# Lecture Notes in Physics

**Editorial Board**

R. Beig, Wien, Austria
J. Ehlers, Potsdam, Germany
U. Frisch, Nice, France
K. Hepp, Zürich, Switzerland
W. Hillebrandt, Garching, Germany
D. Imboden, Zürich, Switzerland
R. L. Jaffe, Cambridge, MA, USA
R. Kippenhahn, Göttingen, Germany
R. Lipowsky, Golm, Germany
H. v. Löhneysen, Karlsruhe, Germany
I. Ojima, Kyoto, Japan
H. A. Weidenmüller, Heidelberg, Germany
J. Wess, München, Germany
J. Zittartz, Köln, Germany

Springer
*Berlin*
*Heidelberg*
*New York*
*Barcelona*
*Hong Kong*
*London*
*Milan*
*Paris*
*Tokyo*

**Physics and Astronomy**   ONLINE LIBRARY

http://www.springer.de/phys/

## Editorial Policy

The series *Lecture Notes in Physics* (LNP), founded in 1969, reports new developments in physics research and teaching -- quickly, informally but with a high quality. Manuscripts to be considered for publication are topical volumes consisting of a limited number of contributions, carefully edited and closely related to each other. Each contribution should contain at least partly original and previously unpublished material, be written in a clear, pedagogical style and aimed at a broader readership, especially graduate students and nonspecialist researchers wishing to familiarize themselves with the topic concerned. For this reason, traditional proceedings cannot be considered for this series though volumes to appear in this series are often based on material presented at conferences, workshops and schools (in exceptional cases the original papers and/or those not included in the printed book may be added on an accompanying CD ROM, together with the abstracts of posters and other material suitable for publication, e.g. large tables, colour pictures, program codes, etc.).

## Acceptance

A project can only be accepted tentatively for publication, by both the editorial board and the publisher, following thorough examination of the material submitted. The book proposal sent to the publisher should consist at least of a preliminary table of contents outlining the structure of the book together with abstracts of all contributions to be included.
Final acceptance is issued by the series editor in charge, in consultation with the publisher, only after receiving the complete manuscript. Final acceptance, possibly requiring minor corrections, usually follows the tentative acceptance unless the final manuscript differs significantly from expectations (project outline). In particular, the series editors are entitled to reject individual contributions if they do not meet the high quality standards of this series. The final manuscript must be camera-ready, and should include both an informative introduction and a sufficiently detailed subject index.

## Contractual Aspects

Publication in LNP is free of charge. There is no formal contract, no royalties are paid, and no bulk orders are required, although special discounts are offered in this case. The volume editors receive jointly 30 free copies for their personal use and are entitled, as are the contributing authors, to purchase Springer books at a reduced rate. The publisher secures the copyright for each volume. As a rule, no reprints of individual contributions can be supplied.

## Manuscript Submission

The manuscript in its final and approved version must be submitted in camera-ready form. The corresponding electronic source files are also required for the production process, in particular the online version. Technical assistance in compiling the final manuscript can be provided by the publisher's production editor(s), especially with regard to the publisher's own Latex macro package which has been specially designed for this series.

## Online Version/ LNP Homepage

LNP homepage (list of available titles, aims and scope, editorial contacts etc.):
http://www.springer.de/phys/books/lnpp/
LNP online (abstracts, full-texts, subscriptions etc.):
http://link.springer.de/series/lnpp/

D. Blaschke  N. K. Glendenning
A. Sedrakian  (Eds.)

# Physics of Neutron Star Interiors

Springer

**Editors**

David Blaschke
Fachbereich Physik
Universität Rostock
18051 Rostock, Deutschland

Armen Sedrakian
Institut de Physique Nuclaire
Groupe de Physique Theorique
15, rue Georges Clemenceau
91406 Orsay Cedex, France

Norman K. Glendenning
Institute for Nuclear
and Particle Astrophysics
Nuclear Science Division
Lawrence Berkeley National Lab.
Berkeley, CA 94720, USA

---

*Cover Picture*: see figure 7, page 350, contribution by H.-Th. Janka in this volume

Die Deutsche Bibliothek - CIP-Einheitsaufnahme

Physics of neutron star interiors / D. Blaschke ... (ed.). - Berlin ;
Heidelberg ; New York ; Barcelona ; Hong Kong ; London ; Milan ; Paris ;
Tokyo : Springer, 2001
   (Lecture notes in physics ; 578)
   (Physics and astronomy online library)
   ISBN 3-540-42340-0

ISSN 0075-8450
ISBN 3-540-42340-0 Springer-Verlag Berlin Heidelberg New York

This work is subject to copyright. All rights are reserved, whether the whole or part of the material is concerned, specifically the rights of translation, reprinting, reuse of illustrations, recitation, broadcasting, reproduction on microfilm or in any other way, and storage in data banks. Duplication of this publication or parts thereof is permitted only under the provisions of the German Copyright Law of September 9, 1965, in its current version, and permission for use must always be obtained from Springer-Verlag. Violations are liable for prosecution under the German Copyright Law.

Springer-Verlag Berlin Heidelberg New York
a member of BertelsmannSpringer Science+Business Media GmbH

http://www.springer.de

© Springer-Verlag Berlin Heidelberg 2001
Printed in Germany

The use of general descriptive names, registered names, trademarks, etc. in this publication does not imply, even in the absence of a specific statement, that such names are exempt from the relevant protective laws and regulations and therefore free for general use.

Typesetting: Camera-ready by the authors/editors
Camera-data conversion by Steingraeber Satztechnik GmbH Heidelberg
Cover design: *design & production*, Heidelberg

Printed on acid-free paper
SPIN: 10845224      54/3141/du - 5 4 3 2 1 0

# Preface

The theory of neutron stars, which along with white dwarfs and black holes form the family of astrophysical compact objects, involves an intimate interplay between diverse branches of theoretical physics. It covers a range from the theory of microscopic nuclear forces to general relativistic gravity, from the particle physics of the radiation of light and neutrinos to the low-temperature physics of superfluids, from the solid-state physics of highly compressed matter to the atomic physics in ultra-high magnetic fields. Hardly in any other physical context do all the forces of nature – the electroweak, strong, and gravitational – emerge as equally important ingredients in the physical picture. It is this diversity of fields and the uniqueness of their interplay that makes the study of neutron stars both exciting and challenging.

The idea of neutron stars has it roots in the 1930s when it was realized that self-gravitating matter can support itself against gravitational contraction by the degeneracy pressure of fermions obeying the Pauli principle. Thus, unlike ordinary stars, which are stabilized by their thermal pressure, neutron stars owe their very existence to the quantum nature of matter. When this idea was combined with the newly developed theory of general relativity neutron stars were born – in theory. It was not until 1967, when the remarkable discovery of pulsars by J. Bell and A. Hewish gave a second birth to neutron stars, that their observational studies became a reality.

The past four decades have seen a dramatic increase in the theoretical activity in this field. Many factors have contributed to the progress. On the observational front the discoveries of neutron stars in X-ray binaries, millisecond pulsars, binary pulsars, and highly magnetized neutron stars (magnetars) have opened new channels of information on these objects. Then, too, the exploration of the nature of interactions among the strongly and weakly interacting constituents of matter at terrestrial accelerators impacted on our conception of superdense matter, its strangeness content, the quark degrees of freedom, phase transitions and reactions involving neutrinos. Another factor is the increase in computational capabilities.

This book is a collection of lectures given at the ECT* (European Centre for Theoretical Studies in Nuclear Physics and Related Areas) in June and July 2000 and covers the theory of neutron star interiors at the forefront of active research. It includes reviews of the traditional material (e.g. the equation of state of superdense matter, the thermal evolution) and, as well, it contains lectures on new issues, for example the recent developments in QCD at finite density, and the possible astrophysical manifestations of the QCD deconfinement phase transition. The choice of topics included in this book was selective. Clearly it is not possible to cover all the current problems of neutron star theory in a single volume; we have provided a list of monographs on the subject for further reference. Naturally enough, the level of presentation throughout the book is uneven; nevertheless, these pedagogical lectures are intermediate between what can be found in the standard texts on the subject and the current research literature; they should be a useful guide to those who wish to enter the field and to those who are actively working in the field.

**Acknowledgements** The editors of this volume express their appreciation for the support and facilities of the ECT* at Trento, Italy, and to its former Director, Rudi Malfliet, for his hearty encouragement of the workshop in the summer of 2000 from which this volume was conceived.

Rostock, Berkeley, Paris  
January 2001

*David Blaschke*  
*Norman K. Glendenning*  
*Armen Sedrakian*

# Contents

**Microscopic Theory of the Nuclear Equation of State and Neutron Star Structure**
*Marcello Baldo, Fiorella Burgio* ........................................ 1

**Superfluidity in Neutron Star Matter**
*Umberto Lombardo, Hans-Josef Schulze* ............................. 30

**Relativistic Superfluid Models for Rotating Neutron Stars**
*Brandon Carter* ........................................................ 54

**The Tensor Virial Method and Its Applications to Self-Gravitating Superfluids**
*Armen Sedrakian, Ira Wasserman* .................................... 97

**Neutron Star Crusts**
*Paweł Haensel* ........................................................ 127

**Kaon Condensation in Neutron Stars**
*Angels Ramos, Jürgen Schaffner-Bielich, Jochen Wambach* ......... 175

**Phases of QCD at High Baryon Density**
*Thomas Schäfer, Edward Shuryak* .................................. 203

**Diquarks in Dense Matter**
*Marten B. Hecht, Craig D. Roberts, Sebastian M. Schmidt* ......... 218

**Color Superconductivity in Compact Stars**
*Mark Alford, Jeffrey A. Bowers, Krishna Rajagopal* ................ 235

**Strange Quark Stars: Structural Properties and Possible Signatures for Their Existence**
*Ignazio Bombaci* ..................................................... 253

**Phase Diagram for Spinning and Accreting Neutron Stars**
*David Blaschke, Hovik Grigorian, Gevorg Poghosyan* .............. 285

**Signal of Quark Deconfinement in Millisecond Pulsars and Reconfinement in Accreting X-ray Neutron Stars**
*Norman K. Glendenning, Fridolin Weber* ........................... 305

**Supernova Explosions and Neutron Star Formation**
*Hans-Thomas Janka, Konstantinos Kifonidis, Markus Rampp* ............ 333

**Evolution of a Neutron Star from Its Birth to Old Age**
*Madappa Prakash, James M. Lattimer, Jose A. Pons, Andrew W. Steiner, Sanjay Reddy* ........................................................ 364

**Neutron Star Kicks and Asymmetric Supernovae**
*Dong Lai* ................................................................ 424

**Spin and Magnetism in Old Neutron Stars**
*Monica Colpi, Andrea Possenti, Serge Popov, Fabio Pizzolato* ............ 440

**Neutrino Cooling of Neutron Stars: Medium Effects**
*Dmitri N. Voskresensky* .................................................. 467

**Books for Further Study** ........................................... 503

**Index** ................................................................ 505

# List of Contributors

**Mark Alford**
Department of Physics and Astronomy
University of Glasgow
G12 8QQ, United Kingdom
m.alford@physics.gla.ac.uk

**Marcello Baldo**
INFN, Sezione di Catania
Corso Italia 57
95129 Catania, Italy
marcello.baldo@ct.infn.it

**David Blaschke**
Fachbereich Physik,
Universität Rostock
Universitätsplatz 3
18051 Rostock, Germany
david.blaschke@physik.uni-rostock.de

**Ignazio Bombaci**
Universita' di Pisa
Dipartimento di Fisica and
INFN Sezione di Pisa
Via Buonarroti 2
56127 Pisa, Italy
bombaci@pi.infn.it

**Jeffrey A. Bowers**
Massachusetts Institute of Technology
77 Massachusetts Ave.
Cambridge, MA 02139, USA
jbowers@mit.edu

**Fiorella Burgio**
INFN Sezione di Catania
57 Corso Italia
95129 Catania, Italy
fiorella.burgio@ct.infn.it

**Brandon Carter**
CNRS
Observatoire de Paris - Meudon
92195 Meudon, France
Brandon.Carter@obspm.fr

**Monica Colpi**
Universita' di Milano
Department of Physics G. Occhialini
Via Celoria 16
20133 Milano, Italy
colpi@mi.infn.it

**Norman K. Glendenning**
Nuclear Science Division and
Institute for Nuclear and Particle
Astrophysics
Lawrence Berkeley Natl. Laboratory
University of California
Berkeley, CA 94720, USA
NKGlendenning@lbl.gov

**Hovik Grigorian**
Yerevan State University
Alex Manoogyan 1
375025, Yerevan, Armenia
hovik@darss.mpg.uni-rostock.de

**Pawel Haensel**
Copernicus Astronomical Center
Bartycka 18
00-716 Warszawa, Poland
haensel@camk.edu.pl

**M.B. Hecht**
Physics Division
Argonne National Laboratory
Argonne, IL 60439-4843, USA
hecht@theory.phy.anl.gov

**Hans-Thomas Janka**
MPI für Astrophysik
Karl-Schwarzschild-Str. 1
85740 Garching, Germany
thj@mpa-garching.mpg.de

**Konstantinos Kifonidis**
MPI für Astrophysik
Karl-Schwarzschild-Str. 1
D-85740 Garching, Germany
kkifonidis@mpa-garching.mpg.de

**Dong Lai**
Cornell University
Space Sciences Building
Ithaca, NY 14853 USA
dong@spacenet.tn.cornell.edu

**James Lattimer**
Department of Physics and Astronomy
State University of New York
Stony Brook, NY 11794-3800, USA
lattimer@astro.sunysb.edu

**Umberto Lombardo**
Dipartimento di Fisica
Universita di Catania and
INFN-LNS
57 Corso Italia
95129 Catania, Italy
umberto.lombardo@ct.infn.it

**Fabio Pizzolato**
Universita' degli Studi di Milano
Dipartimento di Astrofisica
Via Celoria 16
20133, Milano, Italy
fabio@pccolpi.uni.mi.astro.it

**Gevorg S. Poghosyan**
Yerevan State University
Alex Manoogian 1
375025 Yerevan, Armenia
gevorg@ysu.am

**Jose A. Pons**
Department of Physics and Astronomy
State University of New York
Stony Brook, NY 11794-3800, USA
jpons@quark.ess.sunysb.edu

**Sergei B. Popov**
Moscow State University
Universitetskii pr. 13
119899 Moscow, Russia
polar@sai.msu.ru

**Andrea Possenti**
Osservatorio Astronomico di Bologna
Via Ranzani 1
40127, Bologna, Italy
l_possenti@astbo3.bo.astro.it

**Madappa Prakash**
Department of Physics and Astronomy
State University of New York
Stony Brook, NY 11794-3800, USA
prakash@nuclear.physics.sunysb.edu

**Krishna Rajagopal**
Massachusetts Institute of Technology
77 Massachusetts Ave.
Cambridge, MA 02139 USA
krishna@ctp.mit.edu

**Angels Ramos**
University of Barcelona
Departament E.C.M.
Facultat de Fisica
Av. Diagonal 647
08028-Barcelona, Spain
ramos@ecm.ub.es

**Markus Rampp**
MPI für Astrophysik
Karl-Schwarzschild-Str. 1
D-85740 Garching, Germany
mrampp@mpa-garching.mpg.de

**Sanjay Reddy**
Institue for Nuclear Theory
University of Washington
Physics-Astronomy Bldg.
Box 351550
Seattle, WA 98195-1550, USA
reddy@phys.washington.edu

**Craig Roberts**
Physics Division
Argonne National Laboratory
Argonne, IL 60439-4843, USA
cdroberts@anl.gov

**Thomas Schäfer**
Department of Physics
State University of New York
Stony Brook, NY 11794 , USA
schaefer@nuclear.physics.sunysb.edu

**Jürgen Schaffner-Bielich**
Physics Department
RIKEN BNL Research Center
Upton, NY 11973 USA
JSchaffner@bnl.gov

**Hans-Josef Schulze**
E.C.M. Facultat de Fisica
Universitat de Barcelona
Av. Diagonal 647
08028 Barcelona, Spain
schulze@ecm.ub.es

**Sebastian M. Schmidt**
Physics Division
Argonne National Laboratory
Argonne, IL 60439-4843, USA
basti@darss.mpg.uni-rostock.de

**Armen Sedrakian**
Groupe de Physique Theorique
Institut de Physique Nucleaire
91406 Orsay Cedex, France
sedrakia@ipno.in2p3.fr

**Edward Shuryak**
Department of Physics and Astronomy
State University of New York
Stony Brook, NY 11794-3800, USA
shuryak@dau.physics.sunysb.edu

**Andrew W. Steiner**
Department of Physics and Astronomy
State University of New York
Stony Brook, NY 11794-3800, USA
steiner@nuclear.physics.sunysb.edu

**Dmitry Voskresensky**
Moscow Institute for Physics and Engineering
Kashirskoe Shosse 31
115409 Moscow, Russia
d.voskresensky@gsi.de

**Jochen Wambach**
Institut für Kernphysik
TU Darmstadt
Schlossgartenstr. 9
64289 Darmstadt, Germany
wambach@physik.tu-darmstadt.de

**Ira Wasserman**
Cornell University
Department of Astronomy
626 Space Sciences Bldg.
Ithaca, NY 14853, USA
ira@spacenet.tn.cornell.edu

**Fridolin Weber**
University of Notre Dame
Department of Physics
226 Nieuwland Science Hall
Notre Dame, IN 46556-5670, USA
fweber@darwin.helios.nd.edu

# Microscopic Theory of the Nuclear Equation of State and Neutron Star Structure

Marcello Baldo and Fiorella Burgio

Istituto Nazionale di Fisica Nucleare, Sez. Catania, and Universitá di Catania, Corso Italia 57, 95129 Catania, Italy

**Abstract.** The Bethe-Brueckner-Goldstone many-body theory of the Nuclear Equation of State is reviewed in some details. In the theory, one performs an expansion in terms of the Brueckner two-body scattering matrix and an ordering of the corresponding many-body diagrams according to the number of their hole-lines. Recent results are reported, both for symmetric and for pure neutron matter, based on realistic two-nucleon interactions. It is shown that there is strong evidence of convergence in the expansion. Once three-body forces are introduced, the phenomenological saturation point is reproduced and the theory is applied to the study of neutron star properties. One finds that in the interior of neutron stars the onset of hyperons strongly softens the Nuclear Equation of State. As a consequence, the maximum mass of neutron stars turns out to be at the lower limit of the present phenomenological observation.

## 1 Introduction

It is believed that macroscopic portions of (asymmetric) nuclear matter form the interior bulk part of neutron stars, commonly associated with pulsars. Despite the fact that infinite nuclear matter is obviously an idealized physical system, the theoretical determination of the corresponding Equation of State is an essential step towards the understanding of the physical properties of neutron stars. On the other hand, the comparison of the theoretical predictions on neutron stars with the experimental observations can provide serious constraints on the Nuclear Equation of State. Unfortunately, neutron stars are elusive astrophysical objects, and only indirect observations of their structure, including their sizes and masses, are possible. However, the astrophysics of neutron stars is rapidly developing, in view of the observations coming from the new generation of artificial satellites, and one can expect that it will be possible in the near future to confront the theoretical predictions with more and more stringent phenomenological data.

Heavy ion reactions is another field of research where the nuclear Equation of State (EOS) is a relevant issue. In this case, the difficulty of extracting the EOS is due to the complexity of the processes, since the interpretation of the data is necessarily linked to the analysis of the reaction mechanism. An enormous amount of work has been done in the last two decades in the field, but clear indications about the main characteristics of the EOS have still to come. Furthermore, the typical time scale of heavy ion reactions is enormously different from the typical neutron star time scale, and this can prevent a direct link between the two field of research. In particular, nuclear matter inside neutron stars is completely catalyzed, namely it is quite close to the ground state, reachable also by weak processes. In heavy ion reactions the evolution is too rapid to allow weak processes to relax the system towards such a catalyzed state, and therefore the tested Equation of State can differ from the neutron star one, especially at high density.

On the theoretical side, the main general difficulty is the treatment of the strong repulsive core, which dominates the short range behavior of the nucleon-nucleon (NN) interaction, typical of the nuclear system, but which is common to other systems like liquid helium. Simple perturbation theory cannot of course be applied, since the matrix elements of the interaction are too large. One way of overcoming this difficulty is to introduce the two-body scattering G-matrix, which has a much smoother behavior even for strong repulsive core. It is possible to rearrange the perturbation expansion in terms of the reaction G-matrix, in place of the original bare NN interaction, and this procedure is systematically exploited in the Bethe-Brueckner-Goldstone (BBG) expansion [1]. In this contribution we present the latest results on the nuclear EOS based on BBG expansion and their applications to the physics of neutron stars.

## 2 The BBG expansion and the nuclear EOS

The BBG expansion for the ground state energy at a given density, i.e. the EOS at zero temperature, can be ordered according to the number of independent hole-lines appearing in the diagrams representing the different terms of the expansion. This grouping of diagrams generates the so-called hole-line expansion [2]. The smallness parameter of the expansion is assumed to be the "wound parameter" [2], roughly determined by the ratio between the core volume and the volume per particle in the system. It gives an estimate of the decreasing factor introduced by an additional hole-line in the diagram series. The parameter turns out to be small enough up to 2-3 times nuclear matter saturation density. The diagrams with a given number $n$ of hole-lines are assumed to describe the main contribution to the $n$-particle correlations in the system. At the two hole-line level of approximation the corresponding summation of diagrams produces the Brueckner-Hartree-Fock (BHF) approximation, which incorporates the two particle correlations. The BHF approximation includes the self-consistent procedure of determining the single particle auxiliary potential, which is an essential ingredient of the method. Once the auxiliary self-consistent potential is introduced, the expansion is implemented by introducing the set of diagrams which include "potential insertions". To be specific, the introduction of the auxiliary potential can be formally performed by splitting the Hamiltonian in a way which modified from the usual one as

$$H = T + V = T + U + (V - U) \equiv H_0' + V' \tag{1}$$

where $T$ is the kinetic energy and $V$ the nucleon-nucleon interaction. Then one consider $V' = V - U$ as the new interaction potential and $H_0'$ as the new single particle hamiltonian. Then, the single particle energy $e(k)$ is given by

$$e(k) = \frac{\hbar^2 k^2}{2m} + U(k) \qquad (2)$$

while $U$ must be chosen in such a way that the new interaction $V'$ is, in some sense, "reduced" with respect to the original one $V$, so that the expansion in $V'$ should be faster converging. The introduction of the auxiliary potential turns out to be essential, otherwise the hole-expansion would be badly diverging. The total energy $E$ can then be written as

$$E = \sum_k e(k) + B \qquad (3)$$

where $B$ is the interaction energy due to $V'$.

The BHF sums the so called "ladder diagrams". Some of them are depicted in Fig. 1. One has to consider this set of diagrams where one, two, three, and so one, two-body interactions $v$ appear, including exchange terms. Special care must be used in counting correctly the diagrams which give the same contribution.

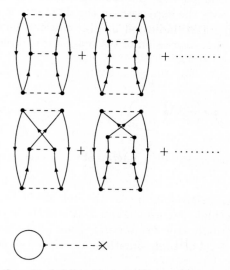

**Fig. 1.** Third and forth order ladder diagrams in the bare interaction (dashed lines) and first order potential insertion (bottom).

The repeated action of the two-body potential $v$ clearly describes the scattering of two nucleons which lie above the Fermi sphere. The summation of the ladder diagrams can be performed by solving the integral equation for the Brueckner G-matrix

$$\langle k_1 k_2 | G(\omega) | k_3 k_4 \rangle = \langle k_1 k_2 | v | k_3 k_4 \rangle +$$
$$+ \sum_{k'_3 k'_4} \langle k_1 k_2 | v | k'_3 k'_4 \rangle \frac{(1-\Theta_F(k'_3))(1-\Theta_F(k'_4))}{\omega - e_{k'_3} - e_{k'_4}} \langle k'_3 k'_4 | G(\omega) | k_3 k_4 \rangle \qquad (4)$$

where $\Theta_F(k) = 1$ for $k < k_F$ and is zero otherwise, $k_F$ being the Fermi momentum. The product $Q(k, k') = (1 - \Theta_F(k))(1 - \Theta_F(k'))$, appearing in the kernel of (4), enforces the scattered momenta to lie outside the Fermi sphere, and it is commonly referred as the "Pauli operator". This G-matrix can be viewed as the in-medium scattering matrix between two nucleons. It has to be stressed that the scattering G-matrix depends parametrically on the entry energy $\omega$, namely it is defined in general also off-shell, as the usual scattering matrix in vacuum. The self-consistent single particle potential $U(k)$ is determined by the equation

$$U(k) = \sum_{k' < k_F} \langle kk'|G(e_{k_1} + e_{k_2})|kk'\rangle_A \tag{5}$$

with $|kk'\rangle_A = |kk'\rangle - |k'k\rangle$.
According to the definition of (2), (5) implies an implicit self-consistent procedure.

Summing up the ladder diagrams to all orders, one then gets the two diagrams, direct and exchange, of Fig. 2, where a wavy lines indicates a Brueckner G-matrix. Indeed, if one expands the G-matrix from (4), in terms of the bare interaction $v$, and inserts the expansion in the diagrams of Fig. 2, one gets the full sets of ladder diagrams, indicated in Fig. 1. More details on the rules for writing down the explicit expression of the diagrams can be found in ref. [1].

**Fig. 2.** The two hole-line contribution in terms of the Brueckner G-matrix (wavy line).

The first potential insertion diagram, at the bottom of Fig. 1, cancels out the potential part of the single particle energy of (2), in the expression for the total energy $E$. This is actually true for any definition of the auxiliary potential $U$. At the two hole-line level of approximation, one therefore gets

$$\begin{aligned} E &= \sum_{k<k_F} \frac{\hbar^2 k^2}{2m} + \frac{1}{2} \sum_{k,k'<k_F} \langle kk'|G(e_k + e_{k'})|kk'\rangle_A \\ &\equiv \sum_{k<k_F} \frac{\hbar^2 k^2}{2m} + \frac{1}{2} \sum_{k<k_F} U(k) \end{aligned} \tag{6}$$

where, in the last equality, the definition of (5) has been adopted. The result, that only the unperturbed kinetic energy appears in the expression for $E$ and all the correlations are included in the potential energy part, holds true to all orders and it is a peculiarity of the BBG expansion. Of course, the modification of the momentum distribution, and therefore of the kinetic energy, is included in the interaction energy part, but it is treated on the same footing as the other correlation effects. This seems to present a noticeable advantage. In fact, the modification of the kinetic energy in itself is quite large and, of course, positive and should be therefore compensate by an extremely accurate calculations of the (negative) correlation energy. On the other hand, putting the two effects on the same footing, one can expect that strong cancellation occur order by order.

Let us now discuss the choice of the single particle potential $U$. As it was discussed in connection with (1), the potential $U$ is in principle arbitrary, and it is used only as a tool for speed up the convergence of the expansion. However, physical considerations suggest the self-consistent procedure defined by (5) to obtain the potential $U$. The self-consistency condition is clearly non-perturbative and it is a generalization of the usual Hartree-Fock (HF) approximation, namely the Brueckner G-matrix is used in place of the bare NN interaction $v$. For nuclear matter the HF approximation would produce unrealistic results, because of the strong repulsive core. The G-matrix takes into account the short range correlations between pairs of nucleons, and therefore it gives a much improved balance between attractive and repulsive contributions. The approximation of (6), together with (2), (5), is usually referred to as the Brueckner-Hartree-Fock (BHF) approximation. This definition of $U$ corresponds to the diagrams of Fig. 3.

**Fig. 3.** The direct and exchange parts of the auxiliary potential $U$ in terms of the Brueckner G-matrix.

It has to be noticed that the G-matrix appearing in the diagrams are calculated on-shell, according to (5), i.e. its entry energy is equal to the energy of the two particles with the two entry momenta. Therefore the total energy at the BHF level of approximation can be written also in terms of the potential $U$, as in the second line of (6).

In the general BBG expansion, in all the higher order diagrams, beyond the BHF approximation, the same definition of $U$ is kept and the bare NN potential is replaced by the G-matrix by performing the corresponding ladder sums whenever it is possible. In this way the diagrammatic expansion is rearranged in terms of the Brueckner G-matrix, in place of the bare NN interaction, with the only obvious prescription that no ladder sums can now appear in the diagrams, in order to avoid double counting.

We have seen that the ladder sum at the BHF level introduce on-shell G-matrices only. This is not necessarily the case if the ladder sum is performed inside a generic higher order energy diagram, since then the entry energy of the resulting G-matrix depends in general on the rest of the diagrams. The energy denominators appearing in the BBG expansion include, in fact, all the particle and hole energies across the diagram. This point will be discuss later and we will see that some exceptions to this expectation can occur.

Another strong reason in favor of keeping the BHF definition for the single particle potential $U$ in the general BBG expansion is the occurrence of cancellation between diagrams including three hole-lines, thus reducing the relevance of higher order contributions. This is true for the two diagrams shown in Fig. 4. The diagram (b) in the right side of the figure is a potential insertion diagram, where the dashed line with the cross indicates a multiplication by a factor $U(k)$, $k$ being the momentum of the hole-line to which the potential is attached. The rule for writing down the potential insertion diagrams can also be found in ref. [1]. The diagram (a) in the left side of Fig. 4 contains a G-matrix loop in place of the potential $U$. If the G-matrix is on-shell, in view of the definition of (5) and the graphical rules, one can easily see that the two diagrams cancel out exactly.

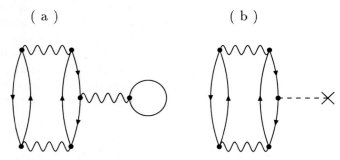

**Fig. 4.** Lowest order three hole-line diagram (a) and the corresponding potential insertion diagram (b).

At first site the G-matrix of diagram (a) should be not calculated on-shell. However, it has been shown in ref. [3] that, if the ladder sums included in the diagram contain bare interactions which appear in all possible positions along the diagram, then their overall contribution reduces indeed to the diagram (a) of Fig. 4, with the G-matrix calculated on-shell, and the above mentioned cancellation holds true.

The definition of (5) does not specify completely the single particle potential $U(k)$. For momenta $k > k_F$ the value of the potential $U(k)$ does not appear explicitly in the energy expression of (6) at the BHF level. In old BHF calculations the potential $U(k)$ was then taken identically zero above the Fermi momentum, with the justification that the interaction between particles above $k_F$ is expected to be small and will, anyway, only slightly affecting the total energy. Within this choice, usually referred to as "standard choice", the potential has then a jump at $k_F$. For this reason it is also often called "gap choice". Most modern BHF calculations adopt a potential $U(k)$ which is defined by extending the definition of (5) also above $k_F$, thus making $U(k)$ continuous across the Fermi sphere. This definition modifies the self-consistent equation and therefore also the potential for $k < k_F$. As a consequence, this different choice, usually called "continuous choice", modifies indirectly also the value of the BHF energy of (6). There are some arguments in favor of the continuous choice. Since $U(k)$ has the physical meaning of single particle potential, it is intimately related to the single particle self-energy. Indeed, one can show [4] that $U(k)$ is the on-shell self energy to first order in the hole expansion. As such, the potential $U(k)$ must be a continuous function of the momentum. Another point to be considered is related to the two other three hole-line diagrams depicted in Fig. 5. They can be obtained from the diagrams of Fig. 4 just by attaching the intermediate G-matrix (diagram $a$) and the potential $U$ (diagram $b$) to the particle-line instead of the hole-line. Diagram ($a$) is usually called the "bubble diagram". In this case the G-matrix is not calculated on-shell, since the argumentation of ref. [3] does not apply, and no exact cancellation can occur between the two diagrams. Actually, in the standard choice the potential insertion diagram $b$ is identically zero, since for this $U(k)$ vanishes for $k > k_F$. On the contrary, in the continuous choice, the potential insertion diagram does not vanish, and some degree of cancellation can be expected, despite the fact that the G-matrix is calculated off-shell, which reduces, also in this case, the contribution from higher order diagrams.

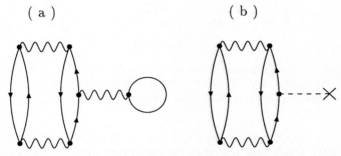

**Fig. 5.** Bubble three hole-line diagram (a) and the corresponding potential insertion diagram (b)

At first sight it can be surprising that the final result for the nuclear matter EOS could depend on the choice of the single particle potential $U$, since the splitting of (1) is a trivial identity and the final result should be independent on the particular choice of $U(k)$. This is of course true only if the full BBG expansion to all orders could be summed up exactly. If the expansion is truncated at a given order, the results can show still a dependence on the choice of $U(k)$, and this dependence will be stronger if the expansion is further away from a reasonable convergence. One can, therefore, take the degree of the dependence on $U(k)$ as a measure of the degree of convergence reached at a given order of the expansion. The gap and continuous choices can be considered two opposite cases for the potential $U(k)$, since any other reasonable choice would modify mainly its definition for $k > k_F$ and would be intermediate between these two cases. In fact, the exact cancellation between the two diagrams of Fig. 4 occurs only with the definition of (5) for $k < k_F$, and it appears inconvenient to adopt a choice for $U(k)$ which does not include the cancellation. However, other choices are surely possible, and one should check also in those other cases the degree of convergence reached at a given level of the expansion. In the sequel we will restrict ourselves to the gap and continuous choices for checking the convergence of the expansion.

Let us consider the symmetric nuclear matter EOS at the BHF level of approximation. The results for the two choices for $U(k)$ are reported in Fig. 6, where the Argonne $v_{14}$ [5] is used for the bare NN potential.

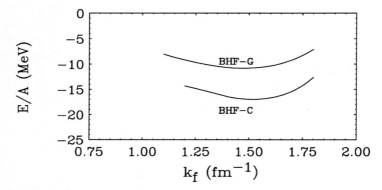

**Fig. 6.** Nuclear matter saturation curve for the Argonne $v_{14}$ NN potential. The solid lines indicate the results at the Brueckner (two hole-lines) level for the standard (BHF-G) and the continuous choices (BHF-C) respectively.

It is apparent from the figure that the degree of convergence is not yet satisfactory at the BHF level. The difference for the energy per particle is of about 4-5 MeV in the considered density range. It has to be kept in mind, however, that the potential energy part of the binding energy of 6 is about -40 MeV around saturation density, and therefore the discrepancy between the two choices is of about 10%. This is the expected degree of convergence at the BHF level, according to the above discussed criterion.

The BHF results imply that, for a check of convergence, it is mandatory to consider the three hole-line diagrams contribution. According to the BBG expansion, this set of diagrams describes the irreducible three-nucleon correlations, i.e. the three-body correlations which cannot be reduced to a product of two-body correlations, already introduced at the BHF level. Let us consider in some detail how the three hole-line diagrams can be summed up exactly, in analogy to the summation of the ladder two hole-line diagrams of the BHF approximation. Indeed, since the two hole-line contribution has been summed up by introducing the G-matrix, which is the in-medium two-body scattering matrix, it is therefore conceivable that the three hole-line diagrams could be summed up by introducing some similar generalization of the scattering matrix for three particles. The three-body scattering problem has a long history by itself, and has been given a formal solution by Faddeev [6]. For three distinguishable particles the three-body scattering matrix $T^{(3)}$ is expressed as the sum of three other scattering matrices, $T^{(3)} = T_1 + T_2 + T_3$. The scattering matrices $T_i$ satisfy a system of three coupled integral equations. The kernel of this set of integral equations contains explicitly the two-body scattering matrices pertaining to each possible pair of particles. Also in this case, therefore, the original two-particle interaction disappears from the equations in favor of the two-body scattering matrix. For identical particles the three integral equations reduce to one due to the symmetry. In fact, the three functions $T_i$ must coincide within a change of variable with a unique function, which we can still call $T^{(3)}$. The analogous equation and scattering matrix in the case of nuclear matter (or other many-body systems in general) has been introduced by Bethe [7,8]. The integral equation, the Bethe–Faddeev equation, reads schematically

$$T^{(3)} = G + G X \frac{Q_3}{e} T^{(3)}$$

$$\langle k_1 k_2 k_3 | T^{(3)} | k'_1 k'_2 k'_3 \rangle = \langle k_1 k_2 | G | k'_1 k'_2 \rangle \delta_K(k_3 - k'_3) + \qquad (7)$$

$$+ \langle k_1 k_2 k_3 | G_{12} X \frac{Q_3}{e} T^{(3)} | k'_1 k'_2 k'_3 \rangle \ .$$

The factor $Q_3/e$ is the analogous of the similar factor appearing in the integral equation for the two-body scattering matrix $G$, see (4). Therefore, the projection operator $Q_3$ imposes that all the three particle states lie above the Fermi energy, and the denominator $e$ is the appropriate energy denominator, namely the energy of the three-particle intermediate state minus the entry energy $\omega$, in close analogy with the equation for the two-body scattering matrix $G$ of (4). The real novelty with respect to the two-body case is the operator $X$. This operator interchanges particle 3 with particle 1 and with particle 2, $X = P_{123} + P_{132}$, where $P$ indicates the operation of cyclic permutation of its indices. It gives rise to the so-called "endemic factor" in the Faddeev equations, since it is an unavoidable complication intrinsic to the three-body problem in general. The reason for the appearance of the operator $X$ in this context is that no two successive $G$ matrices can be present in the same pair of particle lines, since the $G$ matrix already sums up all the two-body ladder processes. In other words, the $G$ matrices must alternate from one pair of particle lines to another, in all possible ways, as it is indeed apparent from the expansion by iteration of (7), which is represented in Fig. 7.

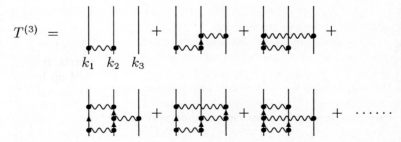

**Fig. 7.** The first few terms in the expansion of the Bethe-Faddeev integral equation.

Therefore, both cyclic operations are necessary in order to include all possible processes. Adding all terms with an arbitrary number of G-matrices, one gets a generalized ladder series for three-particles, analogous to the ladder series introduced for the two particles case in defining the G-matrix. Indeed, this is the basis for the integral equation (7). In the structure of (7) the third particle, with initial momentum $k_3$, is somehow singled out from the other two. This choice is arbitrary, but it is done in view of the use of the Bethe-Faddeev equation within the BBG expansion.

In order to see how the introduction of the three-body scattering matrix $T^{(3)}$ allows to sum up the three hole-line diagrams, we first notice, following B. D. Day [9], that this set of diagrams can be divided into two distinct groups. The first one includes the graphs where two hole-lines, out of three, originate at the first interaction of the graph and terminate at the last one without any further interaction in between. Schematically the sum of this group of diagram can be represented as in part (a) of Fig. 8.

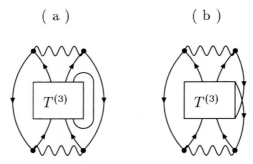

**Fig. 8.** Schematic representation of the direct (a) and exchange (b) three hole-line diagrams.

The third hole-line has been explicitly indicated, out from the rest of the diagram. The remaining part of the diagram describes the rescattering, in all possible way, of three particle-lines, since no further hole-line must be present in the diagram. This part of the diagram is indeed the three-body scattering matrix $T^{(3)}$, and the operator $Q_3$ in (7) assures, as already mentioned, that only particle lines are included.

The second group includes the diagrams where two of the hole-lines enter their second interaction at two different vertices in the diagram, as represented in part (b) of Fig 8. Again the remaining part of the diagram is $T^{(3)}$, i.e. the sum of the amplitudes for all possible rescattering process of three particles. It is easily seen that no other structure is possible. The set of diagrams indicated in part (b) can be obtained by the ones of part (a) by simply interchanging the final (or initial) point of one of the "undisturbed" hole-line with the final (or initial) point of the third hole-line. This means that one can obtain each graph of the group depicted in Fig. 8b by acting with the operator $X$ on the bottom of the corresponding graph of Fig. 8a. In this sense the diagrams of Fig. 8b can be considered the "exchange" diagrams of the ones in Fig. 8a (not to be confused with the term "exchange" introduced previously for the matrix elements of $G$). If one inserts the terms obtained by iterating (7) inside these diagrams in substitution of the scattering matrix $T^{(3)}$ (the box in Fig. 8), the first diagram, coming from the inhomogeneous term in (7), is just the bubble diagram of Fig. 5. The corresponding exchange diagrams is the so called "ring diagram" of Fig. 9.

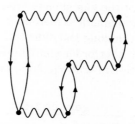

**Fig. 9.** The ring diagram, belonging to the set of three hole-line diagrams. It can be considered the exchange diagram of the bubble diagram.

It is easy to draw the remaining series of diagram which one obtains by going on with the iterations.

Once the Bethe-Faddeev equations are solved, the contribution of the direct three hole-line diagrams of Fig. 8a can be written as

$$E_{3h}^{dir} = \tfrac{1}{2} \sum_{k_1,k_2,k_3 \leq k_F} \sum_{\{k'\},\{k''\} \geq k_F} \langle k_1 k_2 | G | k'_1 k'_2 \rangle_A \cdot$$
$$\cdot \tfrac{1}{e} \langle k'_1 k'_2 k'_3 | X T^{(3)} X | k''_1 k''_2 k''_3 \rangle \tfrac{1}{e'} \langle k''_1 k''_2 | G | k_1 k_2 \rangle_A \quad , \tag{8}$$

In (8) the denominator $e = e_{k'_1} + e_{k'_2} - e_{k_1} - e_{k_2}$, and analogously $e' = e_{k''_1} + e_{k''_2} - e_{k_1} - e_{k_2}$. The exchange diagrams of Fig. 8b can be obtained by multiplying the same expression by a further factor $X$. In summary, the entire set of three hole-line diagrams can be obtained by multiplying the expression of (8) by $1 + X$.

It has been recognized a long ago [8] that the summation of all three-hole diagrams is essential, since individual three-hole diagram can be quite large, but strong cancellation occurs among the different contributions. This is particularly true for the bubble diagram of Fig. 5a and the ring diagram of Fig. 9, which turn out to be quite large but of opposite sign. As already mentioned, the potential insertion diagram of Fig. 5b is different from zero in the continuous choice and it turns out to be essential in compensating the contribution of both bubble and ring diagrams. A scheme of approximation was first devised by B. D. Day [9] within the gap choice for the single particle potential. In this scheme the bubble and ring diagrams are indeed singled out from the whole set of three hole-line diagrams, while the remaining series of diagrams is summed up by solving the Bethe-Faddeev integral equation. The bubble diagram requires special numerical treatment, since very large partial waves contribute to the intermediate G-matrix. Once the bubble and ring diagrams are subtracted from the Bethe-Faddeev equation, the resulting integral equation for the whole set of the higher order diagrams turns out to be much less sensitive to the larger partial waves. We will refer to this contribution as the "higher order" contribution. The numerical solution of the Bethe-Faddeev integral equation is delicate. The main difficulty is the large matrix to be inverted to get the scattering matrix $T^{(3)}$. This difficulty can be overcome by introducing a separable representation of the G-matrix appearing in the kernel of the integral equation, as already performed by B.D. Day [9] in the case of the gap choice. We refer to this reference and to ref. [1] for other details of the numerical methods.

The degree of cancellation among the different terms is apparent in Fig. 10, where the bubble, ring and higher order contributions are displayed [10] in the case of the gap choice and the Argonne $v_{14}$ NN potential.

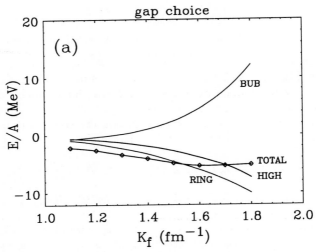

**Fig. 10.** The contributions of the bubble (BUB), ring (RING) and higher order (HIGH) diagrams to the binding energy of symmetric nuclear matter as a function of Fermi momentum, calculated within the gap choice. The line denoted by TOTAL is the sum of all these contributions and gives the overall three hole-line contribution to the EOS.

The final result, denoted as "total", is relatively small and much smaller in size than the individual contributions. The corresponding results for the continuous choice are displayed in Fig. 11. In this case the additional contribution (BUBU) of the potential insertion diagram in Fig. 5b must be considered. One can see the relevance of this term in comparison with the others and its role in determining the size of the total three hole-line contribution. The latter turns out to be much smaller in the continuous choice than in the gap choice.

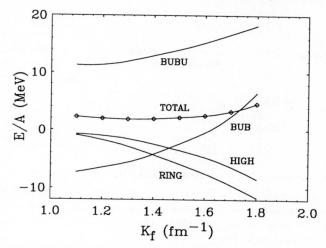

**Fig. 11.** The same as in Fig. 10, but within the continuous choice. Here the line denoted by BUBU is the contribution of the potential insertion diagram of Fig. 5b.

The final Equation of State obtained by adding the three hole-line contribution is reported in Fig. 12, both for the gap choice (squares) and the continuous choice (stars), again for the Argonne $v_{14}$ potential, for a much wider range of densities than in Fig. 6. For comparison, the EOS at the two hole-line level in the continuous choice is also again reported (solid line) from Fig. 6. Two conclusions can be drawn from these results.

i) The two saturation curves in the gap and continuous choices, with the inclusion of the three hole-line diagrams, tend now to collapse in a single EOS, with some deviations only at the highest density. This is a strong indication that a high degree of convergence has been reached at this level of the expansion, according to the criterion discussed above. Notice that the saturation curves extend from a density which is about one half of saturation density to about five times saturation density, and, therefore, it appears unlikely that the agreement between the two choices can be considered as a fortuitous coincidence.

ii) The Brueckner two hole-line EOS within the continuous choice turns out to be already close to the full EOS, since in this case the three hole-line contribution is quite small. In first approximation one can adopt the BHF results with the continuous choice as the nuclear matter EOS.

The phenomenological saturation point for symmetric nuclear matter is, however, not reproduced, which confirms the finding of ref. [9]. The binding energy per particle at the minimum of the saturation curve turns out to be close to the empirical value of about -16 MeV, but the corresponding density comes out about 30-40 % larger than the empirical one. Usually this drawback is corrected by introducing three-body forces in the nuclear Hamiltonian, and indeed all realistic two-nucleon forces, which fit the experimental two-nucleon phase shifts and deuteron data, are not able to reproduce the empirical saturation point. In other words, the results indicate that the missing of the saturation point is not

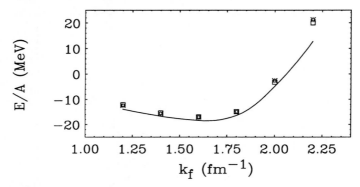

**Fig. 12.** The Nuclear Equation of State including the three hole-line contribution within the gap choice (squares) and the continuous choice (stars), for the Argonne $v_{14}$ potential. For comparison, the EOS at the two hole-line level in the continuous choice is also reported (solid line).

due to a lack of accuracy in the treatment of the nuclear many-body problem, but to a defect of the nuclear Hamiltonian. The need of three-body forces in nuclear matter is consistent with the findings in the study of few nucleon systems, where also the binding energy and radii, as well as scattering data, cannot be reproduced with only two-body forces. Not surprisingly, the effects of three-body forces seem to be more pronounced in nuclear matter than in few body systems.

The standard NN interaction models are based on the meson–nucleon field theory, where the nucleon is considered an unstructured point-like particle. The Paris, the Argonne $v_{14}$ (with the improved version $v_{18}$ [11]), and the set of Bonn potentials [12] fall in this category. In the one-boson exchange potential (OBEP) model one further assumes that no meson–meson interaction is present and each meson is exchanged in a different interval of time from the others. However, the nucleon is a structured particle - it is a bound state of three quarks with a gluon-mediated interaction, according to Quantum Chromodynamics (QCD). The absorption and emission of mesons can be accompanied by a modification of the nucleon structure in the intermediate states, even in the case of NN scattering processes, in which only nucleonic degrees of freedom are present asymptotically. A way of describing such processes is to introduce the possibility that the nucleon can be excited ("polarized") to other states or resonances. The latter can be the known resonances observed in meson–nucleon scattering. At low enough energy the dominant resonance is the $\Delta_{33}$, which has the smallest mass. If the internal nucleon state can be distorted by the presence of another nucleon, the interaction between two nucleons is surely altered by the presence of a third one. This effect produces clearly a definite three-body force, which is absent if the nucleons are considered unstructured. The simplest of such process is depicted in Fig. 13b.
Such a process can be interpreted in different but equivalent ways. One way is to view the pion (meson) coming from the first nucleon to polarize the second one, which therefore interacts with a third one as a $\Delta_{33}$ resonance, surely in a different way than if it had remained a nucleon, like in Fig. 13a. The process of Fig. 13a is

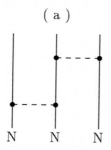 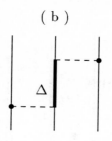

**Fig. 13.** An interaction process among three nucleons with only two-body force (a), and a process involving a genuine three-body force (b).

not indeed a three-nucleon force, but just a repetition of a two-nucleon force. The introduction of a three-nucleon interaction is a consequence of viewing processes like the one of Fig. 13b as an effective interaction among three nucleons, which eventually will be medium-dependent. The genuine three-nucleon forces can be extracted from processes like the one of Fig. 13b by projecting out the $\Delta_{33}$ (or other resonances) degrees of freedom in some approximate way. The theory of three-nucleon forces has a very long history, and it started to be developed since the early stage [13] of the theory of nuclear matter EOS, as well as of few nucleon systems [14]. The most extensive study of the three-nucleon forces (TNF) has been pursued by Grangé and collaborators [15]. Fig. 14, reproduced from [16], indicates some of the processes which can give rise to TNF. Graph of Fig. 14a is a generalization of the process of Fig. 13b, where other nucleon resonances (e.g. the Roper resonance) can appear as intermediate virtual excitation and other exchanged mesons can be present. Graph 14b includes possible non-linear meson-nucleon coupling, as demanded by the chiral symmetry limit [16]. Graph 14c is the simplest one which includes meson-meson interaction. Other processes of this type are of course possible [15,16], which involves other meson-meson couplings, and they should be included in a complete treatment of TNF. Diagram 14d describes the effect of the virtual excitation of a nucleon-antinucleon pair, and it is therefore somehow of different nature from the others. It gives an important (repulsive) contribution and it has been shown [17] to describe the relativistic effect on the EOS to first order in the ratio $U/m$ between the single particle potential and the nucleon rest mass.

The $\sigma$ meson, appearing in some of the diagrams, is a hypothetical scalar meson, believed to be responsible for the intermediate attraction in the two-nucleon interaction, whose mass and coupling constant are treated as parameters. One should therefore be careful, as discussed in [15], to be at least consistent between the treatments of the two-nucleon and the three-nucleon forces. A complete calculations of the TNF in the framework of the meson-nucleon theory, i.e. the calculation of the "best" TNF, is not yet available.

A simpler possibility is to adopt a more phenomenological approach, like the one followed by the Urbana group [18]. Since the EOS obtained with only two-body forces seems to need additional attraction at lower density and an

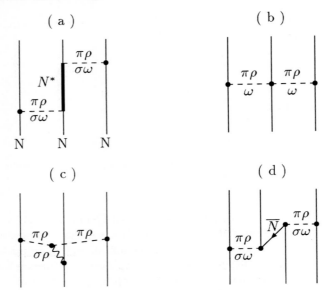

**Fig. 14.** Some of the processes which can produce a genuine three-body force.

additional repulsion at higher density, it is therefore conceivable that the main effect of TNF can be schematized by one attractive and one repulsive term, as representative of the whole set of three-nucleon processes. Actually, once the usual static approximation is made for the nucleons and the resonances in calculating the meson exchange process, the structure of the different three-body forces turns out to be quite similar. Since the strengths of the different vertex appearing in these diagrams cannot be considered fairly well known, one can treat the strengths of the two representative terms as free parameters to be fitted to some known physical quantities. More explicitly, the TNF is written as

$$V_{ijk} = V_{ijk}^{2\pi} + V_{ijk}^{R} \quad . \tag{9}$$

The first (attractive) contribution is a cyclic sum over the nucleon indices $i$, $j$, $k$ of products of anticommutator $\{,\}$ and commutator $[,]$ terms

$$V_{ijk}^{2\pi} = A \sum_{cyc} \Big( \{X_{ij}, X_{jk}\}\{\tau_i \cdot \tau_j, \tau_j \cdot \tau_k\} \\ + \tfrac{1}{4}[X_{ij}, X_{jk}][\tau_i \cdot \tau_j, \tau_j \cdot \tau_k] \Big) \ , \tag{10}$$

where

$$X_{ij} = Y(r_{ij})\sigma_i \cdot \sigma_j + T(r_{ij})S_{ij} \tag{11}$$

is the one–pion exchange operator, $\sigma$ and $\tau$ are the Pauli spin and isospin operators, and $S_{ij} = 3[(\sigma_i \cdot r_{ij})(\sigma_j \cdot r_{ij}) - \sigma_i \sigma_j]$ is the tensor operator. $Y(r)$ and $T(r)$ are the Yukawa and tensor functions, respectively, associated to the one–pion exchange, as in the two–body potential. The repulsive part is taken as

$$V_{ijk}^R = U \sum_{cyc} T^2(r_{ij}) T^2(r_{jk}) \quad . \qquad (12)$$

The strengths $A$ ($< 0$) and $U$ ($> 0$) can be fitted to reproduce the ground state energy of both three nucleon systems (triton and $^3He$), and the empirical nuclear matter saturation point.

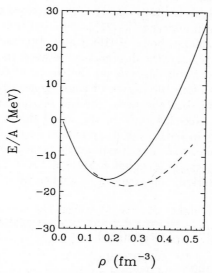

**Fig. 15.** Saturation curve for symmetric nuclear matter in the Brueckner approximation without (dashed curve) and with (full line) three-body forces.

One such a fit, within the Brueckner approximation, is reported in Fig. 15. The empirical saturation point is now reproduced and the EOS become much more repulsive at high density. Of course, the higher density region, needed e.g. in neutron star studies, is obtained by extrapolating the TBF from the region around saturation where they are actually adjusted. This EOS can therefore be inaccurate at the higher densities. One can see indeed that the contribution of the three-body forces is substantial at high density, and therefore an accurate inclusion of the three-body forces is highly demanded.

More detail on the use of phenomenological three-body forces will be given in the Section 4.

The symmetric nuclear matter EOS obtained within the BBG expansion, with only two-body forces included, turns out to be in fair agreement with the corresponding variational calculations of ref. [19,20]. Only at higher density, above 0.6 fm$^{-3}$, the variational results seem to indicate a stronger repulsion. The same trend is seen also when phenomenological three-body forces are included. Furthermore, additional repulsion appears due to the relativistic boost corrections, not included in the present BBG approach.

## 3 The EOS for pure neutron matter

In this Section we will extend the analysis to pure neutron matter EOS, which is more appropriate for neutron star studies, at densities up to about five times the saturation one. Moreover, we consider the calculations for two nucleon-nucleon potentials, the $Av_{14}$ and the $Av_{18}$, in order to analyze the dependence of the results on the nuclear interaction.

We will not give details about contributions of different diagrams, but simply illustrate the results for the neutron matter EOS, obtained by including only two-body forces. The neutron matter EOS [21] is reported (full lines) in Figs. 16 and 17, both for the continuous choice (BHFC) and standard choice (BHFG). As for symmetric nuclear matter, the discrepancy between the two curves indicates to what extent the EOS still depends on the choice of the auxiliary potential at BHF level, and therefore the degree of convergence. The EOS for the $Av_{18}$ appears more repulsive, but the trend for the two potentials is similar. The discrepancy does not exceed 4 MeV in the whole density range for the $Av_{14}$ potential, and it is tiny in the case of $Av_{18}$, except for the highest densities. It is also substantially smaller than in the symmetric nuclear matter case, where the discrepancy is large as much as about 8 MeV at $k_F = 1.8 fm^{-1}$ for the $Av_{14}$.

According to our criterion, this suggests a smaller value of the three hole-line

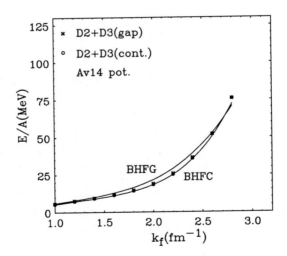

**Fig. 16.** Equation of state of pure neutron matter for the $Av_{14}$ nucleon-nucleon potential. The two full lines correspond to the Brueckner-Hartree-Fock approximation, in the gap (BHFG) and continuous choice (BHFC) respectively. The addition of the three-hole contribution $D_3$ gives the total equation of state for the gap (stars) and continuous choice (open circles) respectively.

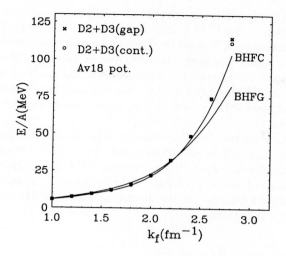

**Fig. 17.** The same as in Fig. 16, but for the Argonne $v_{18}$ potential

contribution and is in agreement with the smaller value in neutron matter of the "wound parameter", which is the smallness parameter of the expansion and should give a rough estimate of the ratio between the three hole-line and the two hole-line contributions (in general between two successive order contributions). It can be estimated by the average depletion of the momentum distribution below the Fermi momentum. Indeed, at densities around the saturation value the wound parameter turns out to be close to 0.1 in neutron matter [22] and about 0.25 in symmetric nuclear matter [23].

These expectations are indeed confirmed by the calculations of the three hole-line contributions. The inclusion of the three hole-line contributions results in the two final EOS depicted in Figs. 16 and 17, where the points marked by stars and the circles correspond to the standard and continuous choices, respectively. For both NN potentials, the very close agreement between the two EOS is a strong evidence that the expansion has reached convergence. Notice that, at Brueckner (two hole) level, the EOS for the standard and continuous choices cross at some value of the density, and at that point the overall three hole-line contribution has the same value in both choices. Furthermore, in the continuous choice the three hole-line contribution is substantially smaller, and it is actually negligible to a first approximation. It appears that the corresponding values of the wound parameter, which is close to 0.1 in neutron matter [22], give an upper limit for the ratio between three and two hole-line contributions.

The final EOS appears more repulsive at high density for the $Av_{18}$ than for the $Av_{14}$ potential. At lower density, up to about $k_F = 2.0 fm^{-1}$, the two potentials produce very similar EOS. This is not surprising, since both potentials fit the NN experimental phase shifts up to 350 MeV Lab energy, which indeed corresponds to a relative momentum of about $2.0 fm^{-1}$. Above this density, the main contribution to the EOS comes from values of the relative NN momentum which need extrapolation beyond the region where the potentials have been fitted to the empirical data. It is likely that different extrapolations are obtained from different potentials in general, and therefore the EOS at high density is largely dependent on the NN potential model, even without the inclusion of three-body forces. The inclusion of three-body forces in pure neutron matter is discussed in the next Section.

The pure neutron matter EOS obtained within the BBG expansion, with only two-body forces included, turns out to be substantially more repulsive at increasing density than the corresponding variational EOS of ref. [19,20]. The reason of these discrepancies has not yet been clarified. However, when three-body forces are included in both treatments, the discrepancies are substantially reduced, and a fair agreement is obtained.

## 4 Neutron Star Structure

Once the Nuclear Equation of State, both for symmetric and pure neutron matter, has been established on a firm basis, one can try to study the structure of neutron star (NS) interior. It is indeed believed that the interior of neutron stars contains mainly asymmetric nuclear matter with density increasing towards the center of the star. The matter in the outer crust, at sub-nuclear densities, exists in a different state, namely, it forms crystal structures of atomic species, whose mass number increases with increasing density. In the region where neutrons start to drip out of nuclei, the crystalline structure is probably mixed with a neutron gas; this phase persist up to the densities where nuclei merge to form uniform asymmetric nuclear matter. This outer region is the place where many interesting phenomena occur. However, to the extent that the analysis is restricted to the mass and radius of the star, the main contribution is coming from the interior, where nuclear matter sets in. One can hope, therefore, that neutron stars could be a testing ground for the Nuclear Equation of State. Up to now the masses of few neutron stars have been accurately determined. Only recently some indirect indications of neutron star radii have been reported, and, as already noticed, the astrophysics of neutron stars is rapidly developing. An accurate enough measurement of both mass and radius of a neutron star is expected to produce an enormous advancement in our knowledge of the nuclear Equation of State.

The observed neutron star masses are $\approx (1-2)M_\odot$ (where $M_\odot$ is the mass of the sun, $M_\odot = 1.99 \times 10^{33}$g). Typical radii of NS are thought to be of order 10 km, and the central density is a few times normal nuclear matter density ($\rho_0 \approx 0.17$ fm$^{-3}$). This requires a detailed knowledge of the EOS for densities $\rho \gg \rho_0$. This is a very hard task from the theoretical point of view. In fact, whereas at densities $\rho \approx \rho_0$ the matter consists mainly of nucleons and leptons, at higher densities several species of particles may appear due to the fast rise of the baryon chemical potentials with density. Among these new particles are strange baryons, namely, the $\Lambda$, $\Sigma$ and $\Xi$ hyperons. Due to its negative charge, the $\Sigma^-$ hyperon is the first strange baryon expected to appear with increasing density in the reaction $n + n \to p + \Sigma^-$, in spite of its substantially larger mass compared to the neutral $\Lambda$ hyperon ($M_{\Sigma^-} = 1197$ MeV, $M_\Lambda = 1116$ MeV). Other species in stellar matter may appear, like $\Delta$ isobars along with pion and kaon condensations. It is therefore mandatory to generalize the study of nuclear EOS with the inclusion of the possible hardens, other than nucleons, which can spontaneously appear in the inner part of a NS, just because their appearance is able to lower the ground state energy of the nuclear matter dense phase. In the following we will concentrate on the production of strange baryons and assume that a carbonic description of nuclear matter holds up to densities comparable to those encountered in the cores of neutron stars.

As we have seen from the previous Sections, the nuclear EOS can be calculated with good accuracy in the Brueckner two hole-line approximation within the continuous choice for the single particle potential, since the results in this scheme are quite close to the full convergent calculations which include also the three hole-line contribution. It is then natural to include the hyperon degrees of freedom within the same approximation to calculate the nuclear EOS needed to describe the NS interior. For this purpose, one needs also a nucleon-hyperon (NY) and a hyperon-hyperon (YY)interaction. In the following this interaction will be taken as the Nijmegen soft-core model [24]. In the calculations the hyperon-hyperon interaction will be neglected in first approximation. We will comment on this point in the sequel. With these NN and NY potentials, the various $G$ matrices are evaluated by solving numerically the Brueckner equation, which can be written in operator form as

$$G_{ab}[W] = V_{ab} + \sum_c \sum_{p,p'} V_{ac} |pp'\rangle \frac{Q_c}{W - E_c + i\epsilon} \langle pp'| G_{cb}[W], \qquad (13)$$

where the indices $a, b, c$ indicate pairs of baryons and the Pauli operator $Q$ and energy $E$ determine the propagation of intermediate baryon pairs. In a given nucleon-hyperon channels $c = (NY)$ one has, for example,

$$E_{(NY)} = m_N + m_Y + \frac{k_N^2}{2m_N} + \frac{k_Y^2}{2m_Y} + U_N(k_N) + U_Y(k_Y). \qquad (14)$$

The hyperon single-particle potentials within the continuous choice are given by

$$U_Y(k) = \text{Re} \sum_{N=n,p} \sum_{k' < k_F^{(N)}} \left\langle kk' \middle| G_{(NY)(NY)}\left[E_{(NY)}(k,k')\right] \middle| kk' \right\rangle \qquad (15)$$

and similar expressions of the form

$$U_N(k) = \sum_{N'=n,p} U_N^{(N')}(k) + \sum_{Y=\Sigma^-,\Lambda} U_N^{(Y)}(k) \qquad (16)$$

apply to the nucleon single-particle potentials. The nucleons feel therefore direct effects of the other nucleons as well as of the hyperons in the environment, whereas for the hyperons there are only nucleonic contributions, because of the missing hyperon-hyperon potentials. The equations (13–16) define the BHF scheme with the continuous choice of the single-particle energies. Due to the occurrence of $U_N$ and $U_Y$ in (14) they constitute a coupled system that has to be solved in a self-consistent manner for several Fermi momenta of the particles involved. Once the different single-particle potentials are known, the total nonrelativistic carbonic energy density, $\epsilon$, and the total binding energy per baryon, $B/A$, can be evaluated

$$\frac{B}{A} = \frac{\epsilon}{\rho_n + \rho_p + \rho_{\Sigma^-} + \rho_\Lambda}, \tag{17}$$

$$\epsilon = \sum_{i=n,p,\Sigma^-,\Lambda} \int_0^{k_F^{(i)}} \frac{dk\, k^2}{\pi^2} \left( m_i + \frac{k^2}{2m_i} + \frac{1}{2} U_i(k) \right) \tag{18}$$

As we have seen, nonrelativistic calculations, based on purely two-body interactions, fail to reproduce the correct saturation point of symmetric nuclear matter, and three-body forces among nucleons are needed to correct this deficiency. In the sequel the so-called Urbana model will be used, which consists, as we have already seen, of an attractive term due to two-pion exchange with excitation of an intermediate $\Delta$ resonance, and a repulsive phenomenological central term. We introduced the same Urbana three-nucleon model within the BHF approach (for more details see [25]). In our approach the TBF is reduced to a density dependent two-body force by averaging on the position of the third particle, assuming that the probability of having two particles at a given distance is reduced according to the two-body correlation function. The corresponding nucleon matter EOS (no hyperon) satisfies several requirements, namely (i) it reproduces correctly the nuclear matter saturation point, (ii) the incompressibility is compatible with values extracted from phenomenology, (iii) the symmetry energy is compatible with nuclear phenomenology, (iv) the causality condition is always fulfilled.

If leptons, namely electrons and muons, and hyperon are introduced, the general EOS can be calculated for a given composition of the baryon components. This allows the determination of the chemical potentials (by simple numerical derivatives of the energy) of all the species, carbonic and leptonic, which are the fundamental input for the equations of chemical equilibrium. The latter determines the actual detailed composition of the dense matter and therefore the EOS to be used in the interior of neutron stars. Indeed, at high density the matter composition is constrained by three conditions: i) chemical equilibrium among the different species, ii) charge neutrality, and iii) baryon number conservation. At density $\rho \approx \rho_0$ the stellar matter is composed of a mixture of neutrons, protons, electrons, and muons in $\beta$-equilibrium [electrons are ultrarelativistic at these densities, $\mu_e = (3\pi^2 \rho x_e)^{1/3}$]. In that case the equations for chemical equilibrium read

$$\mu_n = \mu_p + \mu_e, \tag{19}$$

$$\mu_e = \mu_\mu. \tag{20}$$

Since we are looking at neutron stars after neutrinos have escaped, we set the neutrino chemical potential equal to zero. Strange baryons appear at density $\rho \approx (2-3)\rho_0$ [26], mainly in carbonic processes like $n + n \to p + \Sigma^-$ and $n + n \to n + \Lambda$. The equilibrium conditions for those processes read

$$2\mu_n = \mu_p + \mu_\Sigma ,  \qquad (21)$$

$$\mu_n = \mu_\Lambda . \qquad (22)$$

The other two conditions of charge neutrality and baryon number conservation allow a unique solution of a closed system of equations, yielding the equilibrium fractions of the baryon and lepton species for each fixed baryon density. They read

$$\rho_p = \rho_e + \rho_\mu + \rho_\Sigma , \qquad (23)$$

$$\rho = \rho_n + \rho_p + \rho_\Sigma + \rho_\Lambda . \qquad (24)$$

Finally, from the knowledge of the equilibrium composition one determines the equation of state, i.e., the relation between pressure P and baryon density $\rho$. It can be easily obtained from the thermodynamical relation

$$P = -\frac{dE}{dV} . \qquad (25)$$

being E the total energy and V the total volume. Equation (25) can be explicitly worked out in terms of the carbonic and leptonic binding energies, respectively $B$ and $E_L$,

$$P = -\frac{dE}{dV} = -\frac{d}{dV}(B + E_L) = P_B + P_L , \qquad (26)$$

$$P_B = \rho^2 \frac{d(B/A)}{d\rho} = \rho^2 \frac{d}{d\rho}\left[(x_n + x_p)\frac{\epsilon_{NN}}{\rho_N} + x_\Sigma \frac{\epsilon_{N\Sigma}}{\rho_\Sigma} + x_\Lambda \frac{\epsilon_{N\Lambda}}{\rho_\Lambda}\right] , \qquad (27)$$

$$P_L = \rho^2 \frac{d(E_L/A)}{d\rho} = \rho^2 \frac{d}{d\rho}\left[x_{e^-}\frac{\epsilon_{e^-}}{\rho_{e^-}} + x_{\mu^-}\frac{\epsilon_{\mu^-}}{\rho_{\mu^-}}\right] . \qquad (28)$$

In the above equations $x_i$ represent the baryon fraction of each species. As far as the leptons are concerned, at those high densities electrons are a free ultra-relativistic gas, whereas muons are relativistic. Therefore their energy densities $\epsilon_L$ are well-known from textbooks, see e.g. ref.[27]. In order to construct models of neutron stars, one needs to calculate the total mass-energy density $\mathcal{E}$ as well. This can be easily obtained just by adding the mass-energy densities of each species $\mathcal{E}_i$

$$\mathcal{E} = \mathcal{E}_N + \mathcal{E}_\Sigma + \mathcal{E}_\Lambda + \mathcal{E}_{e^-} + \mathcal{E}_{\mu^-} , \qquad (29)$$

While the electron and muon contributions, respectively $\mathcal{E}_{e^-}$ and $\mathcal{E}_{\mu^-}$, are known from textbooks, the carbonic contribution is given by

$$\mathcal{E}_N = \frac{1}{c^2}(\epsilon_{NN} + m_N \rho_N) , \qquad (30)$$

$$\mathcal{E}_\Sigma = \frac{1}{c^2}\left(\epsilon_{N\Sigma} + m_\Sigma \rho_\Sigma\right), \tag{31}$$

$$\mathcal{E}_\Lambda = \frac{1}{c^2}\left(\epsilon_{N\Lambda} + m_\Lambda \rho_\Lambda\right). \tag{32}$$

$m_i$ being the rest mass and $c$ the speed of light. For more details, the reader is referred to ref. [26] and references therein.

In figure 18 we show the chemical composition of $\beta$-stable and asymmetric nuclear matter containing hyperons. In the upper panel we display the case when only two-body nucleonic forces are present, whereas in panel b) nucleonic TBF's are included. We observe that the inclusion of TBF's shifts the hyperon onset points down to $\rho \simeq 2-3$ times normal nuclear matter density, since some additional repulsion is now present. Moreover, an almost equal percentage of nucleons and hyperons are present in the stellar core at high densities. Such a low threshold for hyperons is in agreement with other approaches [28,29]. A strong deleptonization of matter takes place, since it is energetically convenient to maintain charge neutrality through hyperon formation than $\beta$-decay. This can have far reaching consequences for the onset of kaon condensation. The main physical features of the nuclear EOS which determine the resulting compositions are essentially the symmetry energy of the nucleon part of the EOS and the hyperon single particle potentials inside nuclear matter. Since at low enough density the nucleon matter is quite asymmetric, the small percentage of protons feels a deep single particle potential, and therefore it is energetically convenient to create a $\Sigma^-$ hyperon since then a neutron must be converted into a proton. The deepness of the proton potential is mainly determined by the nuclear matter symmetry energy. Furthermore, the potential felt by the hyperons can shift substantially the threshold density at which each hyperon sets in. This points are illustrated in Fig. 19, where different single particle potentials are plotted at a given nucleon density. For simplicity, neutron and proton densities are fixed at $\rho_N = 0.4\,\text{fm}^{-3}$ and $\rho_p/\rho_N = 0.1$, and the $\Sigma^-$ density is varied. Under these conditions the $\Sigma^-$ single-particle potential is sizably repulsive, while $U_\Lambda$ is still attractive (see also [26]) and the nucleons are much more strongly bound. The $\Sigma^-$ single-particle potential has a particular shape with an effective mass $m^*/m$ slightly larger than 1, whereas the lambda effective mass is typically about 0.8 and the nucleon effective masses are much smaller.

The resulting Equation of State is displayed in Figure 20. The dotted line represents the case when only two-body forces are present, whereas the solid line shows the case when TBF's are included. The upper curves show the equation of state when stellar matter is composed only by nucleons and leptons. We mainly observe a stiffening of the equation of state because of the repulsive contribution coming from the TBF's. The inclusion of hyperons (lower curves) produces a soft equation of state which turns out to be very similar to the one obtained without TBF's. This is quite astonishing because, in the pure nucleon case, the repulsive character of TBF at high density increases the stiffness of the EOS, thus changing dramatically the equation of state. However, when hyperons are included, the presence of TBF's among nucleons enhances the population of $\Sigma^-$ and $\Lambda$ because of the increased nucleon chemical potentials with respect to the

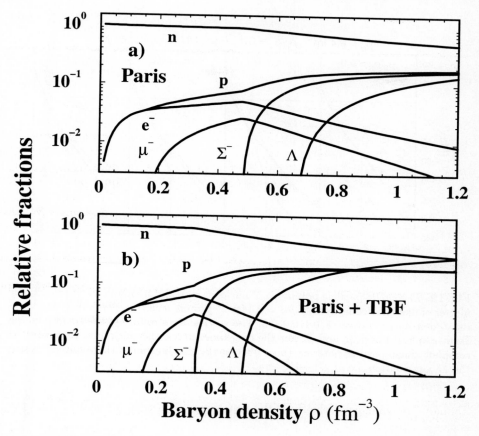

**Fig. 18.** The equilibrium composition of asymmetric and $\beta$-stable nuclear matter containing $\Sigma^-$ and $\Lambda$ hyperons is displayed. In the upper panel only two-body nucleonic forces are present, whereas in the lower panel TBF's have been included.

case without TBF, thus decreasing the nucleon population. The net result is that the equation of state looks very similar to the case without TBF, but the chemical composition of matter containing hyperons is very different when TBF are included. In the latter case, the hyperon populations are larger than in the case with only two-body forces. This has very important consequences for the structure of the neutron stars. Of course, this scenario could partly change if hyperon-hyperon interactions were known or if TBF would be included also for hyperons, but this is beyond our current knowledge of strong interactions.

## 5 Equilibrium configurations of neutron stars

We assume that a star is a spherically symmetric distribution of mass in hydrostatic equilibrium. The equilibrium configurations are obtained by solving the

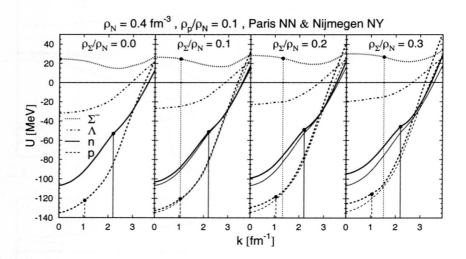

**Fig. 19.** The single-particle potentials of nucleons $n$, $p$ and hyperons $\Sigma$, $\Lambda$ in carbonic matter of fixed nucleonic density $\rho_N = 0.4\,\text{fm}^{-3}$, proton density $\rho_p/\rho_N = 0.1$, and varying $\Sigma$ density $\rho_\Sigma/\rho_N = 0.0, 0.1, 0.2, 0.3$. The vertical lines represent the corresponding Fermi momenta of $n$, $p$, and $\Sigma$. For the nucleonic curves, the thick lines represent the complete single-particle potentials $U_N$, whereas the thin lines show the values excluding the $\Sigma$ contribution, i.e., $U_N^{(n)} + U_N^{(p)}$.

Tolman-Oppenheimer-Volkoff (TOV) equations [27] for the pressure $P$ and the enclosed mass $m$,

$$\frac{dP(r)}{dr} = -\frac{Gm(r)\mathcal{E}(r)}{r^2}\frac{\left[1+P(r)/\mathcal{E}(r)\right]\left[1+4\pi r^3 P(r)/m(r)\right]}{1-2Gm(r)/r}, \quad (33)$$

$$\frac{dm(r)}{dr} = 4\pi r^2 \mathcal{E}(r), \quad (34)$$

being $G$ the gravitational constant. Starting with a central mass density $\mathcal{E}(r=0) \equiv \mathcal{E}_c$, we integrate out until the pressure on the surface equals the one corresponding to the density of matter composed of iron. This gives the stellar radius $R$ and the gravitational mass is then

$$M_G \equiv m(R) = 4\pi \int_0^R dr \; r^2 \mathcal{E}(r). \quad (35)$$

For the outer part of the neutron star we have used the equations of state by Feynman-Metropolis-Teller [30] and Baym-Pethick-Sutherland [31], and for the medium-density regime we use the results of Negele and Vautherin [32]. For density $\rho > 0.08\,\text{fm}^{-3}$ we use the microscopic equations of state obtained in the BHF approximation described above. For comparison, we also perform

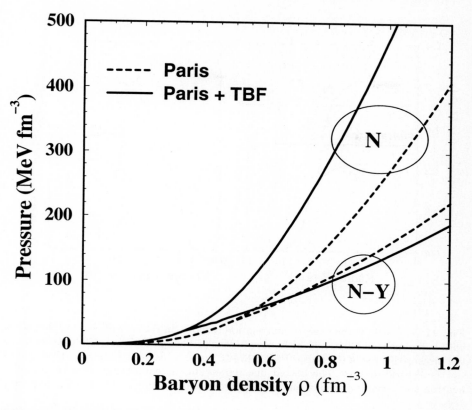

**Fig. 20.** The pressure is displayed vs. the baryon density for hyperon-free (upper curves) and hyperon-rich (lower curves) matter. The solid (dashed) lines represent the case when nucleonic TBF's are (are not) included.

calculations of neutron star structure for the case of asymmetric and $\beta$-stable nucleonic matter. The results are plotted in Fig. 21. We display the gravitational mass $M_G$ (in units of the solar mass $M_o$) as a function of the radius $R$ (panel (a)) and central baryon density $n_c$ (panel (b)). We note that the inclusion of hyperons lowers the value of the maximum mass from about 2.1 $M_o$ down to 1.26 $M_o$. This value lies below the value of the best observed pulsar mass, PSR1916+13, which amounts to 1.44 solar masses. However the observational data can be fitted if rotations are included, see dotted line in panel (b). In this case only equilibrium configurations rotating at the Kepler frequency $\Omega_K$ are shown. However, $\Omega_K$ is much larger than the rotational frequency of that pulsar, and therefore rotation probably does not play any role.

In conclusion, the main finding of our work is the surprisingly low value of the maximum mass of a neutron star, which hardly comprises the observational data. This fact indicates how sensitive the properties of the neutron stars are

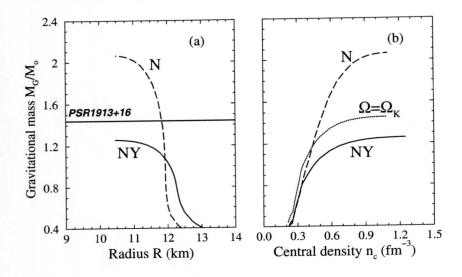

**Fig. 21.** In panel (a) the mass-radius relation is shown in the case of beta-stable matter with hyperons (solid line) and without hyperons (dashed line). The thick line represents the measured value of the pulsar PSR1913+16 mass. In panel (b) the mass is displayed vs. the central density. The dotted line represents the equilibrium configurations of neutron stars containing nucleons plus hyperons and rotating at the Kepler frequency $\Omega_K$.

to the details of the interaction. In particular our result calls for the need of including realistic hyperon-hyperon interactions. However, the use of the available hyperon-hyperon interactions seem to introduce only minor changes in the results [33]. Despite the uncertainty on the NY and YY interactions, it is unlikely that one can obtain a neutron star mass substantially larger. The possible occurrence of a quark core is usually assumed to further soften the EOS and lower the maximum mass. However, this is not necessarily true, since at large density the quark pressure could rise fast enough to increase the stability of the system. In any case, the possible quark core is not expected to change dramatically the critical neutron star mass. Even if an explicit analysis of the quark core has still to be worked out, it is fair to say that the observation of a neutron star with a mass much larger than 1.4-1.5 solar mass would indicate that indeed some basic ingredient is missing in our understanding of neutron star structure.

## Acknowledgments

The material presented in this contribution is the result of a fruitful collaboration, lasting for several years, with a number of people. Special thanks are due to Dr. I. Bombaci, Prof. L.S. Ferreira, Dr. G. Giansiracusa, Prof. U. Lombardo, Dr. H.-J. Schulze and Prof. H. Q. Song.

## References

1. For a pedagogical introduction, see *Nuclear Methods and the Nuclear Equation of State*, Edited by M. Baldo, World Scientific, Singapore, International Review of Nuclear Physics Vol. 9, 1999.
2. B.D. Day, *Brueckner–Bethe Calculations of Nuclear Matter*, Proceedings of the School E. Fermi, Varenna 1981, Course LXXIX, ed. A. Molinari, (Editrice Compositori, Bologna, 1983), p. 1–72; *Rev. Mod. Phys.* **39**, 719 (1967).
3. H.A. Bethe, B.H. Brandow, A.G. Petschek: Phys. Rev. **129**, 225 (1962)
4. J. Hüfner, C. Mahaux: Ann. Phys. (N. Y.) **73**, 525 (1972)
5. R.B. Wiringa, R.A. Smith and T.L. Ainsworth: Phys. Rev. **C29**, 1207 (1984).
6. L.D. Faddeev, *Mathematical Aspects of the Three-Body Problem in Quantum Scattering Theory*, Davey, New York 1965.
7. H.A. Bethe: Phys. Rev. **138**, 804 (1965)
8. R.Rajaraman and H.Bethe: Rev. Mod. Phys. **39**, 745 (1967)
9. B.D. Day: Phys. Rev. **C24**, 1203 (1981); Phys. Rev. Lett. **47**, 226 (1981)
10. H.Q. Song, M. Baldo, G. Giansiracusa U. Lombardo: Phys. Rev. Lett. **81**, 1584 (1998)
11. R.B. Wiringa, V.G.J. Stocks,R. Schiavilla: Phys. Rev. **C51**, 38 (1995)
12. R. Machleidt: Adv. Nucl. Phys. **19**, 189 (1989)
13. J. Fuyita, H. Miyazawa: Progr. Theor. Phys., **17**, 360 (1957)
14. Ch. Hadjuk, P.U. Sauer, W. Streuve: Nucl. Phys. **A405**, 581 (1983)
15. P. Grangé, A. Lejeune, M. Martzolff, J.-F. Mathiot: Phys. Rev **C40**, 1040 (1989), and references therein.
16. J.-F. Mathiot: Phys. Rep. **173**, 63 (1989)
17. G.E. Brown, W. Weise, G. Baym, J. Speth: *Comm. Nucl. Part. Phys.* **17**, 39 (1987)
18. J. Carlson, V.R. Pandharipande, R.B. Wiringa: Nucl. Phys. **A401**, 59 (1983)
19. R.B. Wiringa, V. Fiks and A. Fabrocini, Phys. Rev. **C38**, 1010 (1988)
20. A. Akmal, V.R. Pandharipande and D.G. Ravenhall, Phys. Rev. **C58**, 1804 (1998)
21. M. Baldo, G. Giansiracusa, U. Lombardo. H. Q. Song: *Phys. Lett.* B**473** (2000) 1
22. W. Zuo, G.Giansiracusa, U. Lombardo, N. Sandulesco, H.-J. Schulze: Phys. Lett. **B421**, 1 (1998)
23. W. Zuo, U. Lombardo, H.-J. Schulze: Phys. Lett. **B432**, 241 (1998)
24. P. Maessen, Th. Rijken, J. de Swart: *Phys. Rev.* C**40**, 2226 (1989)
25. M. Baldo, I. Bombaci, G. F. Burgio: Astron. and Astrophys. **328**, 274 (()1997)
26. M. Baldo, G. F. Burgio, H.-J. Schulze: Phys. Rev. C **61**, 055801-1 (()2000)
27. S. L. Shapiro and S. A. Teukolsky, *Black Holes, White Dwarfs and Neutron Stars* (John Wiley & Sons, New York, 1983).
28. N. K. Glendenning, Astrophys. J. **293** 470 (1985).
29. N. K. Glendenning and S. A. Moszkowski, Phys. Rev. Lett. **67** 2414 (1991).
30. R. Feynman, F. Metropolis, E. Teller: Phys. Rev. C **75**, 1561 (()1949)
31. G. Baym, C. Pethick, D. Sutherland: Astrophys. J. **170**, 299 (()1971)
32. J. W. Negele, D. Vautherin: Nucl. Phys. A **207**, 298 (1973).
33. I. Vidaña, A. Polls, A. Ramos, L. Engvik, M. Hjorth-Jensen: preprint, University of Barcellona, 1999.

# Superfluidity in Neutron Star Matter

Umberto Lombardo[1] and Hans-Josef Schulze[2]

[1] Dipartimento di Fisica, Università di Catania,
   Corso Italia 57, I-95129 Catania, Italy
[2] Departament d'Estructura i Constituents de la Matèria, Universitat de Barcelona,
   Av. Diagonal 647, E-08028 Barcelona, Spain

## 1 Introduction, General Formalism

The research on the superfluidity of neutron matter can be traced back to Migdal's observation that neutron stars are good candidates for being macroscopic superfluid systems [1]. And, in fact, during more than two decades of neutron-star physics the presence of neutron and proton superfluid phases has been invoked to explain the dynamical and thermal evolution of a neutron star. The most striking evidence is given by post-glitch timing observations [2,3], but also the cooling history is strongly influenced by the possible presence of superfluid phases [4,5]. On the theoretical side, the onset of superfluidity in neutron matter or in the more general context of nuclear matter was investigated soon after the formulation of the Bardeen, Cooper, and Schrieffer (BCS) theory of superconductivity [6] and the pairing theory in atomic nuclei [7,8].

The peculiar feature of a nucleon system is that it is a strongly interacting Fermi system with a force which has a short-range repulsive component and a long-range attractive one. The first question raised by scientists was whether or not the strongly repulsive core might prevent the formation of a superfluid state. But it was indeed shown [9] that the BCS approach, based on the mechanism of Cooper pairs, can be successfully extended to nuclear matter and that superfluid states could in fact exist for a wide class of nucleon–nucleon potentials [10]. The second question is related to the fact that the superfluid state of nuclear matter is a self-sustaining state in the sense that nucleons participating to the pairing coupling also screen the pairing itself. From this point of view one expects the strong correlations to play an important role in delimiting the magnitude of the pairing gap. Therefore it appears necessary to go beyond the pure BCS approach and properly add the effects of the medium polarization as self-energy and vertex corrections.

Since the neutron star "laboratory" can only provide an indirect evidence of the nucleon pairing in infinite matter and its relation with the pairing in nuclei is still too much model dependent, we need to rely on very accurate quantitative theoretical predictions of its properties such as energy gap, superfluidity density domain, critical temperature, and other physical quantities associated with the various superfluid states. These quantities may only be obtained from *ab initio* calculations, i.e., microscopic approaches using as input the bare nucleon-nucleon interaction, because we are exploring a density domain much wider than the saturation region where phenomenological interactions such as Skyrme forces are well suited. There is a more basic reason to refrain from using effective interactions, that is a double counting of the particle-particle (p-p) correlations incorporated in the effective interaction, but also in the gap equation.

Fortunately, for more than two decades *realistic* potentials, based on field-theoretical approaches, have been supplied to describe the bare nucleon-nucleon interaction. The term 'realistic' means that the parameters contained in such potentials are adjusted to simultaneously reproduce the experimental phase shifts of nucleon-nucleon scattering and the binding energies of the lightest nuclei.

In this chapter the problem of superfluidity in neutron matter is surveyed and special emphasis is devoted to new theoretical developments and calculations. In the following section the general formalism of pairing in a strongly interacting Fermi system is presented. In Sect. 2 the possible superfluid states of nucleon matter are described within the BCS theory extended to non-zero angular momentum. In the last section some aspects of the generalized gap equation will be discussed, including the medium polarization effects at very low density (Sect. 3.1), the induced interaction approach (Sect. 3.2), and the role of self-energy corrections (Sect. 3.3).

We mention three previous works for a comprehensive study of superfluidity in nuclear matter: the early papers based on the generalized BCS-Bogolyubov theory [11,12], which already give a systematic survey of most superfluid states of nuclear matter; the second one [13], based on the method of the correlated basis functions, mainly focussed on the $^1S_0$ pairing, but containing a wide discussion of the important medium correlation effects; and lastly, a recent more general overview of pairing in nuclear matter [14].

## 1.1 Green Function Formalism, Generalized Gap Equation

In this section we briefly review the main points of the treatment of a superfluid Fermi system within the Green function formalism. A detailed account is given in various textbooks [15–20].

The principal equations describing a superfluid system are the Gorkov equations, that can be considered a generalization of the Dyson equation for a normal Fermi system. A diagrammatic representation of the Gorkov equations is shown in Fig. 1(a). They express the relation between normal and anomalous propagators $G$ and $F$, defined by

$$G(1,2) = \frac{1}{i}\langle T(\Psi_1 \Psi_2^\dagger)\rangle, \tag{1a}$$

$$F(1,2) = \frac{1}{i}\langle T(\Psi_1 \Psi_2)\rangle, \tag{1b}$$

and the self-energy $\Sigma$ and gap function $\Delta$. In a homogeneous system, these four quantities depend only on a four-vector $k = (k_0, \boldsymbol{k})$ and can be written as $2 \times 2$ matrices in spin space, for example,

$$\boldsymbol{\Delta}(k) = \begin{pmatrix} \Delta_{\uparrow\uparrow} & \Delta_{\uparrow\downarrow} \\ \Delta_{\downarrow\uparrow} & \Delta_{\downarrow\downarrow} \end{pmatrix}(k). \tag{2}$$

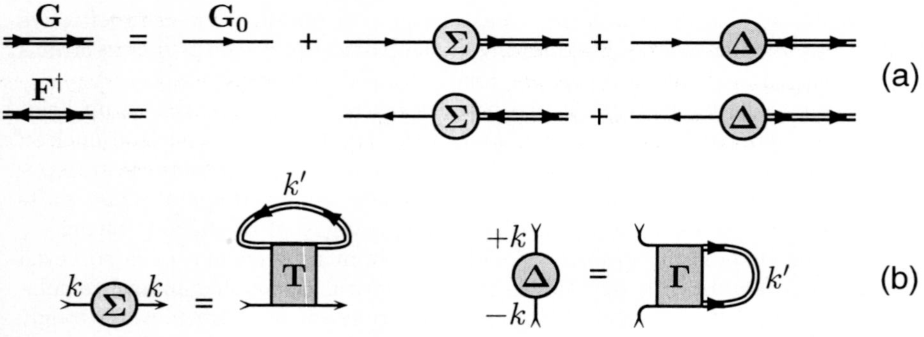

**Fig. 1.** (a) Diagrammatic representation of the Gorkov equations. (b) Equations for the self-energy $\Sigma$ and gap function $\Delta$

Using the free fermion propagator

$$G_0(k) = \frac{e^{i0k_0}}{k_0 - \mathbf{k}^2/2m + \mu + i0k_0}, \qquad (3)$$

and defining

$$\varepsilon(k) = \frac{\mathbf{k}^2}{2m} + \Sigma(k_0, \mathbf{k}) - \mu, \qquad (4)$$

one can write the system of equations explicitly as

$$\begin{pmatrix} [k_0 - \varepsilon(+k)]\mathbf{1} & \mathbf{\Delta}(k) \\ \mathbf{\Delta}^\dagger(k) & [k_0 + \varepsilon(-k)]\mathbf{1} \end{pmatrix} \begin{pmatrix} \mathbf{G}(k) \\ \mathbf{F}^\dagger(k) \end{pmatrix} = \begin{pmatrix} \mathbf{1} \\ \mathbf{0} \end{pmatrix}, \qquad (5)$$

where $\mathbf{1}$ denotes the two-dimensional unit matrix. In order to take into account at the same time pairing correlations $\Delta_S$ with spin $S = 0$ and $S = 1$, one can make the ansatz

$$\mathbf{\Delta} = \begin{pmatrix} 0 & +\Delta_0 + i\Delta_1 \\ -\Delta_0 + i\Delta_1 & 0 \end{pmatrix}, \qquad (6)$$

and equivalently for $\mathbf{F}^\dagger$. The self-energy $\mathbf{\Sigma}$ and $\mathbf{G}$ are diagonal in the spin indices. If the ground state is assumed to be time-reversal invariant, the gap function has in general the structure of a unitary triplet state [21,22], i.e., it fulfills

$$\mathbf{\Delta}^\dagger(k)\mathbf{\Delta}(k) = \Delta(k)^2 \mathbf{1}, \qquad (7)$$

where by $\Delta(k)^2$ we denote the determinant of $\mathbf{\Delta}$ in spin space.

The system (5) can then be inverted with the solution

$$G(k) = \frac{k_0 + \varepsilon(-k)}{D(k)}, \qquad (8a)$$

$$F_S^\dagger(k) = \frac{\Delta_S(k)}{D(k)}, \qquad (8b)$$

where
$$D(k) = [k_0 - \varepsilon(+k)][k_0 + \varepsilon(-k)] - \Delta_0(k)^2 - \Delta_1(k)^2 , \qquad (9)$$
that expresses the propagators $G$ and $F^\dagger$ in terms of $\Sigma$ and $\Delta$, respectively.

In order to determine uniquely the four quantities one needs two more equations, which relate $\Sigma$ and $\Delta$ to the interaction. These equations are displayed in Fig. 1(b) and read explicitly

$$\Sigma_{\alpha\alpha}(k) = \frac{1}{i}\int \frac{d^4k'}{(2\pi)^4} \sum_\beta \langle k\alpha, k'\beta|T|k\alpha, k'\beta\rangle G_{\beta\beta}(k') , \qquad (10a)$$

$$\Delta_{\alpha\beta}(k) = i\int \frac{d^4k'}{(2\pi)^4} \sum_{\alpha',\beta'} \langle k\alpha, -k\beta|\Gamma|k'\alpha', -k'\beta'\rangle F_{\alpha'\beta'}(k') , \qquad (10b)$$

where greek letters denote spin indices and $T$ and $\Gamma$ are the scattering matrix and the irreducible interaction kernel, respectively. Clearly these equations cannot be solved in full generality, but one has to recur to some approximation at this stage. The simplest, very common, BCS approximation, is to replace $T$ and $\Gamma$ by the leading term, namely the bare interaction $V$. In this case the interaction is energy independent and the $k_0$ integration in (10a,10b) can be carried out trivially, leading to

$$\Sigma(k) = \sum_{k'} \frac{v_{k'}^2}{2}\Big[\langle k, k'|V_0 + 3V_1|k, k'\rangle \\ - \langle k, k'|3V_1 - V_0|k', k\rangle\Big] , \quad v^2 = \frac{1}{2}\left(1 - \frac{\varepsilon}{E}\right) , \qquad (11a)$$

$$\Delta_S(k) = \sum_{k'} (u_S v)_{k'} \langle +k', -k'|V_S|+k, -k\rangle_a , \quad u_S v = \frac{-\Delta_S}{2E} , \qquad (11b)$$

where
$$E^2 = \varepsilon^2 + \Delta_0^2 + \Delta_1^2 , \quad \varepsilon = \frac{k^2}{2m} + \Sigma(k) - \mu . \qquad (12)$$

Together with an equation fixing the chemical potential $\mu$ for given density $\rho$,
$$\rho = 2\sum_k v_k^2 , \qquad (13)$$

this is the coupled set of equations that needs to be solved in order to find the Hartree-Fock self-energy $\Sigma_{\rm HF}(k)$ and the BCS gap function $\Delta_{\rm BCS}(k)$ in a superfluid system. In contrast to a normal Fermi system, the smooth occupation numbers $v_k^2$ instead of the Fermi function $\theta(k_F - |k|)$ appear in the HF equation.

## 2 BCS Approximation

In this section we present the solutions of the BCS gap equation

$$\Delta_{TS}(k) = -\sum_{k'} \langle k|V_{TS}|k'\rangle \frac{\Delta_{TS}(k')}{2E(k')} , \qquad (14a)$$

$$\rho = \frac{k_F^3}{3\pi^2} = 2\sum_k \frac{1}{2}\left[1 - \frac{\epsilon(k)}{E(k)}\right] , \qquad (14b)$$

where
$$E(\bm{k})^2 = \epsilon(\bm{k})^2 + \sum_{T,S=0,1} \Delta_{TS}(\bm{k})^2 , \quad \epsilon(\bm{k}) = e(\bm{k}) - \mu \tag{15}$$

with $\mu$ being the chemical potential and $e(\bm{k})$ the single-particle spectrum. Different realistic nucleon-nucleon potentials $V$ [23–28] will be used as input. The equations are valid for pure neutron matter ($T = 1$) and also for *symmetric* nuclear matter ($T = 0, 1$), if the derivation in the previous section is extended to include isospin quantum number $T$ in analogy to the spin $S$.

## 2.1 Pairing in Different Partial Waves

In order to reduce the three-dimensional integral equation (14a) to a set of one-dimensional ones, it is advantageous to perform partial wave expansions of the potential and the gap function. In this way one arrives at separate equations in the different ($TSLL'$) channels of the interaction, provided an angle-average approximation is made by replacing $\Delta(\bm{k})^2 \to \int d\hat{\bm{k}}/4\pi\, \Delta(\bm{k})^2$ in (15). The following equations for the partial wave components of the gap function are then obtained [12,29–33]:

$$\Delta_{TSL}(k) = -\frac{1}{\pi} \int_0^\infty dk' k'^2 \sum_{L'} \frac{V_{LL'}^{TS}(k,k')}{\sqrt{\epsilon(k')^2 + \Delta(k')^2}} \Delta_{TSL'}(k') , \tag{16}$$

where
$$\Delta(k)^2 = \sum_{T,S,L} \Delta_{TSL}(k)^2 , \tag{17}$$

and with the matrix elements of the bare potential in momentum space

$$V_{LL'}^{TS}(k,k') = \int_0^\infty dr\, r^2\, j_{L'}(k'r)\, V_{LL'}^{TS}(r)\, j_L(kr) . \tag{18}$$

It should be noted that the different equations are still coupled due to the fact that the total gap appearing in the denominator on the r.h.s. of (16) is the r.m.s. value of the gaps in the different partial waves. The gap equation allows in principle the coexistence of pairing correlations with different quantum numbers ($TS$), even though the different ($TS$) channels are not mixed by the interaction. In practice, however, so far no such mixed solutions of the gap equation have been found: even if at a given density two or more uncoupled solutions exist, the strong nonlinear character of the gap equation prohibits a coupled solution. (In finite nuclei such mixed solutions seem to exist under certain conditions [34]). This means that in practice this ($TS$)-coupling can be neglected and that at a given density the solution of the uncoupled gap equation with the largest gap is the energetically flavored one.

The only case when it is clearly necessary to keep the coupled equations is the mixing of partial waves due to the tensor potential. In this case the gap equation can be written in matrix form (for given $S = 1, T, L$):

$$\begin{pmatrix} \Delta_L \\ \Delta_{L+2} \end{pmatrix}(k) = -\frac{1}{\pi} \int_0^\infty dk' k'^2 \frac{1}{E(k')} \begin{pmatrix} V_{L,L} & V_{L,L+2} \\ V_{L+2,L} & V_{L+2,L+2} \end{pmatrix}(k,k') \begin{pmatrix} \Delta_L \\ \Delta_{L+2} \end{pmatrix}(k') \tag{19}$$

with

$$E(k)^2 = [e(k) - \mu]^2 + \Delta_L(k)^2 + \Delta_{L+2}(k)^2 \ . \tag{20}$$

This is relevant equation for the $^3SD_1$ ($T = 0$) and $^3PF_2$ ($T = 1$) channels, for example.

Let us finally mention that usually, apart from the $^3SD_1$ channel, the two equations (14a) and (14b) can be decoupled by setting $\mu = e(k_F)$. The reason is the small value of the ratio $\Delta/\mu$, so that a Fermi surface is still quite well defined.

We come now to the presentation of the results that are obtained by solving the previous equations numerically, using a kinetic energy spectrum $e(k) = k^2/2m$ for the moment. In practice, one finds in pure neutron matter ($T = 1$) gaps only in the $^1S_0$ [12,13,35–38] and $^3PF_2$ [12,29–33] partial waves. They are reported in Figs. 2 and 3, respectively. It can be observed that the maximum pairing gap is about 3 MeV in the $^1S_0$ channel and of the order of 1 MeV in the $^3PF_2$ wave. It is remarkable that solutions obtained with different nucleon-nucleon potentials are nearly indistinguishable in the $^1S_0$ case, whereas in the $^3PF_2$ wave such good agreement can only be observed up to $k_F \approx 2\,\text{fm}^{-1}$ (with the exception of the Argonne $V_{14}$ potential that is not very well fitted to the

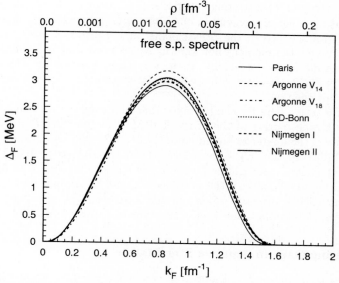

**Fig. 2.** $^1S_0$ gap evaluated in BCS approximation with free single-particle spectrum and different potentials

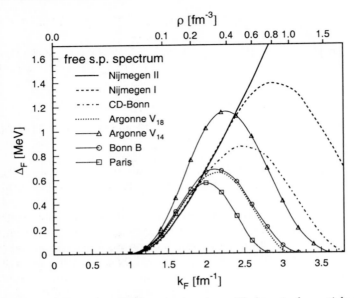

**Fig. 3.** $^3PF_2$ gap evaluated in BCS approximation with free single-particle spectrum and different potentials

phase shifts), from where on the predictions start to diverge from each other. The reason [32,38] is the fact that the various potentials are constrained by the phase shifts only up to scattering energies $E_{\text{lab}}$ of about 350 MeV, which roughly corresponds to a Fermi momentum of $k_F \approx \sqrt{mE_{\text{lab}}/2} \approx 2\,\text{fm}^{-1}$. Thus even on the BCS level the gap in the $^3PF_2$ channel at a neutron density higher than $\approx 0.3\,\text{fm}^{-3}$ is at the moment not known. Apart from that it is clear that the BCS approximation is not reliable at the very large densities for which a gap is predicted in Fig. 3. However, nobody has so far attempted to include polarization effects in this channel.

Let us mention here for completeness that in symmetric nuclear matter one finds very strong pairing of the order of 10 MeV in the $^3SD_1(T=0)$ channel, reminiscent of the deuteron bound state [12,21,22,39–46], and also a gap of the order of 1 MeV in the $^3D_2$ wave [12,47]. This, however, is probably not very relevant for neutron star physics, since a prerequisite for this $T=0$ neutron-proton pairing to take place is the existence of (nearly) isospin symmetric nuclear matter, the pairing correlations being rapidly destroyed by increasing asymmetry [48–50]. We will therefore not discuss this type of pairing further on.

## 2.2 Pairing Gaps in Neutron Star Matter

Let us now come to the $T=1$ gaps that can be expected in isospin asymmetric (beta-stable and charge neutral) neutron star matter. On the BCS level, the only influence of isospin asymmetry on these gaps is via the neutron single-particle

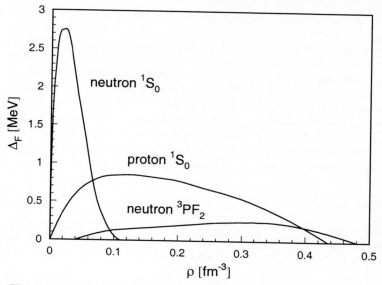

**Fig. 4.** The different $T = 1$ gaps in neutron star matter as a function of total nucleonic density

energy $e(k)$ appearing in the gap equation. Calculations have been performed using $e(k)$ determined in the BHF approximation extended to asymmetric nuclear matter [30,31,51–56], and typical results are shown in Fig. 4. One observes that $^1S_0$ pairing can take place independently in the neutron and in the proton component of the matter. If plotted as a function of total baryon density, the neutron pairing occurs naturally at lower density and with a larger amplitude than the proton pairing, because at the higher density the proton effective mass is smaller and the pairing therefore more reduced. The same is true for the $^3PF_2$ pairing in the neutron component, which is strongly reduced with respect to the calculation with a free spectrum shown in Fig. 3 above. We stress that the results displayed in Fig. 4 can only be qualitative, because they clearly depend on the details (in particular the proton fraction) of the equation of state that is used. The results shown were obtained with a BHF EOS based on the Argonne $V_{14}$ potential and involving $n, p, e, \mu$ components [52].

## 3 Beyond BCS

In the previous section we have presented many results that were all obtained within the BCS approximation. However, as has been explained in the introduction, this approximation amounts to a mean-field approach, equivalent to and consistent with the Hartree-Fock approximation in a normal Fermi system. More precisely, the BCS approximation neglects completely any contribution beyond the bare potential to the interaction kernel $\Gamma$ appearing in the general gap equation (10b).

Going consistently beyond the BCS approximation is however a very difficult task and has been only partially achieved so far. We will review in the following sections some aspects of these extensions. We begin with a discussion of the situation at extremely low density, where certain analytical results are known. Following that, the general framework at more relevant densities will be set up, and we will briefly present the results that have been obtained so far by various authors. Finally, in the last section, we will focus on a certain part of the problem that has recently been tackled, namely the treatment of the energy dependence of the self-energy that appears in the gap equation.

## 3.1 Low Density

In order to derive an exact analytical result for the pairing gap including polarization effects that is valid at very low density (more precisely, for $k_F \ll 1/|a|$, where $a$ is the relevant scattering length), we begin again with the BCS gap equation,

$$\Delta_k = -\sum_{k'} V_{kk'} \frac{1}{2E_{k'}} \Delta_{k'} \,, \quad E_k = \sqrt{(e_k - e_F)^2 + \Delta_k^2} \,. \tag{21}$$

It is then useful [36,54,57] to introduce a modified interaction $T$ that is given by the solution of the integral equation

$$T_{kk'} = V_{kk'} - \sum_{k''} T_{kk''} F_{k''} V_{k''k'} \,, \tag{22}$$

where $F_k$ is for the moment an arbitrary function. (We have used the symbol $T$, although in general this quantity is not to be identified with the scattering matrix). Making use of this equation, the gap equation is transformed into

$$\Delta_k = -\sum_{k'} T_{kk'} \left( \frac{1}{2E_{k'}} - F_{k'} \right) \Delta_{k'} \,. \tag{23}$$

Of particular interest is now the choice $F_k = \text{sgn}(k - k_F)/2E_k$, which leads to the set of equations

$$\Delta_k = -2 \sum_{k' < k_F} T_{kk'} \frac{1}{2E_{k'}} \Delta_{k'} \,, \tag{24}$$

$$T_{kk'} = V_{kk'} - \sum_{k''} T_{kk''} \frac{\text{sgn}(k'' - k_F)}{2E_{k''}} V_{k''k'} \,. \tag{25}$$

Therefore, in the limit $\Delta/e_F \to 0$ that is approached with vanishing density, $T$ does become identical to the free scattering matrix, because $E_k \to |e_k - e_F|$ in this situation. At the same time, in the gap equation (24) the interaction $T$

is now cut off at $k' = k_F$ (at the cost of introducing a factor 2), so that with vanishing density it is ultimately sufficient to use the low-energy result [58] for the $T$-matrix:

$$T_{kk'} \to T_{00} = \frac{4\pi a_{nn}}{m}, \qquad (26)$$

where $a_{nn} = -18.8\,\text{fm}$ is the neutron-neutron scattering length. This yields finally the gap equation

$$1 = -\frac{4k_F a_{nn}}{\pi} \int_0^1 dx \frac{x^2}{\sqrt{(1-x^2)^2 + (\Delta/e_F)^2}}, \qquad (27)$$

which in the limit $\Delta/e_F \to 0$ is solved by [37,59–61]

$$\Delta(k_F) \stackrel{k_F \to 0}{\longrightarrow} \Delta_0(k_F) = \frac{8}{e^2} \frac{k_F^2}{2m} \exp\left[\frac{\pi}{2k_F a_{nn}}\right]. \qquad (28)$$

This is the universal asymptotic result for the BCS pairing gap in a low-density Fermi system with negative scattering length. Unfortunately its validity is limited to the region $k_F \ll 1/|a_{nn}| \approx 0.05\,\text{fm}^{-1}$, far below the densities of interest for neutron star physics or even pairing in finite nuclei.

Going now beyond the BCS approximation, in the low-density limit one should take into account the corrections to the interaction kernel that are of leading order in density. Diagrammatically these are the polarization diagrams of first order (i.e., comprising one polarization "bubble") that are displayed in Fig. 5. The interaction appearing in these diagrams is in the present case the free scattering matrix $T$. It can be shown [59] that in the low-density limit it is again sufficient to neglect the momentum dependence of the $T$-matrix, as in (26). One obtains then for the lowest-order polarization interaction

$$W_{^1S_0}(k,k') = -\frac{T_{00}^2}{8\pi} \frac{1}{2kk'} \int_{|k-k'|}^{k+k'} dq\, q\, \Pi(q), \qquad (29)$$

where

$$\Pi(q) = -\frac{mk_F}{\pi^2} \left[\frac{1}{2} + \frac{1-x^2}{4x} \ln\left|\frac{1+x}{1-x}\right|\right], \quad x = \frac{q}{2k_F} \qquad (30)$$

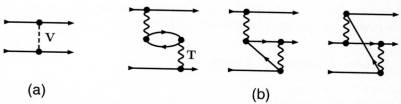

**Fig. 5.** (a) Bare potential and (b) first-order (direct and exchange) polarization diagrams contributing to the interaction kernel in the low-density limit

is the static Lindhard function [58]. Thus the lowest-order polarization modifies the BCS interaction kernel (18) by a (repulsive) term proportional to $k_F$, whereas any other polarization diagram contributes only in higher order of $k_F$.

We can use this result and insert it into the previously obtained approximation in the BCS case:

$$\Delta(k_F) \stackrel{k_F \to 0}{\longrightarrow} \frac{8}{e^2} \frac{k_F^2}{2m} \exp\left[\frac{\pi/2}{\kappa + c\kappa^2}\right] , \quad \kappa = k_F a_{nn} , \tag{31}$$

where

$$c = -\frac{2\pi}{mk_F} \int_0^{2k_F} \frac{dq\, q}{2k_F^2} \Pi(q) = \frac{2}{3\pi}(1 + 2\ln 2) \approx 0.506 \tag{32}$$

accounts for the polarization effects to first order. Expanding now the argument of the exponential up to second order in $\kappa$, one obtains for the ratio relative to the BCS value, (28),

$$\frac{\Delta(k_F)}{\Delta_0(k_F)} = \exp\left[-\frac{\pi}{2}c\left[1 - c\kappa + \mathcal{O}(\kappa^2)\right]\right] \tag{33}$$

$$\approx \left[\frac{1}{(4e)^{1/3}}\right]^{(1-c\kappa)} \tag{34}$$

$$\stackrel{k_F \to 0}{\longrightarrow} \frac{1}{(4e)^{1/3}} . \tag{35}$$

Let us stress that the above "derivation" of this result can only be considered heuristic. A rigorous proof was given originally in Ref. [59].

Therefore, one arrives at the striking conclusion that in the low-density limit the polarization corrections suppress the BCS gap by a factor $(4e)^{-1/3} \approx 0.45$, independent of the strength of the interaction $a_{nn}$. This is quite surprising, since the polarization interaction clearly vanishes with vanishing density; its effect on the pairing gap, however, does not. The reason is the nonanalytical dependence of the gap on the interaction strength, as expressed by (28).

All this means that the BCS approximation cannot even be trusted at very low density, and that one can in general expect quite strong modifications due to polarization effects at higher density as well.

This low-density behavior of neutron pairing could be met in the study of exotic nuclei with a long density tail or in nuclei embedded in a neutron matter environment, as occurring in the neutron star crust.

## 3.2 General Polarization Effects

To go beyond the simple low-density approximations derived before requires considerable effort and has in fact so far not been accomplished in a satisfactory manner, so that ultimate results cannot be presented here. The reason is that many effects that could be neglected in the low-density limit become important now, and that on the other hand the pairing gap is extremely sensitive to even slight changes of the interaction kernel.

First, outside the low-density region $k_F \ll 1/|a_{nn}|$, the interaction kernel has to be extended beyond the lowest-order polarization diagrams. This means summing up polarization diagrams of all orders in the particle-particle channel, but also including them in the particle-hole channel, replacing the $T$-matrix by the general particle-hole interaction $F$. In this way a self-consistent scheme is established that is depicted in Fig. 6. It requires as input the Brueckner $G$-matrix

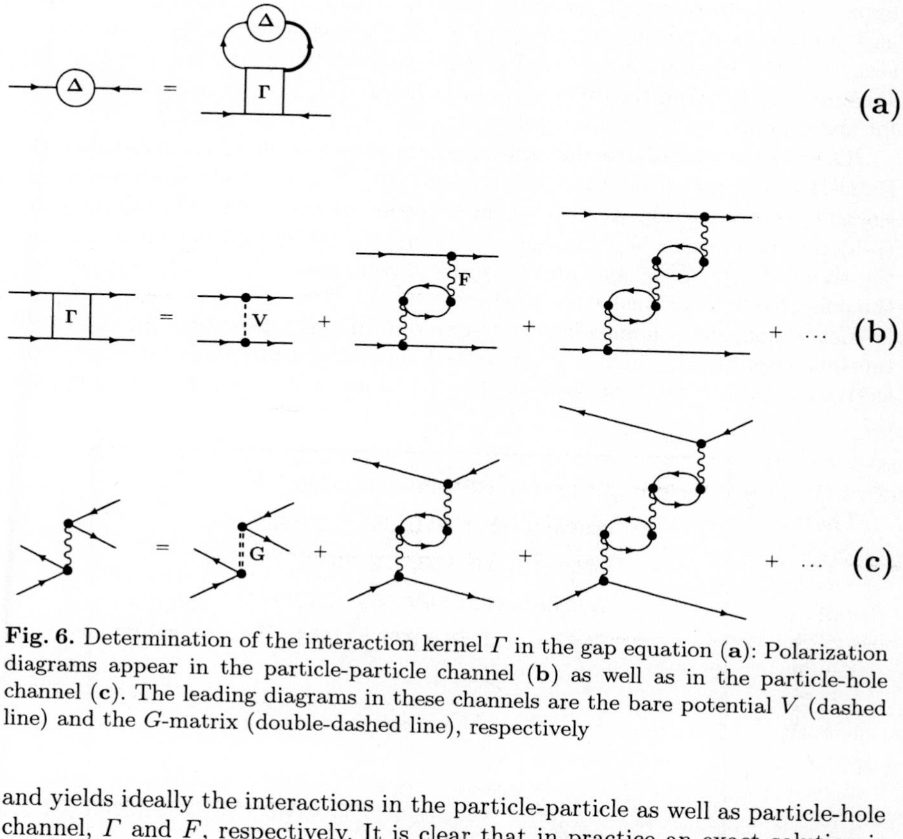

**Fig. 6.** Determination of the interaction kernel $\Gamma$ in the gap equation (a): Polarization diagrams appear in the particle-particle channel (b) as well as in the particle-hole channel (c). The leading diagrams in these channels are the bare potential $V$ (dashed line) and the $G$-matrix (double-dashed line), respectively

and yields ideally the interactions in the particle-particle as well as particle-hole channel, $\Gamma$ and $F$, respectively. It is clear that in practice an exact solution is impossible, but that usually crude approximations have to be performed that cast a doubt on the reliability of the results that are obtained. We will later discuss this point in some more detail.

Second, related to the previous item, the energy dependence of the full gap equation [see (10b)] needs to be taken into account. This is obvious, since the interaction kernel becomes now a complex, energy-dependent quantity. To our knowledge, this problem has so far not been studied in detail in the literature. It is therefore not known how far the gap could be changed by this more elaborate treatment of the equations.

Third, and in connection with the two previous points, the choice of a particular interaction kernel requires also the choice of a compatible self-energy appearing in the gap equation. This will be explained in more detail in the following section, as it has recently been addressed in the literature.

It should be clear by now that the influence of medium effects on pairing constitutes an extremely difficult problem. Consequently the results that can be found in the literature [13,62–66] addressing this task in certain approximations agree only on the fact that generally a strong reduction with respect to the BCS gap is obtained. A collection of these results is displayed in Fig. 7. It can be seen that the precise amount and density dependence of the suppression vary substantially between the different approaches and must be considered unknown for the time being.

The most advanced description of medium polarization effects is based on the Babu-Brown induced interaction model [67–72]. The microscopic derivation of the effective interaction starts from the following physical idea: The particle-hole (p-h) interaction can be considered as made of a *direct* component containing the short-range correlations and an *induced* component due to the exchange of the collective excitations of the medium.

Let us consider a homogeneous system of fermions interacting via an instantaneous potential $V$, which is also translationally and rotationally invariant. Collective excitations are described by the ring series, which can easily be summed

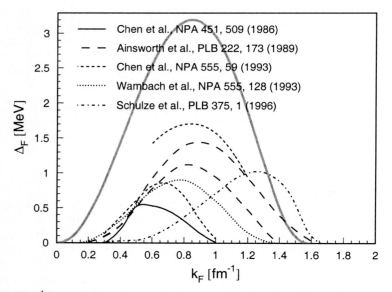

**Fig. 7.** The $^1S_0$ gap in pure neutron matter predicted in several publications taking account of polarization effects. The curve in the background shows the BCS result

up [58], and the p-h interaction can be written as (disregarding for the moment spin degrees of freedom)

$$V_{\text{ph}}(q) = V(q) + \frac{\Pi(q)V(q)^2}{1 - \Pi(q)V(q)} . \tag{36}$$

In this simple case the interaction itself plays the role of the direct term and the sum of the ring series that of the induced term.

In the nuclear case the presence of the hard core imposes the bare interaction $V$ to be renormalized in order to incorporate the short-range correlations. This goal is reached by introducing the $G$-matrix, which sums particle-particle (p-p) ladder diagrams to all orders, and the diagrammatic expansion can be recast just replacing $V$ by $G$. But now the new ring series cannot be summed up any longer, mainly since the $G$-matrix is nonlocal. An averaging procedure has been devised to bring the $G$-matrix into a local form $G(q)$ [66]. Then, in analogy to (36), the p-h interaction $F$ is given by

$$F(q) = G(q) + \frac{\Pi(q)G(q)^2}{1 - \Pi(q)G(q)} . \tag{37}$$

In a simplified version of the theory the direct term now coincides with the $G$-matrix and the induced term is the approximate sum of the renormalized ring diagrams. In Fig. 6 the direct term is represented by the first diagram on the r.h.s. of the series (c). The next diagrams form the ring (or bubble) series.

But, since the RPA series with the $G$-matrix produces a too strong polarizability of nuclear matter, in the Babu-Brown approach it has been proposed to include in the RPA series the full p-h interaction itself, since the particle (hole) coupling vertex with the p-h bubble can indeed be identified with the irreducible p-h interaction. The series (c) of Fig. 6 is, in fact, the final result of such a procedure, where the wiggles represent the effective p-h interaction $F$ on either side. If we denote by $F_d$ the direct interaction ($G$-matrix in our case) and by $F_i$ the induced interaction, the effective p-h interaction can be written in the form (at the Landau limit, in which $p_1$ and $p_2$ are restricted to the Fermi surface)

$$F(\boldsymbol{p}_1, \boldsymbol{p}_2) = F_d(\boldsymbol{p}_1, \boldsymbol{p}_2) + F_i(\boldsymbol{p}_1, \boldsymbol{p}_2; F) , \tag{38}$$

where $F_i$ is, as said before, the RPA series of Fig. 6(c) with the $G$-matrix replaced by $F$ itself. The previous equation clearly entails a self-consistent procedure to determine the interaction $F(\boldsymbol{p}_1, \boldsymbol{p}_2)$. Referring to [66] for details, one may reduce (38) to a numerically tractable form

$$F(q) = G(q) + \frac{\Pi(q)F(q)^2}{1 - \Pi(q)F(q)} , \tag{39}$$

which corresponds to replacing the $G$-matrix by $F$ in the induced term.

Once the irreducible p-h interaction $F$ has been determined, one can construct the irreducible p-p interaction $\Gamma$ by performing the transformation of the matrix elements of the interaction from the p-h to the p-p channel. However, this is not enough, since in the p-p channel a set of additional diagrams must be added arising from p-h diagrams which are reducible in that representation. After including these terms, the interaction contains both direct and exchange terms, which guarantees antisymmetry and Landau sum rules and, in addition, it should simultaneously make the nuclear matter Landau parameter $F_0$ less negative so that the stability condition is satisfied. The first contributions to the p-p interaction are depicted in line (b) of Fig. 6. We stress once more that they describe the influence of the medium polarization on the nucleon-nucleon interaction within the induced interaction model of Babu-Brown.

We discuss now the effects of medium polarization on the superfluidity of neutron stars in the channel $^1S_0$. First of all, the irreducible p-p interaction to be used in the pairing problem must not include any ladder sum already included in the gap equation, and therefore the first term in line (b) of Fig. 6 is the bare neutron-neutron interaction $V$. The next terms include the irreducible p-h interaction in the vertices of the p-h bubble. If their momentum dependence is neglected, these vertices can be identified with the Landau parameters [69].

In neutron matter the polarization of the medium is due to density fluctuations and spin-density fluctuations given by

$$\delta\rho_{\mathbf{k}} = \delta\rho_{\mathbf{k}\uparrow} + \delta\rho_{\mathbf{k}\downarrow}, \tag{40a}$$

$$\delta\rho_{\mathbf{k}} = \delta\rho_{\mathbf{k}\uparrow} - \delta\rho_{\mathbf{k}\downarrow}, \tag{40b}$$

respectively. Solving the Babu-Brown self-consistent equation, (39), with the $G$-matrix as direct interaction and the renormalized RPA series as induced interaction, one determines the p-h interaction and eventually, after the Landau angle expansion, the lowest-order Landau parameters $F_0$, related to the nuclear compression modulus, and $G_0$, related to the spin waves. (In the following the interaction will be expressed in terms of these two Landau parameters for simplicity). Then the effective interaction is calculated including the diagrams of Fig. 6(b). The pairing interaction in the $^1S_0$ channel is then given by

$$\Gamma_{^1S_0} = V_{^1S_0} + \frac{1}{2k_F^2} \int_0^{2k_F} dq\, q \left[ \frac{F_0^2 \Pi(q)}{1 - F_0 \Pi(q)} - \frac{3G_0^2 \Pi(q)}{1 - G_0 \Pi(q)} \right]. \tag{41}$$

This equation shows that the medium screening effect is determined by the competition between the attractive term induced by density fluctuations and the repulsive term induced by spin-density fluctuations (first and second term in the bracket, respectively).

As depicted in Fig. 8, the effect of the medium polarization is an overall suppression of the gap due to the prevalence of the spin-density fluctuations over the density fluctuations. A similar effect is found in a less crude calculation [66], as shown in Fig. 7, where the peak value is shifted to higher density. Such a suppression is common to all calculations existing in the literature even if, as to its magnitude, the different predictions do not agree with each other, as shown in Fig. 7.

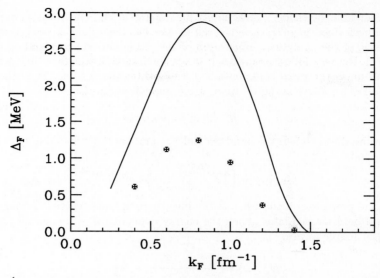

**Fig. 8.** $^1S_0$ pairing gap in neutron matter as a function of the Fermi momentum $k_F$. The full curve corresponds to using the bare $V_{14}$ potential as direct interaction. The symbols show the effect of the medium polarization described in terms of the Landau parameters, according to (41)

### 3.3 Self-energy Effects

Dynamical effects of the interaction on the gap function have been completely left aside from the discussion on the medium polarization. So far no solution of the gap equation has been attempted considering the irreducible interaction block as an energy-dependent quantity. On the other hand the energy dependence in the self-energy can affect deeply the magnitude of the energy gap in a strongly correlated Fermi system such as nucleon matter [73,74].

To discuss self-energy effects we come back to the generalized gap equation presented in section 1.1. Let us rewrite (10b) in the following form (for the $^1S_0$ channel)

$$\Delta_k(\omega) = -\int \frac{d^3 k'}{(2\pi)^3} \int \frac{d\omega'}{2\pi i} \Gamma_{k,k'}(\omega, \omega') \frac{\Delta_{k'}(\omega')}{D_{k'}(\omega')}, \qquad (42)$$

with [cf. (8a) and (9)]

$$-D_k(\omega) = [G_k(-\omega) G_k^s(\omega)]^{-1} = G_k^{-1}(\omega) G_k^{-1}(-\omega) + \Delta_k^2(\omega). \qquad (43)$$

The functions $G_k(\omega)$ and $G_k^s(\omega)$ are the nucleon propagators of neutron matter in the normal state and in the superfluid state, respectively. The $\omega$-symmetry in the two propagators is to be traced to the time-reversal invariance of the Cooper pairs. The effective interaction $\Gamma$ is the block of all irreducible diagrams of the interaction. Since we want to focus only on the self-energy effects, we assume the interaction to be the bare interaction $V_{k,k'}$, as in BCS. Then the pairing gap does not depend on the energy (static limit), i.e., $\Delta_k(\omega) \equiv \Delta_k$, and the $\omega$-integration can be performed in (42), once the self-energy has been determined. A general discussion of the analytic $\omega$-integration of the gap equation has been is given in Ref. [74]. Here we follow a simplified treatment based on the fact that, at each momentum $k$, the main contribution to the $\omega$-integration comes from the pole of $G_k(\omega)$. This latter is the solution of the implicit equation

$$\omega_k = k^2/2m + \Sigma_k(\omega_k) - \mu \,. \tag{44}$$

Expanding the self-energy around the pole $\omega_k$ amounts to expanding $G^{-1}$ itself, which yields

$$G_k^{-1}(\omega) \approx \left(1 - \left.\frac{\partial \Sigma}{\partial \omega}\right|_{\omega=\omega_k}\right)(\omega - \omega_k)\,, \tag{45}$$

where the prefactor on the r.h.s. is the inverse of the quasiparticle strength $Z_k$ that will be discussed later. Then the energy dependence of the kernel $D_k^{-1}(\omega)$ takes the simple form

$$D_k^{-1}(\omega) = \frac{Z_k^2}{\omega^2 - \omega_k^2 - (Z_k \Delta_k)^2}\,, \tag{46}$$

and the $\omega$-integration can be performed in the usual way, leading to

$$\Delta_k = -\int \frac{d^3 k'}{(2\pi)^3}\, (Z_k V_{k,k'} Z_{k'})\, \frac{\Delta_{k'}}{2\sqrt{\omega_{k'}^2 + \Delta_{k'}^2}}\,. \tag{47}$$

Comparing with the BCS result, (14a), the new gap equation contains the quasiparticle strength $Z$ to the second power. Since $Z$ is significantly less than one in a strongly correlated Fermi system, a substantial suppression of the energy gap is to be expected.

Moreover $Z$ deviates from unity only in a narrow region around the Fermi surface and hence it is not a severe approximation to restrict the $\omega$-integration to only the pole part at the Fermi energy. Expanding then $\Sigma_k(\omega)$ around the Fermi surface ($\omega = 0$ and $k = k_F$), (44) is easily solved and we get $\omega_k \approx (k^2 - k_F^2)/2m^*$, where $m^*$ is the effective mass at $k_F$ [see (53)]. The gap equation becomes

$$\Delta_k = -Z_F^2 \int \frac{d^3 k'}{(2\pi)^3}\, \frac{V_{k,k'} \Delta_{k'}}{2\sqrt{[(k'^2 - k_F^2)/2m^*]^2 + \Delta_{k'}^2}}\,, \tag{48}$$

where $Z_F^2$ is the quasiparticle strength at the Fermi surface.

As is well known the pairing modifies the chemical potential which is calculated self-consistently with the gap equation from the closure equation for the average density of neutrons

$$\rho = 2 \int \frac{d^3 k}{(2\pi)^3} \int \frac{d\omega}{2\pi i}\, G_k^s(\omega^+) \tag{49}$$

$$\approx Z_F \int \frac{d^3 k}{(2\pi)^3} \left[1 - \frac{(k^2 - k_F^2)/2m^*}{\sqrt{[(k^2 - k_F^2)/2m^*]^2 + Z_F^2 \Delta_k^2}}\right]\,. \tag{50}$$

The latter approximation is sufficient to investigate the self-energy effects.

Before we present the predictions based on (48), we discuss some properties of the self-energy $\Sigma_k(\omega)$ of neutron matter. In the Brueckner approach [75] the perturbative expansion of $\Sigma$ can be recast according to the number of hole lines as follows

$$\Sigma_k(\omega) = \Sigma_k^{(1)}(\omega) + \Sigma_k^{(2)}(\omega) + \dots . \qquad (51)$$

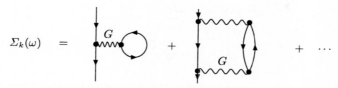

**Fig. 9.** Hole-line expansion of the self-energy

The on-shell values of $\Sigma^{(1)}$ represent the Brueckner-Hartree-Fock (BHF) mean field (first diagram in Fig. 9); the ones of $\Sigma^{(2)}$ represent the so-called rearrangement term (second diagram in Fig. 9), which is the largest contribution due to ground-state correlations. The off-shell values of the self-energy are required to solve the generalized gap equation. Fig. 10 displays a typical result for the off-shell neutron self-energy $\Sigma_k(\omega)$ calculated up to the second order of the hole-line expansion. The calculations are based on Brueckner theory adopting the continuous choice as auxiliary potential [76].

Since we are interested in the behavior of the self-energy around the Fermi surface ($k = k_F$ and $\omega = 0$), we may use the expansion

$$\Sigma_k(\omega) \approx \Sigma_{k_F}(0) + \left.\frac{\partial \Sigma}{\partial \omega}\right|_F \omega + \left.\frac{\partial \Sigma}{\partial k}\right|_F (k - k_F) . \qquad (52)$$

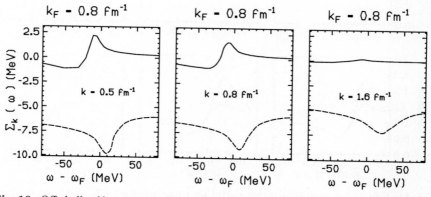

**Fig. 10.** Off-shell self-energy in neutron matter at $k_F = 0.8\,\mathrm{fm}^{-1}$. *Solid curves*: BHF approximation $\Sigma^{(1)}$. *Dashed curves*: rearrangement contribution $\Sigma^{(2)}$. The Argonne $V_{14}$ potential [24] has been used in the Brueckner calculations

In this case the quasiparticle energy takes the simple form

$$\omega_k \approx \frac{k^2 - k_F^2}{2m} \frac{1 + (m/k_F)(\partial \Sigma/\partial k)|_F}{1 - (\partial \Sigma/\partial \omega)|_F} = \frac{k^2 - k_F^2}{2m^*}. \tag{53}$$

In the previous equation we have introduced the effective mass $m^*/m$ as the product of the $e$-mass $m_e$ and the $k$-mass $m_k$, which are defined respectively as follows [75]

$$\frac{m_e}{m} = 1 - \left.\frac{\partial \Sigma}{\partial \omega}\right|_F = \frac{1}{Z_F}, \tag{54a}$$

$$\frac{m_k}{m} = \left[1 + \frac{m}{k_F}\left.\frac{\partial \Sigma}{\partial k}\right|_F\right]^{-1}. \tag{54b}$$

The partial derivatives are evaluated at the Fermi surface. The $k$-mass is related to the non-locality of the mean field and, in the static limit ($\omega = \omega_F$), coincides with the effective mass. This quantity is of great interest in heavy-ion collision physics, since the transverse flows are very sensitive to the momentum dependence of the mean field. The $e$-mass is related to the quasi-particle strength. This latter gives the discontinuity of the neutron momentum distribution at the Fermi surface, and measures the amount of correlations included in the adopted approximation.

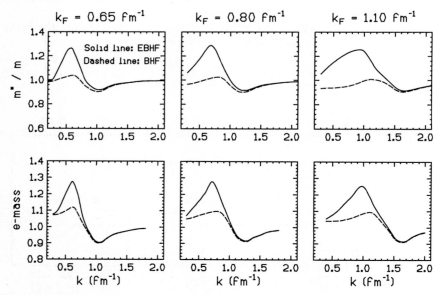

**Fig. 11.** Effective masses in neutron matter for three densities: effective mass $m^*$ (upper panels) and $e$-mass (lower panels). Dashed lines correspond to $\Sigma^{(1)}$ and solid lines to $\Sigma^{(1)} + \Sigma^{(2)}$.

From $\Sigma_k(\omega)$ the effective masses are extracted according to (54a) and (54b). They are depicted in Fig. 11, where the full calculation is compared to that including only the BHF self-energy. We may distinguish two momentum intervals: at $k \approx k_F$ the momentum dependence of the effective mass $m^*$ is characterized by a bump, whose peak value exceeds the value of the bare mass; far above $k_F$ the bare mass limit is approached. The contribution from the rearrangement term exhibits a pronounced enhancement in the vicinity of the Fermi energy, which is to be traced back to the high probability amplitude for p-h excitations near $\epsilon_F$ [76]. At high momenta this contribution vanishes. One should take into account that in this range of $k_F$ the neutron density is quite small ($\rho = 0.074\,\mathrm{fm}^{-3}$ at the maximum $k_F = 1.3\,\mathrm{fm}^{-1}$). This behavior of the effective mass $m^*$ is mostly due to the $e$-mass, as shown in the lower panels of Fig. 11. In all panels of Fig. 11 is also reported for comparison the effective mass in the BHF limit (only $\Sigma^{(1)}$ included), which exhibits a much less pronounced bump at the Fermi energy.

With the off-shell values of the self-energy discussed above as input, the gap equation has been solved in the form of (48), coupled with (50) [77]. This is a quite satisfactory approximation, especially in view of studying the self-energy effects on pairing. The Argonne $V_{14}$ potential has been adopted as pairing interaction, which is consistent with the self-energy data for which the same force has been used. The results are reported in Fig. 12 for a set of different $k_F$-values. The diamonds connected by a solid line represent the solution of (48) replacing the effective mass by the free one, $m^* = m$, together with $Z = 1$. This is very close to the prediction obtained from the BHF approximation for the effective

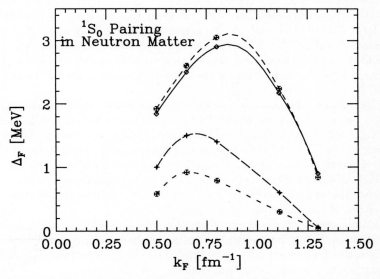

**Fig. 12.** Energy gap in different approximations for the self-energy: free s.p. spectrum (*solid line*); effective mass with $Z_F = 1$ (*upper dashed line*); $Z$ from $\Sigma^{(1)}$ (*long dashed line*); $Z$ from $\Sigma^{(1)} + \Sigma^{(2)}$ (*lower dashed line*)

mass $m^*$, but still keeping $Z = 1$ (fancy crosses connected by the dashed line). This similarity stems from the fact that at the Fermi surface $m^*/m$ from BHF is close to one, as shown in Fig. 11. The self-energy effects are estimated in two approximations. In the first one $m^*$ and the $Z$-factor are calculated from a BHF code. In the considered density domain the $Z$-factor ranges around 0.9. Despite its moderate reduction a dramatic suppression of the gap is obtained, as shown by the long dashed line in Fig. 12. It is due to the exponential dependence of the gap on all quantities. Still a further but more moderate reduction is obtained when the rearrangement term is included in the second approximation. The smaller $Z$-factor ($Z \approx 0.83$ at $k_F = 0.8\,\text{fm}^{-1}$) is to a certain extent counterbalanced by an increase of the effective mass ($m^*/m \approx 1.2$ at the same $k_F$).

From the previous discussion we may conclude that including self-energy effects is an important step forward in understanding the pairing in nucleon matter, but one should also include, on an equal footing, vertex corrections.

## 4 Conclusions

This chapter addressed the present status of the theoretical progress on the superfluidity of neutron matter and the microscopic calculations of nucleonic pairing gaps relevant for the conditions that are encountered in the interior of a neutron star. We have seen that most of the results that can be found in the literature, are obtained within the BCS approximation.

Unfortunately, as has also been pointed out, this approximation cannot be considered reliable in any region of density, since neutron matter is a strongly correlated Fermi system. The same nucleons that undergo pairing instability and form Cooper pairs also act to screen the pairing interaction. This requires necessarily to improve the interaction kernel by in-medium polarization corrections that, at the present time, can be treated with confidence only at very low density, where the ring series can be truncated at the order of the one-bubble term.

The induced interaction approach is a promising candidate to describe these effects, but its predictions are still affected by too severe approximations. The existing calculations addressing this problem point to a substantial reduction of pairing in the $(T = 1)$ $^1S_0$ channel with respect to the BCS results. In the $^3PF_2$ channel, no estimate of polarization effects has been made so far at all.

Also self-energy effects, when properly included in the gap equation, strongly influence the pairing mechanism. We saw in fact that the gap is reduced with the square of the quasi-particle strength $Z$.

In any case it remains the problem of a complete solution of the generalized gap equation, taking into account simultaneously screening and self-energy corrections, which moreover have to be treated on the same footing. This latter goal remains a considerable theoretical challenge for the future.

Due to lack of space and time we have not discussed more speculative subjects like pairing in isospin asymmetric matter, relativistic effects on pairing, hyperon pairing, etc., which might also have relevance for neutron star physics.

# References

1. A. B. Migdal, Soviet Physics JETP **10**, 176 (1960).
2. J. A. Sauls, in *Timing Neutron Stars*, ed. by H. Ögelman and E. P. J. van den Heuvel, (Dordrecht, Kluwer, 1989) pp. 457.
3. *The Structure and Evolution of Neutron Stars*, Proc. US-Japan Joint Seminar, Kyoto, 6-10 November 1990, ed. by D. Pines, R. Tamagaki, and S. Tsuruta, (Addison-Wesley, Reading, 1992).
4. S. Tsuruta, Phys. Rep. **292**, 1 (1998).
5. H. Heiselberg and M. Hjorth-Jensen, Phys. Rep. **328**, 237 (2000).
6. J. Bardeen, L. N. Cooper, and J. R. Schrieffer, Phys. Rev. **108**, 1175 (1957).
7. A. Bohr, B. Mottelson, and D. Pines, Phys. Rev. **110**, 936 (1958).
8. A. Bohr, B. Mottelson, *Nuclear Structure*, Vol. 2 (Benjamin, New York, 1974).
9. L. N. Cooper, R. L. Mills, and A. M. Sessler, Phys. Rev. **114**, 1377 (1959).
10. V. J. Emery and A. M. Sessler, Phys. Rev. **119**, 248 (1960).
11. R. Tamagaki, Prog. Theor. Phys. **44**, 905 (1970).
12. T. Takatsuka and R. Tamagaki, Prog. Theor. Phys. Suppl. **112**, 27 (1993).
13. J. M. C. Chen, J. W. Clark, R. D. Davé, and V. V. Khodel, Nucl. Phys. **A555**, 59 (1993).
14. U. Lombardo, 'Superfluidity in Nuclear Matter', in *Nuclear Methods and Nuclear Equation of State*, ed. by M. Baldo, (World Scientific, Singapore, 1999) pp. 458-510.
15. A. A. Abrikosov, L. P. Gorkov, and I. E. Dzyaloshinskii, *Methods of Quantum Field Theory in Statistical Physics* (Prentice-Hall, Englewood Cliffs, 1963).
16. P. Nozières, *Le problème à N corps* (Dunod, Paris, 1963).
17. P. Nozières, *Theory of Interacting Fermi Systems* (Benjamin, New York, 1966).
18. J. R. Schrieffer, *Theory of Superconductivity* (Addison-Wesley, New York, 1964).
19. A. B. Migdal, *Theory of Finite Systems and Applications to Atomic Nuclei* (Benjamin, New York, 1964).
20. P. Ring and P. Schuck, *The Nuclear Many-Body Problem* (Springer, Berlin, 1980).
21. M. Baldo, I. Bombaci, and U. Lombardo, Phys. Lett. **B283**, 8 (1992).
22. M. Baldo, U. Lombardo, and P. Schuck, Phys. Rev. **C52**, 975 (1995).
23. M. Lacombe, B. Loiseaux, J. M. Richard, R. Vinh Mau, J. Côté, D. Pirès, and R. de Tourreil, Phys. Rev. **C21**, 861 (1980).
24. R. B. Wiringa, R. A. Smith, and T. L. Ainsworth, Phys. Rev. **C29**, 1207 (1984).
25. R. Machleidt, Adv. Nucl. Phys. **19**, 189 (1989).
26. V. G. J. Stoks, R. A. M. Klomp, C. P. F. Terheggen, and J. J. de Swart, Phys. Rev. **C48**, 792 (1993).
27. R. B. Wiringa, V. G. J. Stoks, and R. Schiavilla, Phys. Rev. **C51**, 38 (1995).
28. R. Machleidt, F. Sammarruca, and Y. Song, Phys. Rev. **C53**, 1483 (1996).
29. L. Amundsen and E. Østgaard, Nucl. Phys. **A442**, 163 (1985).
30. M. Baldo, J. Cugnon, A. Lejeune, and U. Lombardo, Nucl. Phys. **A536**, 349 (1992).
31. Ø. Elgarøy, L. Engvik, M. Hjorth-Jensen, and E. Osnes, Nucl. Phys. **A607**, 425 (1996).
32. M. Baldo, Ø. Elgarøy, L. Engvik, M. Hjorth-Jensen, and H.-J. Schulze, Phys. Rev. **C58**, 1921 (1998).
33. V. A. Khodel, V. V. Khodel, and J. W. Clark, Phys. Rev. Lett. **81**, 3828 (1998).
34. A. L. Goodman, Nucl. Phys. **A186**, 475 (1972); Phys. Rev. **C60**, 014331 (1999).
35. L. Amundsen and E. Østgaard, Nucl. Phys. **A437**, 487 (1985).

36. M. Baldo, J. Cugnon, A. Lejeune, and U. Lombardo, Nucl. Phys. **A515**, 409 (1990).
37. V. A. Khodel, V. V. Khodel, and J. W. Clark, Nucl. Phys. **A598**, 390 (1996).
38. Ø. Elgarøy and M. Hjorth-Jensen, Phys. Rev. **C57**, 1174 (1998).
39. T. Alm, G. Röpke, and M. Schmidt, Z. Phys. **A337**, 355 (1990).
40. B. E. Vonderfecht, C. C. Gearhart, W. H. Dickhoff, A. Polls, and A. Ramos, Phys. Lett. **B253**, 1 (1991).
41. T. Alm, B. L. Friman, G. Röpke, and H. Schulz, Nucl. Phys. **A551**, 45 (1993).
42. H. Stein, A. Schnell, T. Alm, and G. Röpke, Z. Phys. **A351**, 295 (1995).
43. T. Alm, G. Röpke, A. Sedrakian, and F. Weber, Nucl. Phys. **A406**, 491 (1996).
44. M. Baldo, U. Lombardo, P. Schuck, and A. Sedrakian, *Condensed Matter Theories*, Vol. 12, ed. by J. W. Clark (Nova Science Publishers, 1997), pp. 265-277.
45. Ø. Elgarøy, L. Engvik, E. Osnes, and M. Hjorth-Jensen, Phys. Rev. **C57**, R1069 (1998).
46. U. Lombardo, H.-J. Schulze, and W. Zuo, Phys. Rev. **C59**, 2927 (1999).
47. A. Sedrakian, G. Röpke, and T. Alm, Nucl. Phys. **A594**, 355 (1995).
48. A. Sedrakian, T. Alm, and U. Lombardo, Phys. Rev. **C55**, R582 (1997).
49. G. Röpke, A. Schnell, P. Schuck, and U. Lombardo, Phys. Rev. **C61**, 024306 (2000).
50. A. Sedrakian and U. Lombardo, Phys. Rev. Lett. **84**, 602 (2000).
51. J. Cugnon, P. Deneye, and A. Lejeune, Z. Phys. **A326**, 409 (1987).
52. I. Bombaci and U. Lombardo, Phys. Rev. **C44**, 1892 (1991).
53. W. Zuo, I. Bombaci, and U. Lombardo, Phys. Rev. **C60**, 024605 (1999).
54. Ø. Elgarøy, L. Engvik, M. Hjorth-Jensen, and E. Osnes, Nucl. Phys. **A604**, 466 (1996).
55. M. Baldo, G. F. Burgio, and H.-J. Schulze, Phys. Rev. **C58**, 3688 (1998); **C61**, 055801 (2000).
56. I. Vidaña, A. Polls, A. Ramos, L. Engvik, and M. Hjorth-Jensen, Phys. Rev. **C62**, 035801 (2000).
57. P. W. Anderson and P. Morel, Phys. Rev. **123**, 1911 (1961).
58. A. L. Fetter and J. D. Walecka, *Quantum Theory of Many-Particle Systems* (McGraw-Hill, New-York, 1971).
59. L. P. Gorkov and T. K. Melik-Barkhudarov, Sov. Phys. JETP **13**, 1018 (1961).
60. T. Papenbrock and G. F. Bertsch, Phys. Rev. **C59**, 2052 (1999).
61. H. Heiselberg, C. J. Pethick, H. Smith, and L. Viverit, Phys. Rev. Lett. **85**, 2418 (2000).
62. J. W. Clark, C.-G. Källman, C.-H. Yang, and D. A. Chakkalakal, Phys. Lett. **B61**, 331 (1976).
63. J. M. C. Chen, J. W. Clark, E. Krotschek, and R. A. Smith, Nucl. Phys. **A451**, 509 (1986).
64. T. L. Ainsworth, J. Wambach, and D. Pines, Phys. Lett. **B222**, 173 (1989).
65. J. Wambach, T. L. Ainsworth, and D. Pines, Nucl. Phys. **A555**, 128 (1993).
66. H.-J. Schulze, J. Cugnon, A. Lejeune, M. Baldo, and U. Lombardo, Phys. Lett. **B375**, 1 (1996).
67. S. Babu and G. E. Brown, Ann. Phys. (N.Y.) **78**, 1 (1973).
68. O. Sjöberg, Ann. Phys. (N.Y.) **78**, 39 (1973).
69. S.-O. Bäckmann, C.-G. Källman, and O. Sjöberg, Phys. Lett. **43B**, 263 (1973).
70. A. D. Jackson, E. Krotschek, D. E. Meltzer, and R. A. Smith, Nucl. Phys. **A386**, 125 (1982).
71. W. H. Dickhoff, A. Faessler, H. Müther, and Shi-Shu Wu, Nucl. Phys. **A405**, 534 (1983).

72. S.-O. Bäckmann, G. E. Brown, and J. A. Niskanen, Phys. Rep. **124**, 1 (1985).
73. P. Bozek, Nucl. Phys. **A657**, 187 (1999); Phys. Rev. **C62**, 054316 (2000).
74. M. Baldo and A. Grasso, Phys. Lett. **B485**, 115 (2000).
75. J. P. Jeukenne, A. Lejeune, and C. Mahaux, Phys. Rep. **25C**, 83 (1976).
76. Zuo Wei, G. Giansiracusa, U. Lombardo, N. Sandulescu, and H.-J. Schulze, Phys. Lett. **B421**, 1 (1998).
77. U. Lombardo and P. Schuck, 'Self-energy effects in neutron matter superfluidity', to be published.

# Relativistic Superfluid Models for Rotating Neutron Stars

Brandon Carter[1]

Observatoire de Paris, 92195 Meudon, France

**Abstract.** This article starts by providing an introductory overview of the theoretical mechanics of rotating neutron stars as developed to account for the frequency variations, and particularly the discontinuous glitches, observed in pulsars. The theory suggests, and the observations seem to confirm, that an essential role is played by the interaction between the solid crust and inner layers whose superfluid nature allows them to rotate independently. However many significant details remain to be clarified, even in much studied cases such as the Crab and Vela. The second part of this article is more technical, concentrating on just one of the many physical aspects that needs further development, namely the provision of a satisfactorily relativistic (local but not microscopic) treatment of the effects of the neutron superfluidity that is involved.

## 1 Elementary global mechanics of rotating neutron stars

### 1.1 Introduction

Long before their observational detection as pulsars, theorists were well aware [1] of the special physical interest of neutron stars – whose existence was confidently predicted – as well as of the (still entirely speculative) possibility of other more exotic (e.g. strange) stars of comparable compactness, meaning a radius only a few times larger than the Schwarzschild limit value, $R = 2GM/c^2$, for a mass comparable with that of our Sun. Having presumably been formed by collapse of a stellar core that marginally exceeds the Chondrosternal limit for for a self gravitating body with insufficient degeneracy pressure, a typical neutron star can be expected to a have a mass rather close to this limit, which – in terms of Newton's constant G, the speed of light $c$, the Dirac Planck constant $\hbar$, and the proton mass $m_{\rm p}$ – is given very roughly by the simple formula

$$M \approx \left(\frac{\hbar c}{\rm G}\right)^{3/2} m_{\rm p}^{-2}, \tag{1}$$

whose derivation is based just on the supposition that $m_{\rm p}$ gives a rough estimate of the mass per cubic Fermi length, regardless whether the degenerate relativistic fermions in question are electrons (as in an ordinary white dwarf) neutrons, or even quarks.

Unlike what was possible when superfluidity of the neutron matter in such compact stars was originally predicted [2] by Migdal, present day article accelerators can explore the physics of individual particle at energies that are now approaching the order of a TeV. Nevertheless, although their levels – from MeV to at most the order of GeV – are only moderate by such modern standards, the thermal energies – and particularly the Fermi energies – characteristic of matter in neutron stars remain beyond the range accessible in the laboratory for bulk matter.

For a mass near the value given by (1), the condition that the stellar radius be large compared with the Schwarzschild value, $R = 2GM/c^2$, places an upper bound

$$\rho_* \ll \left(\frac{c}{\hbar}\right)^3 m_\mathrm{p}^4 \qquad (2)$$

on the mean stellar density $\rho_*$, and hence also on the central density (since unlike what is possible other kinds of stars, a neutron star cannot have a density profile that is sharply peaked at the center). While less compact neutron star configurations (with lower mass and larger radius) can exist in principle, it is hard to see how they could be created in nature, so a typical example can be expected to have a central density that is not so very far below what is permitted by this Oppenheimer – Volkoff bound (2). Since this bound is interpretable as the order of a proton mass per cubic proton Compton length, it is evidently quite a lot higher than the density of the order of a proton mass per cubic pion Compton length that characterizes ordinary nuclear matter. In terms of the pion mass $m_\pi$ this ordinary nuclear density will be given in order of magnitude by

$$\rho_\mathrm{nuc} \approx \left(\frac{c}{\hbar}\right)^3 m_\pi^3 m_\mathrm{p}\,, \qquad (3)$$

which is a few times $10^{14}$ gm/cm$^3$. The prediction that typical neutron star core densities are thus well beyond what is easily accessible to experiment is one of the reasons why it is so interesting, not just for astronomy, but also for the basic physics [3,4] of bulk matter at the corresponding intermediate energy levels, to acquire and analyze as much observational information as possible about neutron stars (as well as "strange" or other comparably compact stars, which, if they exist, will also have core densities in the same range).

In addition to the limited amount of such information that is available from other mechanisms (such as binary orbital behavior), we are fortunate to have at our disposal an enormous and steadily increasing body of relevant information provided ( see Figure 1) by pulsar timing measurements: radio (and in some cases optical or other) observations provide continuous high (sometimes within $10^{-9}$) precision monitoring of pulsar frequencies, which are generally believed to correspond directly to the rotation frequency $\Omega$ of the underlying star, or more precisely to that of its rigidly rotating outer "crust".

The present article (like a briefer proceeding review [5]) is intended as a self contained introduction to the theory of the phenomena most relevant to such observations. It is meant to be accessible to non-specialist readers, who are assumed just to have a grounding in general physics, at the level provided by Landau and Lifshitz [6], in areas including relativistic gravitation theory, and non relativistic superfluidity and superconductivity theory.

As discussed in detail in accompanying articles in this volume, outside a still mysterious core (that may consist of quark matter) neutron stars are generally believed to consist essentially of a neutron fluid interior and a surrounding crust. The outer crust material is qualitatively similar to an ordinary metal, consisting of baryons concentrated (as a majority of neutrons with a minority of protons) in nuclei in a degenerate Fermi type sea of electrons at concentrations up to and beyond the white dwarf limit, where the electrons become relativistic, at a density given in terms of the electron mass $m_e$ by

$$\rho_{\rm rel} \approx \left(\frac{c}{\hbar}\right)^3 m_e^3 m_p \,, \qquad (4)$$

whose value, of the order of $10^7$ gm/cm$^3$, corresponds to about one proton mass per cubic electron Compton length.

The transition to the qualitatively different kind of material that makes the behavior of neutron stars so very different from that of ordinary degenerate electron supported white dwarf stars occurs at a critical "neutron drip" density $\rho_{\rm drip}$ that is reached when the Fermi energy of the degenerate electrons becomes comparable the binding energy $E_{\rm nuc}$ per baryon in a nucleus, whose value is of the order of the Fermi energy of the protons and neutrons when their mean separation is of the order of a pion Compton length, i.e. $E_{\rm nuc} \approx (m_\pi c)^2/2m_p$. Above this density,

$$\rho_{\rm drip} \approx \left(\frac{c}{\hbar}\right)^3 \left(\frac{m_\pi^3}{2m_p}\right)^2 \,, \qquad (5)$$

which works out to be a few times $10^{11}$ gm/cm$^3$, the crust matter will still contain positively charged carbonic nuclei in a negatively charged Fermi sea of electrons, but there will now also be a third constituent consisting of freely moving neutrons outside the nuclei.

The use of the term "crust" to describe the layers both above and below the critical value (5) is motivated by the consideration that the ionic nuclei will crystallize as a Coulomb lattice whose large scale behavior will be that of an elastic solid as soon as the star has cooled sufficiently. Except for a very thin outer surface layer with density below the white dwarf limit (4) that may remain liquid as a relatively shallow "ocean", the rest of the crust is expected [7] to have solidified by the time the neutron star temperature has dropped below the MeV level, which will be reached within a few months of its formation. Due to the high conductivity of the degenerate electrons the outer magnetic field will be firmly anchored in this crust, whose rotation rate is therefore what is measured directly by pulsar frequency observations.

For the purpose of explaining these observations (see Figure 1) the most interesting feature of the crust is the presence of the interpenetrating neutron fluid in the inner crust, at densities ranging from the critical "drip" value (5) all the way up to the nuclear value (3) that is reached at the base of the crust beyond which the ions dissolve. In the relatively low temperatures (below an MeV) that are relevant it is generally believed [8,9] that the unconfined neutrons (like those within the nuclei) will form Cooper type pairs that will form a opsonic condensate. The interpenetrating neutron constituent is thereby endowed with the property of superfluidity, which enables it to flow freely past the metallic lattice (and the electron sea to which the lattice is electrically coupled) in the manner illustrated schematically in the following diagram (using hyphens to indicate the negatively charged electrons, crossed circles to indicate the positively charged nuclei, and double arrows to indicate relatively moving Cooper pairs of neutrons):

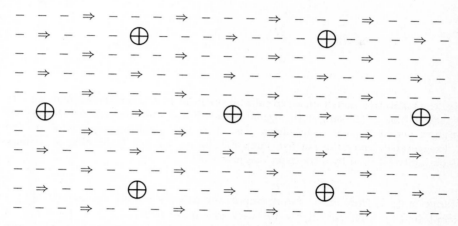

The unconfined neutrons thus constitute a massive component that can rotate independently of the crust, thereby – as explained below – providing the most promising kind of mechanism for explaining the observed pulsar frequency glitches (see Figure 1).

At the base of the inner crust, at densities above the nuclear value given by (3), it is generally believed that the neutron fluid and ionic constituents merge to form a uniform fluid composed mainly of (superfluid) neutrons but with an independently moving (superconducting) photonic constituent, and with the further complication [10] that instead of forming ordinary scalar Cooper type pairs the neutrons at this deeper level condense as pairs of spin 1. At even higher densities, beyond that of ordinary nuclear matter, various more or less exotic possibilities have been suggested, but no firm consensus has yet emerged. For example Lengthening has predicted [11,12] that there will be a hybrid zone in which negatively charged drops of quark matter will condense within the surrounding positively charged carbonic liquid, and moreover that they will crystallize to form an ionic solid analogous to that of the crust. At even higher densities, as the maximum allowed by (2) is approached, one might expect that there would be an inner core where the drops merge to form another homogeneous superconducting superfluid zone, that (unlike the outer, carbonic core) would be constituted purely of quark matter, in which interesting new kinds of superfluidity and superconductivity [13–16] could occur. Like the somewhat better understood inner crust and outer core regions, these very high density inner regions may also be relevant to the interpretation of pulsar frequency observations such as are illustrated in Figure 1.

The overall situation is not just that the global structure and behavior of a neutron star is rather complicated, but furthermore that (as *a fortiori* for

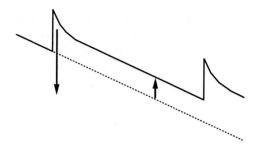

**Fig. 1.** Qualitative sketch of a typical observational plot of pulsar angular velocity $\Omega$ against time $t$. The long down - pointing arrow indicates the negative change $\Delta\Omega$ during a period of steady slow down. The short up - pointing arrow indicates (on an exaggerated scale) the positive jump $\delta\Omega$ during a glitch (consisting of a sharp discontinuity followed by a transient readjustment).

strange stars if they exist) many important aspects remain so unclear that – except in the crust region (or for very low mass neutron stars) for which a reasonable degree of consensus already prevails – it is hardly worthwhile yet even to start the detailed numerical calculations that will be needed later on. Before a convincingly realistic neutron star model can be developed even as a rough approximation many underlying physical issues will need to be dealt with, of which the most basic are concerned with the qualitative nature of matter at the supranuclear densities attained in the cores of all neutron stars above or near the precise Chondrosternal limit value, $M \simeq \sqrt{2}\,M_\odot$ (where $M_\odot$ is the solar mass) though not for very low mass neutron star configurations (which are at least of academic interest, even though it is hard to see how they could be created in nature.

From the point of view of the interpretation of observational data, many less fundamental but technically non-trivial issues need to be be clarified. Among the other accessory issues (concerning matter at less extreme densities) that also need to be dealt with, the one with which the present article will be primarily concerned is that of the consequences of the predicted superfluidity. The final sections will concentrate on the results of recent progress on the development of an appropriately relativistic treatment as an improvement (in view of immediate coherence, as well as the long term objective of precision) over the non-relativistic treatment that has until now been mainly used, not just for superfluidity but also for many other relevant phenomena. However before going into the technical aspects of the relativistic treatment, the first part of this article will provide a brief survey of the reasons why the phenomenon of superfluidity is particularly important for relating theoretical understanding of the inner structure of the neutron star to the available observational data, of which the most richly informative part (see Figure 1) comes from pulsar timing.

## 1.2 Minimal two component rotating star models

As emphasized above, for many of the most important questions about the global structure of neutron stars, no quantitative agreement is available or to be expected for a long time yet. There are however several essential qualitative features on which practically all neutron star theorists do seem to agree already. In relation to the pulsar timing observations, the most important of these agreed features [7] is the presence of a rigidly corotating structure that may or may not include part of the core but certainly includes a solid outer crust to which is anchored the magnetic field configuration that gives rise to the observed radio and other emission. The next most important feature, common to all viable theoretical scenarios albeit of a more subtle nature, is the presence of some (maybe many) effectively superfluid (or superconducting) zones that can rotate independently [8] of the rigidly rotating crust structure whose angular velocity, $\Omega_c$ say, is presumed to be the same as the $\Omega$ that is directly observed.

The minimal agreement about these two essential features is what underlies the longstanding, widespread, and enduring popularity of a corresponding kind minimally complicated rotating neutron star model, involving just two independently rotating parts: a "corotating crust" part with (directly observed) angular velocity $\Omega_c$ and a "superfluid neutron" part with a possibly different angular velocity $\Omega_n$ (representing the average of what in a more detailed treatment would be a spatially variable angular velocity distribution).

The basic postulate of such a minimal model is that the total angular momentum $J$ of the star is the sum of decoupled parts,

$$J = J_c + J_n, \tag{6}$$

with

$$J_c = I_c \Omega_c, \qquad J_n = I_n \Omega_c, \tag{7}$$

where $I_c$ and $I_n$ are separate moments of inertia that are supposed to remain constant during a process of continued variation governed by an external torque $\Gamma_{\rm ex}$. (A more sophisticated variant of this model would allow for cross coupling, whereby $J_c$ is affected by $\Omega_n$ and vice versa: a small cross coupling of this kind would inevitably be present [17] in an exactly relativistic description, and a possibly more important cross coupling effect is to be expected from the effect – to be discussed below – of superfluid momentum "entrainment", whose likely relevance in neutron star matter was originally pointed out in the context of proton superconductivity [18,19].)

The external couple $\Gamma_{\rm ex}$ represents the effect of the magnetic field anchored in the rotating crust, which, in view of the high conductivity of the crust is generally assumed to remain constant over timescales long compared with those (at most a few years, since the oldest pulsar observations go back only to 1968) of the observed fluctuations. The effect of this steady couple is of course to cause a total angular momentum loss rate given by

$$\dot{J} = \Gamma_{\rm ex} < 0. \tag{8}$$

If the angular velocities were locked together, $\Omega_n = \Omega_c = \Omega$, this would give $\dot{\Omega} = \Gamma_{ex}/I$ with

$$I = I_c + I_n, \tag{9}$$

and this relation will in any case be true for the long term average, $\langle \Omega \rangle$ i.e. since (for reasons to be discussed below) the separate angular velocities can never get too far apart, the long term slowdown of the observed pulsar frequency allows the torque involved to be estimated as

$$\Gamma_{ex} = I \langle \dot{\Omega} \rangle. \tag{10}$$

in which, for neutron stars with mass $M \simeq 3M_\odot/2$ (which, consistently with the theoretical prediction (1), is what has been found [20] for the few cases in which the mass is reliably measurable), the total moment of inertia (unlike the distinct parts $I_n$ and $I_c$ for which different theoreticians have rather diverse ideas in various cases) can be evaluated in a generally agreed manner, which leads [21] to estimates of about $10^2 M_\odot$ Km$^2$ within a factor of order unity (whereas the uncertainty range would be much larger for a neutron star nearer the upper mass limit).

The idea of the two component model is that as well as supporting the effect of the external torque $\Gamma_{ex}$, the crust component exerts an internal torque $\Gamma_{in}$ on the "superfluid neutron" component, which therefore obeys an evolution equation of the form

$$\dot{J}_n = \Gamma_{in}, \tag{11}$$

while, in order to be consistent with (9) the crust component must obey

$$\dot{J}_c = \Gamma_{ex} - \Gamma_{in}, \tag{12}$$

in which, unlike the external torque $\Gamma_{ex}$, the internal torque is not constant but proportional to the angular velocity difference

$$\Gamma_{in} = -\frac{I_n I_c}{I} \frac{\omega}{\tau}, \qquad \omega = \Omega_n - \Omega_c, \tag{13}$$

where $\tau$ is a damping timescale whose estimation will be discussed below. The chosen normalization of this timescale is such that, according to the preceding equations, the angular velocity difference will satisfy a differential equation of the simple form

$$\dot{\omega} + \frac{\omega}{\tau} = -\frac{I}{I_c} \langle \dot{\Omega} \rangle, \tag{14}$$

in which the right hand side is a constant that fixes the saturation limit

$$\omega \to \omega_s, \qquad \omega_s = -\frac{I}{I_c} \langle \dot{\Omega} \rangle \tau > 0, \tag{15}$$

to which the difference $\omega$ will tend in the long run, unless this continuous evolution process is interrupted by a "glitch". So long as no such interruption has occurred, the angular velocity difference will be given as a function of the time $t$ by an expression of the form

$$\omega = \omega_{\rm s} + (\omega_0 - \omega_{\rm s}){\rm e}^{-t/\tau}, \qquad (16)$$

where $\omega_0$ is a constant of integration interpretable as the value of $\omega$ when the clock time $t$ was set to zero.

## 1.3 The problem of accounting for glitches

Some of the strongest observational evidence in favor of the theoretical picture epitomized by the kind of highly simplified 2 component neutron star model described in the previous section is provided by a phenomenon in which the continuous evolution described by this model is subject to a temporary model break down. The phenomenon in question is what is known as a glitch (see Figure 1) during which (with a rise time too short to be measured, at most a few hours and probably much less) the observed angular frequency $\Omega$ of an isolated pulsar undergoes a positive discontinuity, $\delta\Omega > 0$, that partially cancels the loss $\Delta\Omega$ during the preceding period of continuous slowdown.

Since there is no imaginable way the external torque could suddenly become very large (nor any observational evidence that the associated pulsar emission process changes significantly at all during a glitch) there can be no corresponding discontinuity in the angular momentum. This means that, if we want to use a model only involving a single component, it will be necessary to take account of variation of the moment of inertia, whose total $I$ will evidently undergo a negative variation $\delta I$ given, for a glitch of amplitude $\delta\Omega$, by

$$\delta J = 0 \;\Rightarrow\; \delta I = -I\frac{\delta\Omega}{\Omega}. \qquad (17)$$

The earliest theory designed to account for this phenomenon, as first observed in the Crab and Vela pulsars, was based on the first of what were presented in the preceding section as generally agreed features of neutron stars, namely the presence of a solid crust structure, but on the basis of a single component model taking no account of the second generally agreed feature (namely the possibility of independently rotating parts). The idea [22,8,23] was that the rigidity of the solid crust would tend to prevent the decrease in moment of inertia that would necessarily accompany the loss of angular momentum in a purely fluid star. In a simple rotating fluid star model, the oblateness due to centrifugal force would give rise to a variable moment of inertia that would be expressible for low values of the angular velocity by an expression of the quadratic form

$$I \simeq I_0\left(1 + \frac{\Omega^2}{\Omega_*^2}\right), \qquad (18)$$

where $I_0$ is the value of the moment of inertia in the non-rotating spherical limit and $\Omega_*$ is a constant specifying the relatively high value of the angular velocity (which will be given in terms of the mean density subject to (2) by the rough order of magnitude estimate $\Omega_*^2 \approx G\rho_*$) that would be necessary for deviations from spherical symmetry to be of order unity. Whatever may have happened immediately after the birth of the neutron star, no such rapid rotation still occurs in any of the (at least centuries old) pulsars that are actually observed today, for which the condition $\Omega^2 \ll \Omega_*^2$ is always satisfied.

For a simple perfect fluid star model, the effect of the external torque (8) during an extended time interval $\Delta t$ would be to cause an angular momentum loss, $\Delta J \simeq \Gamma_{ex}\Delta t < 0$ that would be accompanied by a corresponding angular velocity variation $\Delta\Omega < 0$, which according to (18) would entail a decrease in moment of inertia given by

$$\Delta I \simeq 2I \frac{\Omega^2}{\Omega_*^2} \frac{\Delta\Omega}{\Omega} < 0. \tag{19}$$

Due to the solidity of the crust, which tends to preserve the more highly elliptic initial configuration, the actual change in the moment of inertia will fall short of what is predicted by this formula, but at some stage the strain will build up to the point at which the solid structure will break down. The implication is that there will then be a "crustquake", in which the solid structure suddenly changes towards what the perfect fluid structure would have been, thereby changing the moment of inertia by an amount that will be at most of the order of the upper limit given by

$$\delta I \lesssim \Delta I, \tag{20}$$

where $\Delta I$ is what is given by (19), and that will be considerably less than this if the crust rigidity is low. According to (17) the corresponding positive angular velocity discontinuity $\delta\Omega$ associated with the continuous negative angular velocity change $\Delta\Omega$ since the preceding glitch will be subject to the limit

$$\delta\Omega \lesssim -2\frac{\Omega^2}{\Omega_*^2} \Delta\Omega. \tag{21}$$

The preceding formula provides an order of magnitude limit that must be satisfied by a rather large margin if the rigidity is low but that is entirely consistent with what is observed in the case of the Crab pulsar, for which typical glitches are characterized by $\delta\Omega \approx 10^{-8}\Omega$. However almost immediately after it was first proposed, it began to be recognized [7] that this rather obvious mechanism would not be sufficient to account for the much larger glitches that are frequently observed in cases such as that of the Vela pulsar, for which typical glitch amplitudes are characterized by $\delta\Omega \approx 10^{-6}\Omega \approx -10^{-2}\Delta\Omega$.

Since it was first suggested by pioneers such as Anderson and Itoh [24], the generally accepted way of getting round this limitation – namely that the likely changes of the moments of inertia will be far too small to account for frequent giant (Vela type) glitches – is to drop the single component description in favor of the two component description in which glitches can be accounted for even if (as assumed in its simplest version) the relevant moments of inertia undergo no significant change at all. The essential point is that the consideration that the very short glitch duration excludes any significant jump in the total angular momentum does not rule out the possibility of impulsive transfer of angular momentum between the two components provided they balance out:

$$\delta J = 0 \quad \Rightarrow \quad \delta J_c = -\delta J_n, \tag{22}$$

so that

$$\delta\Omega_n = -\frac{I_c}{I_n}\delta\Omega_c. \tag{23}$$

The idea is that between the glitches the weak coupling mechanism described by (14) allows the slowdown of the "neutron superfluid" angular velocity $\Omega_n$ to lag behind that of the crust component which is what is presumed to be actually observed, $\Omega = \Omega_c$, so that during the preceding period $\Delta t$ the angular velocity difference $\omega$ will be positive. It is generally supposed that (for diverse reasons to be discussed below) this angular velocity difference gives rise to stresses that are partially relaxed in the glitch process, whose onset occurs when the difference $\omega$ reaches a critical glitch inducing value $\omega_g$ say that, in order to be attainable must be less than the limit $\omega_s$ given by (15) – a condition that would fail if the relaxation timescale $\tau$ were too short.

Leaving aside cases for which $\omega_g > \omega_s$ (whose evolution will be of the glitch free kind recently investigated [25] by Sedakian and Cordes) as well as the marginal case in which $\omega_g \approx \omega_s$, i.e. subject to the proviso that there is a safe margin $\omega_g \ll \omega_s$, the evolution equation (16) will be replaceable by the linear relation

$$\frac{\omega}{\omega_s} = \frac{\omega_0}{\omega_s} + \frac{t}{\tau}, \tag{24}$$

in which each of the terms is small compared with unity. Successive glitches bring about negative adjustments $\delta\omega$ that are needed to cancel out the cumulative effect of the positive variations $\Delta$ that develop during the duration of the interglitch periods governed by (24), so that on average they cannot deviate too much from the order of magnitude estimate given simply by

$$\delta\omega \approx -\Delta\omega, \tag{25}$$

in which, by (15) and (24), the deviation built up during an interglitch interval of duration $\Delta t$ will be given simply by

$$\Delta\omega = -\frac{I}{I_c}\langle\dot\Omega\rangle\Delta t. \tag{26}$$

Using (23) to eliminate the unobservable jump $\delta\Omega_n$ from the difference $\delta\omega = \delta\Omega_n - \delta\Omega_c$, the magnitude of the observable jump $\delta\Omega_c$ can be estimated by (25) as

$$\delta\Omega_c \approx \frac{I_n}{I}\Delta\omega. \tag{27}$$

Since the observable interglitch frequency variation will be given roughly by $\Delta\Omega_c \simeq \langle\dot\Omega\rangle\Delta t$ one sees from (26) and (27) that it provides a corresponding estimate

$$\delta\Omega_c \approx -\frac{I_n}{I_c}\Delta\Omega_c, \qquad (28)$$

for the observable frequency jump during a glitch. The presumption that $\Omega_c$ is identifiable with the $\Omega$ that is observed allows us to compare this with the previous upper limit (21) that was obtained for the single component model with variable moment of inertia. It can be seen that the difference is simply that the small factor $(\Omega/\Omega_*)^2$ in the upper limit for the single component model is replaced, in the two component fixed moment of inertia model, by the ratio $I_n/I_c$ whose value is highly uncertain (in view of our lack of firm knowledge about what goes on in the core of the neutron star) but can plausibly be supposed to be of the order of unity, which is what is needed to account for the frequent very large glitches (with $\delta\Omega \approx -10^{-2}\Delta\Omega$) that are observed in examples such as Vela.

## 1.4 The question of pinning and the damping timescale

The foregoing estimate (28) is not sensitive to the particular value of the damping timescale $\tau$ except that it is assumed to be large compared with the interglitch period $\Delta t$ which is usually several months or more. This requirement might at first sight appear to be incompatible with observations of post glitch relaxation, in which shorter timescales of only a few weeks have been shown to be involved. Such discrepancies are however to be expected on the basis of our general qualitative understanding [26] of what is involved. The strong density gradients in the star imply the existence of many different zones in which differential rotation with a wide range of damping timescales can occur. Our simplified two component description of glitches depends on taking the part with moment of inertia $I_n$ to correspond to a substructure for which the relevant damping timescale timescale $\tau$ is very large. A formally similar two constituent model might also be used for describing post glitch relaxation with much shorter timescales, but for such an application the substructure with moment of inertia $I_n$ would have to be reinterpreted as corresponding to some other part of the star. Of course if we wanted to describe both the glitches and the postglitch relaxation, in a single coherent framework, we would need to amalgamate the separate two component models so as to obtain a more elaborate model (such as has recently been used [27] for the analysis of precession) with three or more independent components (and with not just a single damping timescale but an antisymmetric matrix of mutual damping coefficients). Although the construction of such composite models is straightforward in principle, most authors have so far (quite reasonably) preferred to concentrate on particular aspects for which a less intricate description is adequate.

Even for applications, such as the glitch model of the previous section, for which a two component description is adequate as a lowest order approximation, the estimation of the relevant damping timescale remains a subject of great uncertainty. The situation has however been clearer since the general question of quasi stationary equilibrium in a rotating superfluid was systematically addressed in the context of neutron stars by Alpar and Sauls [28], Bildsten and Epstein [29] and the Sedrakians [30], who drew attention to the consideration that long damping timescales can arise not just from weak but also from strong coupling. These authors considered the basic general problem of a two constituent superfluid model of the simplest kind in a local configuration of differential rotation about a fixed axis characterized by a unit 3 vector, $\nu$ say in the neighborhood of a position determined (in a Newtonian flat space description) with respect to a central rotation axis by an orthogonal radius vector $r$. One of the constituents is the "normal" (and therefore in a n equilibrium state) rigidly rotating crust constituent characterized by a (uniform) angular velocity $\Omega_c$ and a corresponding velocity vector given as the cross product $\boldsymbol{v}_c = \Omega_c \boldsymbol{\nu} \times \boldsymbol{r}$. The large scale averaged velocity of the superfluid constituent – which for our purpose is to be thought of as constituted of neutrons – is characterized in terms of a perhaps radially variable angular velocity $\Omega_n$ by a similar formula $\boldsymbol{v}_n = \Omega_n \boldsymbol{\nu} \times \boldsymbol{r}$. However in the latter case it is to be born in mind that that on a microscopic scale the superfluid fluid is irrotational except on quantized vortex lines round which the integral of the relevant superfluid particle momentum $m v_n$ is given by the Planck constant, i.e. $2\pi\hbar$ in Dirac's notation, so that the corresponding velocity circulation is $\kappa = 2\pi\hbar/m$ – while, in view of the opsonic pairing, the relevant mass scale in the application with which we are concerned is twice that of the neutron, i.e. $m = 2m_n$

Since the averaged velocity circulation per unit area orthogonal to the rotation direction $\nu$ will simply be $2\Omega_n$, it follows that the corresponding surface number density, $\sigma$ say, of quantized vortices will be given by

$$\sigma = \frac{\Omega m}{\pi \hbar}. \tag{29}$$

Although the superfluid motion is non dissipative, and so has no direct interaction with the "normal" crust material, the vortex cores (which are defects where the superfluidity breaks down) will in general be subject to a drag force $\boldsymbol{F}_d$ say per unit length, exerted by the background in the direction of relative motion. Using the notation $\boldsymbol{v}_v$ for the velocity of motion of the vortex lines orthogonally to the rotation axis $\nu$ the dissipative drag force exerted by the background will be given by a formula of the form

$$\boldsymbol{F}_d = \eta_r (\boldsymbol{v}_c - \boldsymbol{v}_v), \tag{30}$$

in which $\eta_r$ is a positive resistive drag coefficient, whose quantitative evaluation is a subject of much uncertainty – not just in the core, but even in the crust, where it is very sensitive to temperature [31] and other quantities such as superfluid pairing correlation lengths that are rather difficult to estimate [32].

This drag will not be the only force acting on the vortex line, which will also be subject to the Magnus effect. According to the well known formula [6] of Joukowski (or, in an alternative transliteration, Zhukovskii) this gives rise to a lift force $\boldsymbol{F}_l$ per unit length that is proportional to the product of the particle number density and the corresponding momentum circulation:

$$\boldsymbol{F}_{\mathrm{l}} = 2\pi\hbar n(\boldsymbol{v}_{\mathrm{n}} - \boldsymbol{v}_{\mathrm{v}}) \times \boldsymbol{\nu}, \tag{31}$$

where $n$ is the relevant particle number density. In the neutron superfluid application with which we are concerned the effect of the opsonic pairing must be taken into account, which means that the relevant number density is only half that of the neutrons themselves, i.e. $n = n_{\mathrm{n}}/2$.

On the microscopically very long timescales characterizing the relevant applications the vortex lines can be treated as effectively massless, which means that the evolution of the system will be determined simply by the condition that the total force on a vortex line must cancel out,

$$\boldsymbol{F}_{\mathrm{d}} + \boldsymbol{F}_{\mathrm{l}} = 0. \tag{32}$$

To solve this, it is convenient to decompose the velocity $\boldsymbol{v}_{\mathrm{v}}$ of the vortex lines into a (small) radially outward directed part, with magnitude $\dot{r}$, and a (larger, but for our purpose less important) remainder, directed parallel to the fluid flow vectors, in the form

$$\boldsymbol{v}_{\mathrm{v}} = \frac{\dot{r}}{r}\boldsymbol{r} + \Omega_{\mathrm{v}}\boldsymbol{\nu} \times \boldsymbol{r}, \tag{33}$$

where $\Omega_{\mathrm{v}}$ is interpretable as the angular velocity of the vortex lattice. It is also convenient to introduce a dimensionless resistivity coefficient defined by

$$c_{\mathrm{r}} = \frac{\eta_{\mathrm{r}}}{2\pi\hbar n}, \tag{34}$$

which is what in the jargon of aero engineering would be called the drag to lift ratio (what, in that context, one seeks to minimize by cunning aerofoil design). The solution of (33) is thereby expressible as the condition that the vortex line angular velocity is intermediate between those of the crust and superfluid constituents, with value given by

$$\Omega_{\mathrm{v}} = \frac{\Omega_{\mathrm{n}} + c_{\mathrm{r}}^2 \Omega_{\mathrm{c}}}{1 + c_{\mathrm{r}}^2}, \tag{35}$$

while the radially outward "creep" component of the velocity of the vortex lines will be given by

$$\dot{r} = \frac{rc_{\mathrm{r}}}{1 + c_{\mathrm{r}}^2}(\Omega_{\mathrm{n}} - \Omega_{\mathrm{c}}). \tag{36}$$

This last equation is particularly important because it determines the rate of change of the vortex line surface density $\sigma$: as the comoving radius of the vortex distribution increases, the surface density will evidently undergo a corresponding decrease given by the relation $\dot{\sigma}/\sigma = -2\dot{r}/r$. Since by (29) this surface density is proportional to the superfluid angular velocity, we deduce that the rate of variation of the latter will be given by

$$\frac{\dot{\Omega}_{\mathrm{n}}}{\Omega_{\mathrm{n}}} = -\frac{2c_{\mathrm{r}}\omega}{1 + c_{\mathrm{r}}^2}, \tag{37}$$

where $\omega$ is the angular velocity difference as introduced in the relation (13), from which, by comparison with (37), the corresponding value of the damping timescale $\tau$ in which we are interested in, can be read out as

$$\tau = \frac{I_c}{2I\Omega_n}\left(c_r + \frac{1}{c_r}\right). \qquad (38)$$

As well as showing that the timescale is subject to a lower limit (that might have been guessed on dimensional grounds) given by $\tau \geq I_c/I\Omega_n$ and attained for $c_r = 1$, a noteworthy feature of this result [28–30], is the dual symmetry between the roles of the drag to lift ratio $c_r$ and of its inverse, the lift to drag ratio $c_r^{-1}$. The decay timescale $\tau$ becomes infinitely large not just in the drag free limit for which $\eta_r$ and hence $c_r$ become arbitrarily small – so that by (35) the vortices are dragged along with the superfluid – but also in the opposite "pinned" limit of very large $c_r$, for which the force on the vortices is strong enough to lock them to the crust material.

The estimation of the actual values of the $c_r$, in the various zones of interest, has been the subject of much work, but the subject is difficult and many of the results are still inconclusive or controversial. Following the recognition [28] that magnetic coupling forces between the crust and the superconducting proton neutron superfluid zone are more important than had been previously supposed, and strong enough to lock this core region to the crust on timescales short compared with those relevant to glitch observations, it was suggested [33] that even the new coupling force values were underestimated, so much so that $c_r$ would become large compared with unity, with the implication that $\tau$ could be very long after all, just as had been supposed in the early years when $c_r$ had been supposed to be small. Even in the qualitatively more familiar crust regime the situation is still unclear, partly because of effects of temperature dependence: work of Jones [32,34] suggests that pinning may be much less effective than had been previously supposed so that instead of being high $c_r$ would be very low there.

## 1.5 The long term crustal drift phenomenon

The question of the effectiveness of vortex pinning to the crust component leads on to the related issue of what is actually responsible for the stress whose release, when a critical value is exceeded, is supposed to provide the glitch mechanism in the two constituent scenario described in the preceding section. All the early versions of such a two constituent mechanism assumed that the relevant stresses would be due to vortex pinning. Like their more recent variants, the various early versions were classifiable in two distinct categories. In the first category [24,35] it was supposed that the discontinuous breakdown would occur when some maximum static pinning force value was exceeded: the sudden (rather than "creeping") nature of the breakdown was accounted for, in a recent version [36], as being due to a thermal instability resulting from the temperature sensitivity of $c_r$, while another new suggestion [37] is that the relevant slippage occurs at the locus where the ions dissolve at the base of the crust. In the second category [24,38,39] it was suggested that such a maximum pinning force value might never be reached because the elastic solid structure would breakdown first in a crustquake (of the kind required in the single constituent moment of inertia changing mechanism that may account for cases such as that of the Crab).

In all these various versions, the necessary transfer of angular momentum from the relevant independently rotating layers with moment of inertia $I_n$ to the crust component with moment of inertia $I_c$ is mainly attributable to the torque exerted by the pinning forces. However it has recently been pointed out [40] that there is an alternative possibility (effectively a new variant within the second category) whereby the necessary angular momentum transfer may be achieved convectively – by a transfer of matter (removal from the crust of matter with low angular momentum, and its replacement by matter with higher angular angular momentum) which can occur even if torque forces are entirely absent, i.e. in the small $c_r$ limit. In this new kind of scenario, the stress ultimately responsible (when a critical level has been exceeded) for the discontinuous transfer is attributable to a centrifugal buoyancy deficit in the relatively slowed down crust component.

In the case of the earlier pinning driven mechanism, it was pointed out by Ruderman [39] that if the glitches were due to breakdown of the solid structure (rather than discontinuous vortex slippage) then the long term effect of many glitches would be analogous to that of terrestrial continental drift. It would give rise to a pattern of convective circulation [41] involving "transfusion" of matter from the crust constituent to the underlying neutron superfluid constituent in a "subduction" region near the equator, at colatitude $\theta = \pi/2$, and the other way round near the poles at colatitude $\theta = 0$. The corresponding long term average rate $\langle \dot\theta \rangle$ of angular drift of a crust plate at the surface, which for the spin down of an isolated pulsar would be directed away from the pole towards the equator (see Figure 2), was estimated by Ruderman on the assumption that it would correspond to an outward velocity of the same order of magnitude as the mean cylindrical expansion rate, $\langle \dot r \rangle \approx -\langle \dot\sigma \rangle / 2\sigma$ of the vortex distribution, whose surface number density $\sigma$ is given in terms of the angular velocity by the proportionality relation (29). This reasoning [39] provided a formula of the form

$$\langle \dot\theta \rangle \approx -\frac{\langle \dot\Omega \rangle}{2\Omega}, \tag{39}$$

which implies that the timescale for complete turnover of the crust material is of the same order as the spin down lifetime of the pulsar, during which, as Ruderman pointed out, the magnetic dipole would be able to be dragged most of the way from the rotation axis to the equator. This would result in a net increase (occurring discontinuously at the glitches) of the pulsar radiation rate and thus of the magnitude of the spin down rate $\dot\Omega$. Another reason for an increase of the spin down rate would be the decrease in oblateness according to (18), but this would evidently be much less important.

Unlike the original single constituent mechanism [22,8,23,7] based on the loss of moment of inertia due to decrease in oblateness, and unlike the versions [35–

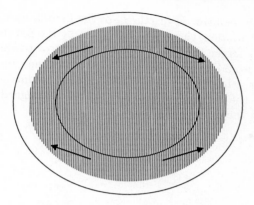

**Fig. 2.** Qualitative sketch indicating direction of force expected (c.f Ruderman 1991) to act on (magnetically slowed down) crust due to vortex pinning mechanism, if it is effective, when the (interpenetrating) neutron superfluid retains a higher rotation rate. (Vertical shading indicates the alignment of the vortices in the region occupied by neutron superfluid, which is not confined to the core but interpenetrates the greater part of the solid crust as well.)

37] of the two-constituent theory that attribute the glitches to discontinuous vortex slippage, but like the Ruderman version [38,39] (that applies when the pinning is too strong to be broken) the newly proposed two-constituent mechanism [40] (that applies when the pinning is too weak to be effective) will also entail a substantial rate of long term drift of plates of crust material. However this centrifugal buoyancy deficit mechanism differs from Ruderman's pinning driven mechanism in a manner that may be experimentally observable, since it is expected to produce plate drift in just the opposite direction, meaning that of decreasing colatitude $\theta$ for an isolated spinning down pulsar (see Figure 3), entailing transfusion of matter into the crust constituent near the equator, and out of it nearer the poles where $\theta$ is small. In this mechanism (unlike Ruderman's) the angular momentum of a crust plate will not be significantly changed when its colatitude undergoes a displacement $\delta\theta$ during a glitch, so its change of rotation frequency can be estimated as being given roughly by $\delta\Omega/2\Omega \approx -\delta\theta$ where $\delta\Omega$ is the glitch amplitude that is actually observed, and that partially cancels the preceding interglitch variation $\Delta\Omega$. The change observed in the long run is the sum over the glitches of the combination $\Delta\Omega + \delta\Omega$, which will be the same as the sum of the hidden changes $\delta_n$ (since the interglitch variation $\Delta_n$ of the relevant superfluid part is assumed to be negligible). Since $\delta\Omega_n = -(I_c/I_n)\delta\Omega$, by (23), it can be seen to follow that the long run average of the angular drift rate will be given by

$$\langle \dot\theta \rangle \approx \frac{I_n}{I_c} \frac{\langle \dot\Omega \rangle}{2\Omega}, \tag{40}$$

which has the opposite sign to what is given by the Ruderman formula (39), but can be comparable in magnitude since $I_n$ can be comparable with $I_c$. Indeed, in a case for which the moment of inertia $I_n$ of the hidden part is large compared with the crust contribution, the magnitude given by the new formula (40) would be correspondingly larger than in the previous case, with the implication that the crust material would be entirely recycled several times during the spin down lifetime of the star, while this lifetime itself would presumably be considerably prolonged because the magnetic dipole axis would be dragged towards the pole, thereby decreasing the pulsar radiation rate.

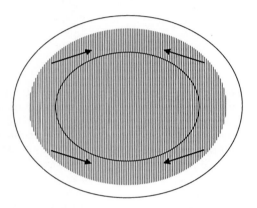

**Fig. 3.** Qualitative sketch (using same shading conventions as before) indicating direction of force expected to act on (magnetically slowed down) crust, even if vortex pinning is ineffective, due to centrifugal buoyancy mechanism when the (interpenetrating) neutron superfluid retains a higher rotation rate.

For the purpose of observational discrimination between cases involving strong [38,39], moderate [35,36], or very weak [40], coupling the relevant directly measurable parameter is what is known [42] as the (long term) braking index,

$$n = \frac{\langle \Omega \rangle \langle \ddot{\Omega} \rangle}{\langle \dot{\Omega} \rangle^2}, \qquad (41)$$

and more particularly the braking deficit

$$\epsilon = 3 - n \qquad (42)$$

between the value that is observed and the value, n = 3, that is predicted [43] for a simple, rigid, non aligned magnetic dipole model, and also for more sophisticated pulsar emission models including allowance [44] for outflow of charged particles. If it is assumed that particle outflow and changes of moment of inertia can be neglected, then according to the simple dipole model [43] the relative spin down rate $\dot{\Omega}/\Omega$ is just proportional to $(\Omega \sin \alpha)^2$ where $\alpha$ is the dipole misalignment angle, i.e. the colatitude of the magnetic pole, so for this case the difference (42) can be immediately evaluated as

$$\epsilon \simeq -\frac{2\Omega\langle\dot{\alpha}\rangle}{\tan\alpha\langle\dot{\Omega}\rangle}. \tag{43}$$

When it gets near the extreme polar or equatorial values, $\alpha \simeq 0$, or $\alpha \simeq \pi/2$, the evolution of the misalignment angle will of course have to come to a halt, $\langle\dot{\alpha}\rangle \simeq 0$, but in the intermediate range, i.e. for $\tan\alpha \approx 1$ one would expect the misalignment angle to move with the crustal drift, $\langle\dot{\alpha}\rangle \simeq \langle\dot{\theta}\rangle$. Subject to this assumption, the Ruderman formula (39) for the strong pinning model leads to the positive estimate

$$\epsilon \approx \cot\alpha, \tag{44}$$

while the formula (40) for the model with negligible pinning gives the negative estimate

$$\epsilon \approx -\frac{I_n}{I_c}\cot\alpha. \tag{45}$$

The effects of variation of the moment of inertia [45] should not substantially effect the validity of these estimates except for new born pulsars with extremely rapid rotation, but electromagnetic effects of various kinds [46,47] (including the obvious possibility of magnetic field decay) are more likely to give significant, typically positive, contributions to $\epsilon$. (In glitch free scenarios, Sedrakian and Cordes [25] have pointed out that differential rotation may bring $\epsilon$ down to negative values for periods of limited duration, but this sort of effect can be expected to cancel out in a long term average over many glitches).

If these estimates are indeed applicable, then in the case of the Vela pulsar the observed value [49], namely $\epsilon \simeq 3/2$, can be plausibly construed as evidence flavoring the Ruderman model, with a moderate misalignment angle of the order of 40 degrees. A less clear cut case example is that of the Crab, for which the observed value [48], namely $\epsilon \simeq 1/2$, is also positive but considerably smaller, which suggests that this may be another instance to which Ruderman model applies, though with a relatively high misalignment angle. However in view of the above mentioned likelihood [46,47] of other positive contributions to $\epsilon$, this evidence is too inconclusive to exclude the possibility that the Crab glitches may, after all, be attributable a slippage mechanism [35–37] of the first category, or even to the original simple oblateness mechanism subject to (21).

What transpires from all the work that has been rather rapidly surveyed in the preceding sections is that the available theory of the internal structure of neutron stars seems to provide all the elements needed to account for the accumulated pulsar frequency data within the framework of scenarios in which superfluidity and differential rotation commonly play an essential role. However we are still a long way short of reaching any consensus about the detailed modelling of the many different kinds of behavior observed in particular cases such as the Crab and Vela pulsars. Before any definitive understanding can be reached it will be necessary to carry out much more work on the technicalities of basic physical processes, particularly those involving electromagnetic effects, which were barely mentioned in the preceding overview, but that are extremely important for the detailed estimation of important quantities such as the drag to lift ratio, $c_r$.

The remainder of this article will be concerned with just one of the many technical problems that needs to be dealt with before a satisfactorily complete understanding can be achieved. This is the problem of developing an appropriately relativistic treatment of the superfluidity that has been seen to play such an essential role in accounting for cases such as that of Vela, and even the less extreme case of the Crab.

## 2 Essentials of relativistic superfluid mechanics

### 2.1 Motivation and background

In calculations of global quantities such as the mass and radius of a neutron star with a given baryon number, it has been known since before the earliest pulsar observations in 1968 that a fully relativistic treatment is indispensable for even a minimally acceptable level of accuracy. It is fortunate that quantities such as this [3] can be obtained within the framework of an exactly spherical perfect fluid description for which a fully relativistic treatment is easily applicable and has always been used. This contrasts with what has been done about secondary effects such as precession [50,27,51], involving mechanisms such as elasticity (for which a relativistic treatment has long been available [52,53] but is difficult to apply) and superfluidity (for which the relevant macroscopic treatment [54–56] is relatively new). Even in the relatively tractable context of stationary axisymmetric problems, or in contexts involving the (intrinsically relativistic) effect of gravitational radiation, superfluidity and superconductivity have nearly always been dealt with using a non-relativistic Newtonian, even in relatively recent work [57–59]

There are two essentially different reasons why it is worthwhile to try to do better. One is of course that a fully relativistic treatment should in principle be more accurate, and will no doubt become necessary for this purpose sooner or later. However in the short run this is not always what is most important, since the errors inherent in the use of a purely Newtonian treatment range typically from a few per cent to a few tens of percent which is not very significant compared with the order of unity (or worse) uncertainties about many of the physical quantities involved. The other kind of reason (which some readers may find surprising) is that for many purposes the use of a relativistic treatment is actually easier. In many cases the advantage of a relativistic treatment is due to the fact that the Lorentz group is in the technical sense semi-simple, whereas the Galilei group unfortunately is not. However, whether or not it is intrinsically simpler, the use of a relativistic treatment will usually be more convenient in practice whenever one wishes to use the commonly appropriate strategy [21,60,61] of working in terms of perturbations of the available spherically symmetric perfect fluid neutron star models. This is just because (as remarked above) all the best of these models (the only ones that are commonly taken seriously) are already formulated exclusively in a (general) relativistic framework. The same consideration applies to the perturbations of the relativistic axisymmetric rapidly rotating star models that have recently [62–66] been a subject of rapid development, particularly in relation to the question of bar mode instabilities that may be significant as a source of gravitational radiation.

The treatment provided here will be limited to the case of scalar (spin 0) models such as are appropriate for the experimentally familiar example of helium - 4 (though not [67] helium - 3) and also for the mesoscopic description of the neutron fluid that (as discussed above) is predicted to interpenetrate the ions in the lower crust of a neutron star, and that is believed [68] to condense as a superfluid in which the relevant bosons are scalar Cooper type pairs of neutrons. For the mesoscopic (intervortex) treatment of the mixed proton neutron superfluid below the crust a scalar description has commonly been employed [18,69] in a Newtonian treatment, and an analogous relativistic description [70,71] has recently been made available. However for an exact description of such a mixed proton neutron superfluid, in which it is predicted [68] that the neutrons pair up as bosons of spin 1, it would be necessary to use a more elaborate treatment that has yet to be developed).

What is actually needed for the analysis of large scale effects (such as were considered in the preceding sections) is not just a mesoscopic treatment of the superfluid on scales small compared with the spacing between interpenetrating ionic nuclei and the vortices where the irrotationality condition breaks down, but a macroscopic average over much larger scales. An appropriate macroscopic theory of the kind that is needed has recently been developed [56] and is described in the final subsections of this article. The treatment presented here differs from the relativistic analogue [54,55] of the earlier non relativistic description [72] of the averaged effect of vortices in neglecting the small anisotropy due to their effective tension, but instead it includes allowance for what in the long run is likely to be a more important effect, namely the "transfusion" of matter (for the reasons discussed in the previous subsections) between the superfluid constituent and the normal background, which is to be interpreted as representing not just thermal excitations (as in ordinary liquid helium - 4) but the entire crust component. A more elaborate treatment would include allowance for anisotropy of the crust constituent which, as noted above, will be cold enough (except very near the surface) to behave as an elastic solid: the way to do this has been indicated elsewhere [73], but such a mixed fluid solid description has not yet been developed in detail, and will not be dealt with in the introductory treatment provided here.

As a preliminary to the construction of model [56] that is actually needed for the macroscopic treatment of neutron star matter, this presentation starts by recapitulating the long well known essentials of the relativistic version of the single constituent kind of superfluid model that is appropriate for the description of Helium 4 at zero temperature, and of of the more recently developed generalization [74–76] to a two constituent model (of the kind whose non-relativistic analogue was originally developed by Landau) in which the second constituent represents a gas of phonon excitations.

## 2.2 Single constituent perfect fluid models

Before getting into the specific technicalities of superfluidity, it is worthwhile to start by recapitulating the relevant properties of ordinary barotropic fluid models, which includes the category of single constituent (scalar) superfluid models (representing the zero temperature limit of Landau type 2 constituent models) as the special case in which the vorticity is zero. The vorticity, in this context, is to be interpreted as the meaning the exterior derivative of the relevant momentum covector which will be formally defined below, so the vanishing of the vorticity is the condition for this momentum to be the gradient of a scalar potential, which in the superfluid case is to be understood to be proportional to the phase angle of an underlying opsonic quantum condensate. The qualification that this (zero temperature limit) model is barotropic simply means that there is only one independent state function such as the conserved (e.g. baryon) number density $n$ or the mass density $\rho$ (which are proportional in Newtonian theory but non-linearly related in relativistic theory) on which all the other state functions, such as the pressure $P$ are dependent. The equation of state giving $P$ as a function of $\rho$ will also determine a corresponding speed $c_I$ say, of ordinary "first" sound, that will be given by the familiar formula

$$c_I^2 = dP/d\rho, \qquad (46)$$

and that must be subluminal, $c_I^2 \leq c^2$ (where $c$ is the speed of light) in order for the usual causality requirement to be respected.

Before proceeding it is desirable to recall the essential elements of the relativistic kinematics and dynamics that will be required. This is particularly necessary in view of the regrettable tradition in non-relativistic fluid theory – and particularly in non relativistic superfluid theory – of obscuring the essential distinction between velocity (which formally belongs in a tangent bundle) and momentum (which formally belongs in a cotangent bundle) despite the fact that the distinction is generally respected in other branches of non-relativistic condensed matter theory, such as solid state physics, where the possibility of non-alignment between the 3-velocity $v^a$, and the effective 3-momentum $p_a$ of an electron travelling in a metallic lattice is well known. Failure to distinguish between contravariant entities (with "upstairs" indices) such as the velocity $v^a$ and covariant entities (with "downstairs" indices) such as the momentum $p_a$ is something that one can get away with in a non-relativistic treatment only at a price that includes restriction to strictly Cartesian (rather than for example cylindrical or comoving) coordinates.

In a relativistic treatment, even using coordinates $x^\mu \leftrightarrow \{t, x^a\}$ of Minkowski type, with a flat spacetime metric $g_{\mu\nu}$ whose components are of the fixed standard form diag$\{-c^2, 1, 1, 1\}$, the necessity of distinguishing between raised and lowered indices is inescapable. Thus for a trajectory parameterized by proper time $\tau$, the corresponding unit tangent vector

$$u^\mu = \frac{dx^\mu}{d\tau} \qquad (47)$$

is automatically, by construction, a contravariant vector: its space components, $u^a = \gamma v^a$ with $\gamma = (1 - v^2/c^2)^{-1/2}$ will be unaffected by the index lowering operation $u^\mu \mapsto u_\mu = g_{\mu\nu} u^\nu$, but its time component $u^o = dt/d\tau = \gamma$ will differ in sign from the corresponding component $u_0 = -\gamma c^2$ of the associated covector $u_\mu$. On the other hand the 3-momentum $p_a$ and energy $E$ determine a 4-momentum covector $\mu_\nu$ with components $\pi_a = p_a$, $\mu_o = -E$ that are intrinsically covariant. The covariant nature of the momentum can be seen from the way it is introduced by the defining equation,

$$\mu_\nu = \frac{\partial L}{\partial u^\nu}, \qquad (48)$$

in terms of the relevant position and velocity dependent Lagrangian function $L$, from which the corresponding equation of motion is obtained in the well known form

$$\frac{d\mu_\nu}{d\tau} = \frac{\partial L}{\partial x^\nu}. \qquad (49)$$

In the case of a free particle trajectory, and more generally for fluid flow trajectories in a model of the simple barotropic kind that is relevant in the zero temperature limit, the Lagrangian function will have the familiar standard form

$$L = \tfrac{1}{2}\mu g_{\mu\nu} u^\mu u^\nu - \tfrac{1}{2}\mu c^2, \qquad (50)$$

in which (unlike what is needed for more complicated chemically inhomogeneous models [77,73]) it is the same scalar spacetime field $\mu$ that plays the role of mass in the first term and that provides the potential energy contribution in the second term. The momentum will thus be given by the simple proportionality relation

$$\mu_\nu = \mu u_\nu, \qquad (51)$$

so that one obtains the expressions $E = \gamma \mu c^2$, $p_a = \mu \gamma v_a$, in which the field $\mu$ is interpretable as the relevant effective mass.

In the case of a free particle model, the effective mass $\mu$ will of course just be a constant, $\mu = m$. This means that if, as we have been supposing so far, the metric $g_{\mu\nu}$ is that of flat Minkowski type, the resulting free particle trajectories will be obtainable trivially as straight lines. However the covariant form of the equations (47) to (51) means that they will still be valid for less trivial cases for which, instead of being flat, the metric $g_{\mu\nu}$ is postulated to have a variable form in order to represent the effect of a gravitational field, such as that of a Kerr black hole (for which, as I showed in detail in a much earlier Les Houches school [78], the resulting non trivial geodesic equations still turn out to be exactly integrable).

In the case of the simple "barotropic" perfect fluid models with which we shall be concerned here, the effective mass field $\mu$ will be generically non-uniform. In these models the equation of state giving the pressure $P$ as a function of the mass density $\rho$ can most conveniently be specified by first giving $\rho$ in terms of the corresponding conserved number density $n$ by an expression that will be decomposable in the form

$$\rho = mn + \frac{\epsilon}{c^2}, \qquad (52)$$

in which $m$ is a fixed "rest mass" characterizing the kind of particle (e.g. a Cooper type neutron pair) under consideration, while $\epsilon$ represents an extra compression energy contribution. The pressure will then be obtainable using the well known formula

$$P = (n\mu - \rho)c^2, \qquad (53)$$

in which the effective dynamical mass is defined by

$$\mu = \frac{d\rho}{dn} = m + \frac{1}{c^2}\frac{d\epsilon}{dn}. \qquad (54)$$

It is this same quantity $\mu$ (sometimes known as the "specific enthalpy") that is to be taken as the effective mass function appearing in the specification (50) of the relevant Lagrangian function (on what is formally identifiable as the tangent bundle of the spacetime manifold).

When one is dealing not just with a single particle trajectory but a space-filling fluid flow, it is possible and for many purposes desirable to convert the Lagrangian dynamical equation (49) from particle evolution equation to equivalent field evolution equations [77,73]. Since the momentum covector $\mu_\nu$ will be obtained as a field over spacetime, it will have a well defined gradient tensor $\nabla_\rho \mu_\nu$ that can be used to rewrite the right hand side of (49) in the form $d\mu_\nu/d\tau = u^\rho \nabla_\rho \mu_\nu$. Since the value of the Lagrangian will also be obtained as a scalar spacetime field $L$, it will also have a well defined gradient which will evidently be given by an expression of the form

$$\nabla_\nu L = \frac{\partial L}{\partial x^\nu} + \frac{\partial L}{\partial \mu_\rho} \nabla_\nu \mu_\rho.$$

We can thereby rewrite the Lagrangian dynamical equation (49) as a field equation of the form

$$u^\rho \nabla_\rho \mu_\nu + \mu_\rho \nabla_\nu u^\rho = \nabla_\nu L. \qquad (55)$$

An alternative approach is of course to start from the corresponding Hamiltonian function, as obtained in terms of the position and momentum variables (so that formally it should be considered as a function on the spacetime cotangent bundle) via the Legendre transformation

$$H = \mu_\nu u^\nu - L. \qquad (56)$$

In this approach the velocity vector is recovered using the formula

$$\frac{dx^\mu}{d\tau} = \frac{\partial H}{\partial \mu_\nu}, \qquad (57)$$

and the associated dynamical equation takes the form

$$\frac{d\mu_\nu}{d\tau} = -\frac{\partial H}{\partial x^\nu}. \qquad (58)$$

The consideration that we are concerned not just with a single trajectory but with a spacefilling fluid means that, as in the case of the preceding Lagrangian equations, so in a similar way this familiar Hamiltonian dynamical equation can also be converted to a field equation which takes the form

$$2u^\rho \nabla_{[\rho}\mu_{\nu]} = -\nabla_\nu H\,, \tag{59}$$

with the usual convention that square brackets are used to indicate index antisymmetrization. On contraction with $u^\nu$ the left hand side will evidently go out, leaving the condition

$$u^\nu \nabla_\nu H = 0\,, \tag{60}$$

expressing the conservation of the value of the Hamiltonian along the flow lines.

The actual form of the Hamiltonian function that is obtained from the simple barotropic kind of Lagrangian function (50) with which we are concerned will evidently be given by

$$H = \frac{1}{2\mu}g^{\nu\rho}\mu_\nu\mu_\rho + \frac{\mu c^2}{2}\,, \tag{61}$$

in which it is again the same scalar spacetime field $\mu$ that plays the role of mass in the first term and that provides potential energy contribution in the second term.

In order to ensure the proper time normalization for the parameter $\tau$ the equations of motion (in whichever of the four equivalent forms (49), (55), (58), (59) may be preferred) are to be solved subject to the constraint that the numerical value of the Hamiltonian should vanish,

$$H = 0\,, \tag{62}$$

initially, and hence also by (60) at all other times. This is evidently equivalent to imposing the standard normalization condition

$$u^\mu u_\mu = -c^2\,, \tag{63}$$

on the velocity four vector. In more general "non-barotropic" systems, such as are needed for some purposes, the Hamiltonian may be constrained in a non uniform manner [77,73] so that the term on the right of (59) will be non zero, but in the simpler systems that suffice for our present purpose the restraint (62) ensures that this final term will drop out, leaving a Hamiltonian equation that takes the very elegant and convenient form

$$u^\nu w_{\nu\rho} = 0\,. \tag{64}$$

in terms of the relativistic vorticity tensor that is defined as the antisymmetrized ("exterior") derivative of the momentum covector, i.e.

$$w_{\nu\rho} = 2\nabla_{[\nu}\mu_{\rho]}\,. \tag{65}$$

It is an evident consequence (and, as discussed in greater detail in the above cited Les Houches notes [55], would still be true even if (62) were not satisfied) that if $w_{\mu\nu}$ is zero initially it will remain zero throughout the flow, which in this case will be describable as "irrotational".

In cases for which the vorticity is non-zero, the "barotropic" dynamical equation (64) is interpretable as requiring the flow vector $u^\mu$ to be a zero eigenvalue eigenvector of the vorticity tensor $w_{\mu\nu}$, which is evidently possible only if its determinant vanishes, a requirement that is expressible as the degeneracy condition

$$w_{\mu[\nu}w_{\rho\sigma]} = 0. \tag{66}$$

Since the possibility of it having matrix rank 4 is thus excluded, it follows that unless it actually vanishes the vorticity tensor must have rank 2 (since an antisymmetric matrix can never have odd integer rank). This means that the flow vector $u^\mu$ is just a particular case within a whole 2-dimensional tangent subspace of zero eigenvalue vorticity eigenvectors, which (by a well known theorem of differential form theory) will mesh together to form well defined vorticity 2-surfaces as a consequence of the Poincaré closure property,

$$\nabla_{[\mu} w_{\nu\rho]} = 0, \tag{67}$$

that follows from the definition (65).

Although it has long been well known to specialists [82], the simple form (64) of what is interpretable just as the relativistic version of the classical Euler equation is still not as widely familiar as it ought to be, perhaps because its Hamiltonian interpretation was not recognized until relatively recently [77,73]. It does not constitute by itself the complete set of dynamical equations of motion for the perfect fluid, but must be supplemented by a particle conservation equation of the usual form for the particle number current

$$n^\nu = nu^\nu, \tag{68}$$

which must of course satisfy the condition

$$\nabla_\nu n^\nu = 0. \tag{69}$$

A much more widely known, but for computational purposes (particularly in curved spacetime) less useful form of the perfect fluid dynamical equations is to express them in terms of the stress momentum energy density tensor, which is given in terms of the mass density $\rho$ and the pressure $P$ by

$$T^{\mu\nu} = (\rho + \frac{P}{c^2})u^\mu u^\nu + P g^{\mu\nu}, \tag{70}$$

and which must satisfy a so called conservation law of the standard form

$$\nabla_\nu T^{\mu\nu} = 0. \tag{71}$$

Although it is conveniently succinct, a disadvantage of this traditional formulation is that it is directly interpretable as a law of conservation of momentum and energy in the strict sense only in the case of a flat (Minkowski type) spacetime, but not in a curved background such as that of a neutron star. The possibility in the barotropic case (i.e. when $P$ is a function only of $\rho$) of decomposing the combined set of dynamical equations (70) as the combination of the convergence condition (69) (obtained by contracting (71) with $u_\mu$) and the relativistic Euler equation (64), which can be written out more explicitly as

$$n^\nu \nabla_{[\nu} \mu_{\rho]} = 0, \qquad (72)$$

has the advantage that these are interpretable as genuine conservation laws – for particle number flux and vorticity respectively – even in an arbitrarily curved spacetime background.

## 3 Single constituent superfluid models

The simplest superfluid models, namely those pertaining to the zero temperature limit, are just ordinary perfect fluid models subject to the restraint of irrotationality, with a momentum covector given as the gradient

$$\mu_\nu = \nabla_\nu S, \qquad (73)$$

of a scalar field $S$. This scalar field is to interpreted as being proportional to the angle of the mesoscopic phase factor, $e^{i\phi}$ say, of an underlying scalar opsonic condensate, in which the phase angle $\phi$ is given according to the usual correspondence principle by

$$\phi = S/\hbar. \qquad (74)$$

In the most familiar application the bosons are Helium-4 atoms, while between the ions of a neutron star crust below the neutron drip transition they will be Cooper type neutron pairs. (However a less simple description is not sufficient for the spin 1 neutron pairs below the base of the crust, nor in the even more complicated, though experimentally accessible, case [67] of Helium-3, for which a relativistic description is still not available).

In a multiconnected configuration of a classical irrotational fluid the Jacobin action field $S$ obtained from (73) might have an arbitrary periodicity, but in a superfluid there will be a U(1) periodicity quantization requirement that the periodicity of the phase angle $\phi$ should be a multiple of $2\pi$, and thus that the periodicity of the Jacobin action $S$ should be a multiple of $2\pi\hbar$. The simplest configuration for any such superfluid is a uniform stationary state in a flat Minkowski background, for which the phase will have the standard plane wave form

$$S/\hbar = k_a x^a - \omega t, \qquad (75)$$

from which one obtains the correspondence $\mu_\nu \leftrightarrow \{-\hbar\omega, \hbar k_a\}$, which means that the effective energy per particle will be given by $E = \gamma\mu c^2 = \hbar\omega$ and that the 3-momentum will be given by $p_a = \mu\gamma v_a = \hbar k_a$.

It is to be remarked that for ordinary timelike superfluid particle trajectories the corresponding phase speed $\omega/k$ of the wave characterized by (75) will always be superluminal – a fact of which people working with laboratory Helium-4 tend to be blissfully unaware, and can usually safely ignore, since what matters for most practical purposes is not the phase speed but the group velocity of perturbation wave packets.

In the irrotational case characterized by (73) the Euler equation (64) is satisfied automatically, so the only dynamical equation that remains is (69). When the phase scalar is subject to a small perturbation, $\delta\phi = \varphi$ say, it can be seen that the corresponding perturbation of the conservation law (69) provides a wave equation of the form

$$\widetilde{\Box}\varphi = 0, \tag{76}$$

in which $\widetilde{\Box}$ is a modified Dalembertian type operator that is constructed from an appropriately modified space-time metric tensor $\widetilde{g}^{\mu\nu}$ in the same way that the ordinary Dalembertian operator $\Box \equiv \nabla^\mu \nabla_\mu$ is constructed from the ordinary spacetime metric tensor $g^{\mu\nu}$. The appropriately modified spacetime metric, namely the relativistic version of what is known in the context of Newtonian fluid [79,80] and superfluid [81] mechanics as the Unruh metric, can be read out in terms of the light speed $c$ and the (first) sound speed $c_{\rm I}$ given by (46) as

$$\widetilde{g}^{\mu\nu} = \frac{\mu}{n}\left(c_{\rm I}\gamma^{\mu\nu} - c_{\rm I}^{-1}u^\mu u^\nu\right), \tag{77}$$

where $\gamma^{\mu\nu}$ is the spatially projected (positive indefinite) part of the ordinary space time metric, as defined by

$$\gamma^{\mu\nu} = g^{\mu\nu} + c^{-2}u^\mu u^\nu. \tag{78}$$

The quantum excitations of the linearized perturbation field $\varphi$ governed by (76) are what are known as phonons. For such excitations the phase speed and the group velocity are the same, both being given with respect to the unperturbed background by the ordinary ("first") soundspeed, $c_{\rm I}$, as given by (46), which will of course be subluminal. Phonons do nevertheless have a tachyonic aspect of their own, because the fact that their phase speed is subluminal automatically implies that they have a 4-momentum covector that is spacelike, in contrast with that of a ordinary fluid or superfluid particle which is timelike. This means that whereas the effective energy $E$ of an ordinary fluid or superfluid particle is always positive, the effective energy $E$ of a phonon may be positive or negative, depending on whether the frame of reference with respect to which it is measured is moving subsonically or supersonically. The well known implication is that if the superfluid is in contact with a supersonically moving boundary there will inevitably be an instability giving rise to dissipative phonon creation.

Given a dynamical system, one of the first things any physicist is inclined to ask is whether it is derivable from a Lagrangian type variation principle. We have already seen in the previous sections that (64) by itself is obtainable from Lagrangian equations of motion for the individual trajectories, which are of course obtainable from a one dimensional action integral of the form $\int L\, d\tau$ with $L$ as given by (50). The question to be addressed now is how to obtain the complete set of dynamical equations (71), including (64) as well as (69), from an action integral over the 4-dimensional background manifold $\mathcal{S}^{(4)}$ of the form

$$\mathcal{I} = \int \mathcal{L}\, d\mathcal{S}^{(4)}, \qquad d\mathcal{S}^{(4)} = \frac{\|g\|^{1/2}}{c} d^4 x, \qquad (79)$$

for some suitable scalar Lagrangian functional $\mathcal{L}$.

There are several available procedures for doing this for a generic perfect fluid with rotation, involving radically different choices of the independent variables to be varied: although they are all ultimately equivalent "on shell" the "off shell" bundles over which the variations are taken differ not only in structure but even in dimension. These methods notably include the worldline variation procedure (the most economical from a dimensional point of view) developed by Taub [83], and the Clebsch type variation procedures developed by Schutz [84], as well as the more recently developed Kalb-Ramond type method [85,55] that has been specifically designed for generalization [54] to allow for the anisotropy arising from the averaged effect of vortex tension in the treatment of superfluidity at a macroscopic level. None of these various methods is sufficiently simple to have become widely popular.

The problem is much easier to deal with if, to start off with, one restricts oneself to the purely irrotational case characterized by (73), which is all that is needed for the description of zero temperature superfluidity at a mesoscopic level. In this case a very simple and well known procedure is available. In this procedure, the independent variable is taken to be just the Jacobin action $S$, or equivalently in a superfluid context, the phase $\varphi$ as given by (74), and the action is simply taken to be the pressure P expressed as a function of the effective mass $\mu$, with the latter constructed as proportional to the amplitude of the 4-momentum, according to the prescription

$$\mu^2 c^2 = -\mu_\nu \mu^\nu, \qquad (80)$$

with the 4-momentum itself given by the relation (73) that applies in the irrotational case, i.e.

$$\mu_\nu = \hbar \nabla_\nu \phi. \qquad (81)$$

Thus setting

$$\mathcal{L} = P, \qquad (82)$$

and using the standard pressure variation formula

$$\delta P = c^2 n \delta \mu, \qquad (83)$$

one sees that the required variation of the Lagrangian will be given by

$$\delta \mathcal{L} = -n^\nu \delta \mu_\nu = -\hbar n^\nu \nabla_\nu (\delta \phi). \qquad (84)$$

Demanding that the action integral (79) be invariant with respect to infinitesimal variations of $\varphi = \delta\phi$ then evidently leads to the required conservation law (69).

It is to be noted that this variational principle can be reformulated in terms of an independently variable auxiliary field amplitude $\Phi$ and an appropriately constructed potential function $V\{\Phi\}$ as a function of which the action takes the desirably fashionable form

$$\mathcal{L} = -\frac{\hbar^2}{2}\Phi^2(\nabla_\nu \phi)\nabla^\nu \phi - V\{\Phi\}, \qquad (85)$$

which is interpretable as the classical limit of a generalized Landau Ginzburg type model. In this formulation, as discussed in greater detail elsewhere [85,55], the auxiliary amplitude is to be identified as being given by the formula

$$\Phi = \frac{n}{\sqrt{\rho + P/c^2}} = \left(\frac{n}{\mu}\right)^{1/2}, \qquad (86)$$

while the prescription for the corresponding potential energy density function is that it should be given by

$$V = \frac{\rho c^2 - P}{2}. \qquad (87)$$

Having evaluated $V$ as a function of $\Phi$ one can recover the effective mass $\mu$, number density $n$, mass density $\rho$ and pressure $P$ of the fluid using the formulae

$$\mu^2 = \frac{1}{c^2 \Phi}\frac{dV}{d\Phi}, \qquad n = \Phi^2 \mu, \qquad (88)$$

and

$$\rho = \tfrac{1}{2}\Phi^2 \mu^2 + \frac{V}{c^2}, \qquad P = \tfrac{1}{2}\Phi^2 \mu^2 c^2 - V, \qquad (89)$$

which are derivable from (53) and (54). It is to be remarked that the covariant inverse of the generalized Unruh tensor (77) is expressible in this notation as

$$\widetilde{g}^{-1}_{\mu\nu} = \Phi^2(c_{\rm I}^{-1}\gamma_{\mu\nu} - c_{\rm I} c^{-2} u_\mu u_\nu). \qquad (90)$$

A particularly noteworthy example is the conformably invariant special case [55] characterized by a potential function that is homogeneously quartan, $V \propto \Phi^4$, which is what is obtained for a radiation gas type equation of state of the familiar form $P = \rho c^2/3$, and for which the (first) sound speed is given by $c_{\rm I}^2 = c^2/3$.

## 4  Landau type 2-constituent superfluid models

As an intermediate step between the very simple single constituent superfluid models described in the previous section and the more elaborate models needed in the context of neutron stars, the purpose of this section is to describe the relativistic version of the category of non dissipative 2-constituent superfluid that was originally developed by Landau for the description of ordinary superfluid Helium-4 at non-zero temperature. As well as the relevant conserved particle number current $n^\mu$ (representing the flux of Helium atoms in that particular application) such a model involves another independently conserved current vector, $s^\mu$ say, representing the flux of entropy. In the single constituent case characterized by the variation rule (84) we saw how the current vector $n^\nu$ was associated with a dynamically conjugate covector $\mu_\nu$ that is interpretable as representing the effective mean 4-momentum per particle. In a similar way in a 2-constituent model the second current vector $s^\nu$ will be analogously associated with its own dynamically conjugate 4-momentum covector $\Theta_\nu$.

The earliest presentations of the generic category of non-dissipative 2-constituent superfluid were on one hand a generalization [74] of the relativistic Clebsch formulation [84] based on the variation of a generalized pressure function $\Psi$ depending on the 4-momentum convectors $\mu_\nu$ and $\Theta_\nu$ according to the partial differentiation rule

$$d\Psi = -n^\nu d\mu_\nu - s^\nu d\Theta_\nu \,, \tag{91}$$

and on the other hand a generalization [86,73] of the world line variational formulation due to Taub [83] based on the variation of a master function $\Lambda$ depending on the currents $n^\mu$ and $s^\nu$ according to the partial differentiation rule

$$d\Lambda = \mu_\nu \, dn^\nu + \Theta_\nu \, ds^\nu \,. \tag{92}$$

Although they were originally developed independently these alternative formulations were subsequently shown to be equivalent to each other and to an intermediate crossbred version [75] based on a Lagrangian density

$$\mathcal{L} = \Psi + s^\nu \Theta_\nu = \Lambda - n^\mu \mu_\nu \,, \tag{93}$$

depending on the particle 4-momentum covector $\mu_\nu$ and the entropy current $s^\nu$ according to the partial differentiation rule

$$d\mathcal{L} = \Theta_\nu \, ds^\nu - n^\nu \, d\mu_\nu \,. \tag{94}$$

All of these variational formulations are subject to the complication that the allowable field variations are not free but must be suitably constrained to avoid giving overdetermined field equations. Although it violates the symmetry between the two kinds of conserved current $n^\nu$ and $s^\nu$ that are involved, the crossbred formulation characterized by (94) is the one that allows the simplest specification of the constraints required to get the appropriate dynamical equations for the superfluid case. In this formulation [75] the constraint on the particle 4-momentum covector is simply that it should have the same phase gradient form (73) as in the zero temperature limit in which the entropy constituent is absent, namely

$$\mu_\nu = \hbar \nabla_\nu \phi \,. \tag{95}$$

The corresponding constraint on the current vector $s^\nu$ of the "normal" constituent is the not quite so simple Taub type requirement that its variation should be determined by the displacement of the flow lines generated by an arbitrary vector field $\zeta^\nu$ say, which means [73] that it must have the form

$$ds^\nu = \zeta^\rho \nabla_\rho s^\nu - s^\nu \nabla_\rho \zeta^\nu + s^\nu \nabla_\rho \zeta^\nu \,, \tag{96}$$

whose derivation is obtainable by a procedure that will be explained more explicitly in the next Section. Demanding invariance of the volume integral of $\mathcal{L}$ with respect to infinitesimal local variations of the phase variable $\phi$ then gives the usual particle conservation law in the same form (69) as for the single constituent limit, while demanding invariance for an arbitrary local displacement field $\zeta$ gives not only the analogous entropy conservation law

$$\nabla_\nu s^\nu = 0 \,, \tag{97}$$

but also the dynamical equation,

$$s^\nu \nabla_{[\nu} \Theta_{\rho]} = 0 \,, \tag{98}$$

that governs the evolution of the thermal 4-momentum covector in a manner analogous to that whereby the relativistic Euler equation (72) governs the evolution of the momentum covector in an ordinary perfect fluid. These dynamical equations entail (but unlike the single constituent case are not entirely contained in) an energy momentum pseudo-conservation law of the usual form (71) for a stress-momentum-energy density tensor that can be written in the form

$$T^\nu_\rho = n^\nu \mu_\rho + s^\nu \Theta_\rho + \Psi g^\nu_\rho \,, \tag{99}$$

which will in fact (although it is not obvious in this particular expression) be automatically symmetric, $T^{[\nu\rho]} = 0$.

The category of models characterized by the preceding specifications for various conceivable forms of the equation of state specifying $\mathcal{L}$ as a scalar function of $\mu_\nu$ and $s^\nu$ is very large. The use of what is interpretable [87] as a special subcategory therein, on the basis of a particular kind of separation ansatz, was proposed in early work of Israel [88] and Dixon [89] and has been advocated more recently by Olsen [90]. Unfortunately however, the simplification provided by the Israel Dixon ansatz (effectively the relativistic generalization of the obsolete Tisza-London theory that was superseded by that of Landau) is incompatible with the kind of equation of state that is needed for even a minimally realistic treatment of a real superfluid.

A satisfactory treatment of what goes on at temperatures high enough for non-linear "roton" type excitations to be important is not yet available, but in the low temperature "cool" regime, in which only linear "phonon" type excitations are important, it is not difficult to provide a straightforward analytic derivation of the kind of equation of state that is appropriate. Following the lines developed in a non - relativistic context by Landau himself [91] the relativistic version of the appropriate "cool" equation of state has recently been derived [76] by considering perturbations of the single constituent model – with equation of state specified as a pressure function, $P\{\mu\}$ – that describes the relevant zero temperature limit. The result is obtained in an analytically explicit form that (despite the fact that it is not of the separable Israel Dixon kind) can be given a very simple expression in terms of what we referred to as the "sonic" metric, which is specifiable by the conformed relation

$$\mathcal{G}^{\rho\sigma} = \Phi^2 c_I^{-1} \widetilde{g}^{\rho\sigma} \tag{100}$$

in terms the Unruh phonon metric (77) that is associated with the relevant zero temperature limit state as specified by the relevant momentum covector $\mu_\nu$, which by (80) determines the relevant value of the scalar $\mu$ and hence (via the zero temperature equation of state, using the formalism of Section 2.2) also of the relevant phonon speed $c_I$ and field amplitude $\Phi$. While the Unruh metric is more convenient for many purposes, the advantage of the conformed modification we have used, namely

$$\mathcal{G}^{\rho\sigma} = g^{\rho\sigma} + (c^{-2} - c_I^{-2}) u^\rho u^\sigma, \tag{101}$$

is that its spatially projected part agrees with that of the ordinary space metric, from which it differs only in the measurement of time.

The result that is obtained [76] is given by a Lagrangian of the form

$$\mathcal{L} = P - 3\psi \tag{102}$$

in which the deviation from the zero pressure limit value $P\{\mu\}$ is given as a function not just of the particle 4-momentum covector $\mu_\nu$ but also of the entropy flux $s^\nu$ (postulated to be sufficiently weak to be constituted only of phonons) by the formula

$$\psi = \frac{\tilde{\hbar}}{3} c_I^{-1/3} |\mathcal{G}^{-1}_{\rho\sigma} s^\rho s^\sigma|^{2/3}, \tag{103}$$

where $\tilde{\hbar}$ is identifiable to a very good approximation with the usual Dirac-Planck constant $\hbar$, its exact value being given by

$$\tilde{\hbar} = \frac{9}{4\pi} \left(\frac{5\pi}{6}\right)^{1/3} \simeq 0.99 \, \hbar. \tag{104}$$

This is equivalent to taking the generalized pressure function to be

$$\Psi = P + \psi, \tag{105}$$

with

$$\psi = \frac{c_I}{4} \left(\frac{3}{4\hbar}\right)^3 \left(\mathcal{G}^{\rho\sigma} \Theta_\rho \Theta_\sigma\right)^2, \tag{106}$$

in which the effective thermal 4-momentum per unit of entropy is given (according to the partial differentiation formula (94)) by

$$\Theta_\rho = \frac{4\tilde{\hbar}}{3} |c_I \mathcal{G}^{-1}_{\mu\nu} s^\mu s^\nu|^{-1/3} \mathcal{G}^{-1}_{\rho\sigma} s^\sigma, \tag{107}$$

with

$$\mathcal{G}^{-1}_{\rho\sigma} = g_{\rho\sigma} + \left(1 - \frac{c_I^2}{c^2}\right) |\mu^\nu \mu_\nu|^{-1} \mu_\rho \mu_\sigma. \tag{108}$$

An concrete illustration, allowing the explicit evaluation of the relevant quantities, is provided by the polytropic case, as characterized by a (single constituent) equation of state giving the mass density $\rho$ as a function of the number density $n$ in terms of a fixed ("rest") mass per particle $m$, a scale constant $\kappa$ and a fixed dimensionless index $\gamma$ in the form

$$\rho = mn + \kappa n^\gamma \Leftrightarrow \mu = m + \kappa \gamma n^{\gamma-1}, \tag{109}$$

which corresponds to taking the pressure to be given by

$$P = \kappa c^2 (\gamma - 1) n^\gamma = \kappa c^2 (\gamma - 1) \left(\frac{\mu - m}{\kappa \gamma}\right)^{\gamma/(\gamma-1)}, \tag{110}$$

while the corresponding sound speed will be given (independently of $\kappa$) by

$$c_{\rm I}^{\,2} = (\gamma - 1)\left(1 - \frac{m}{\mu}\right) c^2. \tag{111}$$

## 5 Non-conservative model with transfusion and vortex drag

Although the Landau type of model described in the previous section has been found to be very effective for the description of liquid Helium-4 under laboratory conditions, it is not of much use for direct application in neutron star matter because the thermal effects it allows for will in general be less important than other complications whose treatment will require the use of more elaborate models whose relativistic versions are still at a relatively early stage of development and will not be presented here. The most important of these complications, whose treatment in a relativistic framework has been the subject of preliminary work that is discussed elsewhere, are due to the effect of the protons that will be present, either in ionic nuclei that are responsible for the elastic solid behavior [73] of the crust, or as a dissolved superfluid [70,71] at deeper levels. Another complication that is relevant for the macroscopic treatment of a neutron star is the necessity of averaging over an Abrikosov type lattice of quantized vortices (that must be roughly aligned with the rotation axis of the star) whose effective tension entails deviations [54,55] from perfect fluid isotropy.

Like the thermal effect discussed in the preceding section, these various complications can all be provisionally set aside as perturbations to be incorporated at a later stage in a systematic approach whose first stage requires the use only of a relatively crude description in which, except for the superfluid neutrons with baryon number current vector $n_{\rm n}^\nu$ all the other constituents, meaning mainly protons and electrons, move together with the entropy as a single "normal" constituent with baryon number current $n_{\rm c}^\nu$. Whereas the anisotropy arising from vortex tension [54,55] is relatively unimportant, a major role in the long term evolution of the star is likely to be played by the static pinning or dynamical drag forces exerted on the vortices by the composite "normal" background constituent. Another effect that is of importance in the long run is that of "transfusion" whereby – due to the subduction resulting from the drift mechanism whose effect is roughly described by (39) or (40) – the superfluid neutron contribution $n_{\rm n}^\nu$ to the baryon current may undergoes transformation (via weak beta decay type processes) to the "normal" (essentially photonic) constituent, and vice versa, so that only the total baryon current

$$n_{\rm B} = n_{\rm n}^\nu + n_{\rm c}^\nu \tag{112}$$

remains locally conserved throughout:

$$\nabla_\nu n_{\rm B}^\nu = 0 \,. \tag{113}$$

The kind of (non-conservative) 2-constituent model needed for this purpose is obtainable as a generalization of the kind of (conservative) 2-constituent superfluid model discussed in the preceding section, starting from the formulation in terms of a master function $\Lambda$ in which the currents (not momenta) are taken as the independent (but not entirely free) variables.

In a transfusive model of the type set up here, the "normal" constituent is not entirely dependent on (though it does include) entropy, so that it is present even at zero temperature: the primary role of this non - superfluid constituent is to represent the fraction of the carbonic material of the neutron star that is not included in the neutron superfluid, as well as the degenerate electron gas that will be present to neutralize the charge density resulting from the fact that some of these baryons will have the form of protons rather than neutrons. In the solid "crust" layers of a neutron star the protons will be concentrated together with a certain fraction of the neutrons in discrete nuclear type ions, which at the relatively moderate temperatures that are expected to apply will form a solid lattice. In the upper crust the "normal" constituent consisting of the ionic lattice and the degenerate electrons will include everything, but in the lower crust (at densities above about $10^{11}$ gm/cm$^3$) the crust will be interpenetrated by an independently moving neutron superfluid. What we refer to as "transfusion" occurs when compression takes place so that the ionic constituent undergoes a fusion process whereby neutrons are released in the form of newly created superfluid matter, or conversely, when relaxation of the pressure allows excess neutrons to be reabsorbed into the ions.

A more elaborate treatment would specifically allow for the expectation that the protons would form an independently conducting superfluid of their own at very high densities, whereas they will combine with some of the neutrons at intermediate densities, and with all of the neutrons at low densities, to form discrete ions which will tend to crystalize to form a possibly anisotropic lattice. What matters for our present purpose is that regardless of its detailed constitution, all this "normal" matter will in effect be strongly self coupled [19] by short range electromagnetic interactions so that its movement will be describable to a very good approximation as that of a single fluid with a well defined 4-velocity, $u_{\rm C}{}^\mu$ say, the only independent motion being that of the (electromagnetically neutral) neutron superfluid with velocity $u_{\rm N}{}^\mu$ say. The latter will specify the direction of the part of the baryon current,

$$n_{\rm n}{}^\mu = n_{\rm n} u_{\rm N}{}^\mu \,, \tag{114}$$

carried by the neutron superfluid, while the "normal" matter velocity specifies the direction of the remaining *collectively comoving* part,

$$n_c{}^\mu = n_c u_C{}^\mu, \tag{115}$$

of the baryon current.

At densities below the "neutron drip" transition at about $10^{11}$ gm/cm$^3$, the "normal" collectively comoving constituent $n_c{}^\mu$ will of course be identifiable with the total, $n_B{}^\mu$. The reason why the remaining free neutron part $n_n{}^\mu$ – which will always be present at higher densities – is presumed to be in a state of superfluidity is that the relevant condensation temperature, below which the neutrons form opsonic condensate of Cooper type pairs is estimated [92] to be at least of the order of $10^9$ K, while it is expected that a newly formed neutron star will drop substantially below this temperature within a few months [93]. At such comparatively low temperatures the corresponding entropy current $s^\mu$ say will not play a very important dynamical role, but for the sake of exact internal consistency it will be allowed for in the model set up here, in which it will be taken for granted that it forms part of the "normal" collectively comoving constituent so that it will have the form

$$s^\mu = s u_C^\mu. \tag{116}$$

Under conditions of sufficiently slow convection, the transfer needs not involve significant dissipation, so the process should be describable by a Lagrangian scalar, $\Lambda$ say, that will depend just on the currents introduced above, of which the independent components are given just by the vectors $n_c^\mu$ and $n_n^\mu$ and the scalar $s$. As a first approximation (whose accuracy in the various relevant density regimes is a subject that needs much further investigation) one might suppose that the Lagrangian separates in the form $\Lambda = -\rho^c c^2 - \rho^n c^2$ in which $\rho^c$ is a mass density depending only on $s$ and $n_c$, while $\rho^n$ is an another energy mass depending only on $n_n$, but we shall not invoke such a postulate here, i.e. we allow for the likelihood that the properties of "normal" constituent will be affected by the presence of the superfluid constituent and vice versa, which means that there will be an *entrainment* effect [94,18,19,95], whereby for example the velocity of the superfluid neutron current will no longer be parallel to the corresponding momentum. (As an alternative to the more suitable term "entrainment" this mechanism is sometimes referred to in the literatures as "drag", which is misleading because entrainment is a purely conservative, entirely non-dissipative effect, whereas the usual kinds of drag in physics, and in particular the kind of drag to be discussed below, are essentially dissipative processes.)

If we adopted the (gas type) description embodied in the separation ansatz we would have two separate variation laws which in a fixed background would take the form $c^2 \delta \rho^c = \Theta \delta s + c^2 \mu^c \delta n_c$ and $c^2 \delta \rho^n = c^2 \mu^n \delta n_n$, in which $\Theta$ would be interpretable as the temperature, $\mu^c$ would be interpretable as the effective mass per baryon in the "normal" part, and $\mu^n$ would be effective mass per neutron in the superfluid part (which would be equal to its analogue in the "normal" part, i.e. $\mu^{\rm n} = \mu^{\rm c}$, in the particular case of a state of static thermodynamic equilibrium.)

In the less specialized (liquid type) description to be used here, there will just be a single "conglomerated" variation law, whose most general form, including allowance for a conceivable variation of the background metric, will be expressible (correcting one of the copying errors in the originally published version [56]) as

$$\delta\Lambda = -\Theta\delta s + \mu^{\rm c}{}_\nu \delta n^\mu_{\rm C} + \mu^{\rm n}{}_\nu \delta n^\nu_{\rm n} + \tfrac{1}{2}\left(c^{-2}\Theta s u^\mu_{\rm C} u^\nu_{\rm C} + n_{\rm c}{}^\mu \mu^{c\nu} + n_{\rm n}{}^\mu \mu^{n\nu}\right)\delta g_{\mu\nu}\,,\qquad (117)$$

where $\Theta$ is to be interpreted as the temperature and where $\mu^{\rm n}_\mu$ and $\mu^{\rm c}_\mu$ are to be interpreted as the 4-momentum per baryon of the neutron superfluid and the "normal" constituent respectively.

To obtain suitable fluid type dynamical equations from a Lagrangian expressed as above just in terms of the relevant currents, the variation of the latter must be appropriately constrained in the manner[73] that was originally introduced for the case of a simple perfect fluid by Taub [83]. The standard Taub procedure can be characterized as the requirement that the variation of the relevant current three form, which for the "normal" constituents in the present application will be given in terms of the antisymmetric space-time measure tensor $\varepsilon_{\mu\nu\rho\sigma}$ by

$$N_{\mu\nu\rho} = \varepsilon_{\mu\nu\rho\sigma} n^\sigma_{\rm c}\,,\qquad (118)$$

should be given by Lie transportation with respect to an associated, freely chosen, displacement vector field $\zeta^\mu$ say. This ansatz gives the well known result

$$\delta N_{\mu\nu\rho} = \zeta^\lambda \nabla_\lambda N_{\mu\nu\rho} + 3 N_{\lambda[\mu\nu}\nabla_{\rho]}\zeta^\lambda\,.\qquad (119)$$

Although a variation $\delta g_{\mu\nu}$ of the metric has no effect on the fundamental current three form, $N_{\mu\nu\rho}$, it will contribute to the variation of the corresponding vector,

$$n^\mu_{\rm c} = \frac{1}{3!}\varepsilon^{\mu\nu\rho\sigma} N_{\nu\rho\sigma}\,,\qquad (120)$$

for which one obtains

$$\delta n^\mu_{\rm c} = \zeta^\nu \nabla_\nu n^\mu_{\rm c} - n^\nu_{\rm c}\nabla_\nu \zeta^\mu + n^\mu_{\rm c}\left(\nabla_\nu \zeta^\nu - \tfrac{1}{2}g^{\nu\rho}\delta g_{\nu\rho}\right)\,.\qquad (121)$$

(Application of an analogous procedure to the entropy current provides the variation rule (96) that was used in Section 3). In terms of the orthogonally projected metric,

$$\gamma_{\rm c}{}^{\mu\nu} = g^{\mu\nu} + c^{-2} u^\mu_{\rm C} u^\nu_{\rm C}\,,\qquad (122)$$

the corresponding variation of the unit flow vector will be given by

$$\delta u_{\rm C}^\mu = \gamma_{\rm c}{}^\mu{}_\rho\left(\zeta^\nu \nabla_\nu u^\rho_{\rm C} - u^\nu_{\rm C}\nabla_\nu \zeta^\rho\right) + \tfrac{1}{2}c^{-2} u^\mu_{\rm C} u^\nu_{\rm C} u^\rho_{\rm C}\delta g_{\nu\rho}\,,\qquad (123)$$

and the corresponding variation in the current amplitude $n_{\rm c}$ will be

$$\delta n_{\rm c} = \nabla_\nu\left(n_{\rm c}\zeta^\nu\right) + n_{\rm c}\left(c^{-2} u_{\rm C}^\mu u_{\rm C}^\nu \nabla_\mu \zeta_\nu - \tfrac{1}{2}\gamma_{\rm c}{}^{\mu\nu}\delta g_{\mu\nu}\right)\,.\qquad (124)$$

Since the entropy flux is to be considered as comoving with the "normal" constituent, it is subject to a variation given by the same displacement vector $\zeta$, which thus gives

$$\delta s = \nabla_\nu (s\zeta^\nu) + s\big(c^{-2} u_C{}^\mu u_C{}^\nu \nabla_\mu \zeta_\nu - \tfrac{1}{2}\gamma_c^{\mu\nu}\delta g_{\mu\nu}\big). \tag{125}$$

On the other hand for the superfluid constituent there will be an independent displacement vector field $\xi^\mu$ say, in terms of which the analogously constructed variation will be

$$\delta n_n^\mu = \xi^\nu \nabla_\nu n_n^\mu - n_n^\nu \nabla_\nu \xi^\mu + n_n^\mu \big(\nabla_\nu \xi^\nu - \tfrac{1}{2}g^{\nu\rho}\delta g_{\nu\rho}\big). \tag{126}$$

The effect of this variation process on the Lagrangian density $\|g\|^{1/2}\Lambda$ itself can be seen to be expressible in the standard form

$$\|g\|^{-1/2}\delta\big(\|g\|^{1/2}\Lambda\big) = \zeta^\nu f^c{}_\nu + \xi^\nu f^n{}_\nu + \tfrac{1}{2}T^{\mu\nu}\delta g_{\mu\nu} + \nabla_\mu \mathcal{R}^\mu, \tag{127}$$

in which $f^c{}_\nu$ will be interpretable as the force density acting on the "normal" constituent, $f^n{}_\nu$ will be interpretable as the force density acting on the superfluid constituent, and $T^{\mu\nu}$ will be interpretable as the stress momentum energy density of the two constituent as a whole.

By considering the trivial case in which there is no actual physical alteration of the system, but in which the apparent changes are merely due to the displacement of the reference system generated by a vector field $\xi^\nu = \zeta^\nu$, in which case the apparent variation of the metric will be given by $\delta g_{\nu\nu} = 2\nabla_{[\mu}\zeta_{\nu]}$, it can be seen from (127) that the separate forces must automatically satisfy an identity of the form

$$f^c{}_\nu + f^n{}_\nu = f^{\text{ex}}{}_\nu, \tag{128}$$

where $f^{\text{ex}}{}_\nu$ is the conglomerated *external force density* that is defined by

$$f^{\text{ex}}{}_\nu = \nabla_\mu T^\mu{}_\nu. \tag{129}$$

The residual current $\mathcal{R}^\mu$ in the divergence will be of no importance for our present purpose (by Green's theorem it just gives a surface contribution that will vanish by the variational boundary conditions) but it is to be noted for the record that it will have the form

$$\mathcal{R}^\mu = 2\zeta^{[\mu} u_C{}^{\nu]} \big(c^{-2}\Theta s u_{C\nu} + n_c \mu^c_\nu\big) + 2\xi^{[\mu} n_n{}^{\nu]} \mu^n{}_\nu. \tag{130}$$

The conglomerated stress momentum energy density tensor can easily be read out as

$$T^\mu{}_\nu = \Psi g^\mu{}_\nu + c^{-2}\Theta s u_C{}^\mu u_{C\nu} + n_c{}^\mu \mu^c{}_\nu + n_n{}^\mu \mu^n{}_\nu, \tag{131}$$

where

$$\Psi = \Lambda + s\Theta - n_c{}^\nu \mu^c{}_\nu - n_n{}^\nu \mu^n{}_\nu. \tag{132}$$

(Although this expression is not manifestly symmetric, the asymmetric contributions will automatically cancel due to the identity $\mu^{c[\mu} n_c{}^{\nu]} = -\mu^{n[\mu} n_n{}^{\nu]}$).

What matters most for our present purpose is the form of the respective force densities: the force law (i.e. the relevant relativistic generalization of Newton's "second" law of motion) for the "normal" constituent is found to take the form

$$f^c{}_\nu = 2s^\mu \nabla_{[\mu}\bigl(c^{-2}\Theta u_{C\nu]}\bigr) + 2n_c{}^\mu \nabla_{[\mu}\mu^c_{\nu]} + c^{-2}\Theta u_{C\nu}\nabla_\mu s^\mu + \mu^c_\nu \nabla_\mu n^\mu_c, \qquad (133)$$

while the force law for the superfluid component is found to take the simpler form

$$f^n_\nu = f^{ch}_\nu + f^{me}_\nu, \qquad (134)$$

in which the first term is a "chemical" contribution, representing the effect of any neutron superfluid particle creation or destruction, which is given by

$$f^{ch}{}_\nu = \mu^n_\nu \nabla_\mu n^\mu_n. \qquad (135)$$

The last term in (134) is a "mechanical" contribution, allowing for drag or pinning forces exerted on the vortices by the crust and balanced by the Magnus effect, according to the formula

$$f^{me}{}_\nu = n_n{}^\mu w^n_{\mu\nu}, \qquad (136)$$

using the notation

$$w^n_{\mu\nu} = 2\nabla_{[\mu}\mu^n_{\nu]} \qquad (137)$$

for the vorticity 2-form of the superfluid neutrons. It is to be noted that this is not the mesoscopic (intervortex) superfluid vorticity, which simply vanishes, but the average vorticity on a macroscopic scale that is large compared with the spacing (typically a very small fraction of a cm.) between the superfluid vortices. For a very accurate treatment it would be necessary to take account of the macroscopic anisotropy resulting from the effective tension of these vortices, as has already been done [54,55] for the case a single constituent, but for the discussion of global evolution on timescales long compared with the stellar oscillation periods (a small fraction of a second) such an effect seems unlikely to be important.

Although the complete expression (133) is not so simple, it is to be observed that the time component in the "normal" rest frame (representing the rate of working on the "normal" constituent) as obtained by contraction with the relevant unit vector $u_C{}^\nu$ has the comparatively simple form

$$u_C{}^\nu f^c{}_\nu = u_C{}^\nu \mu^c_\nu \nabla_\mu n_c{}^\mu - \Theta \nabla_\mu s^\mu. \qquad (138)$$

If we were to impose the variation principle to the effect that the system should be invariant with respect to arbitrary worldline displacements (as specified by the independent fields $\zeta^\nu$ and $\xi^\nu$) it would follow that each of the forces $f^c_\mu$ and $f^n_\nu$ would have to vanish. However it is evident from the identity (128) that we cannot adopt such a restrictive postulate in a model designed to treat the effect of pulsar slowdown due to a torque attributable to coupling to an external electromagnetic field that is removing angular momentum by radiation to infinity. As well as the intrinsically non-conservative magnetic torque contribution to $f^{ex}_\nu$ it is also important [40] to include a contribution to allow for the effect of the elastic solidity in the crust, which is not incorporated into the simple fluid type model included here (and which would require the use of a much more elaborate model [73] for its detailed evaluation).

Although our ultimate purpose is to allow for a non vanishing external torque force, whatever force law we assume must be such that if the external force $f^{\text{ex}}_\nu$ were somehow switched off so as to leave an effectively isolated system, the second law of thermodynamics (no decrease of entropy in an isolated system) would be respected, i.e. we must have $f^{\text{ex}}_\nu = 0 \Rightarrow \nabla_\nu d^\nu \geq 0$. It can be seen that this is equivalent to the requirement of positivity of the right hand side of the identity

$$\Theta \nabla_\mu s^\mu + u_{\text{C}}{}^\nu f^{\text{ex}}_\nu = u_{\text{C}}{}^\nu (\mu^{\text{n}}{}_\nu - \mu^{\text{c}}{}_\nu) \nabla_\mu n_{\text{n}}{}^\mu + u_{\text{C}}{}^\nu f^{\text{me}}{}_\nu, \qquad (139)$$

that is obtained from (138), taking account of the total baryon conservation law (112). Since they involve very different physical processes, one comes to the conclusion that each of the two terms on the right of (139) must satisfy its own separate positivity condition.

The positivity requirement for the first of these terms is presumably to be attributed to a crust particle creation law of the form

$$\nabla_\mu n_{\text{n}}^\mu = \Xi u_{\text{C}}^\nu (\mu^{\text{n}}{}_\nu - \mu^{\text{c}}{}_\nu) \qquad (140)$$

for some positive coefficient $\Xi$. Such a law is an obviously natural generalization of the kind of creation rate formula that is familiar in chemical physics. In the present context what is involved is conversion of protons to neutrons by weak interactions, and the situation is complicated by the consideration that as far as the large scale mechanics of the neutron star is concerned, the effective rate may depend not just on microscopic processes, but also, when subduction is involved, on the rather messy process whereby the crust is broken up before it ultimately dissolves.

To complete the specification of the system, all that remains is to find the appropriate ansatz for the mechanical force $f^{\text{me}}{}_\nu$. This problem is more delicate than that of the (effectively scalar) chemical case, since as well as the "second law" requirement $u_{\text{C}}^\nu f^{\text{me}}{}_\nu \geq 0$, the answer must respect the nature of the macroscopic vorticity 2-form $w^{\text{n}}_{\mu\nu}$ which although non vanishing (unlike the mesoscopic vorticity between vortices) cannot be arbitrary (as in an ordinary viscous fluid): to be consistent with the underlying superfluid nature of the neutron constituent, it must satisfy an algebraic degeneracy condition of the form (66) in order to be compatible with the existence of a well defined congruence of orthogonal 2-surfaces generated by (non vanishing) tangent vectors, $v^\nu$ say, such that $w^{\text{n}}_{\mu\nu} v^\nu = 0$. It can be seen from the form of the defining relation (136) that the obvious way to obtain this degeneracy property is to take the force law to have the form $f^{\text{me}}{}_\nu = w^{\text{n}}_{\nu\sigma} v^\sigma$ for some suitably chosen vector $v^\nu$ which, to satisfy the "second law" requirement must satisfy $u_{\text{C}}^\nu w^{\text{n}}_{\nu\sigma} v^\sigma \geq 0$. The required ansatz can thus be taken to be given by $v_\mu = \alpha w^{\text{n}}_{\mu\nu} u_{\text{C}}^\nu$ for some positive coefficient $\alpha$. This result is conveniently expressible in terms of the rank-2 tensor $\perp^\mu_\nu$ of orthogonal projection with respect to the vortex 2-surface, which is given by

$$\perp^\mu_\nu = 2(w^{\mathrm{n}\rho\sigma} w^{\mathrm{n}}_{\rho\sigma})^{-1} w^{\mathrm{n}\lambda\mu} w^{\mathrm{n}}_{\lambda\nu}. \tag{141}$$

We end up with an expression taking the form

$$f^{\mathrm{me}}{}_\nu = \eta_{\mathrm{r}} \perp_{\nu\sigma} u_{\mathrm{C}}^\sigma, \tag{142}$$

for a positive resistive drag coefficient $\eta_{\mathrm{r}}$ (given in terms of the previous coefficient $\alpha$ by $2\eta_{\mathrm{r}} = \alpha w^{\mathrm{n}\rho\sigma} w^{\mathrm{n}}_{\rho\sigma}$).

The generic class of dissipative models characterized by finite values of $\Xi$ and $\eta_{\mathrm{r}}$ has four different kinds of non dissipative limit. In the low reactivity limit $\Xi \to 0$ we have the non-transfusive limit characterized by the separate superfluid particle conservation law

$$\nabla_\nu n_{\mathrm{n}}^\nu = 0, \tag{143}$$

whereas in the opposite high reactivity limit $\Xi \to \infty$ we have the chemical equilibrium limit characterized by

$$u_{\mathrm{C}}^\nu (\mu^{\mathrm{c}}_\nu - \mu^{\mathrm{n}}_\nu) = 0. \tag{144}$$

which is what would be expected in cases for which the (continental drift like) crust circulation responsible for the transfusion is characterized by timescales that are very long compared with those [98] of the relevant weak (direct or inverse beta decay) interactions. For each of these conceivably relevant possibilities, we have the drag free limit $\eta_{\mathrm{r}} \to 0$ characterized by the condition of vanishing Magnus force,

$$n_{\mathrm{n}}^\mu w^{\mathrm{n}}_{\mu\nu} = 0, \tag{145}$$

or at the opposite extreme the perfect vortex pinning limit, $\eta_{\mathrm{r}} \to \infty$ characterized by the condition that the vortex worldsheets should be at rest with respect to the "normal" (crust) background,

$$u_{\mathrm{C}}^\mu w^{\mathrm{n}}_{\mu\nu} = 0. \tag{146}$$

The actual evaluation of the drag coefficient $\eta_{\mathrm{r}}$, and the question of whether one or other of these simple extreme limits is realistic depends on delicate technical issues [96,97] whose definitive resolution is not entirely clear. A scenario of the type envisaged by Ruderman [39], as represented by Figure 1, and described by (39) (which seems to be appropriate for Vela) is what would be obtained in the case characterized by (146), whereas the more recently proposed alternative scenario [40] represented by Figure 2, and described by (40), is what would be obtained in the case characterized by (145).

It is of course to be expected that such extreme scenarios will turn out in practise to be oversimplifications of a more complicated reality, whose description is likely to require modelling with not just two [61] but many independently rotating components, to allow for the variation of the chemical and mechanical coefficients $\Xi$ and $\eta_{\mathrm{r}}$ over a wide range of finite values as a function of depth in the star.

## Acknowledgements

The author wishes to thank Silvano Bonazzola, David Langlois, Eric Gourghoulon, Pawel Haensel, Isaac Khalatnikov, Reinhardt Prix, and David Sedrakian for conversations and collaboration.

## References

1. B.K. Harrison, K.S. Thorne, M. Wakano, J.A. Wheeler, *Gravitation theory and gravitational collapse* (University of Chicago Press, Chicago, 1965).
2. A.B. Migdal: *Nucl. Phys.* **13**, 655-674 (1959).
3. N. K. Glendenning, *Compact stars* (Springer, Heidelberg, 1997).
4. F. Weber, "Pulsars as astrophysical laboratories for nuclear and particle physica" (I.O.P. Publishing, Bristol, 1999).
5. D. Langlois:"Superfluidity in relativistic neutron stars". [astro-ph/0008161]
6. L.D. Landau, E.M. Lifshitz, *Course of Theorteical Physics*, Vols. 1 - 9 (Pergamon, Oxford, 1959).
7. G. Baym, D. Pines: *Annals of Physics* **66**, 816-835 (1971).
8. G. Baym, C. Pethick, D. Pines, M. Ruderman: *Nature* **224**, 872-874 (1969).
9. M. Hoffberg, A.E. Glassgold, R.W. Richardson, M. Ruderman: *Phys. Rev. Lett.* **24**, 775-777 (1970).
10. J.A. Sauls, D.L. Stein, J.W. Serene: *Phys. Rev.* **D25**, 967-975 (1982).
11. N.K. Glendenning: *Phys. Rev.* **D46**, 1274-1287 (1992).
12. N.K. Glendenning, S. Pei: *Phys. Rev.* **C52**, 2250-2253 (1995).
13. M. Alford, K. Rajagopal, F. Wilczek: *Nucl. Phys.* **B537**, 443-458 (1999). [hep-th/9804403]
14. D. Blaschke, D.M. Sedrakian, K.M. Shahabasyan: *Astron. Astrophys.* **350**, L47 (1999). [astro-ph/9904395]
15. M. Alford, J. Berges, K. Ragjagopal: "Magnetic fields within colour superconducting neutron star cors". [hep-ph/9910254]
16. D.M. Sedrakian, D. Blaschke, K.M. Shahabasyan, D.N. Voskresensky: "Meissner effect for color superconducting quark matter". [hep-ph/0012383]
17. B. Carter: Annals of Physics, **95**, 53 -73 (1974).
18. G.A. Vardanian, D.M. Sedrakian: *Sov. Phys. J.E.T.P.* **54**(5), 919-921 (1981).
19. M.A. Alpar, S.A. Langer, J.A. Sauls: *Astroph. J.* **282**, 533-541 (1984).
20. S.E. Thorsett, D.Chakrabarty: *Astroph. J.* **512**, 288-299 (1999).
21. E. Chubarian, H. Grigorian, G. Poghosyan, D. Blaschke: *Astron. Astrophys.* **357**, 968-976 (2000). [astro-ph/9903489]
22. M. Ruderman: *Nature* **223**, 597-598 (1969).
23. R. Smoluchowski: *Phys. Rev. Lett.* **24**, 923-925 (1970).
24. P.W. Anderson, N. Itoh: *Nature* **256**, 25-26 (1975).
25. A. Sedrakian, J.M. Cordes: *Astroph. J.* **502**, 378-381 (1989). [astro-ph/9802102]
26. M.A. Alpar, P.W. Anderson, D. Pines, J. Shaham: *Astroph. J. Lett.* **249**, L29 (1981).
27. A. Sedrakian, I. Wasserman, J.M. Cordes: pinning", *Astroph. J.* **524**, 341-360 (1999).
28. M.Q. Alpar, J.A. Sauls: *Astroph. J.* **327**, 723-725 (1988).
29. L. Bildsten, R.I. Epstein: *Astroph. J.*, **342**, 951-957 (1989)
30. A.D. Sedrakian, D.M. Sedrakian: *Sov. Phys. JETP* **75**, 395-399 (1992).

31. M.A. Alpar, P.W. Anderson, D. Pines and J. Shaham: *Astroph. J.* **276**, 325-334 (1984); *Astroph. J.* **278**, 791-805 (1984).
32. P.B. Jones: *Mon. Not. R. Astr. Soc.* **246**, 315-323 (1990).
33. A.D. Sedrakian and D.M. Sedrakian: *Astroph. J.* **447**, 305-323 (1995).
34. P.B. Jones: *Astroph. J.* **372**, 208-212 (1991).
35. B.K. Link, R.I. Epstein: *Astroph. J.* **373**, 592-603 (1991).
36. B.K. Link, R.I. Epstein: *Astroph. J.* **457**, 844-854 (1996).
37. A. Sedrakian, J.M. Cordes: *Mon. Not. R. Astr. Soc.* **307**, 365 (1999). [astro-ph/9806042]
38. M. Ruderman: *Astroph. J.* **203**, 213-222 (1976).
39. M. Ruderman: *Astroph. J.* **366**, 261-269 (1991).
40. B. Carter, D. Langlois, D.M. Sedrakian: *Astron. Astroph* **361**, 795-802 (2000). [astro-ph/0004121].
41. R. Prix: *Astron. Astrophys.* **352**, 623-631 (1999). [astro-ph/9910293].
42. S.L. Shapiro, S.A. Teukolsky, *Black holes, white dwarfs, and neutron stars* (Wiley, New York, 1883).
43. F. Pacini: *Nature*, **216**, 567-568 (1967).
44. P. Goldreich, W.H. Julian: *Astroph. J.* **157**, 869 (1969).
45. N. K. Glendenning, S. Pei, F. Weber: *Phys. Rev. Lett.* **79**, 1603-1606 (1997).
46. V.S. Beskin, A.V. Gurevich, Ya. N. Istomin, *Physics of the pulsar magnetosphere* (C.U.P., Cambridge, 1993).
47. A. Melatos: *Mon. Not. R. Astr. Soc.* **288**, 1049-1059 (1997).
48. A.G. Lyne, R.S. Pritchard, F.G. Smith: *Mon. Not. R. Astr. Soc.* **233**, 667-676 (1988).
49. A.G. Lyne, R.S. Pritchard, F.G. Smith, F. Camilio: *Nature* **381**, 497-498 (1996).
50. J. Shaham: *Astroph. J.* **214**, 251-260 (1977).
51. A. Melatos, "Radiative precession of an isolated neuton star". [astro-ph/0004035]
52. B. Carter, H. Quintana: *Annals of physica* **95**, 74-89 (1975).
53. B. Carter, H. Quintana: *Astroph. J.* **202**, 511-522 (1975).
54. B. Carter, D. Langlois: *Nuclear Physics* **B454**, 402-424 (1995). [hep-th/9611082]
55. B. Carter: in *Topological defects and the non-equilibrium dynamics of symmetry breaking phase transitions* (Les Houches 99), ed. Y.M. Bunkov and H. Godfrin, pp 267-301 (Kluwer, 2000). [gr-qc/9907039]
56. D. Langlois, D. M. Sedrakian, B. Carter: *Mon. Not. Roy. Astr. Soc.* **297**, 1198-1201 (1998). [astro-ph/9711042]
57. L. Lindblom, G. Mendel: *Astroph. J.* **421**, 689-704 (1994).
58. U. Lee: *Astron. Astrop.* **303**, 515-525 [1995).
59. A. Sedrakian, I. Wasserman: *Phys. Rev.* **D63**, 024016 (2000). [astro-ph/0004331]
60. G.L. Comer, D. Langlois, L.M. Lin: *Phys. Rev.* **D60**, 104025 (1999). [gr-qc/9908040]
61. N. Andersson, G.L. Comer, "Slowly rotating general relativistic superfluid neutron stars". [gr-qc/0009089]
62. N. Stergioulas, J. Friedman: *Astroph. J.* **492**, 301-322 (1998). [gr-qc/9705056]
63. S. Bonazzola, J. Frieben, E. Gourgoulhon: *Astron. Astrophys.* **331**, 280-290 (1998). [gr-qc/9710121]
64. S. Morsink, N. Stergioulas, S.R. Blattnig: *Astroph. J.* **510**, 854-861 (1999). [gr-qc/9806008]
65. E. Gourgoulhon, P. Haensel, R. Levine, E. Paluch, S. Bonazzola, J.A. Marck: *Astron. Astroph.* **349**, 851-862 (1999). [astro-ph/9907225]
66. N. Andersson, D.I. Jones, K.D. Kokkotas, N. Stergioulas: *Astroph. J.* **534**, L75 (2000). [astro-ph/0002114]

67. G.E. Volovik, "Exotic properties of superfluid helium 3" (World Scientific, Singapore, 1992).
68. J.A. Sauls "Superfluidity in the interior of neutron stars" in *Timing Neutron Stars, Nato ASI* **C262**, ed. H. Ogelman, E.J.P. van den Heuvel. *pp* 457-490 (Kluwer, Dordrecht, 1988).
69. G. Mendel, L. Lindblom: *Annals of Physics* **205**, 110-129 (1991).
70. B. Carter, D. Langlois: *Nuclear Phys.* **B531**, 478-504 (1998). [gr-qc/9806024]
71. B. Carter: *Gravitation and Cosmology* **6**, Supplement, 204-213 (2000). [astro-ph/0010109]
72. I.L. Bekarevich, I.M. Khalatnikov: *Sov. Phys. J.E.T.P.* **13**, 643-646 (1961).
73. B. Carter: in *Relativistic Fluid Dynamics (C.I.M.E., Noto, May 1987)* ed. A.M. Anile, & Y. Choquet-Bruhat, Lecture Notes in Mathematics **1385** *pp* 1-64 (Springer - Verlag, Heidelberg, 1989).
74. V.V. Lebedev, I.M. Khalatnikov: *Sov. Phys. J.E.T.P.* **56**, 923-930 (1982).
75. B. Carter, I.M. Khalatnikov: *Phys. Rev.* **D45**, 4536-4544 (1992).
76. B. Carter, D. Langlois: *Phys. Rev.* **D51**, 5855-5864 (1995). [hep-th/9507 058]
77. B. Carter: in *Active Galactic Nuclei*, ed. C. Hazard & S. Mitton, *pp* 273-300 (Cambridge U.P., 1979).
78. B. Carter: in *Black Holes* (proc. 1972 Les Houches Summer School), ed. B. & C. DeWitt, *pp* 57-210 (Gordon and Breach, New York, 1973).
79. W. Unruh; *Phys. Rev. Letters* **46**, 1351-1357 (1981).
80. W. Unruh: *Phys. Rev.* **D51**, 2827-2838 (1995). [gr-qc/9409008]
81. C. Barcelo, S. Liberati, M. Visser:, "Covariant Theory of Conductivity in Ideal Fluid or Solid Media", [gr-qc/0011026]
82. A. Lichnerowicz, *Relativistic Hydrodynamics and Magnetohydrodynamics* (Benjamin, New York, 1967).
83. A.H. Taub: *Phys. Rev.* **94**, 1469-1470 (1954).
84. B. Schutz: *Phys. Rev.* **D2**, 2762-2773 (1970).
85. B. Carter: *Class. Quantum Grav.* **11**, 2013-2030 (1994).
86. B.Carter, in *A Random Walk in Relartivity and Cosmology, Proc. Vadya - Raychaudhuri Festschrift, IAGRG 1983*, ed. N. Dadhich, J. Krishna Rao, J.V. Narlikar, C.V. Vishveshwara, *pp* 48 -62 (Wiley Eastern, Bombay, 1985).
87. B. Carter, I.M. Khalatnikov: *Ann. Phys.* **219**, 243-265 (1992).
88. W. Israel: *Physics Letters* **A86**, 79-81 (1981); *Physics Letters* **A92**, 77-78 (1982)
89. W.G. Dixon: *Arch. Rat. Mech. Anal.* **80**, 159 (1982).
90. T.S. Olsen: *Physics Letters* **A149**, 71-75 (1990).
91. L.D. Landau, E.M. Lifshitz, *Statistical Physics* (trad. E. and R.F. Peierls), Section VI (Pergamon, Oxford, 1959).
92. R.I. Epstein: *Astroph. J.* **333**, 880-894 (1988).
93. S. Tsuruta: *Phys. Rep.* **56**, 237-278 (1979).
94. A.F. Andreev, E.P. Bashkin: *Sov. Phys. J.E.T.P.* **42**, 164-167 (1976).
95. O. Sjoberg: *Nucl. Phys.* **A265**, 511-516 (1976).
96. P. B. Jones; *Mon. Not. R. Astr. Soc.*, **257**, 501-506 (1992).
97. B. Link, R.I. Epstein, G. Baym: *Astroph. J.*, **403**, 285-302, (1993).
98. P. Haensel: *Astron. Astroph.*, **262**, 131-137 (1992).

# The Tensor Virial Method and Its Applications to Self-Gravitating Superfluids

Armen Sedrakian[1,2] and Ira Wasserman[3]

[1] Kernfysisch Versneller Instituut, Groningen AA-9747, The Netherlands
[2] Groupe de Physique Theorique, Institut de Physique Nucleaire,
F-91406 Orsay Cedex, France
[3] Center for Radiophysics and Space Research, Cornell University, Ithaca, NY 14853

**Abstract.** This review starts with a discussion of the hierarchy of scales, relevant to the description of superfluids in neutron stars, which motivates a subsequent elementary exposition of the Newtonian superfluid hydrodynamics. Starting from the Euler equations for a superfluid and a normal fluid we apply the tensor virial method to obtain the virial equations of the first, second, and third order and to compute their Eulerian perturbations. Special emphasis is put on the computation of perturbations of the new terms due to mutual gravitational attraction and mutual friction between the two fluids. The oscillation modes of superfluid Maclaurin spheroids are derived from the first and second order perturbed virial equations. We discuss two generic classes of oscillation modes which correspond to the *co-moving* and *relative oscillations* of two fluids. These modes decouple if the normal fluid is inviscid. We also discuss the mixing of these modes (when the normal fluid is viscous) and its effect on the dynamical and secular instabilities of the co-moving modes and their damping.

## 1 Introduction

Radio and x-ray observations of neutron stars provide strong evidence for the superfluidity of neutron star interiors. Perhaps, the most striking manifestations of their superfluidity are the long (on time-scales from several hours to hundreds of days) relaxations that follow the glitches in the spin and spin-down rates of some pulsars. Although the majority of pulsars are very precise clocks, timing observations reveal persistent random fluctuations in times of arrival of radio signals. Some pulsars show long-term periodicities in their spin-characteristics and periodic changes of their pulse shape. However, the relation between the superfluidity of neutron star interiors and these latter anomalies of pulsar timing is not firmly established yet.

Further evidence for the superfluidity of neutron star interiors came in the 1980s with the advent of the orbiting x-ray satellites. The measurements of the thermal radiation from a dozen or so hot neutron stars provided indirect information on the temperature of superfluid phases in neutron stars. Theoretical thermal histories of superfluid neutron stars are consistent with the x-ray data (within the limits of our knowledge of the input physics). Non-superfluid stars, as a rule, cool too fast to below the threshold of detection.

Apart from the radiation in the electro-magnetic spectrum, neutron stars are expected to be primary sources of gravitational wave radiation, which are expected to be detectable by future laser interferometer detectors.

It is hoped that one can probe neutron star interiors and their superfluidity using gravity waves, as their eigen-frequencies and damping may depend on the dissipation in the superfluid, at temperatures below the superfluid phase transition. This review concentrates on the oscillations of superfluid self-gravitating ellipsoids within the tensor virial method. The method was originally introduced by Chandrasekhar and Fermi in the context of magneto-hydrodynamics [1]. It was extensively developed in the 1960s by Chandrasekhar for the study of the ellipsoidal figures of equilibrium and their oscillations. A comprehensive account of this work is contained in Chandrasekhar's monograph *Ellipsoidal Figures of Equilibrium* (hereafter EFE)[2].

The ellipsoidal approximation provides an idealized picture of oscillations of neutron stars. One can think of several arguments in favor of adopting such an approach: first, the combination of the Newtonian gravity and two-fluid hydrodynamics of superfluid defines

an exactly solvable model if we assume that the fluids are incompressible and inviscid; second, past experience with single-fluid self-gravitating ellipsoids shows that most of the qualitative features found for these ellipsoids have their analogs in more "realistic" systems; third, the method is transparent and in many cases analytical results can be obtained which shed light on the underlying physics. The tensor virial method is not the only tool for investigating the properties of ellipsoidal figures. Alternative formulations exist in the literature and we refer to [3–5] for further details; for a pedagogical introduction see the textbook [6]. Note, also, that various formulations of the theory of ellipsoids, to a large extent, are equivalent.

Superfluid oscillations were studied using various methods and approximations in the past decade or so. Epstein pointed out that the superfluidity of neutron star interiors has potentially important effects on the propagation of seismically excited acoustic waves. It allows for additional types of waves to propagate by virtue of doubling of degrees of freedom in a superfluid; superfluid phases create acoustic discontinuities in which wave velocities or polarizations change abruptly on the bounding interfaces [7]. The effects of superfluidity on global hydrodynamic oscillations were investigated by Lindblom and Mendell [8] in a model where the superfluid and the normal fluid are coupled via gravitational attraction and the *entrainment effect*[1]. Their solutions reveal that the lowest frequency pulsations are almost indistinguishable from those derived from the ordinary-fluid hydrodynamics; however, their analytical solutions also reveal the existence of a spectrum of modes which are absent in a single fluid star. Nonradial oscillations of non-rotating superfluid neutron stars were computed by Lee, whose numerical solutions for the radial and non-radial pulsations of

---

[1] The latter effect arises in the layers where neutrons and proton condensates coexist, hence the flow of one condensate is accompanied by the motion of the other. See B. Carter's article in this volume for further details.

the two-fluid stars identified distinct superfluid modes [9]. The effects of shear viscosity of the electron fluid and mutual friction on the r-mode oscillations were studied by Lindblom and Mendell [10] by constructing an energy functional and computing the time-scales associated with the dissipative terms. The oscillation modes of superfluid analogs of the classical Maclaurin, Jacobi and Roche ellipsoids were derived recently by Sedrakian and Wasserman [11] within the tensor virial method. The oscillation modes of superfluid ellipsoids separate into two classes corresponding to relative and co-moving (or center-of-mass, hereafter referred as CM) motions of two fluids. The CM oscillations are identical to the oscillations of single-fluid ellipsoids and are undamped if one ignores the viscosity of the normal fluid. The mutual friction contributes only to the damping of the relative oscillation modes. One important feature of the latter modes is that they do not emit gravitational radiation as there is no mass transport associated with them. Our discussion of the tensor virial method is based on [11].

This review is organized as follows. In the remainder of the Introduction we motivate the averaged superfluid hydrodynamics and identify the relevant scales in the problem. In Sect. 2 we give a tutorial introduction to the Newtonian superfluid hydrodynamics, which is mainly built on the work of Bekarevich and Khalatnikov [12]. Sect. 3 introduces the tensor virial method and illustrates its applications to the superfluid ellipsoidal figures by computing the perturbations of the new terms in virial equations due to their two-fluid nature. We discuss in Sect. 4 the oscillations of superfluid Maclaurin spheroids including the effects of mutual friction and viscosity of the normal fluid. Sect. 5 contains a brief summary. We refer the reader to the accompanying article by B. Carter for a review of relativistic models and an overview of the state of the art of the theory of superfluidity in neutron stars.

## 1.1 Characteristic length scales

The physics of neutron star superfluidity unfolds on a hierarchy of three distinct length-scales. The separation of these scales is useful, as often the physics of a neighboring scale enters a theory at a given scale in the form of phenomenological constants, which can potentially be fixed by comparison with measured observables.

At the *microscopic* level the physical scale of the order of fermi (fm= $10^{-13}$ cm) is set by the nuclear forces. The long-range attractive interaction between nucleons leads to an instability of the normal state of the nuclear matter against Cooper pairing, in analogy to the microscopic Bardeen-Cooper-Schrieffer theory of superconductivity of metals. The nuclear forces control the size of the "elementary bosons" of the theory - the Cooper pairs, which appear as weakly bound states of two fermions near the Fermi surface. The size of a Cooper pair, the *coherence length* $\xi$, is of the order of 10 fm. It sets, obviously, the lower scale on which the hydrodynamic description of superfluids breaks down. On length scales larger than $\xi$, the condensate of Cooper-pairs can be described by a single wave function $\psi(\mathbf{x})$, i.e., the condensate forms a macroscopically coherent state.

At the *local hydrodynamic* level the relevant physical scale is set by the size of vortices - macroscopic quantum objects, whose fundamental property is the quantization of the circulation around a path encircling their core. The circulation is quantized in units of $2\pi\hbar$ since the condensate wave-function must be single valued function at each point of the condensate. On writing $\psi = \psi_0 e^{i\chi}$, the gauge invariant superfluid velocities can be expressed through the gradient of the phase of superfluid order parameter $\chi$ and the vector potential, $\mathbf{A}$

$$\mathbf{v}_\tau = \frac{\hbar}{2m_\tau}\nabla\chi_\tau - \frac{e_\tau}{m_\tau c}\mathbf{A}, \tag{1}$$

where $e_\tau \equiv (e, 0)$ is the electric charge of protons and neutrons respectively, $m_\tau$ is their bare mass; $\tau \in \{n, p\}$, where $n$ and $p$ stand for neutrons and protons respectively. Applying the curl operator to (1) and implementing quantization of the circulation (the phase of the superfluid order parameter changes by $2\pi$ around a closed path) we find

$$\operatorname{curl} \mathbf{v}_\tau = \frac{\pi\hbar}{m_\tau}\boldsymbol{\nu}_\tau \sum_j \delta^{(2)}(\mathbf{x} - \mathbf{x}_{\tau j}) - \frac{e_\tau}{m_\tau c}\mathbf{B} \equiv \boldsymbol{\omega}_\tau, \tag{2}$$

where $\pi\hbar/m_\tau$ is the circulation quantum, $\boldsymbol{\nu}_\tau \equiv \boldsymbol{\omega}_\tau/\omega_\tau$ is a unit vector along the vortex lines, $\mathbf{x}_{\tau j}$ defines the position of a vortex line in the plane orthogonal to the vector $\boldsymbol{\nu}_\tau$, $\delta^{(2)}$ is a two–dimensional delta function in this plane and $\mathbf{B} = \operatorname{curl} \mathbf{A}$ is the magnetic field induction. The $j$–summation is over the sites of vortex lines. Note that (2) treats the vortex cores as singularities in the plane orthogonal to $\boldsymbol{\nu}_\tau$, which is justified on scales larger than the coherence length of a condensate. The crossover from the local hydrodynamic scale to the microscopic scale can be studied within the Ginzburg-Landau theory as we briefly discuss below.

For a single vortex the integral of (2) completely determines the superfluid pattern; as this equation is linear, for larger number of vortices the superfluid pattern is a superposition of the flows induced by each vortex. The resulting net flow, obviously, depends on the arrangement of the vortices. It turns out that the integral of (2) on the local hydrodynamic scale is radically different from the one on the scales involving larger number of vortices in a rotating superfluid.

To appreciate the difference in superfluid patterns on different scales let us look for vortex solutions in a neutral condensate in cylindrical geometry. The condensate wave function has the form $\psi(\mathbf{x}) = f(r)e^{i\theta}$ in the polar-cylindrical coordinates $(r, \theta, z)$; the neutron superfluid velocity upon integrating (2) becomes

$$\mathbf{v}_n = \frac{\hbar}{2m_n r}\hat{\theta}. \tag{3}$$

The divergence of the superfluid velocity when $r \to 0$ is avoided by introducing cut-off on scales of the order of the coherence length. The cut-off scale, for Fermi-liquids, can be understood by noting that an increase of $\mathbf{v}_n$ when $r \to 0$ causes an increase of the kinetic energy of Cooper pairs which eventually becomes larger than the binding energy of a pair. The broken pairs will perform a rigid-body rotation with an angular velocity which scales as $Cr$ and is regular in the $r \to 0$ limit (here $C$ is a constant). This crossover can be seen from the well-known solutions of the Ginzburg-Landau equation for the amplitude $f(r)$:

$$\frac{d^2 f}{d\zeta^2} + \frac{1}{\zeta}\frac{df}{d\zeta} - \frac{1}{\zeta^2}f + f - f^3 = 0, \qquad (4)$$

where $\zeta = r/\xi_n$ and $\xi_n$ is the size of the vortex core. The asymptotic solutions of (4) are

$$f(\zeta) = \begin{cases} C\zeta & \zeta \ll 1, \\ 1 - (2\zeta^2)^{-1} & \zeta \to \infty, \end{cases} \qquad (5)$$

while numerical solutions for intermediate values of $\zeta$ show that the condensate wave function is at half of its value in a homogeneous condensate when $\zeta = 1$. Note the long-range nature of the superfluid vortex velocity, and the resulting slow fall-off of the density perturbation in the condensate. This behavior is specific to neutral condensates; for charged condensates the super-current is screened exponentially on length scales of the order of the penetration depth $\lambda$. The solution of (2) for a charged condensate is

$$\mathbf{v}_p = \frac{\hbar}{2m_p \lambda} K_1\left(\frac{r}{\lambda}\right)\hat{\theta}, \qquad (6)$$

where $K_1$ is the Bessel function of imaginary argument; as for $r \gg \lambda$, $K_1(r/\lambda) \simeq \exp(-r/\lambda)$, therefore the superfluid circulation decays exponentially; in the opposite limit $r \ll \lambda$, $K_1(r/\lambda) \simeq \lambda/r$ and (6) assumes a form identical to (3).

At the *global hydrodynamic* level the relevant scales are of the order of the size of the system, which in neutron stars is of the order of kilometers. On these scales the hydrodynamic and thermodynamic variables are course-grained quantities, i.e. they are averages over a large number of vortices. The solution (3) does not minimize the energy $E - \mathbf{L} \cdot \mathbf{\Omega}$ of a rotating superfluid, where $E$ and $\mathbf{L}$ represent the kinetic energy and the angular momentum respectively. The energy acquires its minimum for a superfluid flow which to a high precision mimics a rigid body rotation i.e., $v_n = \Omega r$, where $r$ is the distance from the rotation axis; (small deviation occur only at the bounding surface of the superfluid). On the global hydrodynamic scales a transition to a continuum vortex distribution can be carried out on the right-hand side of (2) by defining vortex densities $n_\tau = \sum_j \delta^{(2)}(\mathbf{x} - \mathbf{x}_{\tau j})$. Since for rigid-body rotations the curl of $\mathbf{v}_n$ is simply $2\Omega$, the number density of vortices in the neutron superfluid is related to the macroscopic angular velocity of the neutron condensate by the familiar Feynman formula

$$n_n = \frac{2m_n \Omega}{\pi \hbar}. \qquad (7)$$

For typical pulsar periods, $0.05 < P < 0.5$ s, $n_n \simeq 6.3. \times 10^3 P^{-1} \sim 10^4\text{-}10^5$ per cm$^2$. The minute difference between the superfluid and normal angular velocities in a neutron star decelerating under external braking torques is neglected here. For a charged superfluid (2) can be transformed to a contour integral over a path where $\mathbf{v}_p = 0$, as the super-current is screened beyond the magnetic field penetration depth $\lambda$. Again, going over to the continuum vortex limit we find

$$n_p = \frac{B}{\Phi_0} \simeq 5 \times 10^{18} \text{ cm}^{-2}. \tag{8}$$

Note that the number of proton vortices per neutron vortex is $n_p/n_n \sim 10^{13} - 10^{14}$ independent of their arrangement. The energy of a bundle of neutron or proton vortices is minimized by a triangular lattice with a unit cell area

$$n_\tau^{-1} = (\sqrt{3}/2) \, d_\tau^2.$$

The length of a "basis vector" of such a lattice in a neutron condensate (the neutron inter-vortex distance) is

$$d_n = \left(\frac{\pi \hbar}{\sqrt{3}\, m_n\, \Omega}\right)^{1/2}. \tag{9}$$

For the inter-vortex distance in the proton condensate we find

$$d_p = \left(\frac{2\, \Phi_0}{\sqrt{3}\, B}\right)^{1/2}, \tag{10}$$

where $B$ is the mean magnetic field induction. Using the estimates given in (7) and (8) we find that the neutron and proton inter-vortex distances are $d_n \sim 10^{-2} - 10^{-3}$ cm and $d_p \sim 10^{-9}$ cm respectively. For typical values of the microscopic parameters the penetration depth is of the order of 100 fm, therefore the conditions $\xi_n \ll d_n$ and $\xi_p \ll \min(\lambda,\, d_p)$ are satisfied and the use of the hydrodynamics on the local scale is justified. It is also clear that the global hydrodynamics can be applied on scales that are much larger than $d_n$ (a fraction of millimeter).

The remainder of this review concentrates on the physics of the global hydrodynamic scale and on neutral superfluids only, as the dominant fraction of the moment of inertia of a neutron star resides in the neutron fluid and it plays the main role in the hydrodynamic oscillations of the star. Charged superfluids will be absorbed in the normal fluid of the star formally, as they are coupled to the electron liquid via electro-magnetic forces on short time-scales. Note that their role is crucial in controlling the mutual friction on the local hydrodynamic scale; however the physics of this scale will enter the theory on the global scale via phenomenological constants, which we will treat as free parameters of the theory.

## 2  Two-fluid Newtonian superfluid hydrodynamics

The superfluid phases in neutron stars coexist with normal fluids whose interaction with superfluid vortices leads to the effect of the *mutual friction* between a superfluid and normal fluid. A phenomenological description of this effect is based on the two-fluid dissipative hydrodynamics. A particularly simple and transparent formulation which, with some care, can be taken over to describe superfluids in neutron stars, was developed by Bekarevich and Khalatnikov for liquid He$^4$ [12]. It is interesting that the general form of mutual friction forces can be obtained by utilizing only the conservation laws and some reasonable assumptions on the form of the dissipation. Although the superfluids in a neutron star can be assumed to be at zero temperature, i.e., the number of quasi-particle excitations is small, they coexist with a normal liquid of electrons in the core and a nuclear lattice in the crusts. Hence, entropy is irreversibly produced due to various dissipative mechanisms in the normal fluid even though the superfluid matter is effectively at zero temperature. Formally, the electron liquid and the nuclear lattice in the crusts take over the role of quasi-particle excitations in the superfluid hydrodynamics of liquid He$^4$.

The conservation of the combined mass of the two fluids is given by the continuity equation

$$\frac{\partial \rho}{\partial t} + \boldsymbol{\nabla} \cdot \mathbf{J} = 0, \tag{11}$$

where the net mass $\rho = \rho_S + \rho_N$ is the sum of the masses of superfluid and normal fluid, $\mathbf{J} = \rho_N \mathbf{v}_N + \rho_S \mathbf{v}_S$ is the mass current (hereafter the indexes $S$ and $N$ refer to superfluid and normal fluid, respectively.) The total momentum conservation is

$$\frac{\partial J_i}{\partial t} + \frac{\partial P_{ik}}{\partial x_k} = 0, \tag{12}$$

where $P_{ik}$ is the stress energy tensor. The time evolution of the entropy, $S$, of normal fluid can be written as

$$\frac{\partial S}{\partial t} + \boldsymbol{\nabla} \cdot S\mathbf{v}_n = \frac{R}{T}, \tag{13}$$

where $R$ is the dissipative function, $T$ is the temperature; finally, the conservation of the energy, $E$, reads

$$\frac{\partial E}{\partial t} + \boldsymbol{\nabla} \cdot \mathbf{Q} = 0, \tag{14}$$

where $\mathbf{Q}$ is the energy current. Equations above should be supplemented by the Euler equations for the superfluid and the normal fluid

$$\rho_S \left[ \frac{\partial \mathbf{v}_S}{\partial t} + (\mathbf{v}_S \cdot \boldsymbol{\nabla}) \cdot \mathbf{v}_S \right] = -\frac{\rho_S}{\rho} \boldsymbol{\nabla} p - \rho_S \boldsymbol{\nabla} \phi + \mathbf{F}, \tag{15}$$

$$\rho_N \left[ \frac{\partial \mathbf{v}_N}{\partial t} + (\mathbf{v}_N \cdot \boldsymbol{\nabla}) \cdot \mathbf{v}_N \right] = -\frac{\rho_N}{\rho} \boldsymbol{\nabla} p - \rho_N \boldsymbol{\nabla} \phi + \eta_N \boldsymbol{\Delta} \mathbf{v}_N - \mathbf{F}, \tag{16}$$

where $\mathbf{F}$ is the mutual friction force, and $\eta_N$ is the viscosity of the normal fluid, $\phi$ is the Newtonian gravitational potential. To determine the unknowns in the hydrodynamic equations, let us write the total energy of the fluid in the frame in which the normal fluid is at rest as

$$E = \frac{1}{2}\rho v_S^2 + (\mathbf{J} - \rho \mathbf{v}_S) \cdot \mathbf{v}_S + \mathcal{E}, \tag{17}$$

where the internal energy $\mathcal{E}$ is given by the second law of thermodynamics as

$$d\mathcal{E} = TdS + \mu d\rho + (\mathbf{v}_N - \mathbf{v}_S) \cdot d(\mathbf{J} - \rho \mathbf{v}_S) + \Lambda d\boldsymbol{\omega}. \tag{18}$$

The energy due to the vorticity is represented by the term which is proportional to $\boldsymbol{\omega} = \nabla \times \mathbf{v}_S$. Differentiating (17) with respect to time and eliminating the time derivatives using the conservation laws above we recover the conservation of the energy

$$\frac{\partial E}{\partial t} + \nabla \cdot \mathbf{Q} = R + \left(P'_{ik} - \Lambda \omega \delta_{ik} + \Lambda \frac{\omega_i \omega_k}{\omega}\right) \frac{\partial v_{Ni}}{\partial x_k}$$
$$+ (\mathbf{J} - \rho \mathbf{v}_N + \nabla \times \Lambda \boldsymbol{\nu}) \cdot \{\mathbf{F} + [(\mathbf{v}_S - \mathbf{v}_N) \times \boldsymbol{\omega}]\} = 0, \tag{19}$$

where $P'_{ik}$ is the part of the stress tensor associated with the vorticity; the explicit form of the energy current is not indicated since it will not be used in the following. Since the dissipative function $R$ must be positive, the remaining terms on the right-hand side of (19) must be quadratic forms for small deviations from equilibrium. This implies that the most general form of the mutual friction force is

$$\mathbf{F} = -[\boldsymbol{\omega} \times (\nabla \times \Lambda \boldsymbol{\nu})] - \beta\left[\boldsymbol{\nu} \times [\boldsymbol{\omega} \times (\mathbf{v}_N - \mathbf{v}_S - \nabla \times \Lambda \boldsymbol{\nu})]\right]$$
$$- \beta'[\boldsymbol{\omega} \times (\mathbf{v}_N - \mathbf{v}_S - \nabla \times \Lambda \boldsymbol{\nu})] + \beta'' \boldsymbol{\nu} \cdot [\boldsymbol{\omega} \cdot (\mathbf{v}_N - \mathbf{v}_S - \nabla \times \Lambda \boldsymbol{\nu})], \tag{20}$$

where $\beta$, $\beta'$ and $\beta''$ are phenomenological coefficients. On substituting the mutual friction force in the Euler equation for the superfluid, (15), we see that the vorticity propagates with a velocity $\mathbf{v}_L$, that is

$$\frac{\partial \boldsymbol{\omega}}{\partial t} = \nabla \times (\mathbf{v}_L \times \boldsymbol{\omega}), \tag{21}$$

which is defined, assuming $\beta'' \ll \beta, \beta'$, as

$$\mathbf{v}_L = \mathbf{v}_S + \nabla \times \Lambda \boldsymbol{\nu} + \beta'(\mathbf{v}_N - \mathbf{v}_S - \nabla \times \Lambda \boldsymbol{\nu})$$
$$+ \beta[\boldsymbol{\omega} \times (\mathbf{v}_N - \mathbf{v}_S - \nabla \times \Lambda \boldsymbol{\nu})]. \tag{22}$$

The latter equation can be put in a form reflecting the balance of forces acting on a vortex

$$\rho_S\left[(\mathbf{v}_S + \nabla \times \Lambda \boldsymbol{\nu} - \mathbf{v}_L) \times \boldsymbol{\omega}\right] - \eta(\mathbf{v}_L - \mathbf{v}_N) + \eta'\left[(\mathbf{v}_L - \mathbf{v}_N) \times \boldsymbol{\nu}\right] = 0, \tag{23}$$

with the new phenomenological coefficients $\eta$ and $\eta'$ defined as

$$\beta = \frac{\eta \rho_S \omega}{\eta^2 + (\rho_S \omega - \eta')^2}, \quad \beta' = 1 - \frac{\rho_S \omega (\rho_S \omega - \eta')}{\eta^2 + (\rho_S \omega - \eta')^2}. \tag{24}$$

The first term in (23) is a non-dissipative lifting force due to a superflow past the vortex (the Magnus force). The remaining terms reflect the friction between the vortex and the normal fluid. The coefficients $\eta$ and $\eta'$, therefore, measure the friction parallel and orthogonal to the vortex motion in the plane orthogonal to the average direction of the vorticity. A nonzero $\beta''$ implies friction along the *average* direction of the vorticity, which is possible if vortices are oscillating, or are subject to other deformations in the plane orthogonal to the rotation. One may assume, at least under stationary conditions, that $\beta'' \ll \beta, \beta'$.

In the both limits of either *strong coupling* ($\eta \gg \rho_S \omega$) or *weak coupling* ($\eta \ll \rho_S \omega$) between a vortex and the normal fluid, one finds that $\beta \to 0$ as a function of $\eta$, with the maximum $\beta_{\max} = 0.5$ at $\eta = \rho_S \omega$. In the strong coupling limit $\beta'(\eta) \to 1$, while in the opposite weak coupling limit $\beta'(\eta) \to 0$ (generally we assume that the quasi-particle–vortex scattering kinematics implies $\eta' \ll \eta$ and that for the relevant densities $\eta' \ll \rho_S \omega$). Note that $\beta'(\eta)$ approaches its asymptotic strong-coupling values quadratically, while $\beta(\eta)$ does so linearly; the asymptotic behavior for large $\eta$'s, therefore, is dominated by $\beta(\eta)$.

## 3 Virial equations and their perturbations

Virial equations of various order are constructed by taking moments of the hydrodynamic equations. Since the computation of their perturbations is central to the theory of superfluid ellipsoids we review this somewhat technical issue in this section. The reader who is interested only in the physics of superfluid oscillations can proceed to the next section where we discuss the oscillations of Maclaurin spheroids.

The equations of motion (15) and (16), written in a frame rotating with angular velocity $\boldsymbol{\omega}$ relative to some inertial coordinate reference system, can be combined in a single equation

$$\rho_\alpha \left( \frac{\partial}{\partial t} + u_{\alpha,j} \frac{\partial}{\partial x_j} \right) u_{\alpha,i} = -\frac{\partial p_\alpha}{\partial x_i} - \rho_\alpha \frac{\partial \phi}{\partial x_i} + \frac{1}{2} \rho_\alpha \frac{\partial |\boldsymbol{\omega} \times \mathbf{x}|^2}{\partial x_i} + 2 \rho_\alpha \epsilon_{ilm} u_{\alpha,l} \Omega_m + F_{\alpha\beta,i}, \tag{25}$$

where the subscript $\alpha \in \{S, N\}$ identifies the fluid component, and Latin subscripts denote coordinate directions; $\rho_\alpha$, $p_\alpha$, and $\mathbf{u}_\alpha$ are the density, pressure, and velocity of fluid $\alpha$. The two fluids are coupled to one another by mutual gravitational attraction and the mutual friction force $\mathbf{F}_{\alpha\beta}$ [equation (20)]. The gravitational potential $\phi$ is derived from

$$\nabla^2 \phi = \nabla^2 (\phi_S + \phi_N) = 4\pi G [\rho_S(\mathbf{x}) + \rho_N(\mathbf{x})]; \tag{26}$$

the individual fluid potentials $\phi_\alpha$ obey $\nabla^2 \phi_\alpha = 4\pi G \rho_\alpha$. For a normal-superfluid mixture the mutual friction force, written in components, is

$$F_{SN,i} = -\rho_S \omega_S \beta_{ij} (u_{S,j} - u_{N,j}), \tag{27}$$

where the mutual friction tensor is

$$\beta_{ij} = \beta \delta_{ij} + \beta' \epsilon_{ijm} \nu_m + (\beta'' - \beta) \nu_i \nu_j, \tag{28}$$

with $\beta$, $\beta'$ and $\beta''$ being the mutual friction coefficients, and $\boldsymbol{\omega}_S = \boldsymbol{\nu} \omega_S \equiv \nabla \times \mathbf{u}_S$.

The Euler equation (25) can be extended to include external gravitational sources, for example the tidal potential of an external point source of gravity acting on an ellipsoid (Roche ellipsoids). We will not discuss here the stability and oscillations of superfluid counterparts of the classical Roche binaries; their relative oscillation modes are derived in [11].

The general strategy for finding the equilibrium shapes of ellipsoidal figures and modes of their oscillations within the tensor virial method consist of (i) constructing moments of the hydrodynamic equations describing fluid motions in the rotating frame; (ii) computing Eulerian perturbations of the resulting virial equations; (iii) expressing these perturbations in terms of the virials of various order; these are defined as the moments of the Lagrangian displacement $\boldsymbol{\xi}_\alpha$ of fluid $\alpha$:

$$V_{\alpha,i} \equiv \int_{V_\alpha} d^3x\, \rho\, \xi_{\alpha,i}, \quad \text{(first order)} \tag{29}$$

$$V_{\alpha,i;j} \equiv \int_{V_\alpha} d^3x\, \rho\, \xi_{\alpha,i} x_j, \quad \text{(second order)} \tag{30}$$

$$V_{\alpha,i;jk} \equiv \int_{V_\alpha} d^3x\, \rho\, \xi_{\alpha,i} x_j x_k, \quad \text{(third order)} \tag{31}$$

...

The advantage of using the homogeneous ellipsoidal approximation is that the perturbations of the gravitational energy tensor of an ellipsoid can be expressed in terms of the *index symbols* defined as (cf. EFE Chap. 3)

$$A_{ijk...} = a_1 a_2 a_3 \int_0^\infty \frac{du}{\Delta(a_i^2 + u)(a_j^2 + u)(a_k^2 + u)\ldots}, \tag{32}$$

$$B_{ijk...} = a_1 a_2 a_3 \int_0^\infty \frac{u\, du}{\Delta(a_i^2 + u)(a_j^2 + u)(a_k^2 + u)\ldots}, \tag{33}$$

where $\Delta^2 = (a_1^2 + u)(a_2^2 + u)(a_3^2 + u)$ and $a_1$, $a_2$, and $a_3$ are the semi-axis of the ellipsoid. This strategy is described in detail in EFE. Below, we concentrate on its extension to superfluids with an emphasis on the new effects of mutual friction and mutual gravitational attraction of the superfluid and normal fluid.

### 3.1 First order virial equations

On taking the zeroth moment of (25) which amounts to integrating over $V_\alpha$ we obtain the "first order 'virial' equation"[2]

---

[2] The word "virial" is in quotes because the equations are intrinsically dissipative.

$$\frac{d}{dt}\left(\int_{V_\alpha} d^3x \rho_\alpha u_{\alpha,i}\right) = 2\epsilon_{ilm}\Omega_m \int_{V_\alpha} d^3x \rho_\alpha u_{\alpha,l} + (\Omega^2 \delta_{ij} - \Omega_i \Omega_j)\int_{V_\alpha} d^3x \rho_\alpha x_j$$
$$-(1-\delta_{\alpha\beta})\int_{V_\alpha} d^3x \rho_\alpha \frac{\partial \phi_\beta}{\partial x_i} + \int_{V_\alpha} d^3x F_{\alpha\beta,i}. \quad (34)$$

Note that we impose the boundary condition $p_\alpha = 0$ for each fluid on the bounding surface of $V_\alpha$. The fluids are not restricted to occupy the same volume. Apart from simple doubling of the number of the inertial forces, which do not couple the two fluids, there are two forces that do couple them: gravity and friction. The net mutual gravitational force between the fluids vanishes only if they (i) occupy the same volume and (ii) have densities that are proportional to one another (i.e. $\rho_S \propto \rho_N$). The mutual friction force is nonzero as long as the fluids move relative to one another. Although the mutual friction force is nonzero only in the *overlap* volume of the two fluids - a restriction which is necessary to derive conservation of total momentum for the combined fluids - it would be effective throughout the entire volume of fluids because the force is mediated by a macroscopically extended vortex lattice.

For isolated single-fluid ellipsoids the first harmonic oscillations are trivial, since they correspond to motions of the center-of-mass of an ellipsoid and can be eliminated by a transformation to the reference frame whose origin is located at the center-of-mass of the ellipsoid. For two-fluid ellipsoids the fluid motions include the counter-phase (relative) oscillations of two fluids, which can not be eliminated by any transformation. These are the only new type of oscillations for the superfluid Maclaurin and Jacobi ellipsoids (the solitary ellipsoids with vanishing internal motions). In the case of ellipsoids in an external gravitational field (e.g. the Roche ellipsoids), the CM motions are not trivial any more, since the external (inhomogeneous by assumption) source of gravitational field breaks the translational symmetry of the problem. Hence, apart from the relative oscillations of two fluids, the CM oscillations become non-trivial.

Consider first the variation of the first order virial equation under the influence of perturbations. The variations of the inertial terms proceeds in full analogy to EFE. Here we concentrate on variations of the new terms corresponding to the mutual gravitational attraction and mutual friction. The variation of the first force is

$$-\delta \int_{V_\alpha} d^3x \rho_\alpha \frac{\partial \phi_\beta}{\partial x_i} = -\int_{V_\alpha} d^3x \rho_\alpha(\mathbf{x})\xi_{\alpha,l}(\mathbf{x})\frac{\partial}{\partial x_l}\int_{V_\beta} \frac{d^3x' \rho_\beta(\mathbf{x}')(x_i - x_i')}{|\mathbf{x}-\mathbf{x}'|^3}$$
$$+ \int_{V_\beta} d^3x \rho_\beta(\mathbf{x})\xi_{\beta,l}(\mathbf{x})\frac{\partial}{\partial x_l}\int_{V_\alpha} \frac{d^3x' \rho_\alpha(\mathbf{x}')(x_i - x_i')}{|\mathbf{x}-\mathbf{x}'|^3} \quad (35)$$

which is manifestly antisymmetric on $\alpha \leftrightarrow \beta$. Assuming $V_\alpha = V_\beta = V$ and $\rho_\alpha = f_\alpha \rho(\mathbf{x})$ in the background equilibrium, we can simplify this to

$$-\delta \int_{V_\alpha} d^3x \rho_\alpha \frac{\partial \phi_\beta}{\partial x_i} = f_\alpha f_\beta \int_V d^3x \rho(\mathbf{x})[\xi_{\beta,l}(\mathbf{x}) - \xi_{\alpha,l}(\mathbf{x})]$$
$$\times \frac{\partial}{\partial x_l}\int_V \frac{d^2x' \rho(\mathbf{x}')(x_i - x_i')}{|\mathbf{x}-\mathbf{x}'|^3}. \quad (36)$$

Although we simplified the final answer by assuming that the fluids occupy identical volumes and have proportional densities in the background state, we could not have derived the correct perturbation of the first order virial theorem if we had not allowed the volumes to differ.

For uniform ellipsoids, we can simplify the mutual gravitational term further. First, we note that the second integral is simply the derivative of the gravitational potential which at any interior point of a homogeneous ellipsoid is

$$\phi(\mathbf{x}) = -\pi G \rho \left( I - \sum_{k=1}^{3} A_k x_k^2 \right), \tag{37}$$

where $I$ is a constant. Thus, the mutual gravitational contribution to the equation of motion for the perturbed center-of-mass is

$$2\pi G \rho^2 A_i f_\alpha f_\beta \int_V d^3 x (\xi_{\beta,i} - \xi_{\alpha,i}). \tag{38}$$

If in the background state, the two fluids move together or are stationary, the variation of the mutual friction force becomes

$$\delta \int_V d^3 x F_{\alpha\beta,i} = -\mathcal{S}_{\alpha\beta} \frac{d}{dt} \left[ f_S \int_V d^3 x \rho(\mathbf{x}) \omega_S \beta_{ij} (\xi_{\alpha,j} - \xi_{\beta,j}) \right], \tag{39}$$

where $\mathcal{S}_{\alpha\beta} = 1 - \delta_{\alpha\beta}$. Collecting terms we find the first order virial equation

$$f_\alpha \frac{d^2 V_{\alpha,i}}{dt^2} = 2\epsilon_{ilm} f_\alpha \Omega_m \frac{d}{dt} V_{\alpha,l} + (\Omega^2 \delta_{ij} - \Omega_i \Omega_j) f_\alpha V_{\alpha,j}$$
$$- 2\pi G \rho A_i f_\alpha f_\beta (V_{\alpha,i} - V_{\beta,i}) - \mathcal{S}_{\alpha\beta} f_S \omega_S \beta_{ij} (V_{\alpha,j} - V_{\beta,j}). \tag{40}$$

The CM and relative motions can be decoupled by defining

$$V_i \equiv f_S V_{S,i} + f_N V_{N,i}, \qquad U_i \equiv V_{S,i} - V_{N,i}. \tag{41}$$

The CM motions of two fluids are trivial (as they can be eliminated by a linear transformation of the reference frame) and, therefore, $V_i = 0$. The virial equation describing the relative motions of the two fluids is

$$\frac{d^2}{dt^2} U_i = 2\epsilon_{ilm} \Omega_m \frac{d}{dt} U_l$$
$$+ (\Omega^2 \delta_{ij} - \Omega_i \Omega_j) U_j - 2 A_i U_i - 2\Omega \left(1 + \frac{f_S}{f_N}\right) \beta_{ij} \frac{d}{dt} U_j. \tag{42}$$

From the latter equation it is straightforward to compute the first harmonic relative oscillation modes of irrotational ellipsoids, which is done in the next section for Maclaurin spheroids.

## 3.2 Second order virial equations

Taking the first moment of (25) results in the second order 'virial' equation

$$\frac{d}{dt}\left(\int_{V_\alpha} d^3x \rho_\alpha x_j u_{\alpha,i}\right) = 2\epsilon_{ilm}\Omega_m\left(\int_{V_\alpha} d^3x \rho_\alpha x_j u_{\alpha,l}\right) + \Omega^2 I_{\alpha,ij} - \Omega_i\Omega_k I_{\alpha,kj}$$
$$+ 2\mathcal{T}_{\alpha,ij} + \delta_{ij}\Pi_\alpha + \mathcal{M}_{\alpha,ij} + (1-\delta_{\alpha\beta})\mathcal{M}_{\alpha\beta,ij} + \mathcal{F}_{\alpha\beta,ij}, \qquad (43)$$

where

$$I_{\alpha,ij} \equiv \int_{V_\alpha} d^3x\, \rho_\alpha x_i x_j \qquad (44)$$

$$\Pi_\alpha \equiv \int_{V_\alpha} d^3x\, p_\alpha \qquad (45)$$

$$\mathcal{T}_{\alpha,ij} \equiv \frac{1}{2}\int_{V_\alpha} d^3x \rho_\alpha u_{\alpha,i}u_{\alpha,j} \qquad (46)$$

$$\mathcal{M}_{\alpha,ij} \equiv -\frac{G}{2}\int_{V_\alpha} \frac{d^3x\, d^3x'\, \rho_\alpha(\mathbf{x})\rho_\alpha(\mathbf{x}')(x_i - x_i')(x_j - x_j')}{|\mathbf{x} - \mathbf{x}'|^3} \qquad (47)$$

$$\mathcal{M}_{\alpha\beta,ij} \equiv -G\int_{V_\alpha} d^3x \int_{V_\beta} \frac{d^3x'\, \rho_\alpha(\mathbf{x})\rho_\beta(\mathbf{x}')x_j(x_i - x_i')}{|\mathbf{x} - \mathbf{x}'|^3} \qquad (48)$$

$$\mathcal{F}_{\alpha\beta,ij} \equiv \int_{V_\alpha} d^3x\, x_j F_{\alpha\beta,i}. \qquad (49)$$

When there is only one fluid present, this equation reduces to the one found in Chap. 2 of EFE. Again we consider only variations of the new terms in the second order virial equation due to the mutual gravitational attraction ($\mathcal{M}_{\alpha\beta,ij}$) and mutual friction ($\mathcal{F}_{\alpha\beta,ij}$). The first variation is

$$\delta\mathcal{M}_{\alpha\beta,ij} = -Gf_\alpha f_\beta\Bigg\{\int_V d^3x \rho(\mathbf{x})\xi_{\alpha,l}(\mathbf{x})\frac{\partial}{\partial x_l}\int_V \frac{d^3x'\rho(\mathbf{x}')(x_i - x_i')(x_j - x_j')}{|\mathbf{x} - \mathbf{x}'|^3}$$
$$+ \int_V d^3x\rho(\mathbf{x})[\xi_{\alpha,l}(\mathbf{x}) - \xi_{\beta,l}(\mathbf{x})]\frac{\partial}{\partial x_l}\int \frac{d^3x'\rho(\mathbf{x}')(x_i - x_i')x_j'}{|\mathbf{x} - \mathbf{x}'|^3}\Bigg\}, \qquad (50)$$

where we have specialized to backgrounds with proportional densities and identical bounding volumes. The first term in the brackets can be combined with $\delta\mathcal{M}_{\alpha,ij}$. The resulting equation can be written more compactly in terms of the functions

$$\mathcal{B}_{ij} \equiv G\int_V \frac{d^3x'\rho(\mathbf{x})(x_i - x_i')(x_j - x_j')}{|\mathbf{x} - \mathbf{x}'|^3}, \qquad (51)$$

$$\frac{\partial \mathcal{D}_j}{\partial x_i} \equiv -G\int_V \frac{d^3x'\rho(\mathbf{x}')x_j'(x_i - x_i')}{|\mathbf{x} - \mathbf{x}'|^3}, \qquad (52)$$

which are related by

$$\frac{\partial \mathcal{D}_j}{\partial x_i} = \mathcal{B}_{ij} - x_j\frac{\partial \phi}{\partial x_i}; \qquad (53)$$

we find

$$\delta \mathcal{M}_{\alpha,ij} + (1-\delta_{\alpha\beta})\delta \mathcal{M}_{\alpha\beta,ij} = -f_\alpha \int_V d^3x \rho \xi_{\alpha,l} \frac{\partial \mathcal{B}_{ij}}{\partial x_l}$$
$$+ f_\alpha f_\beta \int_V d^3x \rho (\xi_{\alpha,l} - \xi_{\beta,l}) \frac{\partial^2 \mathcal{D}_j}{\partial x_l \partial x_i}. \quad (54)$$

For the uniform ellipsoids [cf. EFE, Chap. 3, (125) and (126)],

$$\frac{\mathcal{D}_j}{\pi G \rho} = a_j^2 x_j \left( A_j - \sum_{k=1}^{3} A_{jk} x_k^2 \right), \quad (55)$$

$$\frac{\mathcal{B}_{ij}}{\pi G \rho} = 2B_{ij} x_i x_j + a_i^2 \delta_{ij} \left( A_i - \sum_{i=1}^{3} A_{ik} x_k^2 \right), \quad (56)$$

$$\frac{1}{\pi G \rho} \frac{\partial^2 \mathcal{D}_j}{\partial x_l \partial x_i} = 2B_{ij}(\delta_{il} x_j + \delta_{jl} x_i) - 2a_i^2 \delta_{ij} A_{il} x_l. \quad (57)$$

Using these results, and defining symmetric in their indexes second order virials as

$$V_{\alpha,ij} = V_{\alpha,i;j} + V_{\alpha,j;i}, \quad (58)$$

we finally obtain

$$\frac{\delta \mathcal{M}_{\alpha,ij} + (1-\delta_{\alpha\beta})\delta \mathcal{M}_{\alpha\beta,ij}}{\pi G \rho} = -f_\alpha \left( 2B_{ij} V_{\alpha,ij} - a_i^2 \delta_{ij} \sum_{l=1}^{3} A_{il} V_{\alpha,ll} \right)$$
$$- a_j^2 f_\alpha f_\beta \left[ 2A_{ij}(V_{\alpha,ij} - V_{\beta,ij}) + \delta_{ij} \sum_{l=1}^{3} A_{il}(V_{\alpha,ll} - V_{\beta,ll}) \right]. \quad (59)$$

For the perturbations of mutual friction force we find

$$\delta \int_{V_\alpha} d^3x x_j F_{\alpha\beta,i} = -\mathcal{S}_{\alpha\beta} f_S \int_V d^3x \rho \omega_S x_j \beta_{ik} \left( \frac{d\xi_{S,k}}{dt} - \frac{d\xi_{N,k}}{dt} \right). \quad (60)$$

For perturbations of uniform ellipsoids, $\omega_S$ and $\rho$ are independent of position in the unperturbed background, and we may also assume that $\beta_{ij}$ is constant; for backgrounds in which there are no fluid motions

$$\delta \int_{V_\alpha} d^3x x_j F_{\alpha\beta,i} = -\mathcal{S}_{\alpha\beta} f_S \rho \omega_S \beta_{ik} \left( \frac{dV_{\alpha,k;j}}{dt} - \frac{dV_{\beta,k;j}}{dt} \right). \quad (61)$$

Thus the second order virial equation for a fluid $\alpha$, in a frame rotating with an angular velocity $\Omega$, is

$$f_\alpha \frac{d^2 V_{\alpha,i;j}}{dt^2} = 2\epsilon_{ilm} \Omega_m f_\alpha \frac{dV_{\alpha,l;j}}{dt} + \Omega^2 f_\alpha V_{\alpha,ij} - \Omega_i \Omega_k f_\alpha V_{\alpha,kj} + \delta_{ij} \delta \Pi_\alpha$$

$$- f_\alpha \pi G\rho \left( 2B_{ij}V_{\alpha,ij} - a_i^2 \delta_{ij} \sum_{l=1}^{3} A_{il}V_{\alpha,ll} \right)$$

$$- a_j^2 f_\alpha f_\beta \pi G\rho \left[ 2A_{ij}(V_{\alpha,ij} - V_{\beta,ij}) + \delta_{ij} \sum_{l=1}^{3} A_{il}(V_{\alpha,ll} - V_{\beta,ll}) \right]$$

$$- \mathcal{S}_{\alpha\beta} f_\alpha \omega_S \beta_{ik} \frac{d}{dt}\left( V_{\alpha,k;j} - V_{\beta,k;j} \right). \tag{62}$$

We can replace these equations with a different set by defining

$$V_{i;j} \equiv f_S V_{S,i;j} + f_N V_{N,i;j} \qquad U_{i;j} \equiv V_{S,i;j} - V_{N,i;j}. \tag{63}$$

In terms of these new quantities we find

$$\frac{d^2 V_{i;j}}{dt^2} = 2\epsilon_{ilm}\Omega_m \frac{dV_{l;j}}{dt} + \Omega^2 V_{ij} - \Omega_i \Omega_k V_{kj} + \delta_{ij}\delta\Pi$$

$$- \pi G\rho \left( 2B_{ij}V_{ij} - a_i^2 \delta_{ij} \sum_{l=1}^{3} A_{il}V_{ll} \right), \tag{64}$$

$$\frac{d^2 U_{i;j}}{dt^2} = 2\epsilon_{ilm}\Omega_m \frac{dU_{l;j}}{dt} + \Omega^2 U_{ij} - \Omega_i \Omega U_{kj} + \delta_{ij}\left(\frac{\delta\Pi_S}{f_S} - \frac{\delta\Pi_N}{f_N}\right)$$

$$- 2\pi G\rho A_i U_{ij} - 2\Omega \beta_{ik}\frac{d}{dt}U_{k;j}. \tag{65}$$

The first equation is identical to the second order virial equation for a normal inviscid fluid. The second equation is specific to superfluids and contains all the new modes of relative oscillations between the normal fluid and superfluid. It is clear that the separation of the oscillation modes in the CM and relative modes is the result of the symmetry of the two-fluid hydrodynamic equations with respect to the interchange $\alpha \leftrightarrow \beta$. If this symmetry is broken the two classes of modes mix. We shall consider below the effect of the viscosity of normal fluid which by definition acts only in the normal component and hence breaks the $\alpha \leftrightarrow \beta$ symmetry.

The second order virial equation for viscous fluids, quite generally, acquires the term

$$\mathcal{P}_{\alpha,ij} = \int_{V_\alpha} P_{\alpha,ij} d\mathbf{x}, \qquad P_{\alpha,ik} \equiv \delta_{\alpha,N}\rho_N \nu \left(\frac{\partial u_{\alpha i}}{\partial x_k} + \frac{\partial u_{\alpha,k}}{\partial x_i} - \frac{2}{3}\frac{\partial u_{\alpha,l}}{\partial x_l}\delta_{ik}\right), \tag{66}$$

which is called the shear-energy tensor; $\nu$ is the kinematic viscosity[3]. For background states which are stationary and without internal motions the variation of the stress-energy tensor is

$$\delta \mathcal{P}_{\alpha,ij} = \delta_{\alpha,N}\int_{V_\alpha} \rho_N \nu \frac{\partial}{\partial t}\left(\frac{\partial \xi_{\alpha,i}}{\partial x_k} + \frac{\partial \xi_{\alpha,k}}{\partial x_i} - \frac{2}{3}\frac{\partial \xi_{\alpha l}}{\partial x_l}\delta_{ik}\right). \tag{67}$$

---

[3] We use the same symbol $\nu$ for the kinematic viscosity and for the unit vector along the vortex circulation; no confusion should arises as the latter quantity does not appear in the virial equations.

It is impossible in general to express the variations of the stress-energy tensor in terms of the virials $V_{\alpha,i;j}$. However, in the low Reynolds-number approximation, this tensor can be approximated in a perturbative manner using as the leading order approximation the proper solutions for the displacements corresponding to the inviscid limit. Since the latter are linear functions of the virials, $\xi_{Ni} = \sum_{k=1}^{3} 5 V_{N,i;k} x_k / M_N a_k^2$, with $M_N$ being the mass in the normal fluid, one finds

$$\delta \mathcal{P}_{\alpha,ij} = 5\nu \delta_{\alpha,N} \frac{d}{dt}\left(\frac{V_{\alpha,i;j}}{a_j^2} + \frac{V_{\alpha,j;i}}{a_i^2}\right) \tag{68}$$

in the incompressible limit. Thus, the second order virial equation which includes the viscosity of the normal matter becomes

$$f_\alpha \frac{d^2 V_{\alpha,i;j}}{dt^2} = 2\epsilon_{ilm}\Omega_m f_\alpha \frac{dV_{\alpha,l;j}}{dt} + \Omega^2 f_\alpha V_{\alpha,ij} - \Omega_i \Omega_k f_\alpha V_{\alpha,kj} + \delta_{ij}\delta\Pi_\alpha$$

$$- f_\alpha \pi G \rho \left(2 B_{ij} V_{\alpha,ij} - a_i^2 \delta_{ij} \sum_{l=1}^{3} A_{il} V_{\alpha,ll}\right)$$

$$- a_j^2 f_\alpha f_\beta \pi G \rho \left[2 A_{ij}(V_{\alpha,ij} - V_{\beta,ij}) + \delta_{ij}\sum_{l=1}^{3} A_{il}(V_{\alpha,ll} - V_{\beta,ll})\right]$$

$$- \mathcal{S}_{\alpha\beta} f_\alpha \omega_S \beta_{ik} \frac{d}{dt}\left(V_{\alpha,k;j} - V_{\beta,k;j}\right) - \delta_{\alpha,N} 5\nu f_\alpha \frac{d}{dt}\left(\frac{V_{\alpha,i;j}}{a_j^2} + \frac{V_{\alpha,j;i}}{a_i^2}\right). \tag{69}$$

Apart from the last term, the remaining terms in (69) manifestly preserve the symmetry with respect to the interchange $\alpha \leftrightarrow \beta$; note that they might have different parities under this transformation. The last term breaks this symmetry as the viscosity acts only in the normal fluid.

### 3.3 Third order virial equations

To obtain the third order virial equation we take the second moment of (25) and integrate over $V_\alpha$:

$$\frac{d}{dt}\left(\int_{V_\alpha} d^3x \rho_\alpha u_{\alpha,i} x_j x_k\right) = 2\epsilon_{ilm}\Omega_m\left(\int_{V_\alpha} d^3x \rho_\alpha u_{\alpha,l} x_j x_k\right) + \Omega^2 I_{\alpha,ijk}$$

$$- \Omega_i \Omega_l I_{\alpha,ljk} + 2(\mathcal{T}_{\alpha,ij;k} + \mathcal{T}_{\alpha,ik;j}) + \delta_{ij}\Pi_{\alpha,k} + \delta_{ik}\Pi_{\alpha,j}$$

$$+ \mathcal{M}_{\alpha\beta,ijk} + \mathcal{F}_{\alpha\beta,ijk}, \tag{70}$$

where

$$I_{\alpha,ijk} \equiv \int_{V_\alpha} d^3x\, \rho_\alpha x_i x_j x_k \tag{71}$$

$$\Pi_{\alpha,i} \equiv \int_{V_\alpha} d^3x\, p_{\alpha,i} \tag{72}$$

$$\mathcal{T}_{\alpha,ij;k} \equiv \frac{1}{2}\int_{V_\alpha} d^3x \rho_\alpha u_{\alpha,i}u_{\alpha,j}x_k \tag{73}$$

$$\mathcal{M}_{\alpha\beta,ijk} \equiv -\int_{V_\alpha} d^3x x_j x_k \rho_\alpha \frac{\partial\phi}{\partial x_i} \tag{74}$$

$$\mathcal{F}_{\alpha\beta,ijk} \equiv \int_{V_\alpha} d^3x\, F_{\alpha\beta,i} x_j x_k. \tag{75}$$

Below, we compute only the perturbations of the tensors in the last line of (70), which correspond to the gravitational potential energy and the mutual friction; the perturbations of the remainder terms is computed in analogy to EFE.

For the Eulerian perturbation of the gravitational potential tensor we have

$$-\delta\int_{V_\alpha} d^3x x_j x_k \rho_\alpha \frac{\partial\phi}{\partial x_i} = -\delta G \int_{V_\alpha} d^3x x_j x_k \rho_\alpha(\mathbf{x}) \left[ \int_{V_\alpha} d^3x' \rho_\alpha(\mathbf{x'}) \frac{(x_i-x'_i)}{|\mathbf{x}-\mathbf{x'}|^3} \right.$$
$$\left. + \int_{V_\beta} d^3x' \rho_\beta(\mathbf{x'}) \frac{(x_i-x'_i)}{|\mathbf{x}-\mathbf{x'}|^3} \right]. \tag{76}$$

Assuming $V_\alpha = V$ and $\rho_\alpha = f_\alpha \rho(\mathbf{x})$ in the background equilibrium, and defining [cf. EFE, Chap. 2, (14) and (22)]

$$\mathcal{B}_{ij} \equiv G\int_V d^3x' \rho(\mathbf{x'}) \frac{(x_i-x'_i)(x_j-x'_j)}{|\mathbf{x}-\mathbf{x'}|^3}, \tag{77}$$

$$\mathcal{D}_{ik;j} \equiv G\int_V d^3x' \rho(\mathbf{x'}) x'_j \frac{(x_i-x'_i)(x_k-x'_k)}{|\mathbf{x}-\mathbf{x'}|^3}, \tag{78}$$

one finds for the 'self-interaction' term that

$$-\delta \int_{V_\alpha} d^3x x_j x_k \rho_\alpha \frac{\partial \phi_\alpha}{\partial x_i} = -\frac{1}{2} f_\alpha^2 \int_V d^3x \rho(\mathbf{x}) \xi_{\alpha,l} \frac{\partial}{\partial x_l}(\mathcal{B}_{ij} x_k + \mathcal{D}_{ij;k})$$
$$-\frac{1}{2}f_\alpha^2 \int_V d^3x \rho(\mathbf{x}) \xi_{\alpha,l} \frac{\partial}{\partial x_l}(\mathcal{B}_{ik} x_j + \mathcal{D}_{ik;j}) \equiv f_\alpha^2(\delta\mathcal{M}_{\alpha,ij;k} + \delta\mathcal{M}_{\alpha,ik;j}). \tag{79}$$

To arrive at the symmetric in the indexes $k,j$ we used the identity

$$\mathcal{B}_{ij} x_k + \mathcal{D}_{ik;j} = \mathcal{B}_{ik} x_j + \mathcal{D}_{ij;k}. \tag{80}$$

The perturbation of mutual interaction term in (76), assuming again $V_\alpha = V_\beta = V$ and $\rho_\alpha = f_\alpha \rho(\mathbf{x})$ in the background equilibrium is

$$-\delta\int_{V_\alpha} d^3x x_j x_k \rho_\alpha \frac{\partial \phi_\beta}{\partial x_i} = -f_\alpha f_\beta \int_V d^3x \rho(\mathbf{x}) \xi_{\alpha,l}(\mathbf{x}) \frac{\partial}{\partial x_l}(\mathcal{B}_{ij} x_k + \mathcal{D}_{ik;j})$$
$$- f_\alpha f_\beta \int_V d^3x \rho(\mathbf{x}) \left[\xi_{\alpha,l}(\mathbf{x}) - \xi_{\beta,l}(\mathbf{x})\right] \frac{\partial^2 \mathcal{D}_{jk}}{\partial x_l \partial x_i} \tag{81}$$

where

$$\mathcal{D}_{jk} = G\int_V d^3x' \frac{\rho(\mathbf{x'}) x'_k x'_j}{|\mathbf{x}-\mathbf{x'}|}. \tag{82}$$

Combining (79) and (81) we find

$$-\delta \int_{V_\alpha} d^3x x_j x_k \rho_\alpha \frac{\partial \phi}{\partial x_i} = -f_\alpha \int_V d^3x \rho(\mathbf{x}) \xi_{\alpha,l} \frac{\partial}{\partial x_l}(\mathcal{B}_{ij}x_k + \mathcal{D}_{ik;j})$$
$$- f_\alpha f_\beta \int_V d^3x \rho(\mathbf{x}) \left[\boldsymbol{\xi}_{\alpha,l}(\mathbf{x}) - \boldsymbol{\xi}_{\beta,l}(\mathbf{x})\right] \frac{\partial^2 \mathcal{D}_{jk}}{\partial x_l \partial x_i}. \quad (83)$$

For the perturbation of the mutual friction tensor, assuming stationary background equilibrium, we find

$$\delta \mathcal{F}_{\alpha\beta,ijk} = \delta \int_{V_\alpha} d^3x \, F_{\alpha\beta,i} x_j x_k$$
$$= -\mathcal{S}_{\alpha\beta} f_S \int_V d^3x x_j x_k \rho(\mathbf{x}) \omega_S \beta_{il} \left(\frac{d\xi_{S,l}}{dt} - \frac{d\xi_{N,l}}{dt}\right). \quad (84)$$

Putting together all the terms we arrive at the third order virial equation

$$f_\alpha \frac{d^2 V_{\alpha,i;jk}}{dt^2} = 2\epsilon_{ilm}\Omega_m f_\alpha \frac{dV_{\alpha,l;jk}}{dt} + \delta_{ij}\delta\Pi_{\alpha,k} + \delta_{ik}\delta\Pi_{\alpha,j}$$
$$+ (\Omega^2 \delta_{il} - \Omega_i\Omega_l)f_\alpha V_{\alpha,ljk} - f_\alpha \int_V d^3x \rho(\mathbf{x})\xi_{\alpha,l}\frac{\partial}{\partial x_l}(\mathcal{B}_{ij}x_k + \mathcal{D}_{ik;j})$$
$$- f_\alpha f_\beta \int_V d^3x \rho(\mathbf{x}) \left[\boldsymbol{\xi}_{\alpha,l}(\mathbf{x}) - \boldsymbol{\xi}_{\beta,l}(\mathbf{x})\right] \frac{\partial^2}{\partial x_l \partial x_i} \mathcal{D}_{jk}$$
$$- \mathcal{S}_{\alpha\beta} f_S \omega_S \beta_{il} \left[\frac{dV_{S,l;jk}}{dt} - \frac{dV_{N,l;jk}}{dt}\right], \quad (85)$$

where the symmetric in its indexes third order virial is defined as

$$V_{\alpha,ijk} = V_{\alpha,i;jk} + V_{\alpha,j;ki} + V_{\alpha,k;ij}. \quad (86)$$

To separate the CM and relative motions of the two fluids introduce the virials

$$V_{i;jk} \equiv f_S V_{S,i;jk} + f_N V_{N,i;jk} \qquad U_{i;jk} \equiv V_{S,i;jk} - V_{N,i;jk}. \quad (87)$$

The new set of equations in terms of these virials is

$$\frac{d^2 V_{i;jk}}{dt^2} = 2\epsilon_{ilm}\Omega_m \frac{dV_{l;jk}}{dt} + \delta_{ij}\delta\Pi_k + \delta_{ik}\delta\Pi_j + (\Omega^2 \delta_{il} - \Omega_i\Omega_l)V_{ljk}$$
$$- \int_V d^3x \rho(\mathbf{x}) \left[f_S \boldsymbol{\xi}_{S,l} + f_N \boldsymbol{\xi}_{N,l}\right] \frac{\partial}{\partial x_l}(\mathcal{B}_{ij}x_k + \mathcal{D}_{ik;j}), \quad (88)$$

and

$$\frac{d^2 U_{i;jk}}{dt^2} = \left[2\epsilon_{ilm}\Omega_m - \left(1 + \frac{f_S}{f_N}\right)\omega_S \beta_{il}\right]\frac{dU_{l;jk}}{dt} + \delta_{ij}\left(\frac{\delta\Pi_{S,k}}{f_S} - \frac{\delta\Pi_{N,k}}{f_N}\right)$$
$$+ \delta_{ik}\left(\frac{\delta\Pi_{S,j}}{f_S} - \frac{\delta\Pi_{N,j}}{f_N}\right) + (\Omega^2 \delta_{il} - \Omega_i\Omega_l)U_{ljk} - 2\pi G \rho A_i U_{ijk}$$
$$- 2\int_V d^3x \rho(\mathbf{x})[\boldsymbol{\xi}_{S,l} - \boldsymbol{\xi}_{N,l}]\frac{\partial}{\partial x_l}(\mathcal{B}_{ij}x_k + \mathcal{D}_{ik;j}). \quad (89)$$

To obtain the last term in (89) we used the relations [cf. EFE, Chap. 2, equations (29) and (28)]

$$\frac{\partial \mathcal{D}_{jk}}{\partial x_i} = \mathcal{D}_{ji;k} + x_j \frac{\partial \mathcal{D}_k}{\partial x_i}, \qquad \frac{\partial \mathcal{D}_k}{\partial x_i} = \mathcal{B}_{ik} + x_k \frac{\partial \phi}{\partial x_i}, \qquad (90)$$

and the explicit expression for the gravitational potential of an ellipsoid, (37). The terms in the last lines of (88) and (89) can be worked out to a form involving linear combinations of virials and index symbols, however the present form already makes clear that they will involve the virials describing the CM and relative motions, respectively.

## 4 Small amplitude oscillations of superfluid Maclaurin spheroid

In this section, we specialize our discussion to Maclaurin spheroids, the equilibrium figures of a self-gravitating fluid with two equal semi-major axis, say $a_1$ and $a_2$, rotating uniformly about the third semi-major axis $a_3$ (i.e. the $x_3$ axis). For these figures in many cases analytical results are available. The superfluid oscillations of more complicated non-axisymmetric figures like the Jacobi and Roche ellipsoids require numerical analysis which is beyond the scope of this review (see [11]). The sequence of quasi-equilibrium figures of Maclaurin spheroids can be parameterized by the eccentricity $\epsilon^2 = 1 - a_3^2/a_1^2$, with (squared) angular velocity $\Omega^2 = 2\epsilon^2 B_{13}$, in units of $(\pi \rho G)^{1/2}$.

Surface deformations related to various modes can be classified by corresponding terms of the expansion in surface harmonics labeled by indexes $l, m$. We shall concentrate below on the first and second harmonic surface deformations correspond to $l = 1, 2$ and $-1 \leq m \leq 1, -2 \leq m \leq 2$ respectively.

### 4.1 First order

If the time-dependence of the Lagrangian displacements is of the form

$$\boldsymbol{\xi}_\alpha(x_i, t) = \boldsymbol{\xi}_\alpha(x_i) e^{\lambda t}, \qquad (91)$$

then the characteristic equation for the first order relative oscillation modes becomes

$$\lambda^2 U_i = 2\epsilon_{ilm} \Omega_m \lambda U_l + (\Omega^2 \delta_{ij} - \Omega_i \Omega_j) U_j - 2 A_i U_i - 2\Omega \tilde{\beta}_{ij} \lambda U_j, \qquad (92)$$

where all frequencies are measured in units $(\pi G \rho)^{1/2}$, $\tilde{\beta}_{ij} \equiv (1 + f_S/f_N)\beta_{ij}$, and, since we assumed no internal motions in the background equilibrium, $\omega_S = 2\Omega$. The CM oscillations are trivial as they can be always eliminated by a transformation to the center-of-mass reference frame.

Assume that the ellipsoid is rotating about the $x_3$ axis. Then, writing (92) in components, we find

$$\left(\lambda^2 + 2\Omega\tilde{\beta}\lambda + 2A_1 - \Omega^2\right)U_1 - 2\Omega(1-\tilde{\beta}')\lambda U_2 = 0, \tag{93}$$

$$\left(\lambda^2 + 2\Omega\tilde{\beta}\lambda + 2A_2 - \Omega^2\right)U_2 + 2\Omega(1-\tilde{\beta}')\lambda U_1 = 0, \tag{94}$$

$$\left(\lambda^2 + 2\Omega\tilde{\beta}''\lambda + 2A_3\right)U_3 = 0. \tag{95}$$

The equations which are even and odd with respect to the index 3 decouple. For the perturbations along $x_3$ the relative displacement vanishes, $U_3 = 0$. (95) (which is odd in index 3) gives, on writing $\lambda = i\sigma$,

$$\sigma_{1,2}^{\text{odd}} = \pm\sqrt{2A_3 - \tilde{\beta}''^2\Omega^2} + i\tilde{\beta}''\Omega. \tag{96}$$

The first order odd parity oscillations are stable; they are purely imaginary if $\tilde{\beta}''^2 \geq 2A_3/\Omega^2$, otherwise they develop a real part. For Maclaurin spheroids $2A_3/\Omega^2 \geq 5.040$; the lower bound corresponds to eccentricity of the ellipsoid $\epsilon = 0.865$. For Jacobi ellipsoids this minimal value is slightly lower, $2A_3/\Omega^2 = 4.148$, and occurs at the point of the bifurcation of the Jacobi sequence from the Maclaurin sequence where the axis-ratio is defined by $\text{Cos}^{-1}(a_3/a_1) = 54.48$. Since the $\tilde{\beta}''$-coefficient is the measure of friction along the average direction of the vorticity, it is reasonable to assume that $\tilde{\beta}'' \ll \tilde{\beta}, \tilde{\beta}'$; and since $\tilde{\beta} \leq 1/2$ and $\tilde{\beta}' \leq 1$.

The characteristic equations for the modes even in index 3 is

$$\lambda^4 + 4\Omega\tilde{\beta}\lambda^3 + 2\left[(A_1+A_2) + \Omega^2(1+2\tilde{\beta}^2 - 4\tilde{\beta}' + 2\tilde{\beta}'^2)\right]\lambda^2$$
$$+4\tilde{\beta}\Omega\left(A_1+A_2-\Omega^2\right)\lambda + (2A_1 - \Omega_2)(2A_2 - \Omega_2) = 0. \tag{97}$$

For Maclaurin spheroids $(A_1 = A_2)$, upon writing $\lambda = i\sigma$, the solution becomes

$$\sigma_{1,2}^{\text{even}} = i\tilde{\beta}\Omega \pm \sqrt{2A_1 - \Omega^2[1-\tilde{\beta}^2 - (1-\tilde{\beta}')^2]}. \tag{98}$$

These modes represent stable, damped oscillations since the inequality $1-\tilde{\beta}^2 - (1-\tilde{\beta}')^2 \leq 2A_1/\Omega^2$ is always fulfilled. Indeed, the left-hand side is always larger than unity, while the maximal value of the right-hand side is $1/2$. The latter upper limit is easy to deduce by minimizing the left-hand side of the inequality with respect to $0 \leq \eta/\rho_S\omega_S \leq \infty$ defined via the relations (24).

## 4.2 Second order

Second order harmonic deformations correspond to $l = 2$ with five distinct values of $m$, $-2 \leq m \leq 2$. Again, let us assume time-dependent Lagrangian displacements to have the form (91). The characteristic equation for the second order oscillation modes become

$$\lambda^2 V_{i;j} = 2\epsilon_{ilm}\Omega_m\lambda + \Omega^2 V_{ij} - \Omega_i\Omega_k V_{kj}$$
$$+ \delta_{ij}\delta\Pi - \pi G\rho\left(2B_{ij}V_{ij} - a_i^2\delta_{ij}\sum_{l=1}^{3}A_{il}V_{ll}\right)$$
$$- 5\nu f_N\lambda\left(\frac{V_{i;j}}{a_j^2} + \frac{V_{j;i}}{a_i^2}\right) + 5\nu f_N f_S\lambda\left(\frac{U_{i;j}}{a_j^2} + \frac{U_{j;i}}{a_i^2}\right), \quad (99)$$

$$\lambda^2 U_{i;j} = 2\epsilon_{ilm}\Omega_m\lambda U_{l;j} + \Omega^2 U_{ij} - \Omega_i\Omega U_{kj}$$
$$+ \delta_{ij}\left(\frac{\delta\Pi_S}{f_S} - \frac{\delta\Pi_N}{f_N}\right) - 2\pi G\rho A_i U_{ij}$$
$$- 2\Omega\tilde{\beta}_{ik}\lambda U_{k;j} + 5\nu\lambda\left(\frac{V_{i;j}}{a_j^2} + \frac{V_{j;i}}{a_i^2}\right) - 5\nu f_S\lambda\left(\frac{U_{i;j}}{a_j^2} + \frac{U_{j;i}}{a_i^2}\right), \quad (100)$$

where the frequencies are measured in the units $(\pi\rho G)^{1/2}$. (99) and (100), which (if written in components) constitute a coupled set of 18 equations each, contain all the second harmonic modes of isolated, incompressible, and irrotational superfluid ellipsoids. In the next sections, we concentrate on solutions of these equations for the special case of Maclaurin spheroids, i.e. the case where the axial symmetry about the axis of rotation is assumed.

## 4.3 Transverse shear modes ($l = 2$, $m = |1|$)

These modes correspond to surface deformations with $|m| = 1$ and represent relative shearing of the northern and southern hemispheres of the ellipsoid. They are determined by the eight components of the (99) and (100) which are odd in index 3; i.e. $V_{3;i}$, $V_{i;3}$, $U_{3;i}$ and $U_{i;3}$, where $i = 1, 2$. The odd equations for the virials describing the CM-motions are

$$\left(\lambda^2 + f_N\nu\lambda + 2B_{13}\right)V_{13} - \left(\lambda^2 + \gamma f_N\nu\lambda + 2B_{13}\right)V_{1;3}$$
$$-f_N f_S\nu\lambda U_{13} + f_N f_S\gamma\nu\lambda U_{1;3} = 0, \quad (101)$$

$$\left(\lambda^2 + f_N\nu\lambda + 2B_{23}\right)V_{23} - \left(\lambda^2 + \gamma f_N\nu\lambda + 2B_{13}\right)V_{2;3}$$
$$-f_N f_S\nu\lambda U_{23} + f_N f_S\gamma\nu\lambda U_{2;3} = 0, \quad (102)$$

$$\left(\lambda^2 - \gamma f_N\nu\lambda\right)V_{1;3} - 2\Omega\lambda V_{2;3} + \left(2B_{13} - \Omega^2 + f_N\nu\lambda\right)V_{13}$$
$$+f_N f_S\gamma\nu\lambda U_{1;3} - f_N f_S\nu\lambda U_{13} = 0, \quad (103)$$

$$\left(\lambda^2 - \gamma f_N\nu\lambda\right)V_{2;3} + 2\Omega\lambda V_{1;3} + \left(2B_{13} - \Omega^2 + f_N\nu\lambda\right)V_{23}$$
$$+f_N f_S\gamma\nu\lambda U_{2;3} - f_N f_S\nu\lambda U_{23} = 0, \quad (104)$$

where $\gamma \equiv 1 - a_1^2/a_3^2$ and we have redefined the kinematic viscosity as $\nu' = 5\nu/a_1^2$ and dropped the prime in the equations above. The relations (58) were used to manipulate the original equations to the form above. For the virials describing the relative motions the odd parity equations are

$$\left(\lambda^2 + 2\Omega\tilde{\beta}''\lambda + f_S\nu\lambda + 2A_3\right)U_{13} - \left(\lambda^2 + 2\Omega\tilde{\beta}''\lambda + \gamma f_S\nu\lambda\right)U_{1;3}$$

$$-\nu\lambda V_{13} + \gamma\nu\lambda V_{1;3} = 0, \tag{105}$$

$$\left(\lambda^2 + 2\Omega\tilde{\beta}''\lambda + f_S\nu\lambda + 2A_3\right)U_{23} - \left(\lambda^2 + 2\Omega\tilde{\beta}''\lambda + \gamma f_S\nu\lambda\right)U_{2;3}$$
$$-\nu\lambda V_{23} + \gamma\nu\lambda V_{2;3} = 0, \tag{106}$$

$$\left(\lambda^2 + 2\Omega\tilde{\beta}\lambda - \gamma f_S\nu\lambda\right)U_{1;3} + (2A_1 - \Omega^2 + f_S\nu\lambda)U_{13} - 2\Omega(1 - \tilde{\beta}')\lambda U_{2;3}$$
$$+\gamma\nu\lambda V_{1;3} - \nu\lambda V_{13} = 0, \tag{107}$$

$$\left(\lambda^2 + 2\Omega\tilde{\beta}\lambda - \gamma f_S\nu\lambda\right)U_{2;3} + (2A_1 - \Omega^2 + f_S\nu\lambda)U_{23} + 2\Omega(1 - \tilde{\beta}')\lambda U_{1;3}$$
$$+\gamma\nu\lambda V_{2;3} - \nu\lambda V_{23} = 0. \tag{108}$$

According to the symmetries of the original virial equation (69), the two sets (101)-(104) and (105)-(108) decouple in the limit $\nu \to 0$, as they should. The dissipation in the first set is driven by the viscosity of the normal matter; the superfluid contributes to the damping of the CM modes indirectly, via their coupling to the relative oscillation modes. In the second set the normal matter viscosity directly renormalizes the mutual friction damping time scale ($2\Omega\tilde{\beta} \to 2\Omega\tilde{\beta} - \gamma f_S\nu$), thus reducing the damping of the relative modes. Note that this renormalization vanishes for a sphere since then $\gamma = 0$. One important feature of each set, which remains preserved when they are coupled, is the balance between the tensors describing the perturbations of the rotational kinetic energy and gravitational energy; in the first set only the two-index symbols enter ($B_{ij}$), while in the second one appear only the one-index symbols ($A_i$). As a results the neutral points (if any) along a sequence of ellipsoids (parameterized in terms of eccentricity) remain unaffected by the coupling between the different sets. This implies that as long as there are no neutral points for the relative transverse-shear modes in the uncoupled case, the conclusion about their stability can not be affected by the viscosity of the normal component. The CM modes do not show neutral points along the Maclaurin sequence and therefore their stability is guaranteed.

In the absence of the viscosity the components of (99), which are odd in index 3, decouple into two separate sets. The first set for virials $V_{ij}$, which describes the CM motions of the fluids is identical to the one found in EFE:

$$\left(\lambda^2 + 2B_{13}\right)V_{13} - \left(\lambda^2 + 2B_{13}\right)V_{1;3} = 0, \tag{109}$$

$$\left(\lambda^2 + 2B_{23}\right)V_{23} - \left(\lambda^2 + 2B_{13}\right)V_{2;3} = 0, \tag{110}$$

$$\lambda^2 V_{1;3} - 2\Omega\lambda V_{2;3} + \left(2B_{13} - \Omega^2\right)V_{13} = 0, \tag{111}$$

$$\lambda^2 V_{2;3} + 2\Omega\lambda V_{1;3} + \left(2B_{13} - \Omega^2\right)V_{23} = 0. \tag{112}$$

The corresponding modes are described in EFE (see also Fig. 1 below). The second set, which describes the relative oscillations of the fluids, is

$$\left(\lambda^2 + 2\Omega\tilde{\beta}''\lambda + 2A_3\right)U_{13} - \left(\lambda^2 + 2\Omega\tilde{\beta}''\lambda\right)U_{1;3} = 0, \tag{113}$$

$$\left(\lambda^2 + 2\Omega\tilde{\beta}''\lambda + 2A_3\right)U_{23} - \left(\lambda^2 + 2\Omega\tilde{\beta}''\lambda\right)U_{2;3} = 0, \tag{114}$$

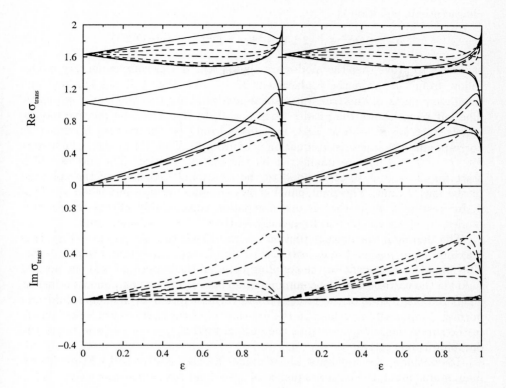

**Fig. 1.** The real (upper panel) and imaginary (lower panel) parts of the CM and relative transverse-shear modes of superfluid Maclaurin spheroid as a function of eccentricity for values of $\eta/\omega_S\rho_S = 0.0$ (*solid line*), 0.5 (*long-dashed line*), 1 (*dashed line*), and 50 (*dashed-dotted line*). The left panel corresponds to the solutions in the inviscid limit $\nu = 0$. The right panel corresponds to the case where $\nu = 4\tilde{\beta}\Omega$. The fraction of the normal fluid $f_N = 0.1$.

$$\left(\lambda^2 + 2\Omega\tilde{\beta}\lambda\right) U_{1;3} + (2A_1 - \Omega^2)U_{13} - 2\Omega(1 - \tilde{\beta}')\lambda U_{2;3} = 0, \quad (115)$$

$$\left(\lambda^2 + 2\Omega\tilde{\beta}\lambda\right) U_{2;3} + (2A_1 - \Omega^2)U_{23} + 2\Omega(1 - \tilde{\beta}')\lambda U_{1;3} = 0. \quad (116)$$

Let us concentrate first on the second set and consider the limit of zero mutual friction (i.e. the case where the two fluids are coupled only by their mutual gravitational attraction). The characteristic equation can be factorized by substituting $\lambda = i\sigma$ to find

$$\sigma\left[\sigma^2 - 2(A_1 + A_3) + \Omega^2\right] \pm 2\Omega(\sigma^2 - 2A_3) = 0. \quad (117)$$

The purely rotational mode $\sigma = \Omega$ decouples only in the spherical symmetric limit where $A_1 = A_3$. If only axial symmetry is imposed then the third order characteristic equation is

$$\sigma^3 \pm 2\Omega\sigma^2 + \left[-2\left(A_1 + A_3\right) + \Omega^2\right]\sigma \mp 4A_3\Omega = 0. \tag{118}$$

It is easy to prove that the modes are always real. Three complementary modes follow from (118) via the replacement $\Omega \to -\Omega$. Fig. 1 shows the real and imaginary parts of the transverse-shear modes along the Maclaurin sequence. The left panel shows the results when $\nu = 0$. In that case the relative modes, which start for $\Omega \to 0$ at 1.63, are affected only by the mutual friction (the corresponding characteristic equation is of order 6). The CM modes, which start for $\Omega \to 0$ at 1.03, are unaffected by the mutual friction. The modes which start for $\Omega \to 0$ at 0, correspond to the rotational frequency of the spheroid in the low-$\Omega$ limit. The right panel shows the same modes but when $\nu/\tilde{\beta}\Omega = 4$. Interestingly, while the viscous dissipation considerably affects the relative modes, its effect on the real frequencies of the CM modes is marginal.

The damping via mutual friction, as seen from the left panel of Fig. 1, is maximal for $\eta/\rho_S\omega = 1$ and decreases to zero for $\eta/\rho_S\omega \to 0$ and $\eta/\rho_S\omega \to \infty$. This behavior is specific to the coupling between the superfluid and the normal fluid via the vortex state; the communication between these components is fastest when the magnitude of the forces on the vortex exerted by the superfluid and normal components are close. In the limiting cases the vortices are locked either in the superfluid ($\eta/\rho_S\omega \to 0$) or the normal fluid ($\eta/\rho_S\omega \to \infty$) and hence the damping is ineffective.

To conclude, the transverse-shear modes for the relative and CM modes remain stable along the entire sequence of superfluid Maclaurin spheroids.

## 4.4 Toroidal modes ($l = 2$, $m = |2|$)

These modes correspond to $|m| = 2$ and the motions in this case are confined to planes parallel to the equatorial plane. The toroidal modes are determined by the even in index 3 components of (99) and (100) for the virials $V_{i;i}$, $V_{i;j}$, $U_{i;i}$ and $U_{i;j}$, where $i, j = 1, 2$. These equations can be manipulated to a set of four equations, which read

$$\left(\lambda^2 + 2f_N\nu\lambda + 4B_{12} - 2\Omega^2\right)V_{12} + \Omega\lambda(V_{11} - V_{22})$$
$$-2f_N f_S\nu\lambda U_{12} = 0, \tag{119}$$

$$\left(\lambda^2 + 2f_N\nu\lambda + 4B_{12} - 2\Omega^2\right)(V_{11} - V_{22})$$
$$-4\Omega\lambda V_{12} - 2f_N f_S\nu\lambda(U_{11} - U_{22}) = 0, \tag{120}$$

$$\left(\lambda^2 + 2\Omega\tilde{\beta}\lambda + 2f_S\nu\lambda + 4A_1 - 2\Omega^2\right)U_{12}$$
$$+\Omega(1 - \tilde{\beta}')\lambda(U_{11} - U_{22}) - 2\nu\lambda V_{12} = 0, \tag{121}$$

$$\left(\lambda^2 + 2\Omega\tilde{\beta}\lambda + 2f_S\nu\lambda + 4A_1 - 2\Omega^2\right)(U_{11} - U_{22})$$
$$-4\Omega(1 - \tilde{\beta}')\lambda U_{12} - 2\nu\lambda(V_{11} - V_{22}) = 0. \tag{122}$$

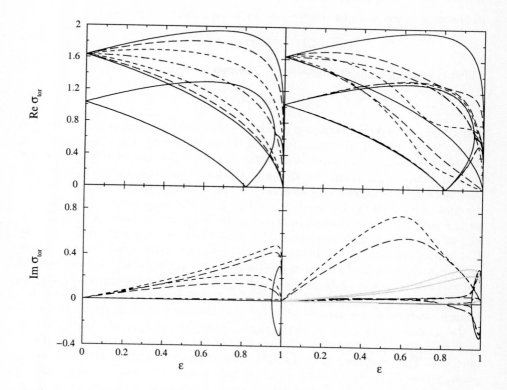

**Fig. 2.** The CM and relative toroidal modes of superfluid Maclaurin spheroid. The imaginary parts shown in *grey* are magnified by factor 10. Other conventions are the same as in Fig. 1. Note the secular instability at the bifurcation point $\epsilon = 0.81$ and the dynamical instability at the point $\epsilon = 0.93$.

In the inviscid limit equations above decouple into separate sets for the CM and relative oscillations. The CM oscillations are described by the equations

$$\left(\lambda^2 + 4B_{12} - 2\Omega^2\right) V_{12} + \Omega\lambda(V_{11} - V_{22}) = 0, \quad (123)$$

$$\left(\lambda^2 + 4B_{12} - 2\Omega^2\right) (V_{11} - V_{22}) - 4\Omega\lambda V_{12} = 0; \quad (124)$$

their solutions are documented in EFE. The relative oscillations are described by the following equations

$$\left(\lambda^2 + 2\Omega\tilde{\beta}\lambda + 4A_1 - 2\Omega^2\right) (U_{11} - U_{22}) - 4\Omega\lambda(1 - \tilde{\beta}')U_{12} = 0, \quad (125)$$

$$\left(\lambda^2 + 2\Omega\tilde{\beta}\lambda + 4A_1 - 2\Omega^2\right) U_{12} + \Omega\lambda(1 - \tilde{\beta}')(U_{11} - U_{22}) = 0. \quad (126)$$

and the characteristic equation for the relative toroidal modes is:

$$\lambda^4 + 4\tilde{\beta}\Omega\lambda^3 + (8A_1 + 4\tilde{\beta}^2\Omega^2 - 8\tilde{\beta}'\Omega^2 + 4\tilde{\beta}'^2\Omega^2)\lambda^2$$
$$+ 8\tilde{\beta}\Omega(2A_2 - \Omega^2)\lambda + 4(2A_1 - \Omega^2)^2 = 0. \quad (127)$$

In the frictionless limit the modes can be found analytically from

$$(\lambda^2 + 4A_1 - 2\Omega^2)^2 + 4\Omega^2\lambda^2 = 0, \tag{128}$$

which is factorized by writing $\lambda = i\sigma$. The two solutions are then

$$\sigma_{1,2} = \Omega \pm \sqrt{4A_1 - \Omega^2}. \tag{129}$$

and there are two complementary modes which are found by substituting $-\Omega$ for $\Omega$. If the mutual friction is included the characteristic equation describing the relative oscillation modes is of order 4; in the presence of viscosity of the normal component, again the CM and relative oscillation modes couple and the characteristic equation is of order 8. The real and imaginary parts of the dissipative toroidal modes are shown in Fig. 2, for the same values of parameters as in Fig. 1. As for the transverse-shear modes, the CM and relative modes start at 1.03 and 1.63, respectively, when $\Omega \to 0$. In the inviscid limit (left panel), the CM modes are unaffected, as they should, while the relative modes are driven against each other and merge in the limit of strong coupling. The damping of the relative modes is finite, while for the CM modes it vanishes exactly up to the point of the onset of the dynamical instability at $\epsilon = 0.95$; beyond this point a mode becomes dynamically (i.e. in the absence of dissipation) unstable. If the kinematic viscosity is finite (right panel), the real parts of the relative modes are strongly affected, while the effect of the viscosity on the CM modes is marginal. The imaginary part of a CM mode changes its sign at the bifurcation point, where $2B_{12} = \Omega^2$ and $\epsilon = 0.813$. This signals the onset of the classical secular instability of the Maclaurin spheroid. The new feature here is that the mutual friction contributes to the secular instability of a CM-mode. On the other hand, ordinary viscosity does not drive the relative modes unstable, in agreement with the fact that there are no neutral points for these modes along the entire Maclaurin sequence. One may conclude that the agents which break the superfluid/normal fluid symmetry can not cause an instability of the relative modes. The only possibility that the relative modes become unstable is a shift of the balance between the kinetic and potential energy perturbations, which might occur for compressible fluids, e.g. polytrops. This problem will be studied elsewhere.

## 4.5 Pulsation mode ($l = 2$, $m = 0$)

Pulsation modes (or breathing modes) are the generalization of the radial pulsation modes of a sphere to the case of rotation. They correspond to $l = 2$ and $m = 0$ indexes in the expansion in spherical harmonics. The pulsation modes are determined by the full set of equations which are even in index 3. By suitable combination of the equations for the virials $V_{i;i}$, $V_{i;j}$, $U_{i;i}$ and $U_{i;j}$, where $i = 1, 2, 3$ and $j = 1, 2$ the original set of equations can be reduced to

$$\left(\lambda^2/2 + f_N\nu\lambda + 4B_{11} - 2B_{13} - \Omega^2\right)(V_{11} + V_{22})$$

$$-\left[\lambda^2 - 2f_N\nu(1-\gamma)\lambda + 6B_{33} - 2B_{13}\right]V_{33}$$
$$+2\Omega\lambda(V_{1;2} - V_{2;1}) - f_N f_S \nu\lambda(U_{11} + U_{22}) - 2f_N f_S \nu(1-\gamma)U_{33} = 0, \quad (130)$$
$$\lambda^2(V_{1;2} - V_{2;1}) - \Omega\lambda(V_{11} + V_{22}) = 0, \quad (131)$$
$$\left(\lambda^2/2 + 2A_1 - \Omega^2 + f_S\nu\lambda + \Omega\tilde{\beta}\lambda\right)(U_{11} + U_{22})$$
$$-\left[\lambda^2 + 2\Omega\tilde{\beta}''\lambda + 4A_3 + 2f_S\nu(1-\gamma)\lambda\right]U_{33}$$
$$+2\Omega(1-\tilde{\beta}')\lambda(U_{1;2} - U_{2;1}) - \nu\lambda(V_{11} + V_{22}) - 2\nu(1-\gamma)\lambda V_{33} = 0, \quad (132)$$
$$(\lambda^2 + 2\Omega\tilde{\beta}\lambda)(U_{1;2} - U_{2;1}) - \Omega(1-\tilde{\beta}')\lambda(U_{11} + U_{22}) = 0. \quad (133)$$

The characteristic equation is found by supplementing these equations by the divergence free constraint on the virials of the CM and relative motions:

$$\sum_{i=1}^{3} \frac{V_{ii}}{a_i^2} = 0, \quad \sum_{i=1}^{3} \frac{U_{ii}}{a_i^2} = 0. \quad (134)$$

An equivalent form of the divergence free constraint for Maclaurin spheroids can be written in terms of eccentricity, $(V_{11} + V_{22})(1 - \epsilon^2) + V_{33} = 0$ and similarly for $U_{ii}$.

In absence of viscosity the equations above decouple into two independent sets for CM and relative oscillations. The CM oscillations are described by

$$\left(\lambda^2/2 + 4B_{11} - 2B_{13} - \Omega^2\right)(V_{11} + V_{22})$$
$$-\left(\lambda^2 + 6B_{33} - 2B_{13}\right)V_{33} + 2\Omega\lambda(V_{1;2} - V_{2;1}) = 0, \quad (135)$$
$$\lambda^2(V_{1;2} - V_{2;1}) - \Omega\lambda(V_{11} + V_{22}) = 0, \quad (136)$$

and coincide with the pulsation modes treated in EFE. The relative oscillation modes are defined by the equations

$$\left(\lambda^2/2 + \Omega\tilde{\beta}\lambda - \Omega^2 + 2A_1\right)(U_{11} + U_{22})$$
$$+2\Omega\lambda(1-\tilde{\beta}')(U_{1;2} - U_{2;1}) - (\lambda^2 + 4A_3 + 2\Omega\tilde{\beta}''\lambda)U_{33} = 0, \quad (137)$$
$$\left(\lambda^2 + 2\Omega\tilde{\beta}\lambda\right)(U_{1;2} - U_{2;1}) - \Omega\lambda(1-\tilde{\beta}')(U_{11} + U_{22}) = 0, \quad (138)$$

which can be combined to:

$$\left[\left(\lambda^2 + 2\Omega\tilde{\beta}\lambda - 2\Omega^2 + 4A_1\right)(\lambda^2 + 2\Omega\tilde{\beta}\lambda) + 4\Omega^2\lambda^2(1-\tilde{\beta}')^2\right](U_{11} + U_{22})$$
$$-2\left[\left(\lambda^2 + 2\Omega\tilde{\beta}\lambda\right)\left(\lambda^2 + 2\Omega\lambda\tilde{\beta}'' + 4A_3\right)\right]U_{33} = 0. \quad (139)$$

The modes are found by supplementing this equation by the divergence free condition (134). The third order characteristic equation in the inviscid limit is

$$(3 - 2\epsilon^2)\lambda^3 + [8\tilde{\beta}\Omega + 4\tilde{\beta}''\Omega - 4(\tilde{\beta} + \tilde{\beta}'')\epsilon^2\Omega]\lambda^2 + [4A_1 + 8(1-\epsilon^2)A_3$$
$$+2\Omega^2 + 4\tilde{\beta}^2\Omega^2 - 8\tilde{\beta}'\Omega^2 + 4\tilde{\beta}'^2\Omega^2 + 8(1-\epsilon^2)\tilde{\beta}\tilde{\beta}''\Omega^2]\lambda$$
$$+8A_1\tilde{\beta}\Omega + 16(1-\epsilon^2)A_3\tilde{\beta}\Omega - 4\tilde{\beta}\Omega^3 = 0, \quad (140)$$

where the trivial mode $\lambda = 0$ is neglected. In the frictionless limit we find ($\lambda = i\sigma$ as before)

$$\sigma = \pm \left[ \frac{2\Omega^2 + 4A_1 + 8A_3(1-\epsilon^2)}{(3-2\epsilon^2)} \right]^{1/2}. \qquad (141)$$

The pulsation modes for a sphere follow in the limit $(\epsilon, \Omega) \to 0$: for a sphere $A_i/(\pi\rho G) = 2/3$, and (141) reduces to $\sigma^2 = 8/3$ [$\sigma$ is given in units of $(\pi\rho G)^{1/2}$]. This result could be compared with the pulsation modes of an ordinary sphere: $\sigma^2 = 16/15$. Thus a superfluid sphere, apart form the ordinary pulsations, shows pulsations at frequencies roughly twice as large as the ordinary ones. In the general case where the viscosity of the normal fluid is taken into account the characteristic equation is of fifth order. The real and imaginary parts of the roots are shown in Fig. 3. In the inviscid limit the CM modes are again unaffected, while the relative modes are suppressed by the mutual friction. The damping of these modes is maximal when $\eta/\rho_s\omega = 1$ and the motions correspond to stable, damped oscillations. In the presence of viscosity, the relative modes are strongly damped and eventually become neutral. The CM modes are weakly affected. The

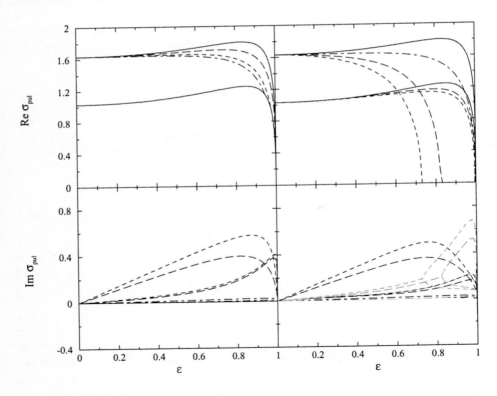

**Fig. 3.** The CM and relative pulsation modes of superfluid Maclaurin spheroid. Conventions are the same as in Fig. 1 and 2.

imaginary parts remain always positive, i.e. one finds stable, damped oscillations. Note that at the point where a relative pulsation mode becomes neutral the number of the imaginary components increases by one (as generally expected for the roots of polynomials), but there are no changes in the sign of the imaginary part of the neutral mode.

## 5  Summary

Let us briefly summarize the main qualitative features of the oscillations of superfluid self-gravitating systems [11]:

- The oscillation modes of the superfluid ellipsoids separate into two generic classes which correspond to *co-moving* and *relative oscillations*. The oscillation frequencies of these two classes have distinct values in the both slow and rapid rotation limits. The first class of oscillations is identical to those of classical single-fluid ellipsoids in the incompressible and inviscid limits. Corresponding modes are undamped if the Euler equations of fluids are symmetric/anti-symmetric with respect to the interchange $\alpha \leftrightarrow \beta$ (the indexes which label the fluids) . When the fluids are coupled by mutual friction and mutual gravitational attraction this symmetry is preserved.
- The second class of oscillations, which is new, corresponds to relative motions of the fluids. The modes are damped by the mutual friction between the superfluid and the normal fluid. These modes correspond to stable oscillations.

- The co-moving (CM) modes emit gravitational radiation and undergo radiation reaction instabilities in full analogy to single-fluid ellipsoids [13]. The relative modes do not emit gravitational radiation at all, since the mass current associated with them is zero. This picture must hold true for a more general class of Chandrasekhar-Friedman-Schutz (CFS) radiation reaction instabilities that are intrinsic to self-gravitating Newtonian fluids [14].
- If the $\alpha \leftrightarrow \beta$ symmetry is broken the two classes of modes mix, for example, when the normal fluid is viscous. The main effect of the mixing is the *renormalization of the mutual friction and viscosity*. The relative modes remain stable as there are no distinct neutral points for these modes along the ellipsoidal sequences. The CM modes become unstable at the classical points of onset of secular/dynamical instabilities, for example at the point of bifurcation of the Maclaurin spheroid into a Jacobi ellipsoid.

These qualitative features are based on general symmetries of underlying hydrodynamic equations *and* the conditions of equilibrium of self-gravitating fluids which are independent of the superfluid nature of underlying fluids (e.g. the existence of bifurcation points). Therefore we may conclude that these features will be preserved in more complex models of oscillations of superfluid neutron stars.

## Acknowledgments

This work was supported in part by FOM at KVI (Groningen) and by CNRS at IPN (Orsay), and by NASA at Cornell.

## References

1. S. Chandrasekhar, E. Fermi: Astrophys. J. **118**, 116 (1953)
2. S. Chandrasekhar, *Ellipsoidal Figures of Equilibrium* (Yale University Press, New Haven, 1969)
3. B. Carter, J. P. Luminet: Mon. Not. R. Astron. Soc. **212**, 23, (1985); J. P. Luminet, B. Carter: Astrophys. J. Suppl. Ser. **61**, 219 (1986)
4. J. R. Ipser, L. Lindblom: Astrophys. J., **355**, 226 (1990); Astrophys. J., **379**, 285 (1991); L. Lindblom, J. R. Ipser, Phys. Rev., **D59** 044009 (1999)
5. D. Lai, F. A. Rasio, S. L. Shapiro: Astrophys. J. Suppl. Ser. **88**, 205 (1993)
6. S. L. Shapiro, S. A. Teukolsky, *Black Holes, White Dwarfs and Neutrons Stars* (Wiley, New York, 1983)
7. R. Epstein: Astrophys. J. **333**, 880 (1988)
8. L. Lindblom, G. Mendell: Astrophys. J. **421**, 689 (1994)
9. U. Lee: Astron. Astrophys. **303**, 515 (1995)
10. L. Lindblom, G. Mendell: Phys. Rev. **D61**, 104003 (2000)
11. A. Sedrakian, I. Wasserman: Phys. Rev. **D63**, 024016 (2000)
12. I. M. Khalatnikov, *Introduction to the Theory of Superfluidity* (Addison Wesley, New York, 1989)
13. S. Chandrasekhar: Phys. Rev. Lett., **24**, 611 (1970)
14. J. L. Friedman, B. F. Schutz: Astrophys. J., **221**, 973 (1978)

# Neutron Star Crusts

Paweł Haensel

N. Copernicus Astronomical Center, Polish Academy of Sciences,
Bartycka 18, PL-00-716 Warszawa, Poland

**Abstract.** The formation, structure, composition, and the equation of state of neutron star crusts are described. A scenario of formation of the crust in a newly born neutron star is considered and a model of evolution of the crust composition during the early neutron star cooling is presented. Structure of the ground state of the crust is studied. In the case of the outer crust, recent nuclear data on masses of neutron rich nuclei are used. For the inner crust, results of different many-body calculations are presented, and dependence on the assumed effective nucleon-nucleon interaction is discussed. Uncertainties concerning the bottom layers of the crust and crust-liquid interface are illustrated using results of various many-body calculations based on different effective nucleon-nucleon interactions. A scenario of formation of a crust of matter-accreting neutron star is presented, and evolution of the crust-matter element under the increasing pressure of accreted layer is studied. Within a specific dense matter model, composition of accreted crust is calculated, and is shown to be vastly different from the ground-state one. Non-equilibrium processes in the crust of mass-accreting neutron star are studied, heat release due to them is estimated, and their relevance to the properties of X-ray sources is briefly discussed. Equation of state of the ground-state crust is presented, and compared with that for accreted crust. Elastic properties of the crust are reviewed. Possible deviations from idealized models of one-component plasmas are briefly discussed.

## 1 Introduction

The crust plays an important role in neutron star evolution and dynamics. Its properties are crucial for many observational properties, despite the fact that the crust mass constitutes only $\sim 1\%$ of neutron star mass, and its thickness is typically less than one tenth of the star radius. The crust separates neutron star interior from the photosphere, from which X-ray radiation is emitted. The transport of heat from neutron star core to the star surface is determined by the thermal conductivity of the outer layers of the crust, which is crucial for determining the relation between observed X-ray flux and the temperature of neutron star core.

Electrical resistivity of the crust is expected to be important for the evolution of neutron star magnetic field. Both thermal conductivity and electrical resistivity depend on the structure of the crust, its nuclear composition, and the presence and number of crystalline defects and impurities. During some stages of neutron star cooling, neutrino emission from the crust may significantly contribute to total neutrino losses from stellar interior. The presence of a crystal lattice of atomic nuclei in the crust is mandatory for modeling of radio–pulsar glitches. Presence of solid crust enables excitation of toroidal modes of oscillations. The toroidal modes in a completely fluid star have all zero frequency, but the presence of a solid crust gives them nonzero frequencies ∼kHz. Presence of the crust can also be important for non-radial pulsations excited in the liquid core, because of specific boundary conditions which are to be imposed at the solid-liquid boundary. Due to its solid character, neutron star crust can be a site of elastic stresses, and can build-up elastic strain during star evolution (cooling, spin-down). In contrast to fluid core, the crust can therefore support deviations of the stellar shape from the axial symmetry, and make from rapidly rotating pulsar an interesting source of gravitational waves. Instabilities in the fusion of light elements, taking place in the outer layers of the crust of an accreting neutron star, are thought to be responsible for the phenomenon of X-ray bursts.

The present review is devoted to the structure, composition, and equation of state of neutron star crust. In Sect.2 we briefly describe formation of the crust of a newly born neutron star. Structure, composition, and equation of state of the outer crust in the ground state approximation is described in Sect.3. Theoretical models of the inner crust in the ground state approximation and with $\rho \lesssim 10^{14}$ g cm$^{-3}$ are presented in Sect. 4. Section 5 is devoted to the presentation of theoretical models of the ground state of the bottom layers of the inner crust, with $\rho \gtrsim 10^{14}$ g cm$^{-3}$, and to determination of the location of and conditions at the bottom edge of the crust. In Sect. 6 we consider a scenario of formation of the crust in accreting neutron star. Then, in Sect. 7 we study non-equilibrium nuclear processes in the crust interior, and derive its structure and nuclear composition. Sect. 8 is devoted to the equation of state of neutron star crust, both in the ground state approximation and in the case of an accreted crust. Elastic properties of the crust are discussed in Sect. 9. Possible deviations from idealized crust models studied in the preceding sections are briefly reviewed in Sect.10.

## 2 Formation of the crust in a newly born neutron star

Neutron star formed in gravitational collapse of a stellar core is initially very hot, with internal temperature $\sim 10^{11}$ K. At such high temperature, the composition and equation of state of the outer envelope of a newly-born neutron star, with $\rho \lesssim 10^{14}$ g cm$^{-3}$ ($n_b \lesssim 0.1$ fm$^{-3}$), is different from that of a one-year old neutron star. This envelope of a newly-born neutron star will eventually become the crust of neutron star.

We will restrict ourselves to the case in which matter is transparent to neutrinos, a condition satisfied for $T \lesssim 10^{10}$ K ($k_B T \lesssim 1$ MeV). Hot envelope is then a mixture of heavy and light nuclei (mostly α-particles, because of their large binding energy of 28.3 MeV), neutrons, protons, electrons, positrons, and photons. At high densities and temperatures the density of nucleons outside nuclei can be large, and a consistent treatment of both nuclei and nucleons is required. Nuclei and nucleons outside them should be described using the same

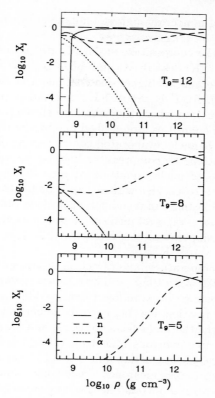

**Fig. 1.** Mass fractions of different constituents of the outer envelope of a newly born neutron star versus matter density, at different temperatures $T_9 = T/(10^9 \text{ K})$. Beta equilibrium assumed. After Haensel et al. [40]. Calculations performed for the Lattimer and Swesty [53] model, with a specific choice of the compression modulus of cold symmetric nuclear matter at the saturation density, $K_0 = 180$ MeV.

nucleon interaction (nucleon hamiltonian), and modifications of nuclear surface properties, and pressure exerted by nucleons on nuclei, have to be calculated in a consistent way. At high densities, where the distance between nuclei is no longer much larger than nuclear size, one has to take into account modification of the nuclear Coulomb energy. Another important complication is that, at temperatures under consideration, excited states of nuclei become populated and must therefore be considered when calculating thermodynamic quantities.

In what follows, we will describe results, obtained using a version of compressible liquid–drop model of nuclei developed by Lattimer and Swesty [53], with a specific choice of compression modulus of symmetric nuclear matter at saturation (equilibrium) density, $K_0 = 180$ MeV. We assume nuclear equilibrium, as well as beta equilibrium of dense hot matter. The assumption of nuclear equilibrium is justified by high temperature. Beta equilibrium is adopted for simplicity;

a very rapid cooling of matter at highest temperatures can produce deviations from beta equilibrium.

In Fig.1 we show the composition of dense, hot matter of neutron star envelope for $T = 5 \times 10^9$ K, $8 \times 10^9$ K, and $1.2 \times 10^{10}$ K. We restrict ourselves to $\rho \lesssim 10^{13}$ g cm$^{-3}$, because at higher densities thermal effects on matter composition are negligible. At $T \gtrsim 5 \times 10^9$ K shell and paring effects, so visible in the $T = 0$ approximation where they show up through jumps in the density dependence of various quantities, are washed out by the thermal effects.

At $T = 1.2 \times 10^{10}$ K nuclei evaporated completely for $\rho \lesssim 10^9$ g cm$^{-3}$. This can be understood within the compressible liquid-drop model of nuclei, which are treated as droplets of nuclear matter. At $\rho \lesssim 10^{11}$ g cm$^{-3}$ these droplets of nuclear matter have to coexist with a vapor of neutrons, protons, and $\alpha$-particles. However, coexistence of two different nucleon phases (denser – nuclear matter liquid, less dense – vapor of nucleons and $\alpha$-particles) is possible only at $T$ lower than critical temperature at given density, $T_{\rm crit}(\rho)$. For $\rho \lesssim 10^9$ g cm$^{-3}$, one has $T_{\rm crit}(\rho) < 1.2 \times 10^{10}$ K.

With decreasing temperature, mass fraction of evaporated nucleons and $\alpha$-particles decreases. At $T = 8 \times 10^9$ K, $\alpha$-particles are present below $10^{10}$ g cm$^{-3}$, while free protons appear below even lower density. Free neutrons are present at all densities, but their fraction does not exceed one percent for $\rho \lesssim 10^{11}$ g cm$^{-3}$.

At $T = 5 \times 10^9$ K the thermal effects are weak, and imply mainly appearance of a small fraction of free neutrons ("neutron vapor") below zero temperature neutron drip density, $\rho_{\rm ND}$; this fraction falls below $10^{-5}$ at $\rho = 10^{10}$ g cm$^{-3}$. Further decrease of $T$ leads to disappearance of neutrons below $\rho_{\rm ND}$, and switching-on of shell effects. Another important effect will be superfluid transition for neutrons (both inside and outside nuclei) and for protons. The composition freezes-out and does not change with further decrease of temperature. A spherical shell of neutron star envelope solidifies if its temperature decreases below the melting point corresponding to local density and composition, $T_{\rm m}$ (see Sect.9).

## 3  Ground state of the matter in the outer crust

The ground state of matter at the densities and pressures, at which all neutrons are bound in nuclei (i.e. below the neutron drip point) can be described by a model formulated in the classical paper of Baym, Pethick, and Sutherland ([4], hereafter referred to as BPS). An essential input for this model are the ground-state masses of atomic nuclei, present in the lattice sites of a crystal. At lowest densities, the relevant nuclei are those whose ground-state masses are determined with high precision by the laboratory measurements. However, at higher densities the nuclei in the ground state of matter become more and more neutron rich. At the time, when the BPS paper was written, the last experimentally studied nucleus, present in the ground state of dense matter, was $^{84}$Se ($Z/A = 0.405$). This nucleus is unstable in laboratory, and its beta-decay half-life time is 3.1 min. The maximum density, at which this experimentally studied nucleus was present, was found to be $8.2 \times 10^9$ g cm$^{-3}$.

During last two decades significant progress has been made in the experimental nuclear physics techniques, and masses of many new neutron rich isotopes have been measured; latest up-to-date results can be found at http://nucleardata.
nuclear.lund.se. As we will see, the last neutron-rich nucleus studied in laboratory, which is expected to be present in the ground state of neutron star crust, is $^{78}$Ni, at about $10^{11}$ g cm$^{-3}$ (for first experimental identification of this nuclide, see Engelmann et al. [32])

We shall assume that matter is in its ground state (complete thermodynamic equilibrium - cold catalyzed matter) and that it forms a perfect crystal with a single nuclear species, (number of nucleons $A$, number of protons $Z$), at lattice sites. Deviations and exceptions from this rule will be discussed later in the present review. At given baryon density, $n_b$, the ground state of matter corresponds to the minimum energy per nucleon $E = \mathcal{E}/n_b$ ($\mathcal{E}$ is energy density, which includes rest energy of constituents of matter). However, $n_b$ (or $\rho = \mathcal{E}/c^2$) is not a good variable to be used in the neutron star interior because it can suffer jumps (discontinuities) at some values of pressure. On the contrary, pressure is strictly monotonic and continuous in the stellar interior, and increases monotonically with decreasing distance from the star center. Therefore, it is convenient to formulate our problem as that of finding the ground state of cold ($T = 0$) matter at given pressure, $P$. This correspond to minimizing the $T = 0$ Gibbs energy per nucleon, $g = (\mathcal{E} + P)/n_b$.

Let us start with $P = 0$, when $g = E = \mathcal{E}/n_b$. The minimum energy per nucleon at zero pressure is reached for a body-centered-cubic (bcc) crystal lattice of $^{56}$Fe, and is $E(^{56}\mathrm{Fe}) = 930.4$ MeV. It corresponds to $\rho = 7.86$ g cm$^{-3}$ and $n_b = 4.73 \times 10^{24}$ cm$^{-3} = 4.73 \times 10^{-15}$ fm$^{-3}$.

The bcc $^{56}$Fe crystal remains the ground state of cold matter up to pressures $\sim 10^{30}$ dyn/cm$^2$, at which matter is compressed to $\sim 10^6$ g cm$^{-3}$ ([81], BPS). At such a high density, matter is a plasma of nuclei and electrons which form a nearly uniform Fermi gas. At given pressure, the values of the average electron density, $n_e$, and the number density of nuclei, $n_\mathcal{N}$, are determined from the relations

$$n_e = Zn_\mathcal{N} , \qquad P = P_e(n_e, Z) + P_L(n_\mathcal{N}, Z) , \qquad (1)$$

where $P_e$ is the electron gas pressure, and $P_L$ is the "lattice" contribution resulting from the Coulomb interactions (see below).

Let us divide the system into electrically neutral unit (Winger-Seine) cells containing one nucleus. The number density of nuclei is $n_\mathcal{N} = n_b/A$, and the volume of each cell $V_c = 1/n_\mathcal{N}$. For a given $A, Z$ nuclide, the Gibbs energy per one unit cell is given by

$$G_\mathrm{cell}(A, Z) = W_\mathcal{N}(A, Z) + W_L(Z, n_\mathcal{N}) + [\mathcal{E}_e(n_e, Z) + P]/n_\mathcal{N} , \qquad (2)$$

where $W_\mathcal{N}$ is the energy of the nucleus (including rest energy of nucleons), $W_L$ is the lattice energy per cell (BPS), and $\mathcal{E}_e$ is the mean electron energy density.

For a bcc lattice one has

$$W_{\rm L} = -0.895929 \frac{Z^2 e^2}{r_c}, \qquad r_c = (4\pi n_{\mathcal{N}}/3)^{-1/3}. \qquad (3)$$

The lattice contribution to pressure, (1), is thus $P_{\rm L} = \frac{1}{3} W_{\rm L} n_{\mathcal{N}}$.

The Gibbs energy per nucleon $g = G_{\rm cell}/A$ is just the baryon chemical potential for a given nuclid, $\mu_{\rm b}(A, Z)$. To find the ground state at given $P$, one has to minimize $\mu_{\rm b}(A, Z)$ with respect to $A$ and $Z$.

At not too high density, the lattice correction to $P$ and $\mu_{\rm b}$ is negligibly small. One can then easily see the reason for matter neutronization using the approximation $\mu_{\rm b}(A, Z) \simeq W_{\mathcal{N}}(A, Z)/A + Z\mu_e/A$ and $P \simeq P_e$. Notice that for $\rho \gg 10^6$ g cm$^{-3}$, electrons are ultrarelativistic and therefore $\mu_e \propto P^{1/4}$. With increasing pressure, it is energetically advantageous to replace $(A, Z)$ by $(A', Z')$ with higher $W_{\mathcal{N}}$ but smaller $Z'/A'$, because increase in $W_{\mathcal{N}}/A$ is more than compensated by the decrease of the $Z\mu_e/A$ term.

We will follow determination of the ground state of cold dense matter by Haensel and Pichon [39] (hereafter referred to as HP). There are some small differences between the approximations used in HP and BPS. In HP, the values of $W_{\mathcal{N}}$ have been obtained from the atomic masses by subtracting not only the electron rest energies, but also removing the atomic electron binding energies. Let us mention, that atomic binding energies were kept in the BPS definition of $W_{\mathcal{N}}$, to simulate the electron screening effects in dense matter. Also, HP used a better approximation for the electron screening effects in dense matter. Their expression for $\mathcal{E}_e$ takes into account deviations of the electron density from uniformity, which result from the electron screening effects. They include also the exchange term in $\mathcal{E}_e$, which was neglected in BPS.

At the pressure $P_i$ at which optimal values $A, Z$ change into $A', Z'$, matter undergoes a density jump, $\Delta \rho$, $\Delta n_{\rm b}$, which to a very good approximation is given by the formula

$$\frac{\Delta \rho}{\rho} \simeq \frac{\Delta n_{\rm b}}{n_{\rm b}} \simeq \frac{Z}{A} \frac{A'}{Z'} - 1. \qquad (4)$$

The above equation results from the continuity of pressure, which in the outer crust is to a very good approximation equal to the electron pressure, $P \simeq P_e$.

Actually, sharp discontinuity in $\rho$ and $n_{\rm b}$ is a consequence of the assumed one-component plasma model. Detailed calculations of the ground state of dense matter by Jog and Smith [45] have shown, that the transition between the $A, Z$ and $A', Z'$ shells takes places through a very thin layer of a *mixed lattice* of these two species. However, since the pressure interval within which the mixed phase exists is typically $\sim 10^{-4} P_i$, the approximation of a sharp density jump is quite a good representation of a nuclear composition of the ground state of matter.

Experimental masses of nuclei in HP were taken from nuclear masses tables of Audi (1992, 1993, private communication) [1] Because of the pairing effect, only

---

[1] Some of masses of unstable nuclei, given in the tables of Audi (1992,1993), were actually semi-empirical evaluations based on the knowledge of masses of neighboring isotopes.

even-even nuclei are relevant for the ground state problem. For the remaining isotopes, up to the last one stable with respect to emission of a neutron pair, HP used theoretical masses obtained using a mass formula of Möller [59] (the description of the formalism can be found in Möller and Nix [60]).

The equilibrium nuclides present in the cold catalyzed matter are listed in Table 1. Only even-even nuclides are present, which results from additional binding due to nucleon pairing (see, e.g., [74]). In the fifth column of this table one finds the maximum density at which a given nuclide is present, $\rho_{max}$. The value of the electron chemical potential, $\mu_e$, at the density $\rho_{max}$, is given in the sixth column. The transition to the next nuclide has a character of a first order phase transition and is accompanied by a density jump. The corresponding fractional increase of density, $\Delta\rho/\rho$, is shown in the last column of Table 1. The last row above the horizontal line, dividing the table into two parts, corresponds to the maximum density, at which the ground state of dense matter contains a nucleus observed in laboratory. The last line of Table 1 corresponds to the neutron drip point in the ground state of dense cold matter. This limiting density can be determined exclusively by the theoretical calculation.

Single-particle energy levels in nuclei are discrete, with large energy gaps between "major shells". The local maxima in the binding energies of nuclei with "magic numbers" $Z = 28$ and $N = 50$, $82$ are associated with filling up these major shells (see, e.g., Preston and Bhaduri [74]). The effect of the closed proton and neutron shells on the composition of the ground state of matter is

**Table 1.** Nuclei in the ground state of cold dense matter. Upper part: experimental nuclear masses. Lower part: from mass mass formula of Möller [59]. Last line corresponds to the neutron drip point. After Haensel and Pichon [39].

| element | $Z$ | $N$ | $Z/A$ | $\rho_{max}$ (g cm$^{-3}$) | $\mu_e$ (MeV) | $\Delta\rho/\rho$ (%) |
|---|---|---|---|---|---|---|
| $^{56}$Fe | 26 | 30 | 0.4643 | 7.96 10$^6$ | 0.95 | 2.9 |
| $^{62}$Ni | 28 | 34 | 0.4516 | 2.71 10$^8$ | 2.61 | 3.1 |
| $^{64}$Ni | 28 | 36 | 0.4375 | 1.30 10$^9$ | 4.31 | 3.1 |
| $^{66}$Ni | 28 | 38 | 0.4242 | 1.48 10$^9$ | 4.45 | 2.0 |
| $^{86}$Kr | 36 | 50 | 0.4186 | 3.12 10$^9$ | 5.66 | 3.3 |
| $^{84}$Se | 34 | 50 | 0.4048 | 1.10 10$^{10}$ | 8.49 | 3.6 |
| $^{82}$Ge | 32 | 50 | 0.3902 | 2.80 10$^{10}$ | 11.44 | 3.9 |
| $^{80}$Zn | 30 | 50 | 0.3750 | 5.44 10$^{10}$ | 14.08 | 4.3 |
| $^{78}$Ni | 28 | 50 | 0.3590 | 9.64 10$^{10}$ | 16.78 | 4.0 |
| $^{126}$Ru | 44 | 82 | 0.3492 | 1.29 10$^{11}$ | 18.34 | 3.0 |
| $^{124}$Mo | 42 | 82 | 0.3387 | 1.88 10$^{11}$ | 20.56 | 3.2 |
| $^{122}$Zr | 40 | 82 | 0.3279 | 2.67 10$^{11}$ | 22.86 | 3.4 |
| $^{120}$Sr | 38 | 82 | 0.3167 | 3.79 10$^{11}$ | 25.38 | 3.6 |
| $^{118}$Kr | 36 | 82 | 0.3051 | (4.33 10$^{11}$) | (26.19) | |

very strong; except for the $^{56}$Fe nucleus, present in the ground state at lowest densities, all nuclides are those with a closed proton or neutron shell (Table 1). A sequence of three increasingly neutron rich isotopes of nickel $Z = 28$ is followed by a sequence of $N = 50$ isotopes of decreasing $Z$, ending at the last experimentally identified $^{78}$Ni. This last nuclid is doubly magic ($N = 50$, $Z = 28$).

At the densities $10^{11}$ g cm$^{-3} \lesssim \rho < \rho_{\rm ND}$ HP get a sequence of $N = 82$ isotopes, of decreasing proton number, from $Z = 44$ down to $Z = 36$, with neutron drip at $\rho_{\rm ND} = 4.3~10^{11}$ g cm$^{-3}$ (Table 1). As shown by HP, results obtained using different mass formula, that of Pearson and collaborators (Pearson, 1993, private communication quoted in [39]) are quite similar to those obtained using the mass formula of Möller [59].

While the persistence of the $N = 50$ and/or $Z = 28$ nuclei in the ground state of the outer crust may be treated as an *experimental fact*, the strong effect of the $N = 82$, dominating at $10^{11}$ g cm$^{-3} \lesssim \rho < \rho_{\rm ND}$, might – in principle – be an artifact of the extrapolation via the semiempirical mass formulae. It should be mentioned, that some many-body calculations of the masses of very neutron rich nuclei suggest, that the effect of the closed $N = 82$ shell might be much weaker, and could be replaced by the strong effect of the closure of the $Z = 40$ subshell [36]. Clearly, there is a need for better understanding of shell effects in nuclei close to the neutron drip.

## 4  Ground state of the matter in the inner crust for $\rho \lesssim 10^{14}$ g cm$^{-3}$

The existence of the inner neutron-star crust, in which very neutron rich nuclei are immersed in a gas of dripped neutrons, has been realized long before the discovery of pulsars (in 1958, [41]). First approach to describe this layer of neutron star envelope consisted in employing a semiempirical mass formula to calculate (or rather estimate) the masses of nuclei, combined with an expression for the energy of neutron gas [41],[42],[88], [51],[6]. It is worth to be mentioned that as early as in 1965 neutron drip density and the density at the bottom edge of the inner crust were estimated as $\rho_{\rm ND} \simeq 3 \times 10^{11}$ g cm$^{-3}$ and $\rho_{\rm edge} \simeq 8 \times 10^{13}$ g cm$^{-3}$ [88], surprisingly close to the presently accepted values of these densities. Further work concentrated on a consistent (unified) description of nuclear matter inside neutron rich nuclei, and of neutron gas outside them, using a single expression for the energy density of nuclear matter as a function of neutron and proton densities and of their gradients [3],[18],[19], [1]. The most ambitious early attempt to calculate the ground state of the inner crust was the Hartree-Fock calculation of Negele and Vautherin [62]. Later work focused on the consistent description of the bottom layers of the crust and included up-dated treatment of both pure neutron matter and effective nucleon-nucleon interaction [43],[66],[67],[55], [87],[28],[31],[30].

In general, calculations of the structure, composition, and equation of state of the inner crust can be divided into three groups, according to the many-body technique used. Full quantum mechanical treatment can be carried out within the Hartree-Fock approximation with an effective nucleon-nucleon interaction. Further approximation of the many-body wave function can be done using semi-classical Extended Thomas-Fermi (ETF) approximation. Basic quantities within the ETF are neutron and proton densities and their spatial gradients. Finally, investigations belonging to the third group use Compressible Liquid Drop Model (CLDM) parameterization for the description of nuclei, with parameters derived within a microscopic nuclear many-body theory (HF or ETF) based on an effective nucleon-nucleon interaction.

## 4.1 Hartree-Fock calculations with effective nucleon-nucleon interaction

Matter is divided into unit cells, which are electrically neutral and contain one nucleus, with cell volume $V_c = 1/n_\mathcal{N}$. Let us assume that a unit cell contains $N$ neutrons and $Z$ protons. The nuclear effective hamiltonian for such a system of $A = N + Z$ nucleons is

$$H_\mathrm{N}^\mathrm{eff} = \sum_{j=1}^{A} t_j + \frac{1}{2} \sum_{j,k=1, j \neq k}^{A} v_{jk}^\mathrm{eff} , \qquad (5)$$

where $t_j$ is the kinetic energy operator of $j$-th nucleon, while $v_{jk}^\mathrm{eff}$ is an operator of effective two-body interaction between the $jk$ nucleon pair. Usually, $v_{jk}^\mathrm{eff}$ contains a component which is an effective two-body representation of the three-body forces, important in dense nucleon medium.

Effective nuclear hamiltonian $H_\mathrm{N}^\mathrm{eff}$ has to reproduce - as well as possible, and within the Hartree-Fock approximation - relevant properties of the ground state of the many-nucleon system, and in particular - ground state energy, $E_0$. This last condition can be written as $\langle \Phi_0 | H_\mathrm{N}^\mathrm{eff} | \Phi_0 \rangle \simeq \langle \Psi_0 | H_\mathrm{N} | \Psi_0 \rangle$, where $\Phi_0$, $\Psi_0$, and $H_\mathrm{N}$ are Hartree- Fock wave function, real wave function, and real nuclear Hamiltonian, respectively.

The complete hamiltonian of a unit cell is $H_\mathrm{cell}^\mathrm{eff} = H_\mathrm{N}^\mathrm{eff} + V_\mathrm{Coul} + H_e$, where $V_\mathrm{coul}$ and $H_e$ are the components corresponding to Coulomb interaction between charged constituents of matter (protons and electrons), and that of a uniform electron gas, respectively. The Hartree-Fock approximation for the many-body nucleon wave function is

$$\Phi_{NZ} = \mathcal{C}_{NZ} \det \left[ \varphi_{\alpha_i}^{(p)}(\xi_k) \right] \det \left[ \varphi_{\beta_j}^{(n)}(\zeta_l) \right] , \qquad (6)$$

where $\varphi_{\beta_j}^{(n)}(\zeta_l)$ and $\varphi_{\alpha_i}^{(p)}(\xi_k)$ are single-particle wave functions (orbitals) for neutrons ($j, l = 1, ..., N$) and protons ($i, k = 1, ..., Z$), respectively, and $\mathcal{C}_{NZ}$ is normalization constant. The space and spin coordinates of $k$-th proton and $j$-th neutron are represented by $\xi_k$ and $\zeta_l$, while $\{\alpha_i\}$ and $\{\beta_j\}$ are sets of quantum numbers of occupied single-particle states for protons and neutrons, respectively.

Further approximation, used by Negele and Vautherin [62], consisted in imposing the spherical symmetry. Unit cell was approximated by a sphere, and quantum numbers of single-particle states were therefore $nlj$. The Hartree-Fock equations for $\varphi^{(p)}$ and $\varphi^{(n)}$ were derived from the minimization of the HF energy functional, at fixed volume of the unit cell, $V_c$,

$$E_{\text{cell}}\left[\varphi_\alpha^{(p)}, \varphi_\beta^{(n)}\right] = \langle \Phi_{NZ}\Phi_e | H_{\text{cell}}^{\text{eff}} | \Phi_{NZ}\Phi_e \rangle = \text{minimum} , \qquad (7)$$

where $\Phi_e$ is the plane-wave Slater determinant for the ultrarelativistic electron gas of constant density $n_e = Z/V_c$. Minimization, performed at fixed $V_c$, corresponds to fixed average neutron and proton densities, $n_n = N/V_c$, $n_p = Z/V_c = n_N Z$. For details concerning actual calculational procedure the reader is referred to the original paper of Negele and Vautherin [62].

Having calculated the HF orbitals, $\varphi_\beta^{(n)}$, $\varphi_\alpha^{(p)}$, one determines the minimum (ground state) value of $E_{\text{cell}}(N, Z)$, filling lowest $N$ neutron states and $Z$ proton states. Then, the absolute ground state configuration is found by minimizing $E_{\text{cell}}(N, Z)$ at fixed $A = N + Z$. Let us notice, that $\alpha_Z$ and $\beta_N$ correspond then to the "Fermi level" for protons and neutrons, respectively. In terms of the single-nucleon orbitals, the neutron drip point corresponds to the threshold density, at which neutron Fermi level becomes *unbound*, i.e., $\phi_{\beta_N}^{(n)}$ extends over the whole unit cell. Even at highest densities considered, no proton drip occurs. As the matter density increases, the neutron gas density outside nuclei increases, and the density of protons within nuclei decreases. As Negele and Vautherin [62] find, at $\rho \gtrsim 8 \times 10^{13}$ g cm$^{-3}$ the differences in energy between various local minima of $E_{\text{cell}}(N, Z)$ become so small, that it is not meaningful to proceed with their calculational scheme to higher density.

One of the most interesting results of Negele and Vautherin [62] was prediction of strong shell effect for protons: it is visualized by persistence of $Z = 40$ (closed proton subshell) from neutron drip point to about $3 \times 10^{12}$ g cm$^{-3}$, and $Z = 50$ (closed major proton shell) for $3 \times 10^{12}$ g cm$^{-3} \lesssim \rho \lesssim 3 \times 10^{13}$ g cm$^{-3}$, Fig.2.

Alas, apart from the work of Negele and Vautherin [62], no other attempt of a Hartree-Fock calculation of nuclear structures in the ground state of the inner crust was carried out. This might result from an unsolved problem of correct treatment of the boundary conditions at the unit cell edge, accompanied by difficulties in finding absolute minimum of the Hartree-Fock energy functional. These problems did not prevent carrying-out Hartree-Fock unit-cell calculations of nuclear structures in hot dense matter, relevant for the equation of state in the gravitational collapse of stellar cores [14], [91]. This seems to be due to the fact that thermal averaging at $k_B T \gtrsim 1$ MeV, as well as much less important role of the nucleon gas outside nuclei, in the relevant case of entropy per nucleon $\sim 1-2$ $k_B$, makes the calculation less dependent on a somewhat arbitrary choice of the boundary condition at the unit cell edge.

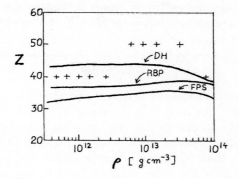

**Fig. 2.** Numbers of protons per nucleus in the ground state of the inner crust, obtained by various authors. Solid lines: RBP - Ravenhall et al. [76]; FPS - as quoted in [72]; DH - Douchin and Haensel [30]. Crosses - Negele and Vautherin [62].

## 4.2 Extended Thomas-Fermi calculations

Above neutron drip, the number of nucleons in the unit cell grows rapidly with increasing density. At $\sim 10^{13}$ g cm$^{-3}$ one has $A_{\rm cell} \sim 1000$ [62], and implementation of the self-consistent HF scheme requires an enormous amount of work and computer time. Large size of nuclei suggests further simplifications of the HF model via *semiclassical approximation*, in which relevant quantities are represented "on the average", with quantum fluctuations (oscillations) being averaged out. The energy of a unit cell is a sum of a nuclear energy $E_{\rm N}$ (which includes nucleon rest energies), Coulomb energy $E_{\rm coul}$, and energy of electron gas, $E_e$. In the Extended TF approximation (see, e.g., [79],[16]), nuclear energy of a unit cell is expressed in terms of energy density functional $\mathcal{E}_{\rm N}$ as

$$E_{\rm N} = \int_{\rm cell} \{\mathcal{E}_{\rm N}\left[n_n(\mathbf{r}), n_p(\mathbf{r}), \nabla n_n(\mathbf{r}), \nabla n_p(\mathbf{r})\right]\} {\rm d}^3 r$$
$$+ \int_{\rm cell} \left[m_n c^2 n_n(\mathbf{r}) + m_p c^2 n_p(\mathbf{r})\right] {\rm d}^3 r \ . \tag{8}$$

The nuclear energy density functional has a non-local character, as it depends on the density gradients. For the ETF approximation to be valid, characteristic length over which density $n_n$ or $n_p$ changes significantly has to be much larger than the mean internucleon distance. One can then restrict to keeping only quadratic gradient terms in $\mathcal{E}_{\rm N}$. To a very good approximation, electron gas is uniform, with $n_e = Z/V_{\rm c}$, and therefore Coulomb energy of a unit cell is given by

$$E_{\rm Coul} = \frac{1}{2} \int_{\rm cell} e\left[n_p(\mathbf{r}) - n_e\right] \phi(\mathbf{r}) {\rm d}^3 r \ , \tag{9}$$

where $\phi(\mathbf{r})$ is the electrostatic potential, to be calculated from the Poisons equation,

$$\nabla^2 \phi(\mathbf{r}) = -4\pi e \left[n_p(\mathbf{r}) - n_e\right] \ , \tag{10}$$

and $e$ is the elementary charge. To calculate the ground state at a given $n_b$, one has to find $n_n(\mathbf{r})$, $n_p(\mathbf{r})$, which minimize $E_{\text{cell}}/V_{\text{c}}$, under the constraints

$$V_c n_b = \int_{\text{cell}} [n_n(\mathbf{r}) + n_p(\mathbf{r})] \, \mathrm{d}^3 r \;, \qquad \int_{\text{cell}} [n_p(\mathbf{r}) - n_e] \, \mathrm{d}^3 r = 0 \;. \qquad (11)$$

The unit cell is approximated by a sphere of radius $r_c = (3V_c/4\pi)^{1/3}$, which simplifies the problem due to spherical symmetry. The boundary conditions are such that far from the nuclear surface (i.e., from the neutron gas–nuclear matter interface) nucleon densities are uniform. This requires that nuclear radius be significantly smaller than $r_c$. The ETF method was first applied to the calculation of the structure of the inner crust by Buchler and Barkat ([18], see also [19] and [2]). In the 1980s the main effort was concentrated on the case of dense and hot matter, relevant for the gravitational collapse of stellar cores and for modeling of type II supernova explosions An exception from this rule is the paper of Ogasawara and Sato [63], who devote Sect.3.1 of their paper to the case of cold catalyzed matter. Their calculational scheme was similar to that used by Barkat et al. [2]. However, Ogasawara and Sato used different models of potential energy of asymmetric nuclear matter. They obtained neutron drip density $3-4 \times 10^{11}$ g cm$^{-3}$ and the values of $Z = 35 - 45$, higher than those of [2]; this difference resulted from different nuclear energy functional models. Results of Ogasawara and Sato were in good agreement with HF results of Negele and Vautherin [62].

Significant progress in the 1980s was achieved in the calculations of the properties of asymmetric nuclear matter and pure neutron matter with realistic bare nucleon-nucleon interactions (see, [33],[90]). On the other hand, calculations using the HF method and its semi-classical simplifications, with new models of effective nucleon-nucleon interaction, reached a high degree of precision in reproducing the properties of atomic nuclei. The ETF calculation in the 1990s focused on detailed investigation of the possibility of appearance of non-spherical nuclei in the densest layer of the crust, which will be described in detail in Sect.5. Oyamatsu [67] studied the ground state of the inner crust within the ETF approximation, with four different energy density functionals $\mathcal{E}_{\text{N}}$. These functionals were constructed so as to reproduce gross properties of laboratory nuclei, and to be consistent with the equation of state of pure neutron matter obtained by Friedman and Pandharipande [33] for realistic bare nucleon-nucleon interaction. Oyamatsu performed explicit minimization of the TF energy functional within a family of *parameterized* $n_n$ and $n_p$ density profiles. Between neutron drip, which takes place at $4 \times 10^{11}$ g cm$^{-3}$, and $10^{14}$ g cm$^{-3}$, Oyamatsu obtained for all four of his models $Z \simeq 40$, in good agreement with HF calculations of Negele and Vautherin [62].

Simultaneously with application of the relativistic Brueckner-Hartree-Fock (RBHF) approach to neutron star matter at supranuclear densities, semi-classical ETF approximation based on the RBHF model was applied for the calculation of the properties of the inner crust. Starting from the RBHF results for bulk asymmetric nuclear matter, Sumiyoshi et al. [87] applied the ETF scheme of Oyamatsu [67], with his parameterization of the nucleon density profiles. The quadratic gradient term in the energy density functional was determined by fitting the properties of terrestrial nuclei. They found neutron drip at $2.4 \times 10^{11}$ g cm$^{-3}$. Their values of $Z$ in the inner crust were systematically lower than those obtained in older work, with $Z \simeq 35$ near neutron drip, decreasing down to about 20 at $\rho \simeq 10^{14}$ g cm$^{-3}$. This may be attributed to the fact that Coulomb energy of nuclei in their model is relatively large, due to smaller nuclear radii. It should be mentioned that their RBHF value of saturation density for symmetric nuclear matter, 0.185 fm$^{-3}$, was significantly larger than the experimental value of 0.16 fm$^{-3}$ and this may explain compactness of their nuclei.

Relativistic Hartree approximation of the ground-state energy functional, calculated in the non-linear relativistic mean field model of dense nucleon matter, can be simplified using the relativistic extended Thomas-Fermi (RETF) approximation proposed by Centelles et al. [24],[25]. In the RETF approximation, one gets $\mathcal{E}_N$ functional containing terms quadratic in $\nabla n_n$, $\nabla n_p$, which are completely determined within the model. The RETF model was applied by Cheng et al. [28] for the calculations of the structure of the ground state of the inner crust, starting from the Boguta and Bodmer [13] nonlinear $\sigma - \omega - \rho$ model Lagrangian. Three sets of the Lagrangian parameters were used in actual calculations. Cheng et al. [28] solved the Euler-Lagrange equations for $n_n(r)$, $n_p(r)$ in the spherical unit cell exactly. They did not give explicitly values of $Z$ as function of the matter density. However, analysis of their figures and tables leads to conclusion that, similarly as Sumiyoshi et al. [87], they get nuclei which are relatively small, and their values of $Z$ are significantly lower than those obtained in non-relativistic calculations.

## 4.3 Compressible liquid drop model

The nature of the HF and ETF calculations does not permit to study separate physical contributions and effects, whose interplay leads to a particular structure of the inner crust. The compressible liquid drop model (CLDM) enables one to separate various terms in $E_{\text{cell}}$, so that their role and mutual interaction can be identified.

There are also practical advantages of using the CLDM. On the one hand, it can be considered as suitable and economical parameterization of results of microscopic calculations of the HF or ETF type. On the other hand, CLDM model avoids technical complications related to the choice of boundary conditions at the edge of the unit cell, plaguing HF approach at highest inner crust densities. Finally, CLDM description allows for thermodynamically consistent and systematic treatment of bulk and finite-size effects, and is particularly convenient for studying phase transitions between different phases of neutron star matter (see Sect.5). In particular, CLDM treats two major effects of the outer neutron gas on nuclei: 1) decrease of the surface tension with growing density, due to increasing similarity of nucleon matter inside and outside nuclei; 2) compression of nuclear matter within nuclei due to the pressure of outer neutron gas.

However, we should stress that all these attractive features of the CLDM model are valid only when finite-size contributions were calculated, in a microscopic HF or ETF approach, from the same effective nucleon hamiltonian as that used for the calculation of the bulk (volume) terms. In particular, only in such a case decrease of the surface tension due to the presence of the outer neutron gas is treated in a correct way.

Within the CLDM, one divides nuclear contribution $E_{\rm N}$ (which excludes Coulomb interactions) to $E_{\rm cell}$ into bulk, $E_{\rm N,bulk}$ and surface, $E_{\rm N,surf}$, terms. Coulomb contributions to the energy of a unit cell are denoted by $E_{\rm Coul}$. Electrons are assumed to form an uniform Fermi gas, and yield the rest and kinetic energy contribution, denoted by $E_e$. Total energy of a unit cell is therefore given by

$$E_{\rm cell} = E_{\rm N,bulk} + E_{\rm N,surf} + E_{\rm Coul} + E_e \ . \quad (12)$$

Here, $E_{\rm N,bulk}$ is the bulk contribution of nucleons, which does not depend on the size and shape of nuclear structures. However, both $E_{\rm N,surf}$ and $E_{\rm Coul}$, which vanish for uniform $npe$ matter, do depend on the size and shape of nuclear structures, formed by denser nuclear matter and the less dense neutron gas. From the point of view of thermodynamics, nucleons are distributed between

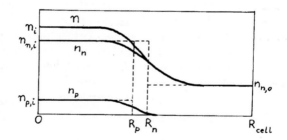

**Fig. 3.** Proton and neutron number density distributions within a spherical unit cell in the inner neutron star crust. Solid lines are actual density profiles, dashed lines correspond to those of the Compressible Liquid Drop Model. $R_n$, $R_p$ are equivalent neutron and proton radii, denoted in the text as $r_n$, $r_p$.

three subsystems: denser nucleon fluid, which will be labeled by "i", less dense neutron fluid, labeled by "o", and nuclear surface (i.e., "i-o" phase interface), labeled by "s". One requires mechanical and chemical equilibrium between these subsystems. Far from the nuclear surface, nucleon densities are constant, and equal to $n_{n,{\rm i}}$, $n_{p,{\rm i}}$ in the denser "i" phase and $n_{n,{\rm o}}$ in the less dense neutron gas. The definition of the surface term is subject to an ambiguity. In the case of spherical nuclei in the inner crust it is convenient to identify it with a sphere of *reference proton radius* $r_p$, such that $\frac{4\pi}{3}r_p^3 n_{p,{\rm i}}$ is equal to the actual $Z$. Such a definition is convenient because of the presence of the Coulomb term in $E_{\rm cell}$, which involves solely proton density distribution. Similarly, neutron radius $r_n$ is defined by $\frac{4\pi}{3}[r_n^3(n_{n,\mathrm{i}} - n_{n,\mathrm{o}}) + r_c^3 n_{n,\mathrm{o}}] = N_{\mathrm{cell}}$ (see Fig.3). In view of a significant neutron excess, the interface includes *neutron skin*, of thickness $s_n = r_n - r_p$, formed by neutrons adsorbed onto the nuclear matter surface.

The nuclear bulk energy of a cell is

$$E_{\mathrm{N,bulk}} = V_c \left[ w \mathcal{E}_{\mathrm{N,i}} + (1-w) \mathcal{E}_{n,\mathrm{o}} \right], \tag{13}$$

where the volume of the cell $V_c = 4\pi r_p^3/3$, $\mathcal{E}_{\mathrm{N,i}}$ is energy density of nuclear matter far from nuclear surface, and $\mathcal{E}_{n,\mathrm{o}}$ is the corresponding quantity for outer neutron gas. The volume fraction occupied by the phase "i" is $w = V_p/V_c = (r_p/r_c)^3$.

The nuclear surface energy term, $E_{\mathrm{N,surf}}$, gives the contribution of the interface between neutron gas and nuclear matter; it includes contribution of neutron skin ([52],[72],[54]),

$$E_{\mathrm{N,surf}} = \mathcal{A}\sigma + N_{\mathrm{s}}\mu_{n,\mathrm{s}}, \tag{14}$$

where $\sigma$ is the *surface thermodynamic potential per unit area*, $\mathcal{A}$ is the area of nuclear surface (in the case of spherical nuclei $\mathcal{A} = 4\pi r_p^2$), $N_{\mathrm{s}}$ is the number of neutrons in neutron skin, and $\mu_{n,\mathrm{s}}$ is the chemical potential of the neutrons adsorbed onto reference proton surface. In the simplest approximation, in which curvature contributions to $E_{\mathrm{N,surf}}$, proportional to $\mathcal{A}/r_p$, are neglected, $\sigma$ is approximated by the *surface tension* $\sigma_{\mathrm{s}}$, and $N_{\mathrm{s}} \simeq (n_{n,\mathrm{i}} - n_{n,\mathrm{o}}) s_n \mathcal{A}$. More precise expression for $E_{\mathrm{N,surf}}$ may be obtained including curvature corrections, which take into account the curvature of the nuclear surface ([54],[29]). In view of the possibility of nuclear structures with infinite volumes (see Sect.5) it is convenient to introduce contribution of neutrons in neutron skin to the total (overall) nucleon density, $n_{\mathrm{s}} = N_{\mathrm{s}}/V_c$ [54].

In order to calculate $E_{\mathrm{Coul}}$, one uses the Winger-Seine approximation. Neglecting diffuseness of the proton distribution one gets

$$E_{\mathrm{Coul}} = \frac{16}{15}\pi^2 (n_{p,\mathrm{i}}e)^2 r_p^5 f_3(w), \quad f_3(w) = \left(1 - \frac{3}{2}w^{1/3} + \frac{1}{2}w\right). \tag{15}$$

At $T=0$, equilibrium can be determined by minimizing total energy density, $\mathcal{E} = E_{\mathrm{cell}}/V_c$, at fixed value of $n_b$. The quantity $\mathcal{E}$ is a function of seven independent variables. A convenient set of variables is: $n_{n,\mathrm{i}}$, $n_{p,\mathrm{i}}$, $n_{\mathrm{s}}$, $n_{n,\mathrm{o}}$, $r_p$, $r_n$, $r_c$; in this way all independent variables will be finite even in the case of infinite nuclear structures, considered in Sect. 5. Imposing fixed $n_b$, and requiring charge neutrality of the cell, we reduce the number of independent variables to five. Therefore, there will be five conditions of equilibrium resulting from the stationarity of $\mathcal{E}$ with respect to variations of thermodynamic variables. Each of these conditions has well defined physical meaning. First condition requires that the neutron chemical potential in the nucleus and in the outer neutron gas be the same. Neglecting curvature corrections, it implies equality of neutron chemical potentials in the bulk phases of nucleon matter, $\mu_{n,\mathrm{i}}^{\mathrm{bulk}} = \mu_{n,\mathrm{o}}^{\mathrm{bulk}}$. Second equation results from minimization with respect to the number of protons, and yields the beta equilibrium condition between neutrons, protons, and electrons. Neglecting curvature corrections, it reads

$$\mu_{n,i}^{\text{bulk}} - \mu_{p,i}^{\text{bulk}} - \mu_e = \frac{8\pi}{5} e^2 n_{p,i} r_p^2 f_3(w) \ . \tag{16}$$

We also need a condition on the number of surface neutrons. It results from the requirement of stationarity with respect to transfer of a neutron from the nucleus interior to the surface, all other particle numbers being fixed. Neglecting curvature corrections, this condition implies that the chemical potential of surface neutrons is equal to the bulk chemical potentials in both phases, $\mu_{n,s} = \mu_{n,i}^{\text{bulk}} = \mu_{n,o}^{\text{bulk}}$.

To these three conditions, expressing chemical equilibrium within the system, we have to add two equations corresponding to mechanical equilibrium. Condition number four results from the requirement of stationarity with respect to change of $r_p$, and expresses the equalities of pressures inside and outside the nucleus. Neglecting curvature corrections, condition number four reads

$$P_i^{\text{bulk}} - P_o^{\text{bulk}} = \frac{2\sigma_s}{r_p} - \frac{4\pi}{15} e^2 n_{p,i}^2 r_p^2 (1-w) \ , \tag{17}$$

where $P_j^{\text{bulk}} = n_j^2 \partial (\mathcal{E}_j^{\text{bulk}}/n_j)/\partial n_j$.

The last fifth equation determines the equilibrium size of the cell. It results from the condition of stationarity with respect to the variation spatial scale of the cell, while $w$ and all densities including $n_s$ are kept constant. Notice, that because $w$ is kept constant, this condition involves only the finite-size terms in $E_{\text{cell}}$. Within our approximation (no curvature corrections), the last condition can be written as

$$E_{\text{N,surf}} = 2E_{\text{Coul}} \ . \tag{18}$$

This the "virial theorem" of the simplified Compressible Liquid Drop Model with no curvature corrections (Baym, Bethe, and Pethick [3], hereafter referred to as BBP), which enables one to express $r_p$ in terms of remaining variables. Generalization of "virial theorem" to the case of nonstandard nuclear shapes will be discussed in Sect.5.

Let us write an explicit expression for nuclear component of the energy density, neglecting for simplicity curvature corrections in $E_{\text{N,surf}}$. Both surface tension, $\sigma \simeq \sigma_s$, and thickness of neutron skin, $s_n = r_n - r_p$, are calculated under the condition of thermodynamic and mechanical equilibrium of the semi-infinite "i" and "o" phases, separated by a plane interface. Therefore, $\sigma$ and $s_n$ depend on only one thermodynamic variable, e.g., proton fraction in the bulk "i" phase, $x_i = n_{p,i}/n_i$, where $n_i = n_{n,i} + n_{p,i}$. The formula for the energy density $\mathcal{E}_N$ reads then

$$\mathcal{E}_N = w\mathcal{E}_{N,i} + (1-w)\mathcal{E}_{n,o} + \frac{3w}{r_p}[\sigma_s + (n_{n,i} - n_{n,o})s_n\mu_n] \ . \tag{19}$$

Let us remind, that in equilibrium chemical potential of neutrons adsorbed onto nuclear surface is equal to the common value of $\mu_n$ in both bulk phases. Possibility of non-spherical shapes of nuclei will be considered in the next Section.

Historically, first CLDM calculations of the structure of the inner crust were performed in the classical paper of Baym, Bethe, and Pethick [3](BBP). The BBP paper formulated the foundations of the subsequent CLDM calculations of the structure of the inner neutron star crust. Unfortunately, BBP used oversimplified estimates of the reduction of $\sigma$ with increasing density, based on dimensional arguments; this resulted in rapid increase of $Z$ with increasing density, corrected in subsequent calculations [76]. Most recent CLDM calculation of the ground state structure of the inner crust were performed by Lorenz [54] and by Douchin and Haensel ([29], [31],[30]). These calculations were based of effective nucleon-nucleon interactions, which were particularly suitable for strongly asymmetric nuclear systems. Lorenz used FPS model (**F**riedman **P**andharipande **S**kyrme [70]), consistent with results of many-body calculations of dense asymmetric nuclear matter with realistic bare nucleon-nucleon interaction and a phenomenological three-nucleon force, performed by Friedman and Pandharipande [33]. Douchin and Haensel used the SLy (**S**kyrme **Ly**on, [26],[27]) effective forces, adjusted to the properties of neutron-rich nuclei, and adjusted also, at $\rho > \rho_0$, to the results of many-body dense asymmetric nuclear matter calculations of Wiringa et al. [90], which were based on bare two-nucleon interaction AV14 and phenomenological UVII three-nucleon interaction.

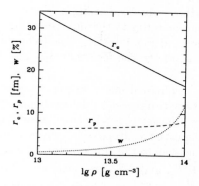

**Fig. 4.** Radius of spherical unit (Winger-Seine) cell, $r_c$, the proton radius of spherical nuclei, $r_p$, and fraction of volume filled by protons, $w$ (in percent), versus matter density $\rho$. Based on Douchin and Haensel [30].

In what follows, we will illustrate CLDM results for spherical nuclei by those of Douchin and Haensel [30]. Geometrical parameters characterizing nuclei in the inner crust, up to $10^{14}$ g cm$^{-3}$, are shown in Fig. 4. Here, $w$ is the fraction of volume occupied by nuclear matter (with our definition of nuclear matter volume equal to that occupied by protons).

More detailed information on neutron-rich nuclei, present in the ground state of the inner crust at $\rho < 10^{14}$ g cm$^{-3}$, can be found in Fig.5. Number of nucleons in a nucleus, $A$, grows monotonically, and reaches about 300 at $10^{14}$ g cm$^{-3}$, where $A_{\rm cell} \simeq 1000$. However, the number of protons changes rather weakly,

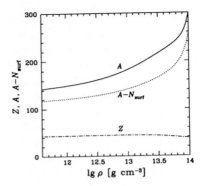

**Fig. 5.** Mass number of spherical nuclei, $A$, and their proton number, $Z$, versus average matter density $\rho$. Dotted line corresponds to number of nucleons after deducing neutrons belonging to neutron skin ($N_{\text{surf}}$ corresponds to $N_{\text{s}}$ in the text). Based on Douchin and Haensel [30].

increasing slightly from $Z \simeq 40$ near neutron drip to somewhat above forty at $10^{13}$ g cm$^{-3}$, and then decreasing to $Z \simeq 40$ at $10^{14}$ g cm$^{-3}$. Results for $Z$ of spherical nuclei are quite similar to those obtained in ([67],[76]), but are somewhat higher than those obtained using a relativistic mean-field model in ([87],[28]). An interesting quantity is the number of neutrons forming neutron skin, $N_{\text{s}}$. As one can see from Fig.5, for $\rho \gtrsim \frac{1}{3}\rho_0$ the value of $N_{\text{s}}$ decreases with increasing density; this is due to the fact that $n_{n,\text{i}}$ and $n_{n,\text{o}}$ become more and more alike.

For $\rho \simeq 10^{14}$ g cm$^{-3}$, spherical nuclei are very heavy, $A \simeq 300$, and doubts concerning their stability with respect to deformation and fission arise.

Originally, the Bohr-Wheeler condition for fission has been derived for isolated nuclei, which were treated as drops of incompressible, charged nuclear matter (see, e.g., [74]). Let us denote the Coulomb and surface energy of a spherical nucleus *in vacuum* by $E_{\text{Coul}}^{(0)}$ and $E_{\text{surf}}^{(0)}$, respectively. The Bohr-Wheeler conditions states that for $E_{\text{Coul}}^{(0)} \geq 2 E_{\text{surf}}^{(0)}$ a spherical nucleus is unstable with respect to small quadrupolar deformations, and is therefore expected to deform spontaneously and fission into smaller drops (fragments). In the case of nuclei in the neutron-star crust one has to include corrections to the Bohr-Wheeler condition, resulting from the presence of electron background and of other nuclei. Such corrections were calculated by Brandt (1985; quoted in [72]). The leading corrections were found to be of the order of $(r_p/r_c)^3$. This is to be contrasted with corrections in $E_{\text{Coul}}$, where the leading correction term is linear in $r_p/r_c$ [see (15)]. Keeping only leading correction to Coulomb energy, one can rewrite the equilibrium condition, (18), in an approximate form

$$E_{\text{surf}} \simeq 2 E_{\text{Coul}}^{(0)} \left(1 - \frac{3}{2}\frac{r_p}{r_c}\right) . \qquad (20)$$

Within the linear approximation, $E_{\rm Coul}^{(0)}$ is the Coulomb energy of the the nucleus itself (self-energy). The quantity $E_{\rm Coul}^{(0)}$ is larger than the actual $E_{\rm Coul}$, which is equal to the half of $E_{\rm surf}$. As the density increases, $E_{\rm Coul}^{(0)}$ can become sufficiently large for the Bohr-Wheeler condition to be satisfied. Within linear approximation, this would happen for $r_p/r_c > 1/2$, i.e., when nuclei fill more than $(1/2)^3 = 1/8$ of space. As one sees from Fig.4, this does not happen for spherical nuclei at $\rho < 10^{14}$ g cm$^{-3}$ for the particular Douchin and Haensel [30] model of the inner crust. However, at $10^{14}$ g cm$^{-3}$ the value of $w = (r_p/r_c)^3$ is rather close to the critical value of $1/8$.

## 5 Ground state of the matter in the bottom layers of the crust

For $\rho \lesssim 10^{14}$ g cm$^{-3}$ ground state of the inner crust contains spherical nuclei; as we will see in the present section, such a structure is stable with respect to transition into different nuclear shapes, or into a uniform $npe$ matter. Of course, as long as $r_p \ll r_c$, we *expect* nuclei in the ground state of dense cold matter to be spherical (or quasispherical). This is particularly clear within the CLDM, where for $r_p \ll r_c$ it is the spherical shape which minimizes the shape-dependent (finite-size) contribution $E_{\rm N, surf} + E_{\rm Coul}$. However, the situation at $\rho \gtrsim 10^{14}$ g cm$^{-3}$, where $r_p/r_c \gtrsim 0.5$, is far from being obvious.

In the present section we will study, in the ground state approximation, the structure and equation of state of the inner crust at $\rho \gtrsim 10^{14}$ g cm$^{-3}$. In particular, we will discuss possible unusual (exotic) shapes of nuclei present in the bottom layers of the crust. We will also study transition between the crust and the liquid neutron star core.

### 5.1 Unusual nuclear shapes

Long ago it has been pointed out that when the fraction of volume occupied by nuclear matter exceeds 50%, nuclei will turn "inside-out", and spherical bubbles of neutron gas in nuclear matter will become energetically preferred (BBP). [2] Generally, in the process of minimization of energy, nuclear shape has to be treated as a thermodynamic variable: the actual shape of nuclei in the ground state of the bottom layer of the inner crust has to correspond to the minimum of $\mathcal{E}$ at a given $n_b$. Historically, first studies along these lines were connected with structure of matter in gravitational collapse of massive stellar cores. Calculations performed within the CLDM for dense hot matter, with $T > 10^{10}$ K and entropy per nucleon $1 - 2$ $k_{\rm B}$ indicated, that before the transition into uniform plasma, matter undergoes a series of phase transitions, accompanied by a change of nuclear shape [77]. These authors considered a basic set of spherical, cylindrical, and planar geometries, corresponding to *dimensionality* $d = 3, 2, 1$. For

---

[2] This result of BBP was obtained within the Liquid Drop Model, neglecting curvature contribution to the surface thermodynamical potential.

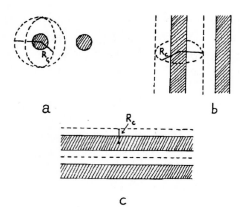

**Fig. 6.** Unit cells for a set of nuclear shapes (spheres, rods, plates) in the inner crust. The radius of the unit cell is denoted by $R_c$ (notation in the text: $r_c$). Hatched regions correspond to nuclear matter, blank to neutron gas. In the case of the "bubbular phase" (tubes, spherical bubbles) one has to exchange the roles of the blank and hatched regions.

each dimensionality, they restricted to simplest shapes with a single curvature radius (i.e., maximal symmetry). So, for $d = 3$ Ravenhall et al. [77] considered spherical nuclei in nucleon gas and spherical bubbles in denser nuclear matter, referred to as 3N and 3B, respectively. In the case of $d = 2$, nuclear structures were cylindrical nuclei (rods, 2N), and cylindrical holes in nuclear matter, filled with nucleon gas (tubes, 2B). Finally, for $d = 1$ they considered parallel plates of nuclear matter separated by nucleon gas; in this case "bubbular" and "nuclear" phases coincide, and were denoted by 1N. [3] Ravenhall et al. [77] found a sequence of phase transitions 3N $\longrightarrow$ 2N $\longrightarrow$ 1N $\longrightarrow$ 2B $\longrightarrow$ 3B, which preceded transition into uniform plasma. These transitions were accompanied by increase of the fraction of volume occupied by denser (nuclear matter) phase.

One of the virtues of the CLDM is its flexibility as far as the shape of nuclei is concerned. The terms $\mathcal{E}_{\rm N,bulk}$ and $\mathcal{E}_e$ are shape independent. The surface and Coulomb terms do depend on the shape of nuclei, but can easily be calculated if one neglects the curvature corrections. In what follows we will describe the formulae for $\mathcal{E}_{\rm N,surf}$ and $\mathcal{E}_{\rm Coul}$ within this simple approximation. For the sake of completeness, we will include also previously considered case of spherical nuclei (phase 3N). Using elementary considerations, one may show that the general formula for the surface energy contribution is

$$\mathcal{E}_{\rm N,surf} = \frac{wd}{r_p} \left[ (n_{n,\rm i} - n_{n,\rm o}) \mu_n s_n + \sigma_{\rm s} \right] , \qquad (21)$$

---

[3] For obvious reasons, culinary terms are also frequently used to denote various phases. So, 3B, 2N, and 1N are referred to as *swiss cheese*, *spaghetti*, and *lasagna* phases, respectively.

where dimensionality the $d = 3$ for 3N, 3B phases, $d = 2$ for the 2N, 2B phases, and $d = 1$ for the 1N phase. The filling factor $w$ is given by a simple formula $w = (r_p/r_c)^d$.

The case of the Coulomb contribution is more complicated, but the result can be also represented by a universal expression, obtained in [77]. The calculation is based on the Winger-Seine approximation. The unit cells for the 3N, 2N and 1N phases are visualized in Fig. 6. In the case of rods, the unit cell is approximated by a cylinder, coaxial with the rod, of radius $r_c$. The number of rods per unit area of the plane perpendicular to rods is $1/(\pi r_c)^2$. In the case of plates, the boundary of the unit cell consists of two planes parallel to the nuclear matter slab, at distance $r_c$ from the slab symmetry plane. For the phases of spherical nuclei (3N), nuclear matter rods (2N) and plates (1N) one obtains then

$$\mathcal{E}_C = \frac{4\pi}{5} (n_{p,i} e r_p)^2 f_d(w) , \qquad (22)$$

where

$$f_d(w) = \frac{5}{(d+2)} \left[ \frac{1}{d-2} \left( 1 - \frac{1}{2} d w^{1-2/d} \right) + \frac{1}{2} w \right] . \qquad (23)$$

In the case of $d = 2$ (rods) one has to take the limit of $d \longrightarrow 2$, in order to get a more familiar expression

$$f_2(w) = \frac{5}{8} \left( \ln \frac{1}{w} - 1 + w \right) . \qquad (24)$$

These formulae hold also in the case of the neutron gas tubes (2B) and neutron gas bubbles (3B) but one has then to replace $w$ by $1 - w$. Of course, $r_p$ is then the radius of the tubes or the bubbles.

The virial theorem, which states that in equilibrium $\mathcal{E}_{N,\text{surf}} = 2\mathcal{E}_{\text{Coul}}$, remains valid for any phase. It is a consequence of scaling of the Coulomb and surface energy density with respect to the value of $r_p$ ($E_{N,\text{surf}} \propto r_p^{-1}$, $E_C \propto r_p^2$), and simultaneous invariance in the case of $d = 2$ and $d = 1$ with respect to the change of the scale in the remaining one and two dimensions. In the case of bubbular phases (bubbles, tubes), one has to replace $w$ by $1 - w$.

Beautiful simplicity of the formulae is lost when one introduces "curvature corrections" to the finite-size terms. In the case of the surface terms, they result from the fact that the energy of the nuclear surface depends on its curvature, which in the case of the five nuclear shapes under consideration is given by $\kappa = (d-1)/r_p$ for the phases 3N, 2N and $\kappa = -(d-1)/r_p$ for the 3B, 2B ones, respectively. Surface thermodynamic potential, calculated including lowest order curvature correction, is then given by $\sigma = \sigma_s + \kappa \sigma_c$. It should be stressed, that in contrast to *surface tension* $\sigma_s$, the *curvature tension* $\sigma_c$ does depend on the choice of the "reference surface", which in our case is taken at $r = r_p$ (see, e.g., [48], [30]). In the case of the Coulomb energy, curvature corrections appear when we include the diffuseness of the proton surface. These corrections were studied in detail by Lorenz [54](see also [29]).

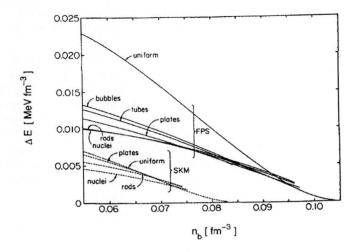

**Fig. 7.** Energy density of a given phase of inner-crust matter minus that of the bulk two-phase nuclear matter-neutron gas-electron gas system, as a function of the average baryon density $n_b$. Label "uniform" corresponds to the case of the uniform $npe$ matter. Calculations performed for the FPS and SkM effective nucleon-nucleon interactions. After [55].

First detailed calculations of the structure of the inner crust at $\rho \gtrsim 10^{14}$ g cm$^{-3}$, performed within the CLDM by Lorenz et al. [55], indicated that the presence or absence of unusual nuclear shapes before transition to uniform $npe$ matter depends on the assumed model of effective nucleon-nucleon interaction. For the FPS model of effective N-N interaction, they found a sequence of 3N ⟶ 2N ⟶ 1N ⟶ 2B ⟶ 3B phase transition, which started at 0.064 fm$^{-3}$ ≃ $\frac{1}{3}n_0$ ($1.1 \times 10^{14}$ g cm$^{-3}$), and ended at $n_{\text{edge}} = 0.096$ fm$^{-3}$ ($\rho_{\text{edge}} = 1.6 \times 10^{14}$ g cm$^{-3}$) with a transition from the 3B phase to uniform $npe$ matter. All phase transitions were very weakly first-order, with relative density jump below 1%. It should be stressed that in the relevant density region the differences between $\mathcal{E}(n_b)$ for various shapes is very small and amounts typically to less than 0.001 MeV/fm$^3$. This is to be compared with $\mathcal{E}(\text{crust; shape}) - \mathcal{E}(\text{uniform}) \simeq 0.01 - 0.02$ MeV/fm$^3$ (see Fig.7).

As Lorenz et al. [55] have shown, the very presence of unusual shapes depends on the assumed model of $v_{\text{NN}}^{\text{eff}}$. In the case of the SkM force (used by Bonche and Vautherin [14],[15] in their dense and hot matter studies) spherical nuclei were energetically preferred down to the bottom edge of the crust, found at significantly lower density $n_{\text{edge}} = 0.074$ fm$^{-3}$ ($\rho_{\text{edge}} = 1.2 \times 10^{14}$ g cm$^{-3}$).

Further calculations confirmed this unfortunate ambiguity, resulting from dependence on $v_{\text{NN}}^{\text{eff}}$. Using parameterized density profiles in the ETF energy density functional, Oyamatsu [67] found complete sequence of phase transitions in the density range $1.0-1.5 \times 10^{14}$ g cm$^{-3}$. Similar sequence of phase transitions was found by Sumiyoshi et al. [87], with however a much narrower range of existence of unusual nuclear shapes, $0.050-0.058$ fm$^{-3}$ ($0.83-0.97 \times 10^{14}$ g cm$^{-3}$), before final transition to uniform $npe$ matter. On the contrary, Cheng et al. [28], using unconstrained relativistic ETF approach, found that spherical nuclei persist in the ground state of the crust down to $n_{\text{edge}}$, which depending on the parameters of their relativistic $\sigma - \omega - \rho$ Lagrangian ranged from 0.058 fm$^{-3}$ to 0.073 fm$^{-3}$. Similarly, calculations performed by Douchin and Haensel [30] with SLy4 effective N-N force indicated absence of unusual nuclear shapes. They found transition to uniform $npe$ matter at $n_{\text{edge}} = 0.078$ fm$^{-3}$ ($\rho_{\text{edge}} = 1.3 \times 10^{14}$ g cm$^{-3}$).

While the presence of unusual nuclear shapes for $\rho \lesssim \rho_{\text{edge}}$ depends on effective nuclear interaction model used, some general qualitative statements, based on existing calculations, can still be made. The very phenomenon of phase transitions between various shapes results from the interplay of three quantities: finite-size (surface and Coulomb) term in $E_{\text{cell}}$, the dominating bulk energy term, and the volume fraction of the denser nucleon fluid, $w$. If finite-size terms are small, then $\rho_{\text{edge}}$ is reached at relatively low value of $w$. However, unusual (non-spherical) shapes become energetically advantageous only at sufficiently large value of $w$. Therefore, small surface tension may prohibit the appearance of unusual shapes before $\rho_{\text{edge}}$ is reached (this is the case of the SLy4 and SkM forces. It should be stressed, however, that phase transitions themselves result from very small energy differences (see Fig. 7) of energy densities: finite-size terms in the relevant density range are very small compared to $E_{\text{bulk}} = E_{\text{N,bulk}} + E_e$.

In the case of the CLDM, one should stress very important role of the curvature term in $E_{\text{N,surf}}$, which should therefore be included in any CLDM calculations of the crust-liquid core transition. As we already mentioned, introducing curvature corrections in the finite-size terms complicates the analysis of the unusual shape problem. In the absence of the curvature correction to $E_{\text{N,surf}}$, it is possible to show that the 3N⟶2N transition has to occur at $w = 0.2$ [66]. However, in the presence of the curvature correction the 3N phase can persist at larger values of $w$.

In actual CLDM or ETF calculations, the change of nuclear shape in the ground state of the inner crust is accompanied by a very small (less than one percent) density jump; it has therefore the character of a very weak first-order phase transition [55],[67],[87]. The equation of state in the region in which the nuclear shape transitions occur is obtained using Maxwell construction at the transition pressures.

The CLDM model is *par excellence* classical. Also, the ETF scheme is a semiclassical approximation to a quantum-mechanical many-body problem. As the differences of energy densities between phases with different nuclear shapes are very small, one may worry about possible importance of neglected quantum effects. In the case of terrestrial nuclear physics, there exists a systematic procedure of adding quantum (shell) corrections to the smooth liquid drop model energies of nuclei (Strutinsky method, see, e.g., [74]). Energy correction, resulting from the quantum shell effects for protons, and for various nuclear shapes, has been calculated by Oyamatsu and Yamada [68]. They found, that with inclusion of proton shell effects changes of nuclear shapes occur at higher densities than those obtained using semiclassical ETF calculation.

Another quantum effect neglected in the CLD or ETF model is pairing of nucleons. However, because of large numbers of nucleons in a unit cell, pairing contribution to the energy is negligible. Larger effects, which clearly need a careful investigation, may result from using the Winger-Seine approximation at $\rho \simeq 10^{14}$ g cm$^{-3}$.

## 5.2 Reaching the bottom edge of the crust from the denser side

The method of the determination of the bottom edge of the crust, based on the CLDM of nuclei, requires a very high precision of the calculation of the finite-size contribution term, $E_{\rm N,surf} + E_{\rm coul}$, in $E_{\rm cell}$. One has to construct a CLDM model of the ground state of the inner crust, and then find the density of the crust–liquid core transition from the condition of the thermodynamic phase equilibrium. This method requires that one uses the same nuclear hamiltonian for the crust and for the liquid core phase. It requires also very precise many–body method for the description of nuclear structures within the bottom layers of the crust, which is a rather difficult task (see Fig. 7). Luckily enough, calculations described in the previous subsection show that the crust–liquid core phase transition is *very weakly first-order* (i.e., the relative density jump at the crust-liquid core interface is very small). Therefore, one can locate the crust-core interface using completely different method, which is based on a well known technique used in the theory of phase transitions in condensed matter. This can be an independent test of precision of the CLDM calculation of $\rho_{\rm edge}$, described in the previous subsection. We will locate the edge of the crust by checking the stability of the uniform *npe* matter, starting from the higher density side where we know that the homogeneous phase is indeed stable with respect to formation of spatial inhomogeneities (BBP, Pethick et al. [71]). By lowering the density, we will eventually find the threshold density, at which the uniform *npe* matter becomes unstable for the first time. As we will see, this threshold density gives a very good approximation of the actual density of the crust edge density, $\rho_{\rm edge}$.

At a given $n_b$, the ground state of a homogeneous *npe* matter corresponds to the minimum of the energy density $\mathcal{E}(n_n, n_p, n_e) = \mathcal{E}_0$, under the constraints of fixed baryon density and electric charge neutrality, $n_p + n_n = n_b$ and $n_e = n_p$, respectively. This implies beta equilibrium between the matter constituents and ensures vanishing of the first variation of $\mathcal{E}$ due to small perturbations $\delta n_j(\mathbf{r})$ (where $j = n, p, e$) of the equilibrium solution (under the constraints of constant total nucleon number and global charge neutrality within the volume $V$ of the system). However, this does not guarantee the stability of the spatially homogeneous state of the *npe* matter, which requires that the second variation of $\mathcal{E}$ (quadratic in $\delta n_j$) be positive.

The expression for the energy functional of slightly inhomogeneous neutron-star matter can be calculated using the semi-classical ETF treatment of the kinetic and the spin-gradient terms in nucleon contribution to $\mathcal{E}$ [16]. Assuming that the spatial gradients are small, we keep only the quadratic gradient terms in the ETF expressions. This approximation is justified by the fact that characteristic wavelengths of periodic perturbations will turn out to be much larger than the internucleon distance. With these approximations, the change of the energy (per unit volume) implied by the density perturbations can be expressed, keeping only second order terms (BBP, [71]),

$$\mathcal{E} - \mathcal{E}_0 = \frac{1}{2} \int \frac{d\mathbf{q}}{(2\pi)^3} \sum_{j,k} F_{jk}(\mathbf{q}) \delta n_j(\mathbf{q}) \delta n_k(\mathbf{q})^* , \quad (25)$$

where we used the Fourier representation

$$\delta n_j(\mathbf{r}) = \int \frac{d\mathbf{q}}{(2\pi)^3} \delta n_j(\mathbf{q}) e^{i\mathbf{q}\mathbf{r}} . \quad (26)$$

The Hermitian $F_{ik}(\mathbf{q})$ matrix determines the stability of the uniform state of equilibrium of the *npe* matter with respect to the spatially periodic perturbations of wavevector $\mathbf{q}$. Due to the isotropy of the homogeneous equilibrium state of the *npe* matter, $F_{ik}$ depends only on $|\mathbf{q}| = q$. The matrix elements $F_{ik}$ are calculated from the second variation of the microscopic energy functional $\mathcal{E}[n_n, n_p, n_e, \nabla n_n, \nabla n_p, \nabla n_e]$ (BBP, [71]).

The condition for the $F_{ij}$ matrix to be positive-definite is equivalent to the requirement that the determinant of the $F_{ij}$ matrix be positive (Pethick et al. 1995). At each density $n_b$, one has thus to check whether $\det[F_{ij}(q)] > 0$. Let us start with some $n_b$, at which $\det[F_{ij}(q)] > 0$ for any $q$. By decreasing $n_b$, we find eventually a wavenumber $Q$ at which stability condition is violated for the first time; this happens at some density $n_Q$. For $n_b < n_Q$ the homogeneous state is no longer the ground state of the *npe* matter since it is unstable with respect to small periodic density modulations.

Calculations performed with several effective nuclear Hamiltonians indicate that $n_Q \simeq n_{\text{edge}}$, within a percent or better [71],[30]. For the ETF approximation to be correct, the value of the characteristic wavelength of critical density perturbations, $\lambda_Q = 2\pi/Q$, must be significantly larger than the mean internucleon distance. The critical wavenumbers $Q$ are typically $\sim 0.3$ fm$^{-1}$. Therefore, despite a small proton fraction (about 3–4% at $n_Q$), $\lambda_Q \sim 20$ fm is typically four times higher than the mean distance between protons $r_{pp} = (4\pi n_p/3)^{-1/3}$; for neutrons this ratio is typically about eight.

The instability at $n_Q$ signals a phase transition with a loss of translational symmetry of the *npe* matter, and appearance of nuclear structures. The agreement of $n_Q$ and $n_{\text{edge}}$ is a good test of the precision of determination of $n_{\text{edge}}$. It implies also that the spherical unit cell approximation for 3N or 3B phases is valid even close to $\rho_{\text{edge}}$. This agreement means also that restriction to linear curvature correction in $\sigma$ within the CLDM is sufficiently precise. Finally, it is a convincing argument for the validity of the CLDM at very large neutron excess.

## 6 Formation of accreted crust and crustal non-equilibrium processes

While a newly born neutron star is clearly made of hot matter in nuclear equilibrium, its subsequent evolution can lead to formation of regions in which matter is far from it. Such a situation may take place in the neutron star crust, where the reshuffling of nucleons necessary for the formation of large nuclei characteristic of cold catalyzed matter may be prohibited due to the high Coulomb barriers. This is the case of an old accreting neutron star. For the accretion rate of the order of $10^{-10} M_\odot/$y typical temperature in the neutron star interior does not exceed $10^8$ K [34]. Let us consider standard scenario connected with phenomenon of X-ray bursts. Explosive burning of the helium layer leads to formation of matter consisting mainly of $^{56}$Ni, which transforms into $^{56}$Fe. The growing layer of processed accreted matter pushes down the original crust. The original catalyzed (ground state) outer crust, which consisted of nuclei embedded in electron gas, is replaced by a new, non-catalyzed one in $\sim 10^5$ y. In view of low temperature ($T \lesssim 10^8$K) the only processes which can take place in crystallized matter when it sinks inwards are: electron captures, neutron emission or absorption and, at sufficiently high density, pycnonuclear fusion. Detailed study of the processes taking place in the crust of an accreting neutron star has been done by Sato [83], who considered several scenarios with different initial composition of matter, and by Haensel and Zdunik [37] (see also Bisnovatyi-Kogan and Chechetkin [8], and references therein).

Non-catalyzed neutron star crust represents a source of energy. The energy release takes place due to the non-equilibrium processes in the crust of an accreting neutron star. Some aspects of this problem have been considered by Vartanyan and Ovakimova [89] using an unrealistic model of neutron star matter. Detailed study of non-equilibrium processes, and resulting crustal heating was presented by Haensel and Zdunik [37].

The non-equilibrium processes lead to the appearance of spherical (or more generally - quasi-spherical) surfaces, on which heat is produced at a rate proportional to accretion rate. As Haensel and Zdunik [37] have shown, the resulting total heat release in the solid crust can be larger than the original inward heat flow resulting from the steady hydrogen burning between the helium flashes [34].

### 6.1 A model of accreted neutron star crust

We assume that at a given pressure, $P$, the neutron star crust is a body-centered cubic crystal lattice of a single species of atomic nucleus $(A, Z)$, immersed in an electron gas, and, above neutron drip point, also in a neutron gas. The maximum temperature in the crust of accreting neutron star can be as high as $10^8$K [34]. Therefore, we can expect that some part of the neutron star crust will be in a liquid phase. While the transport properties of dense matter such as heat conductivity depend sensitively on whether matter is in a liquid or a crystallized phase, melting of the crust introduces only minor corrections to thermodynamic potentials. The latent heat of crystallization is of the order of less than

0.1 keV per one accreted nucleon (c.f.,[49]) and thus negligible. Generally, for $\rho \gtrsim 10^8$ g cm$^{-3}$ and $T \lesssim 10^8$ K thermal contributions to thermodynamic potentials can be safely neglected and the composition and equation of state of dense matter can be calculated using the $T = 0$ K approximation.

In the case of neutron star matter below the helium layer we have $\rho \gtrsim 10^7$ g cm$^{-3}$ and $T \lesssim 10^8$ K and we may therefore calculate all the thermodynamic potentials in the $T = 0$ K approximation. Before the pycnonuclear fusion becomes possible, the unit (W-S) cell contains a fixed number of nucleons, $A_{\text{cell}}$, equal to the mass number of the nucleus produced by explosive helium burning. In other words, the number of nuclei in an evolving neutron star matter element is then fixed.

In what follows, we will describe a scenario developed by Haensel and Zdunik [37],[38]. Before the neutron drip point $A_{\text{cell}} = A$. At given pressure the equilibrium value of $Z$ is determined from the condition that the Gibbs energy of the unit cell, (2), be minimum. Experimental values of $W_\mathcal{N}(A, Z)$ are used whenever they were available. For the nuclei for which no experimental data exist Haensel and Zdunik [37] used a theoretical compressible liquid drop model (CLDM) of Mackie and Baym [56]. A few phenomenological parameters of this model have been fitted to the experimental masses of the atomic nuclei without introducing any shell correction term. Haensel and Zdunik [37], [38] used this model in its original form, which gives the best fit to nuclear masses. Thus, the CLDM formula for $W_\mathcal{N}(A, Z)$ includes the phenomenological even - odd pairing term, which makes even-even nuclei more bound, and odd-odd nuclei less bound than the odd-even ones.

Above neutron drip point, $P > P_{\text{ND}} \equiv P(\rho_{\text{ND}})$, neutrons are present in two phases: bound in nuclei and as a neutron gas outside nuclei. In what follows, we use the formalism and notation applied previously for the determination of the neutron drip point in Sect. 3. The Gibbs energy of the W-S cell is then written in the form

$$G_{\text{cell}}(A, Z) = W_\mathcal{N}(A, Z, n_n) + W_L(n_\mathcal{N}, Z)$$
$$+[\mathcal{E}_e(n_e) + (1 - n_\mathcal{N} V_\mathcal{N}) \, \mathcal{E}_n(n_n) + P]/n_\mathcal{N} \, , \qquad (27)$$

where $\mathcal{E}_n$ is the energy density of neutron gas (including neutron rest energy) and $V_\mathcal{N}$ is the volume of the nucleus. At given $(A, Z)$ the values of $n_\mathcal{N}$, $n_e$, $n_n$ are determined from the system of three equations,

$$P = P_e(n_e) + P_L(n_\mathcal{N}, Z) + P_n(n_n) \, , \qquad n_e = Z n_\mathcal{N}$$
$$A_{\text{cell}} = A + n_n(1/n_\mathcal{N} - V_\mathcal{N}) \, , \qquad (28)$$

supplemented by the condition of mechanical equilibrium of the surface of the nucleus under the external pressure of neutron gas. This last condition, applied to the compressible liquid drop model of Mackie and Baym [56] for $W_\mathcal{N}$ which takes into account the influence of the neutron gas on the nuclear surface energy and on the nuclear radius, yields the equilibrium value of $V_\mathcal{N}$.

The model described above enables one to calculate, at a given pressure, the ground state of a matter element under an additional constraint of a fixed number of nuclei. However, we should remind that our task is to follow the ground state of a matter element as it descends deeper and deeper into the neutron star interior under the pressure of accreted matter. This process is taking place at rather low temperature. In practice, this means that matter element sits at the local minimum of $G_{\rm cell}$, and that this local ground state can change only after the corresponding energy barrier vanishes. [4] In view of the characteristic behavior of $W_N$ as a function of $A$ and $Z$, resulting from the pairing of nucleons in nuclei, this leads to non-equilibrium character of processes which change $Z$ (and $A$) during the evolution of the neutron star crust.

## 6.2 Evolution of matter element

In what follows we will study evolution of a matter element with initial density $\sim 10^8$ g cm$^{-3}$, under a gradual compression up to the density $\sim 10^{13}$ g cm$^{-3}$, the temperature of matter not exceeding significantly $10^8$ K. In order to estimate the timescales for such a compression process, let us consider a 1.4 $M_\odot$ neutron star, with a medium-stiff EOS of the liquid core. Calculation of the density profile of such a neutron star shows, that in order to compress a matter element, initially at $\rho = 10^8$ g cm$^{-3}$, to the density $\rho = 6 \times 10^{11}$ g cm$^{-3}$, which as we will see corresponds to the neutron drip point for our specific scenario, the star should accrete a mass of $3 \times 10^{-5} M_\odot$. This would take $3 \times 10^5/\dot{M}_{-10}$ years, where $\dot{M}_{-10}$ is the accretion rate in the units of $10^{-10} M_\odot/y$. Compression up to $1.2 \times 10^{13}$ g cm$^{-3}$ (this is maximum density which we will consider) would require accretion of $\sim 5 \times 10^{-4} M_\odot$ and thus would take $\sim 5 \times 10^6/\dot{M}_{-10}$ years. After such a time the whole outer part of the neutron star crust with $\rho < 1.2 \times 10^{13}$ g cm$^{-3}$ would consist of non-catalyzed matter, studied in the present section.

Let us follow the evolution of an element of matter produced in the explosive helium burning, as it undergoes compression due to accretion of matter onto stellar surface. Let us start with a pressure close to that just below the helium layer. We have there $A=56$, $Z=26$. We shall follow possible transformations taking place in the unit cell during its travel to the deep layers of the neutron star crust. For pressures corresponding to $\rho < \rho_1 = 5.852 \times 10^8$ g cm$^{-3}$ the minimum of $G_{\rm cell}$ corresponds to $^{56}$Fe. For pressure just above $P_1 = P(\rho_1)$ the minimum is obtained for $^{56}$Cr. However, direct transition $^{56}$Fe$\to ^{56}$Cr would require an extremely slow double electron capture.

In view of the extreme slowness of the $ee$ capture $2e^- + ^{56}$Fe $\longrightarrow ^{56}$Cr$ + 2\nu_e$, reaction

$$^{56}\text{Fe} + e^- \longrightarrow ^{56}\text{Mn} + \nu_e \tag{29}$$

must proceed first. With increasing $P$, the two-step electron capture reactions occur each time when the threshold for a single electron capture is reached,

---

[4] Strictly speaking, even at $T = 0$ *quantum tunneling* through energy barrier is possible. Therefore, strict condition for the possibility of leaving the local minimum is that the energy barrier becomes sufficiently low (or thin) so that the timescale for tunneling is short compared to matter element compression timescale.

according to a general scheme

$$(A, Z) + e^- \longrightarrow (A, Z-1) + \nu_e,$$
$$(A, Z-1) + e^- \longrightarrow (A, Z-2) + \nu_e. \tag{30}$$

Usually, the first step takes place very (infinitesimally) close to the threshold and therefore is accompanied by a very small (infinitesimal) energy release (quasi-equilibrium process). An exception from this rule is the case in which, due to the selection rules, first electron capture must proceed into an excited state of the daughter nucleus. This is the case of reaction (29). Notice, that because of low temperature, the nucleus undergoing an electron capture should always be considered as being in its ground state. If the daughter nucleus is produced in an excited state, then it de-excites by gamma emission before next electron capture. This leads to the heat release $Q_1 = E_{\text{exc}}$ per cell. Second electron capture proceeds always in a non-equilibrium way, because $P_2$ is significantly above the threshold pressure for the electron capture on the odd-odd $(A, Z-1)$ nucleus. Mechanical equilibrium requires that this process takes place at constant pressure, $P_2$. On the other hand, because of very high thermal conductivity of matter, resulting from the presence of degenerate electrons, and a very slow accretion rate, reactions occur at constant temperature, $T$. Thus, the total heat release per one W-S cell, accompanying second capture, (30), is given by the change of the Gibbs energy of the cell (chemical potential of the cell), $Q_2 = G_{\text{cell}}(A, Z-1) - G_{\text{cell}}(A, Z-2)$ (see, e.g., Prigogine [75]). On average, most of the released heat is radiated away by neutrinos, $E_\nu = \frac{5}{6}(\mu_e - \Delta)$, where $\Delta$ is the threshold energy for the second (non-equilibrium) electron capture [7].

The effective *deposited in matter* heat release per one unit cell is thus estimated as

$$Q_{\text{cell}} \simeq Q_1 + \frac{1}{6} Q_2. \tag{31}$$

Generally, $Q_1 \ll Q_{\text{cell}}$.

At $\rho_{\text{ND}} = 6.11 \times 10^{11}$ g cm$^{-3}$ neutrons drip out of the nuclei, which are then $^{56}$Ar. This process, occurring at constant pressure $P_{\text{ND}}$, proceeds in five steps, and is initiated by an electron capture,

$$\begin{aligned}
^{56}\text{Ar} + e^- &\longrightarrow {}^{56}\text{Cl} + \nu_e, \\
^{56}\text{Cl} &\longrightarrow {}^{55}\text{Cl} + n, \\
^{55}\text{Cl} + e^- &\longrightarrow {}^{55}\text{S} + \nu_e, \\
^{55}\text{S} &\longrightarrow {}^{54}\text{S} + n, \\
^{54}\text{S} &\longrightarrow {}^{52}\text{S} + 2n.
\end{aligned} \tag{32}$$

The whole chain of reactions (which we will call *non-equilibrium process*) can be symbolically written as $^{56}\text{Ar} \longrightarrow {}^{52}\text{S} + 4n - 2e^- + 2\nu_e$.

For $P > P_{\text{ND}}$ electron captures induce non-equilibrium neutron emissions, the general rule being that an even number of electron captures is accompanied by emission of an even total number of neutrons. When determining the path the system follows during nuclear transformations one uses a simple rule: if both electron capture and neutron emission are energetically possible, neutron emission - which is more rapid - goes first. However, in order to calculate the effective heat release we have to consider a detailed sequence of reactions, taking place at the threshold pressure for the first "trigger" reaction of electron capture.

As the element of matter moves deeper and deeper into the neutron star interior, each time the threshold density for the electron capture is crossed, a chain of electron captures and neutron emission follows. While the "trigger" reaction produces virtually no (or very little) energy release, subsequent non-equilibrium transformations lead to a significant heat production, due mainly to the downscattering of neutrons and de-excitation of nuclei. This is possible because emitted neutrons have energies well above the Fermi surface of superfluid neutron liquid.

Due to electron captures, the value of $Z$ systematically decreases. In consequence, the lowering of the Coulomb barrier for the nucleus-nucleus reaction, combined with decrease of the separation between nuclei and a simultaneous increase of the energy of the quantum zero-point vibrations around the lattice sites opens a possibility of pycnonuclear reactions (for an introduction, see [85]).

In their calculation of the pycnonuclear reaction rate per unit volume, $r_{\rm pyc}$, Haensel and Zdunik [37] used the formulae of Salpeter and Van Horn [82] (see [37] for details). The pycnonuclear timescale is defined as

$$\tau_{\rm pyc} = \frac{n_\mathcal{N}}{r_{\rm pyc}} \ . \tag{33}$$

The quantity $\tau_{\rm pyc}$ is a sensitive function of $Z$ and of the density, so that the pressure at which pycnonuclear fusion starts can be quite easily pointed out.

The electron capture on $^{40}$Mg nucleus, taking place at $\rho = 1.45 \times 10^{12}$ g cm$^{-3}$, initiates the reaction

$$^{40}{\rm Mg} \longrightarrow {}^{34}{\rm Ne} + 6n - 2e^- + 2\nu_e \ . \tag{34}$$

The subsequent pycnonuclear fusion of the $^{34}$Ne nuclei ($Z = 10$) takes place on a timescale of months, much shorter than the time needed for a significant compression due to accretion. The fusion reaction can be written symbolically as

$$^{34}{\rm Ne} + {}^{34}{\rm Ne} \longrightarrow {}^{68}{\rm Ca} \ . \tag{35}$$

After the pycnonuclear fusion has been completed, the number of nuclei is decreased by a factor of two. Further evolution of the element of matter takes place at a fixed number of nucleons in the unit cell, doubled with respect to the initial one, $A_{\rm cell} = 112$. Pycnonuclear fusion is accompanied by a significant energy release in the form of the excitation energy of the final nucleus. The energy release resulting from pycnonuclear fusion represents an important source of heat within the crust. Results concerning the energy release will be presented in the next subsection.

## 6.3 Non-equilibrium processes and crustal heating

Detailed results describing the non-equilibrium reactions in the crust of an accreting neutron star are shown in Tables 2, 3. In Table 2 we show results for the

**Table 2.** Non-equilibrium processes in the outer crust. Temperature effects are neglected. $P$ and $\rho$ are the threshold pressure and density for reactions initiated by the electron capture. Relative density jump at the threshold pressure is denoted by $\Delta\rho/\rho$. Last two columns give the total energy release, $q_{\text{tot}}$, and heat deposited in matter, $q$, both per one accreted nucleon, accompanying non-equilibrium reactions. After [37].

| $P$ (dyn cm$^{-2}$) | $\rho$ (g cm$^{-3}$) | process | $\Delta\rho/\rho$ | $q_{\text{tot}}$ (MeV) | $q$ (MeV) |
|---|---|---|---|---|---|
| 7.23 10$^{26}$ | 1.49 10$^{9}$  | $^{56}$Fe $\to$ $^{56}$Cr $- 2e^- + 2\nu_e$       | 0.08 | 0.04 | 0.01 |
| 9.57 10$^{27}$ | 1.11 10$^{10}$ | $^{56}$Cr $\to$ $^{56}$Ti $- 2e^- + 2\nu_e$       | 0.09 | 0.04 | 0.01 |
| 1.15 10$^{29}$ | 7.85 10$^{10}$ | $^{56}$Ti $\to$ $^{56}$Ca $- 2e^- + 2\nu_e$       | 0.10 | 0.05 | 0.01 |
| 4.78 10$^{29}$ | 2.50 10$^{11}$ | $^{56}$Ca $\to$ $^{56}$Ar $- 2e^- + 2\nu_e$       | 0.11 | 0.05 | 0.01 |
| 1.36 10$^{30}$ | 6.11 10$^{11}$ | $^{56}$Ar $\to$ $^{52}$S $+ 4n - 2e^- + 2\nu_e$   | 0.12 | 0.06 | 0.05 |

**Table 3.** Non-equilibrium processes in the inner crust. Notation as in Table 2. Neutron fraction in the total number of nucleons, in the layer just above the reaction surface, is denoted by $X_n$. After [37].

| $P$ (dyn cm$^{-2}$) | $\rho$ (g cm$^{-3}$) | process | $X_n$ | $\Delta\rho/\rho$ | $q$ (MeV) |
|---|---|---|---|---|---|
| 1.980 10$^{30}$ | 9.075 10$^{11}$ | $^{52}$S $\to$ $^{46}$Si $+ 6n - 2e^- + 2\nu_e$   | 0.07 | 0.13 | 0.09 |
| 2.253 10$^{30}$ | 1.131 10$^{12}$ | $^{46}$Si $\to$ $^{40}$Mg $+ 6n - 2e^- + 2\nu_e$  | 0.18 | 0.14 | 0.10 |
| 2.637 10$^{30}$ | 1.455 10$^{12}$ | $^{40}$Mg $\to$ $^{34}$Ne $+ 6n - 2e^- + 2\nu_e$  | 0.39 | 0.16 | 0.12 |
| 3.204 10$^{30}$ | 1.951 10$^{12}$ | $^{34}$Ne $+ ^{34}$Ne $\to$ $^{68}$Ca <br> $^{68}$Ca $\to$ $^{62}$Ar $+ 6n - 2e^- + 2\nu_e$ | 0.39 | 0.09 | 0.40 |
| 3.216 10$^{30}$ | 2.134 10$^{12}$ | $^{62}$Ar $\to$ $^{56}$S $+ 6n - 2e^- + 2\nu_e$   | 0.45 | 0.09 | 0.05 |
| 3.825 10$^{30}$ | 2.634 10$^{12}$ | $^{56}$S $\to$ $^{50}$Si $+ 6n - 2e^- + 2\nu_e$   | 0.50 | 0.09 | 0.06 |
| 4.699 10$^{30}$ | 3.338 10$^{12}$ | $^{50}$Si $\to$ $^{44}$Mg $+ 6n - 2e^- + 2\nu_e$  | 0.55 | 0.09 | 0.07 |
| 6.044 10$^{30}$ | 4.379 10$^{12}$ | $^{44}$Mg $\to$ $^{36}$Ne $+ 8n - 2e^- + 2\nu_e$ <br> $^{36}$Ne $+ ^{36}$Ne $\to$ $^{72}$Ca <br> $^{72}$Ca $\to$ $^{66}$Ar $+ 6n - 2e^- + 2\nu_e$ | 0.61 | 0.14 | 0.28 |
| 7.233 10$^{30}$ | 5.665 10$^{12}$ | $^{66}$Ar $\to$ $^{60}$S $+ 6n - 2e^- + 2\nu_e$   | 0.70 | 0.04 | 0.02 |
| 9.238 10$^{30}$ | 7.041 10$^{12}$ | $^{60}$S $\to$ $^{54}$Si $+ 6n - 2e^- + 2\nu_e$   | 0.73 | 0.04 | 0.02 |
| 1.228 10$^{31}$ | 8.980 10$^{12}$ | $^{54}$Si $\to$ $^{48}$Mg $+ 6n - 2e^- + 2\nu_e$  | 0.76 | 0.04 | 0.03 |
| 1.602 10$^{31}$ | 1.127 10$^{13}$ | $^{48}$Mg $+ ^{48}$Mg $\to$ $^{96}$Cr             | 0.79 | 0.04 | 0.11 |
| 1.613 10$^{31}$ | 1.137 10$^{13}$ | $^{96}$Cr $\to$ $^{88}$Ti $+ 8n - 2e^- + 2\nu_e$  | 0.80 | 0.02 | 0.01 |

outer neutron star crust, where matter consists of nuclei immersed in electron gas. Non-equilibrium electron captures generate heat on the spherical surfaces (actually: in very thin shells) at pressures indicated in the first column of Table 2. The density of matter just above the reaction surface is given in the second column. Density of matter undergoes a jump at the reaction surface. This results from the fact that reactions take place at a constant pressure, $P \simeq P_e$, which is determined mainly by the electron density $n_e = Zn_\mathcal{N}$. At constant pressure, a decrease in $Z$, implied by the double electron capture, is thus necessarily accompanied by the baryon density and mass density increase, with $\Delta\rho/\rho \simeq 2/(Z-2)$.

In the fifth column of Table 2 we give the total heat release in a unit cell, accompanying a non-equilibrium reaction, divided by the number of nucleons in the cell. As we have shown in the previous subsection, on average only $\sim \frac{1}{6}$ of this heat is deposited in matter, the remaining part being radiated away with neutrinos. In the last column we give the effective (deposited in matter) heat per nucleon in a non-equilibrium process, $q$. In the steady thermal state of an accreting neutron star, effective heat release per unit time on the $i$-th reaction surface, $Q_i$, is proportional to the mass accretion rate, $\dot{M}$. This relation can be written in a suitable form

$$Q_i = 6.03 \cdot \left(\frac{\dot{M}}{10^{-10} M_\odot/\mathrm{y}}\right) \cdot \left(\frac{q_i}{1\ \mathrm{MeV}}\right) 10^{33}\ \mathrm{erg/s}\ . \tag{36}$$

Let us notice that the heat release on the neutron drip surface exceeds the total remaining heat release in the outer crust. This is due to the fact that non-equilibrium neutron emission represents a very efficient channel of matter heating.

In Table 3 we collected results referring to the inner crust of accreting neutron star. The fraction of nucleons in the neutron gas phase is denoted by $X_n$ and refers to the crust shell laying just above the reaction surface. For the sake of simplicity, the description of non-equilibrium processes is largely symbolic.

Results presented in Table 3 show that when a chain of non-equilibrium processes includes pycnonuclear fusion, heat production may be more than an order of magnitude larger than in the case involving only electron captures and neutron emission.

For $\rho > 1.2\ 10^{13}$ g cm$^{-3}$ the energy release per nucleon, due to non-equilibrium processes, is rather small compared to that in the $\rho_{\mathrm{ND}} < \rho < 1.2\ 10^{13}$g cm$^{-3}$ layer. To some extent this is due to the fact that atomic nuclei immersed in a dense neutron gas contain then only a small fraction of the total number of nucleons. On the other hand, being more and more neutron rich, these nuclei become less and less dense and less and less bound.

The validity of the Haensel and Zdunik [37],[38] model becomes questionable at the densities a few times $10^{13}$g cm$^{-3}$. However, one may expect that at such a high density properties of non-catalyzed matter become rather simple. Calculation shows, that if a matter element produced originally in a helium flash could reach a density $\sim 10^{14}$g cm$^{-3}$, we should expect it to contain only $\sim 10\%$ of nucleons bound in atomic nuclei. In view of this, pressure of matter in the shells of constant $(A, Z)$ (as well as many other properties) at this (and higher) density may be expected to be dominated by non-relativistic neutrons.

## 6.4 Astrophysical consequences

Non-catalyzed matter in the crust of an accreting neutron stars turns out to be an important reservoir of energy, which is partly released in the non-equilibrium reactions involving electron captures, neutron emissions and pycnonuclear fusion. The total heat release due to non-equilibrium reactions in the neutron star crust is larger than a typical inward heat flow produced by the steady thermonuclear burning of accreted matter between the helium flashes. Detailed calculations of the steady thermal state of neutron stars, accreting at rates $10^{-11} \lesssim \dot{M}/(M_\odot/y) \lesssim 10^{-9}$, taking due account of non-equilibrium heat sources in the crust, and with various models of neutron star core, were performed in [58].

Many neutron stars in close X-ray binaries are transient accretors (transients). Such neutron stars exhibit X-ray outbursts separated by long periods (months or even years) of quiescence. It is believed that quiescence corresponds to a low-level, or in extreme case halted, accretion onto neutron star. During high-accretion episodes, heat is deposited by non-equilibrium processes in the deep layers ($10^{12} - 10^{13}$ g cm$^{-3}$) of accreted crust. This has been shown to be possible mechanism to maintain temperature of neutron star interior sufficiently high to explain thermal X-ray radiation in quiescence [17],[80].

## 7 Composition of accreted crust

Many neutron stars may have accreted crust. For example, consider the millisecond pulsars. They are thought to be old neutron stars, spun up by the accretion, via an accretion disk, of $\sim 0.1 M_\odot$ from their companion in a close binary system (see, e.g., [44]). Clearly, if such scenario is correct, the whole crust of a typical millisecond pulsar is built of accreted, non-catalyzed matter.

Composition of accreted crust in the "single nucleus" approximation, discussed in detail in the preceding section, was calculated by Haensel and Zdunik [38]. These authors used the same compressible liquid drop model of nuclei as that applied in their study of non-equilibrium processes in accreting crust. In Table 4 we list the nuclides present in the crust of an accreting neutron star. In the third, fourth and fifth columns we give the maximum pressure, $P_{\rm max}$, mass density, $\rho_{\rm max}$, and baryon density, $n_{b,\rm max}$, at which the nuclide is present. In the sixth column we give the value of the electron chemical potential (including rest energy), $\mu_e$, at this density. The fraction of nucleons in the neutron gas phase within the layer ending at $P_{\rm max}$, denoted by $X_n$, is shown in the seventh column. Transition to the next nuclide is accompanied by a density jump. In the last column of Table 4 we give the corresponding relative density increase, $\Delta \rho/\rho$. To a very good approximation we have $\Delta n_b/n_b \simeq \Delta \rho/\rho$. Relative density jumps are significantly larger than those in the case of cold catalyzed matter, and exceed 10% above neutron drip point.

As one sees from Table 4, composition of the crust of an accreting neutron star is vastly different from that of a standard neutron star composed of the catalyzed matter. In the case of an accreting neutron star the value of $Z$ is typically $\lesssim 20$, to be compared with $Z = 40 - 50$ for the cold catalyzed matter above the neutron drip point. At $\rho \simeq 10^{13}$ g cm$^{-3}$ the mass number of nuclei in the accreted crust is about 60, to be compared with about 200 in cold catalyzed matter of the same density. The neutron drip occurs at a similar density as in the cold catalyzed matter. At the highest density considered, $\rho \simeq 1.2\ 10^{13}$ g cm$^{-3}$, more than 80% of nucleons form neutron gas outside nuclei. The neutron gas gives there a dominating contribution to the pressure. Another remark to be made is that composition given in Table 4 corresponds to an idealized scenario of formation of accreted crust. Possible deviations from this idealized picture are discussed in Sec.10.

An important remark concerns the values of $Z$. The mean charge of nuclei in the crust of an accreting neutron star turns out to be less than half of that characteristic of the cold catalyzed matter. As pointed out by Sato [83], this will result in a significant reduction of the shear modulus of the crust (see Sect. 9).

## 8 Equation of state of the neutron star crust

The equation of state (EOS) constitutes an essential input for the calculation of the neutron star models. In the present Section, we discuss the EOS of the outer and inner neutron star crust. Two basic models, corresponding to different idealized scenarios of crust formation, will be considered. First model will be based on the ground state approximation, in which the crust is assumed to be built of cold catalyzed matter (structure of the crust in this approximation was discussed in Sections 3, 4, 5). Then we will describe the EOS of accreted crust, assuming formation scenario described in Sect.6, where the structure of the outer and inner crust was derived within the "single nucleus" approximation.

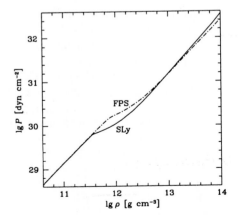

**Fig. 8.** Comparison of the SLy and FPS EOS.

## 8.1 Ground state approximation

The EOS of the outer crust in the ground state approximation is rather well established. Generally, the EOS of Haensel and Pichon [39] is quite similar to the more than two decades older BPS EOS [4]. In some pressure intervals one notices a few percent difference between densities, resulting from the difference in the nuclides present at the same pressures.

As soon as one leaves the region of experimentally known nuclei, the EOS of cold catalyzed matter becomes uncertain. This uncertainty rises above the neutron drip density, where only theoretical models can be used. The properties of nuclei become influenced by the outside neutron gas, which contributes more and more to the total pressure. Therefore, the problem of correct modelling of equation of state of pure neutron gas at subnuclear densities becomes important. The real EOS of cold catalyzed matter stems from a real nucleon Hamiltonian, which is expected to describe nucleon interactions at $\rho \lesssim 2\rho_0$ (at higher densities, non-nucleon degrees of freedom, such as hyperons, quarks (?), meson condensates (?), etc., may become relevant). In practice, in order to make the solution of the many-body problem feasible, the task was reduced to that of finding an *effective nucleon Hamiltonian*, which would enable one to calculate reliably the EOS of cold catalyzed matter for $10^{11}$ g cm$^{-3} \lesssim \rho \lesssim \rho_0$, including therefore the crust-liquid core transition.

Of course, for $\rho \lesssim 4 \times 10^{11}$ g cm$^{-3}$ one can use EOS based on experimental, or semi-empirical nuclear masses, but it is reassuring to check that this EOS is nicely reproduced by a "theoretical EOS", based on an effective nucleon-nucleon interactions FPS and SLy. As one can see in Fig.8, significant differences between the SLy and FPS EOS are restricted to the density interval $4 \times 10^{11} - 4 \times 10^{12}$ g cm$^{-3}$. They result mainly from the fact that $\rho_{\rm ND}({\rm SLy}) \simeq 4 \times 10^{11}$ g cm$^{-3}$

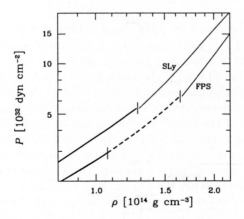

**Fig. 9.** Comparison of the SLy and FPS EOS near the crust-liquid core transition. Thick solid line: inner crust with spherical nuclei. Dashed line corresponds to "exotic nuclear shapes". Thin solid line: uniform *npe* matter.

**Table 4.** Composition of the crust of an accreting neutron star. After [38]. For further explanation see the text.

| Z | A | $P_{max}$ (dyn cm$^{-2}$) | $\rho_{max}$ (g cm$^{-3}$) | $n_{b,max}$ (cm$^{-3}$) | $\mu_e$ MeV | $X_n$ | $\Delta\rho/\rho$ (%) |
|---|---|---|---|---|---|---|---|
| 26 | 56 | 7.235 10$^{26}$ | 1.494 10$^{9}$  | 8.994 10$^{32}$ | 4.59  | 0.00 | 8.2 |
| 24 | 56 | 9.569 10$^{27}$ | 1.1145 10$^{10}$ | 6.701 10$^{33}$ | 8.69  | 0.00 | 8.9 |
| 22 | 56 | 1.152 10$^{29}$ | 7.848 10$^{10}$ | 4.708 10$^{34}$ | 16.15 | 0.00 | 9.8 |
| 20 | 56 | 4.747 10$^{29}$ | 2.496 10$^{11}$ | 1.494 10$^{35}$ | 22.99 | 0.00 | 10.9 |
| 18 | 56 | 1.361 10$^{30}$ | 6.110 10$^{11}$ | 3.651 10$^{35}$ | 29.89 | 0.00 | 12.1 |
| 16 | 52 | 1.980 10$^{30}$ | 9.075 10$^{11}$ | 5.418 10$^{35}$ | 32.78 | 0.07 | 13.1 |
| 14 | 46 | 2.253 10$^{30}$ | 1.131 10$^{12}$ | 6.748 10$^{35}$ | 33.73 | 0.18 | 14.4 |
| 12 | 40 | 2.637 10$^{30}$ | 1.455 10$^{12}$ | 8.682 10$^{35}$ | 34.85 | 0.29 | 17.0 |
| 20 | 68 | 2.771 10$^{30}$ | 1.766 10$^{12}$ | 1.054 10$^{36}$ | 34.98 | 0.39 | 8.3 |
| 18 | 62 | 3.216 10$^{30}$ | 2.134 10$^{12}$ | 1.273 10$^{36}$ | 35.98 | 0.45 | 8.6 |
| 16 | 56 | 3.825 10$^{30}$ | 2.634 10$^{12}$ | 1.571 10$^{36}$ | 37.10 | 0.50 | 9.0 |
| 14 | 50 | 4.699 10$^{30}$ | 3.338 10$^{12}$ | 1.990 10$^{36}$ | 38.40 | 0.55 | 9.3 |
| 12 | 44 | 6.044 10$^{30}$ | 4.379 10$^{12}$ | 2.610 10$^{36}$ | 39.92 | 0.61 | 13.8 |
| 18 | 66 | 7.233 10$^{30}$ | 5.665 10$^{12}$ | 3.377 10$^{36}$ | 39.52 | 0.70 | 4.4 |
| 16 | 60 | 9.2385 10$^{30}$ | 7.041 10$^{12}$ | 4.196 10$^{36}$ | 40.85 | 0.73 | 4.3 |
| 14 | 54 | 1.228 10$^{31}$ | 8.980 10$^{12}$ | 5.349 10$^{36}$ | 42.37 | 0.76 | 4.0 |
| 12 | 48 | 1.602 10$^{31}$ | 1.127 10$^{13}$ | 6.712 10$^{36}$ | 43.41 | 0.79 | 3.5 |
| 24 | 96 | 1.613 10$^{31}$ | 1.137 10$^{13}$ | 6.769 10$^{36}$ | 43.55 | 0.79 | 1.5 |
| 22 | 88 | 1.816 10$^{31}$ | 1.253 10$^{13}$ | 7.464 10$^{36}$ | 43.69 | 0.80 | .... |

(in good agreement with the "empirical EOS" of [39]), while $\rho_{ND}$(FPS) $\simeq$ 6 × 10$^{11}$ g cm$^{-3}$. For $4 \times 10^{12}$ g cm$^{-3} \lesssim \rho \lesssim 10^{14}$ g cm$^{-3}$ the SLy and FPS EOS are very similar, with the FPS EOS being a little softer at highest densities. Detailed behavior of two EOS near crust-liquid core transition can be seen in Fig.9. The FPS EOS is softer than the SLy one.

In the case of the SLy EOS the crust-liquid core transition takes place as a very weak first-order phase transition, with relative density jump of the order of a percent. Let us remind that for this model spherical nuclei persist down to the bottom edge of the crust. As one can see in Fig.9, crust-core transition is accompanied by a noticeable jump of the slope (stiffening) of the EOS. For the FPS EOS, the crust-core transition takes place through a sequence of phase transitions with changes of nuclear shapes. These phase transitions make the crust-core transition smoother than in the SLy case, with a gradual increase of stiffness, which nevertheless suffers a visible jump at the bottom of the bubble-layer edge. All in all, while presence of exotic nuclear shapes is expected to have dramatic consequences for the transport, neutrino emission, and elastic properties of neutron star matter, their effect on the EOS is rather small.

The SLy EOS of the crust, calculated including adjacent segments of the liquid core and the outer crust, is shown in Fig.10. In the outer crust segment, the SLy EOS cannot be graphically distinguished from that of Haensel and Pichon [39], which was based on experimental nuclear masses.

## 8.2 Accreted crust

Equation of state of accreted crust was calculated by Haensel and Zdunik [38], within the "single nucleus" scenario, described in preceding subsection. This EOS is compared with SLy model of cold catalyzed matter in Fig. 11. Up to neutron drip point, both equations of state are quite similar. This is easily understood: for $\rho < \rho_{\rm ND}$ we have $P \simeq P_e$, which in turn depends only the ratio $Z/A$, quite similar for both accreted and ground state EOS.

Significant differences appear for $\rho_{\rm ND} \lesssim \rho \lesssim 10\rho_{\rm ND}$, where EOS of accreted matter is stiffer than that of cold catalyzed matter. Also, one notices well pronounced constant-pressure density jumps in EOS of accreted matter, which are due to discontinuous changes in nuclear composition. These density jumps, accompanying first order phase transitions, are particularly large for $\rho_{\rm ND} \lesssim \rho \lesssim 10\rho_{\rm ND}$, and lead to an overall softening of the EOS of accreted crust. For $\rho \gtrsim 10^{13}$ g cm$^{-3}$ EOS for accreted crust becomes very similar to that of cold catalyzed matter. The EOS of accreted crust is given in the density interval from $\sim 10^8$ g cm$^{-3}$ to $\rho \simeq 1.5 \times 10^{13}$ g cm$^{-3}$. The lower limit corresponds to

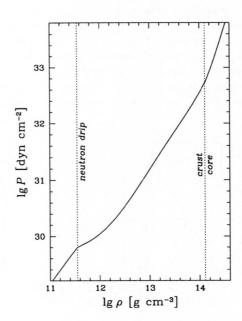

**Fig. 10.** The SLy EOS. Dotted vertical lines correspond to the neutron drip and crust bottom edge.

the minimum density of the processed accreted matter, just below the bottom of the helium layer (matter is there assumed to be composed of $^{56}$Fe). The choice of the upper limit is based on two arguments. Firstly, for $\rho > 10^{13}$ g cm$^{-3}$ our equation of state becomes very similar that of the cold catalyzed matter. Secondly, the validity of the Haensel and Zdunik [38] model of dense matter and, in particular, of the Mackie and Baym [56] model for nuclei, used in their calculations, becomes questionable for the densities much higher than $\sim 10^{13}$ g cm$^{-3}$.

It is therefore fortunate, that the difference between the cold catalyzed matter and accreted crust equations of state decreases for large density and for $\rho > 10^{13}$ g cm$^{-3}$ both curves are very close to each other. This is due to the fact that for such a high density the equation of state in both cases is determined mainly by the properties of neutron gas. In view of this, the use of the equation of state of the catalyzed matter for the calculation of the hydrostatic equilibrium of the high density ($\rho > 10^{13}$ g cm$^{-3}$) interior layer of the crust of an accreting neutron star should give a rather good approximation, as far as the density profile is concerned.

## 9 Elastic properties of neutron star crust

In contrast to the liquid core, solid crust can sustain an *elastic strain*. As neutron stars are relativistic objects, a relativistic theory of elastic media in a curved space-time should in principle be used to describe elastic effects in neutron star structure and dynamics. Such a theory of elasticity has been developed by Carter and Quintana [21] and applied by them to rotating neutron star models in [22],[23]. However, in view of the smallness of elastic forces compared to those of gravity and pressure, we will restrict ourselves, for the sake of simplicity, to the Newtonian version of the theory of elasticity [50].

The state of thermodynamic equilibrium of an element of neutron-star crust corresponds to specific *equilibrium positions* of nuclei, which will be de-

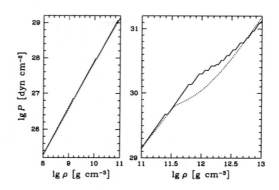

**Fig. 11.** Comparison of the SLy EOS for cold catalyzed matter (dotted line) and the EOS of accreted crust.

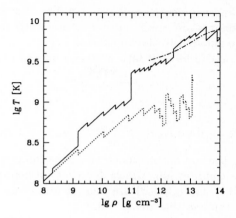

**Fig. 12.** Melting temperature of neutron star crust versus density. Solid line: Melting temperature for the ground state composition of the crust (Haensel and Pichon [39] for the outer crust and Negele and Vautherin [62] for the inner crust). Jumps at some densities correspond to change of the nuclide. Dash-dotted line: melting temperature of the ground state inner crust, based on the compressible liquid drop model calculation of Douchin and Haensel [30]; its smooth behavior results from dense matter model nature. Dotted line corresponds to the accreted crust model of [38].

noted by $\mathbf{r}$. For pure nuclear composition (one-component plasma) at $T = 0$ and $\rho < 10^{14}$ g cm$^{-3}$, $\mathbf{r}$ points to the lattice sites of the bcc lattice of nuclei. (Strictly speaking, $\mathbf{r}$ corresponds to *mean* positions of nuclei, which suffer both quantum zero-point, as well as thermal, oscillations.) Neutron star evolution (e.g., spin-down of rotation, cooling) or some outer influence (tidal forces from a close massive body, accretion of matter, electromagnetic strains associated with strong internal magnetic fields) may lead to *deformation* of the crust. In what follows, we will neglect the thermal contributions to thermodynamic quantities and restrict to the $T = 0$ approximation.

Deformation of a crust element with respect to the ground state configuration implies a *displacement* of nuclei into their new positions $\mathbf{r}' = \mathbf{r} + \mathbf{u}$, where $\mathbf{u} = \mathbf{u}(\mathbf{r})$ is the displacement vector. In the continuum limit, relevant for macroscopic phenomena, both $\mathbf{r}$ and $\mathbf{u}$ are treated as continuous fields. Non-zero $\mathbf{u}$ is accompanied by the appearance of *elastic strain* (i.e., forces which tend to return the matter element to the equilibrium state of minimum energy density $\mathcal{E}_0$), and yields *deformation energy* density $\mathcal{E}_{\mathrm{def}} = \mathcal{E}(\mathbf{u}) - \mathcal{E}_0$.[5] Uniform translation, described by an $\mathbf{r}$-independent displacement field, does not contribute to $\mathcal{E}_{\mathrm{def}}$, and the real (genuine) deformation is described by the (symmetric) *strain tensor*

$$u_{ik} = u_{ki} = \frac{1}{2}\left(\frac{\partial u_i}{\partial x_k} + \frac{\partial u_k}{\partial x_i}\right) , \qquad (37)$$

---

[5] In this section, by "energy" we will always mean energy of a unit volume of matter (i.e., energy density)

where $i, j = 1, 2, 3$, and $x_1 = x$, $x_2 = y$, $x_3 = z$. The above form of $u_{ik}$ is valid when all components of **u** are small, and terms quadratic in the components of **u** can be neglected compared to the linear ones [50].

Each deformation can be split into *compression* and *shear* components,

$$u_{ik} = u_{ik}^{\text{comp}} + u_{ik}^{\text{shear}}, \qquad (38)$$

where

$$u_{ik}^{\text{comp}} = \frac{1}{3}\text{div}\mathbf{u}\,\delta_{ik}, \qquad u_{ik}^{\text{shear}} = u_{ik} - \frac{1}{3}\text{div}\mathbf{u}\,\delta_{ik}. \qquad (39)$$

After a deformation, the volume of a matter element changes according to $dV' = (1 + \text{div}\mathbf{u})dV$. Pure compression, which does not influence a shape of matter element, is described by $u_{ik} = a\delta_{ik}$. Pure shear deformation keeps the volume of matter element constant, so that div **u** = 0.

To lowest order, deformation energy is quadratic in the deformation tensor,

$$\mathcal{E}_{\text{def}} = \frac{1}{2}\lambda_{iklm}u_{ik}u_{lm}, \qquad (40)$$

where summation is assumed over repeated indices. Since deformation energy $\mathcal{E}_{\text{def}}$ is a scalar, $\lambda_{iklm}$ are components of a rank fourth tensor. While the total number of components $\lambda_{iklm}$ is 81, general symmetry relations reduce the maximum number of linearly independent components (elastic moduli) to 21. The number of independent elastic moduli decreases with increasing symmetry of elastic medium, and becomes as small as three in the case of a bcc crystal, and two in the case of an isotropic solid. Elastic stress tensor, $\sigma_{ik}$, is derived from from the deformation energy via $\sigma_{ik} = \partial \mathcal{E}_{\text{def}}/\partial u_{ik}$.

## 9.1 From bcc lattice to isotropic solid

While microscopically the ground state of neutron star crust at $\rho \lesssim 10^{14}$ g cm$^{-3}$ corresponds to a bcc lattice, one usually assumes that its macroscopic properties, relevant for the neutron star calculations, are those of an isotropic bcc polycrystal. Such an assumption is made, because it seems quite probable that neutron star crust is better approximated by a polycrystal than by a monocrystal (see, however, [10]), and also for the sake of simplicity. Elastic properties of an isotropic solid are described by two elastic moduli, and the deformation energy can be expressed as

$$\mathcal{E}_{\text{def}} = \frac{1}{2}K(\text{div }\mathbf{u})^2 + \mu\left(u_{ik} - \frac{1}{3}\delta_{ik}\text{div }\mathbf{u}\right)^2. \qquad (41)$$

Here, $\mu$ is the *shear modulus* and $K$ is the *compression modulus* of isotropic solid. The stress tensor is then calculated as

$$\sigma_{ik} = \frac{\partial \mathcal{E}_{\text{def}}}{\partial u_{ik}} = K\text{div }\mathbf{u}\,\delta_{ik} + 2\mu\left(u_{ik} - \frac{1}{3}\text{div }\mathbf{u}\,\delta_{ik}\right). \qquad (42)$$

Considering pure uniform compression one finds that

$$K = n_b \frac{\partial P}{\partial n_b} = \gamma P ,  \qquad (43)$$

where $\gamma$ is the adiabatic index, $\gamma \equiv (n_b/P)\mathrm{d}P/\mathrm{d}n_b$ .

Detailed calculations of directionally averaged effective shear modulus of a bcc Coulomb solid, appropriate for the polycrystalline crusts of neutron stars, were performed by Ogata and Ichimaru [65]. These authors considered a one component bcc Coulomb crystal, neglecting screening by the degenerate electrons as well as the quantum zero-point motion of the ions about their equilibrium lattice sites. The deformation energy, resulting from the application of a specific strain $u_{ik}$, was evaluated directly through the Monte Carlo sampling.

In the case of an ideal bcc lattice there are only three independent elastic moduli, denoted traditionally as $c_{11}$, $c_{12}$ and $c_{44}$ (see, e.g., [47]). When the crystal is deformed without changing (to lowest order in $u_{ik}$) the volume of matter element, only two independent elastic moduli are relevant, because

$$\mathcal{E}_{\mathrm{def}} = b_{11}(u_{xx}^2 + u_{yy}^2 + u_{zz}^2) + c_{44}(u_{xy}^2 + u_{xz}^2 + u_{yz}^2), \quad \mathrm{for} \;\; \mathrm{div}\mathbf{u} = 0 , \qquad (44)$$

with $b_{11} = \frac{1}{2}(c_{11} - c_{12})$. At $T = 0$, Ogata and Ichimaru (1990) find $b_{11} = 0.0245 n_\mathcal{N}(Ze)^2/r_\mathrm{c}$, $c_{44} = 0.1827 n_\mathcal{N}(Ze)^2/r_\mathrm{c}$. Significant difference between $b_{11}$ and $c_{44}$ indicates high degree of elastic anisotropy of an ideal bcc monocrystal.

While treating neutron star crust as an isotropic solid is a reasonable approximation (ideal long-range order does not exist there, and we are most prob-

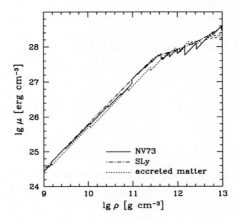

**Fig. 13.** Effective shear modulus $\mu$ versus neutron star matter density, assuming bcc crystal lattice. Solid line - cold catalyzed matter (Haensel and Pichon [39] model for the outer crust, and that of Negele and Vautherin [62] for the inner crust). Dash-dotted line - cold catalyzed matter calculated by Douchin and Haensel [30] (compressible liquid drop model, based on SLy4 effective N-N interaction). Dotted line - accreted crust model of Haensel and Zdunik [38].

ably dealing with a bcc polycrystal), the choice of an "effective" shear modulus deserves a comment. In numerous papers treating the elastic aspects of neutron star dynamics, a standard choice was $\mu = c_{44}$ ([5],[69], [57] and references therein). It is clear, that replacing $\mu$ by a single *maximal* elastic modulus of strongly anisotropic bcc lattice is not appropriate. Correct value of $\mu$ was calculated by Ogata and Ichimaru [65], who performed directional averages over rotations of the Cartesian axes. At $T = 0$, they obtained (neglecting quantum zero-point oscillations of nuclei)

$$\mu = \frac{1}{5}(2b_{11} + 3c_{44}) = 0.1194 \frac{n_N (Ze)^2}{r_c}, \qquad (45)$$

nearly two times smaller than $\mu = c_{44}$ used in ([5],[69],[57]). Dependence of $\mu$ on temperature was studied, using the Monte Carlo sampling method, by Strohmayer et al. [86]. These authors found that their results can be represented via a simple analytic formula

$$\mu(T) = \frac{\mu(0)}{1 + 1.781 \, (100/\Gamma)^2}, \qquad (46)$$

where the ion-coupling parameter $\Gamma = Z^2 e^2/(r_c k_B T)$, and quantum zero-point motion of nuclei has been neglected. Formula (46) fits their numerical results within the estimated numerical error of the Monte Carlo scheme, and reproduces correct $T = 0$ (i.e., $\Gamma = \infty$) limit. As expected, effective shear modulus decreases with increasing temperature.

Let us discuss now qualitative properties of the isotropic neutron star crust. One can easily show, that $\mu \ll K$. This means that neutron star crust is much more susceptible to shear than to compression; its Poisons coefficient $\sigma \simeq 1/2$, while its Young modulus $E \simeq 3\mu$ (for definitions, see [50]).

Strictly speaking, the formulae given in the present subsection hold for the outer crust, where $r_N \ll r_c$. They neglect also the effect of the quantum zero-point vibrations of nuclei around their lattice sites. Therefore, in the case of the inner crust these formulae give only an approximation of the actual values of $\mu$.

## 9.2 Exotic nuclei

Some models of neutron-star matter predict existence of unusual nuclei (rods, plates, tubes, bubbles) in the bottom layer of the crust with $\rho \gtrsim 10^{14}$ g cm$^{-3}$. Possible structure of this bottom layer was discussed in Sect. 5. In what follows we will concentrate on two specific unusual shapes, namely rods and plates, which are expected to fill most of the bottom crust layer. The properties of matter containing rods and plates are intermediate between those of solids and liquids. For example, displacement of an element of plate matter parallel to the plate plane or rod matter in the direction of rods, is not opposed by restoring forces: this is typical property of a liquid. However, elastic strain opposes any bending of planes or rods, a property specific of a solid. Being intermediate between solids and liquids, these kinds of matter are usually referred to as *mesomorphic* phases, or *liquid crystals* (see, e.g., [50],[35]). Elastic properties of rod and plate phases of neutron star matter were studied by Pethick and Potekhin [73] (see also [72]).

## 10 Deviations from idealized models

The ground state crust and accreted crust, described in preceding sections of the present review correspond to idealized perfect one-component plasmas. The real neutron-star crust may be expected to deviate from these idealized models. The practical question to be a asked (and to be answered) is: how much the matter of a real neutron-star crust deviates from a one-component plasma? The knowledge of "imperfections" of the crust is particularly important for its transport properties. The motion of electrons in a significantly disordered ion lattice is qualitatively different from that in a perfect crystal. In the case of a perfect crystal, electrons move in a strictly periodic field, and scatter only on the elementary excitations of the ion lattice - phonons. Disordered ions act as individual scattering centers, strongly limiting electron transport of heat and charge.

### 10.1 Impurities in a crust of a newly-born neutron star

Initial temperature of the outer layers of a newly born neutron star exceeds $10^{10}$ K. Under such conditions, nuclear composition of the matter is characterized by some statistical distribution of $(A, Z)$ nuclei in a hot plasma. Initially, the spread in $(A, Z)$ is rather wide [20]. After solidification of the crust its composition is practically frozen, so that it may be expected to reflect the situation at crystallization point rather than in the ground state. In contrast to the ground state composition of the outer crust, at $T \simeq T_\mathrm{m}$ transitions between shells $(A_1, Z_1)$ and $(A_2, Z_2)$ will be continuous, via a transition layer consisting of a mixture of both nuclides. Only sufficiently far from the transition layer one is dealing with a one-component plasma. Two-component transition layers were studied by De Blasio [11],[12]. The radial width of the transition layers in the outer crust, calculated in [12] for the density range $10^9 - 10^{11}$ g cm$^{-3}$, was $4 - 12$ m.

### 10.2 Non-equilibrium neutrons

Higher temperatures are characterized by larger fraction of evaporated nucleons. The most sensitive region is that around the neutron drip point in cold catalyzed matter, $\rho_\mathrm{ND} \simeq 4 \times 10^{11}$ g cm$^{-3}$. At $T \simeq 5 \times 10^9$ K, there is a non-negligible fraction of free neutrons for $10^{11}$ g cm$^{-3} \lesssim \rho \lesssim \rho_\mathrm{ND}$ (see lower panel of Fig.1). In general, one notices a significant excess of free neutrons for the densities $10^{11}$ g cm$^{-3} \lesssim \rho \lesssim 10^{12}$ g cm$^{-3}$ as compared to the ground-state composition of the crust. With further cooling, there will be a tendency to absorb these excess neutrons by nuclei, which in turn will increase their $A$, and modify their $Z$ due to weak-interaction processes. However, the temperature may be expected to be too low to reach full nuclear equilibrium, mainly because of high Coulomb barriers, and the lack of free protons and $\alpha$-particles. Therefore, one may expect deviations from the ground-state composition (excess of dripped neutrons) in the cooled crust at $10^{11}$ g cm$^{-3} \lesssim \rho \lesssim 10^{12}$ g cm$^{-3}$ [8].

## 10.3 Thermal fluctuations and impurities in the inner crust

General problem of thermal fluctuations of the values of $Z$ and $N_{\text{cell}}$ in the inner crust, at $T \simeq T_{\text{m}} \gtrsim 10^9$ K, was studied by Jones [46]. Detailed calculations [46], performed within the Compressible Liquid Drop Model, and combined with consideration of the shell and pairing effects, suggest a high degree of heterogeneity in $Z$ to be frozen as the temperature falls below $T_{\text{m}}$, with substantial population of two closed $Z$-shells ($Z = 40$ and $Z = 50$). It should be mentioned that high value of $T_{\text{m}}$ (large thermal energies) and large fraction of neutrons (with large fraction of them unbound) in the inner crust are both favorable for impurity fractions higher than those in the outer crust. One has to keep in mind, however, that the kinetics of phase transitions is notoriously difficult for theoretical modelling, especially if approximations used cannot be tested in laboratory. Fortunately, while the "purity" of the crust is of crucial importance for its transport properties, the equation of state is not very sensitive to deviations from the one-nucleus model.

## 10.4 Impurities in accreted crust

If the ashes of this explosive burning are well approximated by pure $^{56}$Fe, then "single-nucleus scenario" described in Sect. 6 may be a valid description. Actually, this is only an approximation; the problem of the detailed outcome of the time-dependent nucleosynthesis during X-ray bursts is very complicated and should be considered as not completely resolved (see, e.g. [78],[84]) The nature of the unstable thermonuclear burning at higher accretion rates $10^{-8}$ M$_\odot$/y $\lesssim \dot{M} \lesssim 10^{-9}$ M$_\odot$/y is not well understood. The ashes from such a burning might contain some admixture of nuclei beyond the iron group, with $A \simeq 60 - 100$ [84]. Of course, even replacing pure $^{56}$Fe by a mix of the iron group elements would substantially complicate the description of the evolution, and would lead to deviation of resulting accreted crust from an idealized model of Sects. 6, 7.

If the starting composition is a mix with significant fractions of different nuclides, one may expect that further evolution will keep heterogeneity of the matter. The thermal and electrical conductivity of such a heterogeneous accreted crust would therefore be drastically lower than that of a perfect crystal. The equation of state would be rather smooth, in contrast to the extreme case of a one-nucleus model with significant density jumps. The values of average $Z$ and $A$ will still be much lower than those characteristic of the ground state of the crust. The number of shells of nonequilibrium processes triggered by electron captures will be much larger, but the total heat release may be expected to be similar to that estimated in Sect. 6.

## 10.5 Other scenarios for accreted crust

Up to now, it has been assumed that neutron star accreted baryon mass $M_{b,acc}$ larger than that of initial "primordial crust" composed of catalyzed matter, $M_{b,crust}^0$. In view of the fact that $M_{b,crust}^0 \sim 0.01$ $M_\odot$, to reach such a situation at constant accretion rate takes $10^8$ years at $\dot{M} = 10^{-10}$ $M_\odot/y$, with required accretion time $\propto (\dot{M})^{-1}$. At earlier times, the crust is composed of an outer layer of accreted and processed matter, of baryon mass $M_{b,acc}$, and an inner layer of baryon mass $M_{b,old} \simeq M_{b,crust}^0 - M_{b,acc}$, composed of compressed, processed primordial matter. Evolution of primordial crust under compression due to the weight of accreted layer can be followed shell by shell, with initial ground state composition of the shell. Such a study for a set of several shells with initial density ranging from $10^{8.9}$ g cm$^{-3}$ to $10^{13.6}$ g cm$^{-3}$ was performed by Sato [83].

## 10.6 Density inversions in accreted crust

They might appear during the evolution of the composition of a "primordial crust" under the weight of accreted layer of matter. In particular, let us focus our attention on the case of the interface between the $^{56}$Fe and $^{62}$Ni layers. Let us denote the ratio of the density of the upper layer to that of the lower one (at the interface) by $r_{u/l}$. This ratio is initially $r_{u/l} = 0.97$ (see Sect. 3). With increasing pressure, first electron capture take place on $^{56}$Fe, $^{56}$Fe $+ e^- \longrightarrow\ ^{56}$Mn $+ \nu_e$, followed by $^{56}$Mn$+e^- \longrightarrow\ ^{56}$Cr$+\nu_e$. The interface $^{56}$Cr/$^{62}$Ni is now characterized by the *density inversion* with $r_{u/l} = 1.05$ [9]. In general, density inversions are expected to appear and disappear at various interfaces during compression of primordial crust (Zdunik 2000, unpublished).

Even more significant density inversions may be expected in the case when accretion is very slow, $10^{-16}$ $M_\odot/y \lesssim \dot{M} \lesssim 10^{-12}$ $M_\odot/y$ (e.g., accretion of interstellar medium). Under such conditions, temperature within the accreted envelope is so low that helium burning takes place in pycnonuclear regime. It starts with $3\alpha$ fusion and typically terminates with $^{12}$C$(\alpha,\gamma)^{16}$O reaction [92]. Further compression of the $^{16}$O/$^{56}$Fe interface, accompanied by electron captures, leads to significant density inversions at the evolving interface. However, the timescales needed to reach such situations might exceed $10^{10}$ yr [9].

If both layers with $r_{u/l} > 1$ were fluid, the interface would be unstable with respect to the Rayleigh-Taylor overturn. However, under typical conditions prevailing at moderate and low accretion rates, $T < T_m$ and both layers are solid. In view of this, when analyzing the stability of the interface with respect to perturbations of its shape, one has to include, an addition to pressure and gravity forces, also elastic forces which are opposing the deformation, and might stabilize the interface [9].

## Acknowledgements

Results obtained for the SLy interaction, and presented in this review, were obtained with F. Douchin in 1998-2000, and I express my gratitude to him for fruitful collaboration. I am very grateful to D. G. Yakovlev and J. L. Zdunik for reading the preliminary version of this review, critical remarks, and helpful comments. I am also very grateful to A. Y. Potekhin for his help in the preparation of figures.

## References

1. J. Arponen: *Nucl. Phys.* **A 191**, 257 (1972)
2. Z. Barkat, J.-R. Buchler, L. Ingber: *Astrophys. J.* **176**, 723 (1972)
3. G. Baym, H.A. Bethe, C. Pethick: *Nucl. Phys.* **A175**, 225 (1971) (**BBP**)
4. G. Baym, C. Pethick, P. Sutherland: *Astrophys. J.* **170**, 299 (1971) (**BPS**)
5. G. Baym, D. Pines: *Ann. Phys. (N.Y.)* **66**, 816 (1971)
6. H.A. Bethe, G. Börner, K. Sato: *Astron. Astrophys.* **7**, 279 (1970)
7. G.S. Bisnovatyi-Kogan, E.F. Seidov,: *Astron. Zh.* **47**, 139 (1970)
8. G.S. Bisnovatyi-Kogan, V.M. Chechetkin: *Uspekhi Fiz. Nauk* **127**, 263 (1979) (English translation: *Sov. Phys. Uspekhi* **22**, 89 (1979))
9. O. Blaes, R. Blandford, P. Madau, S. Koonin: *Astrophys. J.* **363**, 612 (1990)
10. F.V. De Blasio: *Astrophys. J.* **452**, 359 (1995)
11. F.V. De Blasio: *Mon. Not. Roy. Astron. Soc.* **299**, 118 (1998)
12. F.V. De Blasio: *Astron. Astrophys.* **353**, 1129 (2000)
13. J. Boguta, A.R. Bodmer: *Nucl. Phys.* **A 292**, 413 (1977)
14. P. Bonche, D. Vautherin: *Nucl. Phys.* **A 372**, 496 (1981)
15. P. Bonche, D.Vautherin: *Astron. Astrophys.* **112**, 268 (1982)
16. M. Brack, C. Guet, H.-B. Håkansson: *Phys. Rep.* **123**, 275 (1985)
17. E.F. Brown, L. Bildsten, R.E. Rutledge: *Astrophys. J.* **504**, L95 (1998)
18. J.-R. Buchler, Z. Barkat: *Phys. Rev. Letters* **27**, 48 (1971)
19. J.-R. Buchler, Z. Barkat: *Astrophys. Letters* **7**, 167 (1971)
20. A. Burrows, J.M. Lattimer: *Astrophys. J.*, 294 (1984)
21. B. Carter, H. Quintana: *Proc. Roy. Soc. London Ser. A* **331**, 57 (1972)
22. B. Carter, H. Quintana: *Annals of Phys. (N.Y.)* **95**, 74 (1975)
23. B. Carter, H. Quintana: *Astrophys. J.* **202**, 511 (1975)
24. M. Centelles, X. Viñas, M. Barranco, S. Marcos, R.J. Lombard: *Nucl. Phys.* **A 537**, 486 (1992)
25. M. Centelles, X. Viñas, M. Barranco, P. Schuck: *Ann. Phys. (N.Y.)* **221**, 165 (1993)
26. E. Chabanat, P. Bonche, P. Haensel, J. Meyer, R. Schaeffer: *Nucl. Phys.* **A 627**, 710 (1997)
27. E. Chabanat, P. Bonche, P. Haensel, J. Meyer, R. Schaeffer: *Nucl. Phys.* **A 635**, 231 (1998)
28. K.S. Cheng, C.C. Yao, Z.G. Dai: *Phys. Rev.* **C 55 97**, 2092 (1997)
29. F. Douchin, 1999, PhD Thesis, École Normale Supérieure de Lyon, unpublished
30. F. Douchin, P. Haensel: *Phys. Letters* **B 485**, 107 (2000)
31. F. Douchin, P. Haensel, J. Meyer: *Nucl. Phys.* **A 665**, 419 (2000)
32. Ch. Engelmann et al., *Zeitschrift Phys.* **A 352**, 351 (1995)
33. B. Friedman, V.R. Pandharipande: *Nucl. Phys.* **A 361**, 502 (1981)

34. M.Y. Fujimoto, T. Hanawa, I. Iben, Jr., M.B. Richardson: *Astrophys. J* **278**, 813 (1984)
35. P.G. de Gennes, J.Prost, *The physics of liquid crystals*, 2nd ed., Clarendon, Oxford (1993)
36. P. Haensel, J.L. Zdunik, J. Dobaczewski: *Astron. Astrophys.* **222**, 353 (1989)
37. P. Haensel, J.L. Zdunik: *Astron. Astrophys.* **227**, 431 (1990)
38. P. Haensel, J.L. Zdunik: *Astron. Astrophys.* **229**, 117 (1990)
39. P. Haensel, B. Pichon: *Astron. Astrophys.* **283**, 313 (1994)
40. P. Haensel, A.D. Kaminker, D. G. Yakovlev: *Astron. Astrophys.* **314**, 328 (1996)
41. K. Harrison, M. Wakano, J.A. Wheeler, Matter-energy at high density: end point of thermonuclear evolution, in: *La structure et évolution de l'univers*, R. Stoops, Bruxelles, pp.124-140 (1958)
42. B.K. Harrison, K.S. Thorne, M. Wakano, J.A. Wheeler, *Gravitation Theory and Gravitational Collapse*, Chicago University Press, Chicago (1965)
43. M. Hashimoto, H. Seki, M. Yamada: *Prog. Theor. Phys.* **71**, 320 (1984)
44. E.P.J. van den Heuvel, in *IAU Symposium 125, The Origin and Evolution of Neutron Stars*, eds., D.F. Helfand, J.-H. Huang, Reidel, Dordrecht (1987)
45. C.J. Jog, R.A. Smith: *Astrophys. J.* **253**, 839 (1982)
46. P.B. Jones: *Phys. Rev. Letters* **83**, 3589 (1999)
47. C. Kittel, *Introduction to Solid State Physics*, Wiley, New York (1996), Chap. 3
48. K. Kolehmainen, M. Prakash, J.M. Lattimer, J.R. Treiner: *Nucl. Phys. A* **439**, 535 (1985)
49. D.Q. Lamb, H.M. Van Horn: *Astrophys. J.* **200**, 306 (1975)
50. L.D. Landau, E.M. Lifshitz: *Theory of Elasticity*, Pergamon Press, Oxford (1986)
51. W.D. Langer, L.C. Rosen, J.M. Cohen, A.G.W. Cameron: *Astrophys. Space Science* **5**, 259 (1969)
52. J.M. Lattimer, C.J. Pethick, D.G. Ravenhall, D.Q. Lamb: *Nucl. Phys.* **A432**, 646 (1985)
53. J.M. Lattimer, F. Douglas Swesty: *Nucl. Phys.* **A535**, 331 (1991)
54. C.P. Lorenz, PhD Thesis, University of Illinois, unpublished (1991)
55. C.P. Lorenz, D.G. Ravenhall, C.J. Pethick: *Phys. Rev. Letters* **70**, 379 (1993)
56. F.D. Mackie, G. Baym: *Nucl. Phys. A* **285**, 332 (1977)
57. P.N. McDermott, H.M. Van Horn, C.J. Hansen: *Astrophys. J.* **375**, 679 (1988)
58. J. Miralda-Escudé, P. Haensel, B. Paczyński: *Astrophys. J.* **362**, 572 (1990)
59. P. Möller: 1992, unpublished data, (private communication to B. Pichon, see Haensel and Pichon (1994))
60. P. Möller, J.R. Nix: *Atom. Data Nucl. Data Tables* **39**, 213 (1988)
61. W.D. Myers, W.J. Swiatecki: *Nucl. Phys.* **81**, 1 (1966)
62. J.W. Negele, D. Vautherin: *Nucl. Phys. A* **207**, 298 (1973)
63. R. Ogasawara, K. Sato: *Prog. Theor. Phys.* **68**, 222 (1982)
64. R. Ogasawara, K. Sato: *Prog. Theor. Phys.* **70**, 1569 (1983)
65. S. Ogata, S. Ichimaru: *Phys. Rev. A* **42**, 4867 (1990)
66. K. Oyamatsu, M. Hashimoto, M. Yamada: *Prog. Theor. Phys.* **72**, 373 (1984)
67. K. Oyamatsu: *Nucl. Phys.* **A561**, 431 (1993)
68. K. Oyamatsu, M. Yamada: *Nucl. Phys. A* **578**, 181 (1994)
69. V.R. Pandharipande, D. Pines, R.A. Smith: *Astrophys. J.* **208**, 550 (1976)
70. V.R. Pandharipande, D.G. Ravenhall, in: *Proceedings of the NATO Advanced Research Workshop on Nuclear Matter and Heavy Ion Collisions, Les Houches*, ed. by M.Soyeur et al. (Plenum, New York), pp.103 (1989)
71. C.J. Pethick, D.G. Ravenhall, C.P. Lorenz: *Nucl. Phys. A* **584**, 675 (1995)

72. C.J. Pethick, D.G. Ravenhall: *Annu. Rev. Nucl. Part. Sci.* **45**, 429 (1995)
73. C.J. Pethick, A.Y. Potekhin: *Phys. Lett.* B **427**, 7 (1998)
74. M.A. Preston, R. Bhaduri, *Structure of the Nucleus*, Addison-Wesley Publishing Company, Reading, Massachusetts (1975)
75. Prigogine, *Introduction to Thermodynamics of Irreversible Processes*, Interscience, New York (1961)
76. D.G. Ravenhall, C.D. Bennett, C.J. Pethick: *Phys. Rev. Letters* **28**, 978 (1972)
77. D.G. Ravenhall, C.J. Pethick, J.R. Wilson: *Phys. Rev. Letters* **50**, 2066 (1983)
78. F. Rembges, C. Freiburghans, T. Rauscher, F.-K. Thielemann, H. Schatz, M. Wiescher: *Astrophys. J.* **484**, 412 (1997)
79. P. Ring, P. Schuck: *The Nuclear Many-Body Problem*, Springer-Verlag, New York (1980) (Chapter 13).
80. R.E. Rutledge, L. Bilsten, E.F. Brown, G.G. Pavlov, V.E. Zavlin: *Astrophys. J* **514**, 945 (1999)
81. E.E. Salpeter: *Astrophys. J.* **134**, 669 (1961)
82. E.E. Salpeter, H.M. Van Horn: *Astrophys. J.* **155**, 183 (1969)
83. K. Sato: *Prog. Theor. Phys.* **62**, 957 (1979)
84. H. Schatz, L. Bildsten, A. Cumming, M. Wiescher: *Astrophys. J.* **524**, 1014 (1999)
85. S.L. Shapiro, S.A. Teukolsky, *White Dwarfs, Black Holes, and Neutron Stars*, Wiley, New York (1983)
86. T. Strohmayer, S. Ogata, H. Iyetomi, S. Ichimaru, H.M. Van Horn: *Astrophys. J.* **375**, 679 (1991)
87. K. Sumiyoshi, K. Oyamatsu, H. Toki: *Nucl. Phys.* **A 595**, 327 (1995)
88. S. Tsuruta, A.G.W. Cameron: *Can. J. Phys.* **43**, 2056 (1965)
89. Yu.L. Vartanyan, N.K. Ovakimova, *Soobtcheniya Byurakanskoi Observatorii* **49**, 87 (1976)
90. R.B. Wiringa, V. Fiks, A. Fabrocini: *Phys. Rev. C* **38**, 1010 (1988)
91. R.G. Wolff, PhD Thesis, Technische Universität, München (unpublished) (1983)
92. J.L. Zdunik, P. Haensel, B. Paczyński, J. Miralda-Escudé: *Astrophys. J.* **384**, 129 (1992)

# Kaon Condensation in Neutron Stars

Angels Ramos[1], Jürgen Schaffner-Bielich[2], and Jochen Wambach[3]

[1] Departament d'Estructura i Constituents de la Matèria, Universitat de Barcelona, Diagonal 647, 08028 Barcelona, Spain
[2] RIKEN BNL Research Center, Brookhaven National Laboratory, Upton, New York 11973-5000, USA
[3] Institut für Kernphysik, TU Darmstadt, Schlossgartenstr. 9, 64289 Darmstadt, Germany

**Abstract.** We discuss the kaon-nucleon interaction and its consequences for the change of the properties of the kaon in the medium. The onset of kaon condensation in neutron stars under various scenarios as well its effects for neutron star properties are reviewed.

## 1 Introduction – hadrons in dense matter

Due to its non-abelian structure, Quantum Chromodynamics (QCD) becomes very strongly interacting and highly nonlinear at large space-time distances. As a consequence, quarks and gluons condense in the physical vacuum with a gain in condensation energy density of $\Delta\epsilon_0 \sim 500$ MeV/fm$^3$. The condensation of quarks is associated with the spontaneous breaking of chiral symmetry, an additional symmetry of the QCD Lagrangian in the absence of (current) quark masses.

This limit is well justified in the up-down quark sector and to a somewhat lesser extent for the strange quark. There is good evidence that the mechanism for spontaneous chiral symmetry breaking is provided by classical gluon field configurations in euclidean space called 'instantons'. These provide effective quark-(anti)quark interactions which are strong enough to cause a BCS like transition to a condensed state of quarks and antiquarks. It has been shown, that this picture provides an excellent description of hadronic states and correlation functions [1] and it is fair to say that the low-lying hadron spectrum is dominated by spontaneous chiral symmetry breaking with confinement playing a much lesser role.

These observations form the basis for discussing the properties of hadrons, or more precisely hadronic correlation functions, under extreme conditions in temperature and/or density as encountered in the early universe or in the interior of neutron stars. It is well established from lattice QCD that the vacuum undergoes a phase transition (or at least a sharp cross over) when heated. Chiral symmetry is restored in this phase accompanied by a nearly vanishing chiral quark condensate. Though not calculable at present from first principles, the same is expected to happen at finite density. Since light hadrons are dynamically driven by chiral symmetry breaking it is obvious to ask how hadronic properties are related to the vacuum structure and its changes with temperature and density. This is far from trivial and under intense experimental and theoretical scrutiny at present.

As detailed below the most economical way for treating hadrons in matter under extreme conditions is to resort to 'effective field theories' in which hadrons rather than quarks and gluons appear as the fundamental degrees of freedom. Formally one identifies the pertinent quark currents $J_{\Gamma_j}(x) = \bar{q}(x)\Gamma_j q(x)$, $\Gamma_j = 1, \gamma_5, \gamma_\mu,..$ with elementary hadronic fields $\phi_i(x)$ for which the most general effective Lagrangian, consistent with the underlying symmetries and anomaly structure of QCD, is written down. We recall that the spontaneous breaking of chiral symmetry has two important consequences. One is the appearance of (nearly) massless Goldstone bosons (pions, kaons, eta) and the other the absence of parity doublets in the hadron spectrum ($m_\pi \neq m_{f_0}, m_\rho \neq m_{a_1}$ etc). For the present discussion the first is the most relevant. Chiral symmetry does more than just predict the existence of Goldstone bosons. It also prescribes and severely restricts their mutual interactions as well as those with other hadrons. The most rigorous treatment is in terms of 'chiral perturbation theory' [2] but other 'chiral effective Lagrangians' including e.g. vector mesons can be devised [3,4].

To elucidate the connection between the vacuum structure (the chiral condensate) and the properties of light hadrons let us consider a medium of hadronic matter in thermal and chemical equilibrium. This is of course well suited for neutron stars. The QCD partition function is then given in the grand-canonical ensemble as

$$\mathcal{Z}_{QCD}(V,T,\mu_q) = \text{Tr}\exp\{-(\hat{H}_{QCD} - \mu_q \hat{N}_q)/T\} \quad, \tag{1}$$

where $\hat{H}_{QCD}$ denotes the QCD Hamiltonian, $\hat{N}_q$ the quark number operator and $\mu_q$ the quark chemical potential. Statistical expectation of operators are then given as

$$\langle\langle \hat{O} \rangle\rangle = \mathcal{Z}^{-1} \sum_n \langle n|\hat{O}|n\rangle \exp\{-(E_n - \mu_q N_n)/T\} \quad, \tag{2}$$

where $E_n$ are the exact QCD energies (hadrons). The quark condensate $\langle\langle \bar{q}q \rangle\rangle$ can be obtained directly from the free energy density

$$\Omega_{QCD}(T,V) = -\lim_{V\to\infty} \frac{T}{V} \ln \mathcal{Z}_{QCD}(V,T,\mu_q) \tag{3}$$

via the Feynman-Hellmann theorem as

$$\langle\langle \bar{q}q \rangle\rangle = \frac{\partial \Omega_{QCD}}{\partial m_q^\circ} \quad, \tag{4}$$

where $m_q^\circ$ denotes the bare (current) quark mass. An obvious first step is to approximate the free energy density by an ideal gas of hadrons. Using the Gell-Mann–Oakes–Renner relation for the vacuum chiral condensate

$$m_\pi^2 f_\pi^2 = -2 m_q^\circ \langle \bar{q}q \rangle \quad, \tag{5}$$

where $m_\pi$ is the pion mass and $f_\pi$ the pion weak-decay constant, one then finds

$$\frac{\langle\langle \bar{q}q \rangle\rangle}{\langle \bar{q}q \rangle} = 1 - \sum_h \frac{\Sigma_h \varrho_h^s(\mu_q, T)}{f_\pi^2 m_\pi^2} \quad. \tag{6}$$

Here

$$\Sigma_h = m_q^\circ \frac{\partial m_h}{\partial m_h^\circ} \qquad (7)$$

denotes the 'sigma' commutator' (related to the scalar density of quarks in a given hadron) and $m_h$ the vacuum mass of a given hadron. At low temperature and small baryochemical potential ($\mu_B = 3\mu_q$), in which the hadron gas is dominated by thermally excited pions and a free Fermi gas of nucleons, (6) leads to the celebrated leading-order result

$$\frac{\langle\!\langle \bar{q}q \rangle\!\rangle}{\langle \bar{q}q \rangle} = 1 - \frac{T^2}{8f_\pi^2} - 0.3 \frac{\rho}{\rho_0} \qquad (8)$$

where $\rho_0 = 0.16/\text{fm}^3$ is the saturation density of symmetric nuclear matter. This result is model independent and indicates that the mere presence of an ideal gas of hadrons already alters the vacuum structure and leads to a decrease of the condensate, without changing the vacuum properties of the hadrons! Obviously medium-modifications of hadrons and the corresponding non-trivial change of the QCD vacuum has to involve hadronic interactions. They become increasingly important as the medium grows hotter and denser, i.e. as the point of chiral restoration is approached. Thus, the theoretical description in terms of hadrons becomes very complex, involving more and more degrees of freedom. In the vicinity of the restoration transition, in addition, non-perturbative methods are called for which is far from trivial in effective field theories.

In terms of effective fields $\phi_j(x)$ representing the pertinent quark currents $J_{\Gamma_j}(x)$, hadronic correlation functions in a hot and dense environment are defined as the (retarded) current-current correlation functions

$$D_{\phi_j}(\omega, \boldsymbol{q}) = -i \int d^4x \, e^{i q x} \theta(x_0) \langle\!\langle [\phi_j(x), \phi_j(0)] \rangle\!\rangle. \qquad (9)$$

Note that, in contrast to the vacuum, these correlators depend on energy $\omega$ and three-momentum $\boldsymbol{q}$ separately since Lorentz invariance is explicitly broken by the presence of matter. Equation (9) can be rewritten in terms of the self energy $\Sigma_{\phi_j}$ as

$$D_{\phi_j}(\omega, \boldsymbol{q}) = (\omega^2 - \boldsymbol{q}^2 - m_{\phi_j}^2 - \Sigma_{\phi_j}(\omega, \boldsymbol{q}))^{-1} \qquad (10)$$

where $m_{\phi_j}$ denotes the (bare) mass of the field $\phi_j$ and all interaction effects are incorporated via $\Sigma_{\phi_j}$ which depends on $T$ and $\mu_B$. Given the effective Lagrangian, the objective is then to evaluate $\Sigma_{\phi_j}$ as realistic as possible. This is usually attempted by employing 'chiral counting rules' for evaluating loop diagrams contributing to $\Sigma_{\phi_j}$ in the low-density and low-temperature limit and by adjusting the parameters of the effective Lagrangian to as many data for elementary scattering processes as available. If everything is done consistently, chiral symmetry is properly incorporated and the relation between hadronic medium modifications and vacuum changes can be inferred.

In the vacuum, the hadronic correlators $D_{\phi_j}$ are usually dominated by a few fairly sharp hadronic 'resonances'. These are visible as 'peaks' in the corresponding 'spectral functions'

$$\rho_{\phi_j}(\omega, \mathbf{q}) = -\frac{1}{\pi} \mathrm{Im} D_{\phi_j}(\omega, \mathbf{q}) \tag{11}$$

at the hadronic vacuum mass $m_{\phi_j}$. Two things will happen in the interacting medium. On the one hand the mass will change to an effective 'pole-mass'

$$m_{\phi_j}^{*2} = m_{\phi_j}^2 + \mathrm{Re}\Sigma_{\phi_j}(m_{\phi_j}^{*2}, 0) \quad . \tag{12}$$

In fact, one may have *several* solutions and the in-medium spectrum shows more than one 'peak'. As we shall see, this is usually the case. On the other hand, the imaginary part $\mathrm{Im}\Sigma_{\phi_j}$ acquires additional pieces through the interactions with the medium giving rise to an increased width. If the width becomes too large, the peak structure is washed out and the notion of a 'quasiparticle' is lost. This happens when the 'quasiparticle' energy

$$\omega_{\phi_j}^2(\mathbf{q}) = \mathbf{q}^2 + m_{\phi_j}^2 + \mathrm{Re}\Sigma_{\phi_j}(\omega_{\phi_j}^2, \mathbf{q}) \tag{13}$$

is no longer large compared to the width

$$\Gamma_{\phi_j}(\omega_{\phi_j}(\mathbf{q})) = -\frac{1}{2\pi} \mathrm{Im}\Sigma(\omega_{\phi_j}, \mathbf{q}) \quad . \tag{14}$$

This has to be kept in mind when describing in-medium hadrons.

In the following, we will discuss the properties of kaons in the medium and its consequences for the properties of neutron stars. In sect. 6, the elementary kaon-nucleon interaction and the in-medium changes of the kaons as seen in kaonic atoms are described. Chiral effective interactions are constructed which can describe these data and are used to extract the optical potential of kaons at finite density. Section 3 utilizes the results found for the kaon optical potential to apply it to the equation of state for neutron stars. Implications for a first order phase transition to a kaon condensed state are listed. Effects on the onset of kaon condensation by the presence of hyperons in matter are studied. Finally, we summarize in sect. 4 and give an outlook.

## 2 Kaons in dense matter

The properties of kaons and antikaons in the nuclear medium have been the object of numerous investigations since the possibility of the existence of a kaon condensed phase in dense nuclear matter was pointed out by Kaplan and Nelson [5]. If the $K^-$ meson develops sufficient attraction in dense matter it could be energetically more favorable, after a certain critical density, to neutralize the positive charge with antikaons rather than with electrons. A condensed kaon phase would then start to develop, changing drastically the properties of dense neutron star matter [6–15]. In fact, kaonic atom data, a compilation of which is given in [16], favor an attractive $K^-$ nucleus interaction. On the other hand, the enhancement of the $K^-$ yield in Ni+Ni collisions measured recently by the KaoS collaboration at GSI [17] can be explained by a strong attraction in the medium for the K$^-$ [18–21]. However, the antikaons might feel a repulsive potential at the relatively high temperatures attained in heavy-ion reactions, and an alternative mechanism, based on the production of antikaons via an in-medium enhanced $\pi \Sigma \to K^- p$ reaction, has been suggested [22].

The theoretical investigations that go beyond pure phenomenology [23] have mainly followed two different strategies. One line of approach is that of the mean field models, built within the framework of chiral Lagrangians [21,24–26] or based on the relativistic Walecka model which are extended to incorporate strangeness in the form of hyperons [57] or kaons [27] or by using explicitly the quark degrees of freedom [28]. The other type of approach aims at obtaining the in-medium $\bar{K}N$ interaction microscopically by incorporating the medium modifications in a $\bar{K}N$ amplitude using (chiral) effective interactions that reproduces the low energy scattering data and generates the $\Lambda(1405)$ resonance dynamically [29–35]. In this section, we focus on this latter perspective, since it allows one to systematically study the importance of all the different mechanisms that might modify the $\bar{K}N$ interaction in the medium from that in free space.

## 2.1 $\bar{K}N$ interaction in the medium

The free $\bar{K}N$ scattering observables ($\bar{K} = K^-$ or $\bar{K}^0$) are derived from the scattering amplitude, obtained from the Bethe-Salpeter equation

$$T = V + VGT, \qquad (15)$$

which is depicted diagrammatically in Fig. 1. Note that, in the case of $\bar{K}^- p$ scattering, this is a coupled-channel equation involving ten intermediate states, namely $K^- p$, $\bar{K}^0 n$, $\pi^0 \Lambda$, $\pi^+ \Sigma^-$, $\pi^0 \Sigma^0$, $\pi^- \Sigma^+$, $\eta \Lambda$, $\eta \Sigma^0$, $K^+ \Xi^-$ and $K^0 \Xi^0$. The loop operator $G$ stands for the diagonal intermediate meson-baryon ($MB$) propagator and $V$ is a suitable $MB \to M'B'$ transition potential.

A connection with the chiral Lagrangian was established in [3], where the properties of of the $S = -1$ meson-baryon sector were studied in a potential model [3], such that, in Born approximation, it had the same S-wave scattering length as the chiral Lagrangian, including both the lowest order and the

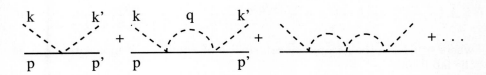

**Fig. 1.** Diagrammatic representation of the Bethe-Salpeter equation for $\bar{K}N$ scattering.

momentum dependent $p^2$ terms. No $\eta\Lambda$, $\eta\Sigma$ or $K\Xi$ channels were considered, on the basis that they were not opened at the $K^-p$ threshold. By fitting five parameters, corresponding to, so far, unknown parameters of the second order chiral Lagrangian plus the range parameters of the potential, the $\Lambda(1405)$ resonance was generated as a quasi-bound meson-baryon state and the $K^-p \to MB$ cross sections, as well as the available branching rations at threshold, were well reproduced. Note that being close to a resonance forces one to resort to nonperturbative approaches, such as summing the infinite Bethe-Salpeter series as represented diagrammatically in Fig. 1. A recent work [36], which shares many points with [3], showed that all the strangeness $S = -1$ meson-baryon scattering observables near threshold could be reproduced using only the lowest order Lagrangian in the Bethe-Salpeter equation and one parameter, the cut-off $q_{\max}$ that regularizes the loop function $G$. All 10 meson-baryon states that can be generated from the octet of pseudoscalar mesons and the octet of ground-state baryons were included, the additional $\eta\Lambda$ state being a quite relevant one. The success of this method in reproducing the scattering observables is analogous to that obtained in the meson-meson sector [37]. An explanation was found in [38] by applying the Inverse Amplitude Method in coupled channels to the same problem, with the lowest and next-to-lowest order meson-meson Lagrangian. It was shown that the selection of an appropriate cut off for a particular $I, J$ channel could minimize the contribution of the next-to-lowest order Lagrangian, reducing the relevant higher order terms to those iterated by the Bethe-Salpeter equation with the lowest order Lagrangian, which is the simplified method followed in [36,37] (see also the discussion in the review [39]).

The most obvious medium effect is that induced by Pauli-blocking on the nucleons in the intermediate states. This makes the $\bar{K}N$ interaction density dependent and modifies, in turn, the $K^-$ properties from those in free space. These effects were already included a long time ago in the context of Brueckner-type many body theory, using a separable $\bar{K}N$ interaction [29]. Some more recent works [31–35] take the $\bar{K}N$ interaction from the chiral Lagrangian in S-wave. However, as seen in the recent work by Tolos et al. [40] which uses the Bonn $\bar{K}N$ potential model [41], the incorporation of higher angular momenta has non-negligible effects on the properties of the antikaon at high momentum, which is of relevance for the analysis of in-medium effects for the $K^-$ in heavy-ion collisions.

In the actual calculations, the effect of Pauli blocking is incorporated by replacing the free nucleon propagator in the loop function $G$ for intermediate $\bar{K}N$ states by an in-medium one of the type

$$A(\sqrt{s} - q^0, -\boldsymbol{q}, \rho) = \frac{1 - n(\boldsymbol{q}_{\text{lab}})}{\sqrt{s} - q^0 - E(-\boldsymbol{q}) + i\epsilon} + \frac{n(\boldsymbol{q}_{\text{lab}})}{\sqrt{s} - q^0 - E(-\boldsymbol{q}) - i\epsilon}, \quad (16)$$

where $n(\boldsymbol{q}_{\text{lab}})$ is the occupation probability of a nucleon of momentum $\boldsymbol{q}_{\text{lab}}$ in the lab frame.

The most spectacular consequence of the Pauli principle is that the blocking of intermediate states shifts the resonance to higher energy. This changes the $\bar{K}N$ interaction at threshold from being repulsive in free space to being attractive in the medium. Therefore, antikaons develop an attractive optical potential which, incorporated again in the $\bar{K}N$ states of the in-medium scattering equation, may compensate the upward shifting of these intermediate states induced by Pauli blocking. This feedback has been recently confirmed by the calculation of Lutz [34], where the dressing of the antikaon is incorporated in the intermediate loops in a self-consistent manner. The $\Lambda(1405)$ resonance remains then unchanged, in qualitative agreement with what was noted in [30] using a constant mean field potential for the $\bar{K}$.

Since the dressing of the antikaon turns out to be so relevant, one might wonder about dressing the other mesons or baryons that play a role in the $\bar{K}N$ system. This has been explored in the recent work [35], where a self-consistent antikaon self-energy is obtained including the dressing of the pions in the $\pi\Lambda$, $\pi\Sigma$ intermediate states, which are the ones that couple strongly to $\bar{K}N$.

Incorporating the medium modified mesons in the calculation is technically achieved by replacing the free meson propagator in the loop function $G$ by

$$D(q^0, \boldsymbol{q}, \rho) = \frac{1}{(q^0)^2 - \boldsymbol{q}^2 - m^2 - \Pi(q^0, \boldsymbol{q}, \rho)} = \int_0^\infty d\omega\, 2\omega\, \frac{S(\omega, \boldsymbol{q}, \rho)}{(q^0)^2 - \omega^2 + i\epsilon}, \tag{17}$$

where $\Pi(q^0, \boldsymbol{q}, \rho)$ is the meson self-energy. The Lehman representation shown in the second equality of (17) introduces the spectral density, $S(\omega, \boldsymbol{q}, \rho) = -\mathrm{Im} D(\omega, \boldsymbol{q}, \rho)/\pi$, which in the case of free mesons reduces to $\delta(\omega - \omega(\boldsymbol{q}))/2\omega(\boldsymbol{q})$. With these modifications the loop integral becomes

$$G(P^0, \boldsymbol{P}, \rho) = \int_{|\boldsymbol{q}|<q_{\max}} \frac{d^3q}{(2\pi)^3} \frac{M}{E(-\boldsymbol{q})} \int_0^\infty d\omega\, S(\omega, \boldsymbol{q}, \rho)$$
$$\times \left\{ \frac{1 - n(\boldsymbol{q}_{\mathrm{lab}})}{\sqrt{s} - \omega - E(-\boldsymbol{q}) + i\epsilon} + \frac{n(\boldsymbol{q}_{\mathrm{lab}})}{\sqrt{s} + \omega - E(-\boldsymbol{q}) - i\epsilon} \right\}, \tag{18}$$

where $(P^0, \boldsymbol{P})$ is the total four-momentum in the lab frame and $s = (P^0)^2 - \boldsymbol{P}^2$. Note that the loop function $G$ does-not contain the $V$ and $T$ amplitudes. This simplification is possible in the chiral approach of [36], where it was shown that the the amplitudes factorize on-shell out of the loop, since the off shell part could be absorbed into a renormalization of the coupling constant $f_\pi$. These arguments can not be applied in potential models and a coupled system of integral equations must be solved. A reasonable simplification is obtained if the dressing of the antikaon in the loop function $G$ is taken into account via an energy-independent self-energy, $\Pi(q^0 = \varepsilon_{qp}(\boldsymbol{q}), \boldsymbol{q}, \rho)$, evaluated at the quasiparticle energy, $\varepsilon_{qp}(\boldsymbol{q})$, which is the solution of the in-medium dispersion relation

$$\varepsilon_{qp}^2(\boldsymbol{q}) = \boldsymbol{q}^2 + m_K^2 + \Pi(q^0 = \varepsilon_{qp}(\boldsymbol{q}), \boldsymbol{q}, \rho). \tag{19}$$

This is the prescription followed in [22,40] and amounts to approximate the actual antikaon spectral function by a symmetric pseudo-Lorentzian peak at the quasiparticle energy given by (19). With this assumption, the loop function $G$ looks like the free one, but replacing the free antikaon energy $\sqrt{\boldsymbol{q}^2 + m_K^2}$ by the complex quasiparticle one, $\varepsilon_{qp}(\boldsymbol{q})$.

One might also include the dressing of the baryons by assuming that the single particle energy $E(-\boldsymbol{q})$ in (16) to contain a mean-field potential of the type $U_0 \times \rho/\rho_0$, as done in [35,40]. For the nucleon, a reasonable depth value is $U_0^N = -70$ MeV, as suggested by numerous calculations of the nucleon potential in nuclear matter. For the $\Lambda$ hyperon, one can take $U_0^\Lambda = -30$ MeV, as implied by extrapolating the experimental $\Lambda$ single particle energies in $\Lambda$ hypernuclei to bulk matter [42]. For the $\Sigma$ hyperon, there is no conclusive information on the potential. Early phenomenological analyzes [43] and calculations [44] found the $\Sigma$ atomic data to be compatible with $U_0^\Sigma \sim -30$ MeV, but more recent analysis indicate a repulsive potential in the nuclear interior [45]. As shown in [35], changing the $\Sigma$ depth from $-30$ to $+30$ MeV does not change the results for the properties of the $K^-$ in the medium considerably.

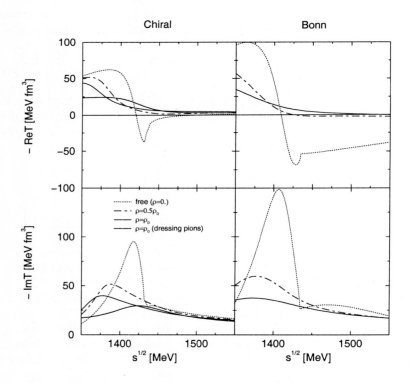

**Fig. 2.** Real and imaginary parts of the $I = 0$ $\bar{K}N$ scattering amplitude as functions of $\sqrt{s}$ for $\mid \boldsymbol{p}_K + \boldsymbol{p}_N \mid = 0$ and several densities (from [35] and [40]).

Using the dressed meson-baryon loop in the coupled-channel Bethe-Salpeter equation, one obtains the in-medium $\bar{K}N$ amplitude $T_{\text{eff}}(P^0, \boldsymbol{P}, \rho)$. In Fig. 2 we compare the free amplitude ($\rho = 0$, dotted line) in the $I = 0$, $L = 0$ channel with that obtained at nuclear densities $\rho = \rho_0 = 0.17$ fm$^{-1}$ (solid line) and

$\rho = 0.5\rho_0$ (dot-dashed line). Two different models are shown, that of [35], based on the lowest order meson-baryon chiral Lagrangian, and that of [40], based on the potential model of the Bonn group [41]. In spite of the appreciable differences seen in the free scattering amplitudes, the medium modified ones show the same qualitative trends. Note also how the real part of the amplitude (upper panels) at the $K^-p$ threshold ($\sqrt{s} = 1433$ MeV) is repulsive in free space and attractive in the medium. The thick solid line on the left panels show the effect on the $K^-p$ amplitude by dressing the pions in the intermediate states [35].

Most of the available models study the in-medium $\bar{K}N$ amplitude in S-wave. However, if one aims at extracting the properties of antikaons through the analysis of heavy-ion collisions, one must keep in mind that they are created at a finite momentum of around $250 - 500$ MeV/c, hence the effect of higher partial waves might be relevant. The meson-exchange $\bar{K}N$ potential of the Jülich group [41] is given in partial waves and the main results from a recent study [40] on the effect of the angular momentum states higher than the commonly considered $L = 0$ one will be summarized in the next subsection. From the chiral perspective, the P-wave amplitudes from the next-to-leading order $\bar{K}N$ chiral Lagrangian have been identified and, recently, the parameters have been fitted to reproduce a large amount of low energy data [46]. However, a nuclear medium application of this model is not available yet.

## 2.2 In medium $\bar{K}$ properties

The $\bar{K}$ self-energy is obtained by summing the in-medium $\bar{K}N$ interaction, $T_{\text{eff}}(P^0, \boldsymbol{P}, \rho)$, over the nucleons in the Fermi sea

$$\Pi_{\bar{K}}(q^0, \boldsymbol{q}, \rho) = 2 \sum_{N=n,p} \int \frac{d^3 p}{(2\pi)^3} n(\boldsymbol{p})\, T_{\text{eff}}(q^0 + E(\boldsymbol{p}), \boldsymbol{q} + \boldsymbol{p}, \rho) \,. \quad (20)$$

Note that a self-consistent approach is required since one calculates the $\bar{K}$ self-energy from the effective interaction $T_{\text{eff}}$ which uses $\bar{K}$ propagators which themselves include the self-energy being calculated.

A P-wave contribution to the $\bar{K}$ self-energy coming from the coupling of the $\bar{K}$ meson to hyperon-hole excitations can be easily included (if it is not already contained in $T_{\text{eff}}$) and the expression can be found in [35]. In that work the pions are also dressed through a pion self-energy that contains one- and two-nucleon absorption and is conveniently modified to include the effect of nuclear short-range correlations (see [47] for details). The resulting pion spectral density in nuclear matter at density $\rho = \rho_0$ is shown in Fig. 3 as a function of the pion energy for several momenta. The strength is distributed over a wide range of energies and, as the pion momentum increases, the position of the peak is increasingly lowered from the corresponding one in free space as a consequence of the attractive pion-nuclear potential. Note that, to the left of the peaks, there appears the typical structure of the $1p1h$ excitations which give rise to $1p1h\Lambda$ and $1p1h\Sigma$ components in the effective $\bar{K}N$ interaction. Although not visible

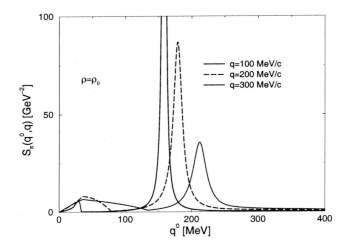

**Fig. 3.** Pion spectral density at $\rho = \rho_0$ for several momenta (from [35]).

in the linear scale used in Fig. 3, there is some additional strength at energies around 300 MeV associated to the excitation of the $\Delta$ resonance.

The spectral function of a $K^-$ meson of zero momentum obtained with the chiral model of [35] is shown in Fig. 4 for various densities: $\rho_0$, $\rho_0/2$ and $\rho_0/4$. The results in the upper panel include only Pauli blocking effects, i.e. the nucleons propagate as in (16) but the mesons behave as in free space. At $\rho_0/4$ one clearly sees two excitation modes. The left one corresponds to the $K^-$ pole branch, appearing at an energy smaller than the kaon mass, $m_K$, due to the attractive medium effects. The peak on the right corresponds to the $\Lambda(1405)$-hole excitation mode, located above $m_K$ because of the shifting of the $\Lambda(1405)$ resonance to energies above the $K^-p$ threshold. As density increases, the $K^-$ feels an enhanced attraction while the $\Lambda(1405)$-hole peak moves to higher energies and loses strength, a reflection of the tendency of the $\Lambda(1405)$ to dissolve in the dense nuclear medium. These features were already observed in [31,33]. The (self-consistent) incorporation of the $\bar{K}$ propagator in the Bethe-Salpeter equation softens the effective interaction, $T_{\text{eff}}$, which becomes more spread out in energies (solid and dot-dashed lines in Fig. 2). The resulting $K^-$ spectral function (middle panel in Fig. 4) shows the displacement of the resonance to lower energies because, as already noted, the attraction felt by the $\bar{K}$ meson lowers the threshold for the $\bar{K}N$ states that had been increased by the Pauli blocking on the nucleons. This has a compensating effect and the resonance moves backwards, slightly below its free space value. The $K^-$ pole peak appears at similar or slightly smaller energies, but its width is larger, due to the strength of the intermediate $\bar{K}N$ states being distributed over a wider region of energies. Therefore the $K^-$ pole and the $\Lambda(1405)$-hole branches merge into one another and can hardly be distinguished. Finally, when the pion is dressed according

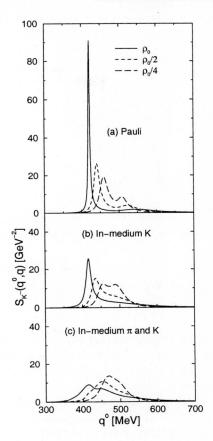

**Fig. 4.** $K^-$ spectral density for zero momentum from the chiral model of ref. [35]

to the spectral function shown in Fig. 3 the effective interaction $T_{\rm eff}$ becomes even smoother (thick solid lines in Fig. 2). The resulting $K^-$ spectral function is displayed in the bottom panel in Fig. 4. Even at very small densities one can no longer distinguish the $\Lambda(1405)$-hole peak from the $K^-$ pole one. As density increases, the attraction felt by the $K^-$ is more moderate and the $K^-$ pole peak appears at higher energies than in the other two approaches. However, more strength is found at very low energies, especially at $\rho_0$, due to the $1p1h$ $2p2h$ components of the pionic strength, which couple the $\bar{K}N$ state to the $1p1h\Sigma$ and $2p2h\Sigma$ ones. It is precisely the opening of the $\pi\Sigma$ channel, on top of the already opened $1p1h\Sigma$ and $2p2h\Sigma$ ones, which causes a cusp structure to appear slightly above 400 MeV.

The calculation of [40] using the Bonn $\bar{K}N$ potential obtains similar results, which are shown in Fig. 5 for the same three densities. We notice some structure of the spectral function to the left of the quasiparticle peak at energies of the $\bar{K}$ around 320 − 360 MeV. This is the in-medium reflection of a singularity in the

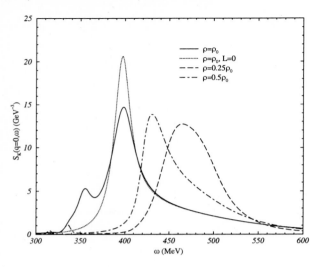

**Fig. 5.** $K^-$ spectral density for zero momentum using the Bonn $\bar{K}N$ potential (from [40]).

$L = 1$, $I = 1$ free space amplitude around the mass of the $\Sigma$ baryon induced by the $\Sigma$-pole diagram present in the Bonn $\bar{K}N$ potential [41]. This peak is therefore indicating the physical excitation of $\Sigma h$ states with antikaon quantum numbers. The dotted line shows the spectral density at $\rho = \rho_0$ but keeping only the $L = 0$ component of the $\bar{K}N$ interaction. In agreement with the behavior of the complex antikaon potential at zero momentum shown below, we observe that the location of the quasiparticle peak (driven by the real part) only moves a few MeV, while the width (driven by the imaginary part) gets reduced by about 30% when including higher partial waves.

One may define a non-relativistic antikaon single-particle potential from the self-energy at the quasiparticle energy via the relation

$$U_K(q) = \frac{\Pi_K(\varepsilon_{qp}(\boldsymbol{q}), \boldsymbol{q}, \rho)}{2m_K} \ . \tag{21}$$

The real and imaginary parts of the antikaon potential at $\rho = \rho_0$, obtained from the chiral model of ref. [35], are shown in Fig. 6 as function of the antikaon momentum for two approximations, one in which only the antikaon self-energy is considered in the intermediate loops (thin solid lines) and another in which the pions are also dressed (thick solide line). Note that the antikaon potential obtained when the pions are also dressed has much less structure. This is due to the smoother in-medium amplitude, but also to the different quasiparticle energy at which the antikaon self-energy is evaluated. This quasiparticle energy is more attractive when only antikaons are dressed and, hence, the amplitude is explored at lower energy regions, closer to the position of the in-medium $\Lambda(1405)$ resonance. Dressing the intermediate pions gives an antikaon potential depth at

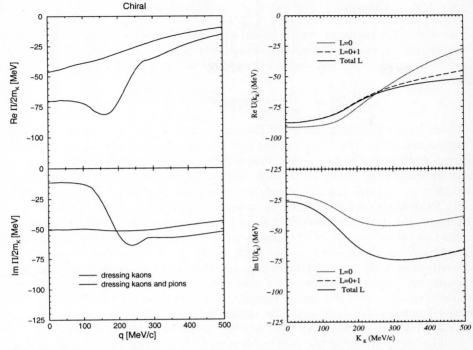

**Fig. 6.** Real and imaginary parts of the $\bar{K}$ optical potential at $\rho = \rho_0$ as functions of the antikaon momentum, as obtained from the chiral model of [35].

**Fig. 7.** The same as Fig. 6, but obtained using the Bonn $\bar{K}N$ potential (taken from [40]).

zero momentum of $-45$ MeV. This is about half the attraction of that obtained with other recent models and approximation schemes [21,26–28,32,33], which give rise to potential depths at the center of the nucleus between $-140$ and $-75$ MeV. Results obtained by the model of [40] using the Bonn $\bar{K}N$ potential [41], where only kaons are dressed, are shown in Fig. 7. In this figure one can also see the effect of including the higher partial waves of the Bonn $\bar{K}N$ interaction. We observe that the antikaon nuclear potential at zero momentum receives some contribution of partial waves higher than $L = 0$, due to the fact that the $\bar{K}$ meson interacts with nucleons that occupy states up to the Fermi momentum, giving rise to finite $\bar{K}N$ relative momenta of up to around 90 MeV/c. Clearly, the effect of the higher partial waves increases with increasing $\bar{K}$ momentum, flattening out the real part of the potential and producing more structure to the imaginary part. At an antikaon momentum of around 500 MeV/c, the inclusion of the higher partial waves practically doubles the size of the antikaon potential with respect to the S-wave value.

## 2.3 Kaonic atoms

Since the $K^-$ in kaonic atoms are bound with small (atomic) energies, their study requires the knowledge of the antikaon self-energy at $(q^0, \boldsymbol{q}) = (m_K, \boldsymbol{0})$. The real and imaginary parts of the isospin averaged in-medium scattering length, defined as

$$a_{\rm eff}(\rho) = -\frac{1}{4\pi}\frac{M}{m_K + M}\frac{\Pi_{\bar{K}}(m_K, \boldsymbol{q}=0, \rho)}{\rho}, \qquad (22)$$

obtained from the chiral model of [35], are shown in Fig. 8 as function of the nuclear density $\rho$. The change of the real part of the effective scattering length Re $a_{\rm eff}$ from negative to positive values indicates the transition from a repulsive interaction in free space to an attractive one in the medium. As shown by the dotted line, this transition happens at a density of about $\rho \sim 0.1\rho_0$ when only Pauli effects are considered, in agreement with what was found in [31,32]. However, this transition occurs at even lower densities ($\rho \sim 0.04\rho_0$) when one considers the self-energy of the mesons in the description, whether one dresses only the $\bar{K}$ meson (dashed line) or both the $\bar{K}$ and $\pi$ mesons (solid line). The deviations from the approach including only Pauli blocking or those dressing the mesons are quite appreciable over a wide range of densities. The thin solid lines show the results obtained with a repulsive $\Sigma$ potential depth of +30 MeV. The deviations from the thick solid line, obtained for an attractive potential depth of $-30$ MeV, are smaller than 10% and only show up at higher densities.

The implications of the density dependent scattering length displayed in Fig. 8 on kaonic atoms have been recently analyzed [48] in the framework of a local density approximation, which amounts to replace the nuclear matter density $\rho$ by the density profile $\rho(r)$ of the particular nucleus. The results displayed in Fig. 9 show that both the energy shifts and widths of kaonic atom states agree well with the bulk of experimental data [49].

Reproducing kaonic atom data with this moderately attractive antikaon nucleus potential of $-45$ MeV is in contrast with the depth of around $-200$ MeV obtained from a best fit to $K^-$ atomic data with a phenomenological potential that includes an additional non-linear density dependent term [16]. A hybrid model, combining a relativistic mean field approach in the nuclear interior and a phenomenological density dependent potential at the surface that is fitted to $K^-$ atomic data, also favors a strongly attractive $K^-$ potential of depth $-180$ MeV [50]. On the other hand, the early Brueckner-type calculations of [29] also obtained a shallow $K^-$-nucleus potential, of the order of $-40$ MeV at the center of $^{12}$C, and predicted reasonably well the $K^-$ atomic data available at that time. Acceptable fits to kaonic atom data have also been obtained using charge densities and a phenomenological $T_{\rm eff}\rho$ type potential with a depth of the order of $-50$ MeV in the nuclear interior [51], which goes down to $-80$ MeV, when matter densities are used instead [16].

A clarifying quantitative comparison of kaonic atom results obtained with various $K^-$-nucleus potentials can be found in [52]. There, the reasonable reproduction of data obtained with the chiral antikaon-nucleus potential of [35], shown in Fig. 9, is quantified with a $\chi^2/d.o.f. = 3.8$. This potential is then

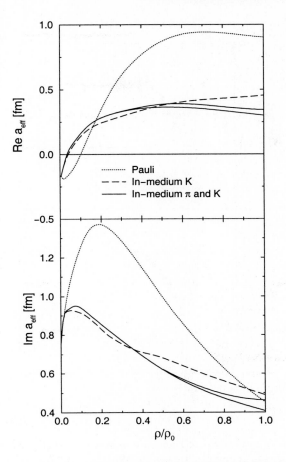

**Fig. 8.** $K^-N$ scattering length as a function of density (from [35]).

modified by an additional phenomenological piece, linear in density, which is fitted to the known data and is able to bring the agreement to the level of $\chi^2/d.o.f. = 1.6$. This results into a final potential which is slightly more attractive ($-50$ MeV at $\rho_0$) and has a reduced imaginary part by about a factor 2. The work [52] reemphasizes that kaonic atoms only explore the antikaon potential at the surface of the nucleus. Therefore, although all models predict attraction for the $K^-$-nucleus potential, the precise value of its depth at the center of the nucleus, which has important implications for the occurrence of kaon condensation, is still not known. It is then necessary to gather more data that could help in disentangling the properties of the $\bar{K}$ in the medium. Apart from the valuable information that can be extracted from the production of $K^-$ in heavy-ion collisions, one could also measure deeply bound kaonic states, which have been predicted to be narrow [48,52,53] and could be measured in $(K^-,\gamma)$ [48] or $(K^-,p)$ reactions [54,55].

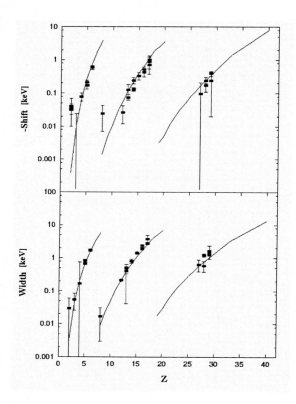

**Fig. 9.** Energy shifts and widths of kaonic atom states (from [48]). The experimental data are taken from the compilation given in [49].

## 3 Kaons in neutron stars

Kaon condensation has first thought to be irrelevant for neutron stars as their mass has to be lowered so drastically to appear in beta-stable neutron star matter [57]. Nevertheless, as demonstrated in the last sections, the in-medium effects for the kaons, especially for the $K^-$ can be quite pronounced which reopened the issue of kaon condensation for neutron stars [5]. The topic has been extensively discussed in the literature over the last years (see e.g. [6–14]). In all these approaches, the antikaon-nucleon interaction has been parameterized in effective field theoretical models which were guided by the investigations of the last sections. So far, a consistent coupled channel calculation incorporating a realistic nucleon-nucleon interactions as well as kaon-nucleon interactions has not been performed due to the complexity of the problem. We will outline in this section, how one can parameterize the antikaon-nucleon interaction in a simple field theoretical model and apply it to the equation of state (EoS) of beta-stable matter and neutron stars.

## 3.1 Effective model of kaon-nucleon interactions

In neutron star matter, only baryon number and charge are conserved. Hence, kaons or antikaons, as well as hyperons, can appear inside neutron stars by strangeness changing processes. The onset of the appearance of the negatively charged $K^-$ is given by the equality of the effective antikaon chemical potential (or effective antikaon energy) in matter with the electrochemical potential

$$\omega_K = \mu_{K^-} = \mu_e \quad . \tag{23}$$

Then processes like

$$e^- \to K^- + \nu_e \qquad n \to p + K^- \tag{24}$$

are energetically allowed. The $K^-$ is replacing electrons from the Fermi surface or equivalently transforming neutrons to protons. The $K^-$ as a boson will form a condensate with zero momenta as the s-wave interaction with nucleons is attractive. The presence of the zero momentum $K^-$'s, compared to the high momenta electrons, will lower the overall energy of the system. Also, the increase in the proton fraction will lower the isospin asymmetry of the matter. As the nuclear asymmetry term is strongly repulsive, $K^-$ can again lower the energy of the system substantially.

Guided by the discussion of the previous sections, we write down now an effective Lagrangian which models the kaon-nucleon interaction:

$$\mathcal{L}_K = \mathcal{D}_\mu^* K^* \mathcal{D}^\mu K - m_K^{*2} K^* K \tag{25}$$

where the vector fields are coupled minimally

$$\mathcal{D}_\mu = \partial_\mu + i g_{\omega K} V_\mu + i g_{\rho K} \boldsymbol{\tau}_K \boldsymbol{R}_\mu \tag{26}$$

and the effective mass of the kaon is defined as a linear shift of the mass term by the scalar field

$$m_K^* = m_K - g_{\sigma K} \sigma \quad . \tag{27}$$

The form of the interaction as mediated by a scalar ($\sigma$) and vector ($V_\mu$, $\boldsymbol{R}_\mu$) meson fields is in close analogy to the relativistic mean-field model which will be used for the baryon-baryon interactions. We will focus now on the $K^-$. The combined equations of motion for the meson fields including nucleons

$$\begin{aligned} m_\sigma^2 \sigma + b\, m_N (g_{\sigma N} \sigma)^2 + c (g_{\sigma N} \sigma)^3 &= g_{\sigma N} \rho_s + g_{\sigma K} \rho_K \\ m_\omega^2 V_0 &= g_{\omega N} (\rho_p + \rho_n) - g_{\omega K} \rho_K \\ m_\rho^2 R_{0,0} &= g_{\rho N} (\rho_p - \rho_n) - g_{\rho K} \rho_K \end{aligned} \tag{28}$$

has a certain simple structure. The scalar densities for nucleons and $K^-$ act as a source term for the scalar field. The vector densities are conserved and build source terms for the corresponding vector fields. The different signs for the source terms of the isospin dependent $R_0$ field reflect the isospin of the hadron. Note, that the kaon scalar and vector density are equal in our approach as the kaon has spin zero, contrary to the nucleon. The dispersion relation for the $K^-$ at zero momentum is given by

$$\omega_K = m_K - g_{\sigma K}\sigma - g_{\omega K}V_0 - g_{\rho K}R_{0,0} \quad . \tag{29}$$

The solution of the equations of motion provides then an EoS of the form

$$\epsilon = \epsilon_N + \epsilon_K + \epsilon_{e,\mu} \tag{30}$$

$$p = p_N + p_{e,\mu} \quad . \tag{31}$$

Note, that the direct contribution of the kaons to the pressure vanishes, as it involves a Bose condensate. The energy contribution of the $K^-$ reads

$$\epsilon_K = m_K^* \rho_K \quad . \tag{32}$$

Now there are two parameters for the $K^-$, which have to be fixed. For the vector coupling constants, we use simple quark counting rules and set

$$g_{\omega K} = \frac{1}{3}g_{\omega N} \quad \text{and} \quad g_{\rho K} = g_{\rho N} \quad . \tag{33}$$

The scalar coupling constant is fixed to the optical potential of the $K^-$ at $\rho_0$:

$$U_K(\rho_0) = -g_{\sigma K}\sigma(\rho_0) - g_{\omega K}V_0(\rho_0) \tag{34}$$

and will be varied between $-140$ and $-80$ MeV according to the results of the coupled channel calculations of the previous section.

## 3.2 Phase transition to kaon condensation

The phase transition to kaon condensed matter can be in principle of any order. If the phase transition is of first order, then Gibbs' general condition for two phases have to be applied. As there are two conserved charges for cold neutron star matter, baryon number and charge, the Gibbs conditions read

$$p^{\mathrm{I}} = p^{\mathrm{II}}, \quad \mu_B^{\mathrm{I}} = \mu_B^{\mathrm{II}}, \quad \mu_e^{\mathrm{I}} = \mu_e^{\mathrm{II}} \tag{35}$$

in kinetic and chemical equilibrium. Note, that the standard Maxwell construction can not be used, as it ensures to conserve only *one* chemical potential.

For a sufficiently large attraction for the $K^-$, $U_K(\rho_0) > -80$ MeV, we find that the phase transition is indeed of first order. Figure 10 shows the EoS of neutron star matter with a kaon condensate. The pure charge neutral nucleon phase is shown by the solid line. The pure charge neutral kaon condensed phase by the dotted lines. A mechanical instability of the latter phase is apparent for some density range as the slope gets negative, $dp/d\epsilon < 0$. The Gibbs construction

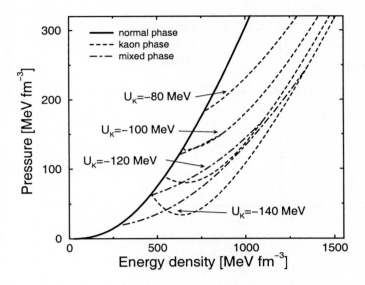

**Fig. 10.** EoS of kaon condensed matter. The physical solution of the mixed phase is given by the Gibbs construction (dash-dotted lines) as taken from [14]

for the mixed phase, plotted by the dash-dotted lines, is mechanical stable and is a continuously rising function of the density. The mixed phase starts at a lower density compared to the onset of the pure charge neutral kaon condensed phase and can extend to rather large densities. The phase transition turns out to be of second order for small values of the optical potential, i.e. $U_K(\rho_0) \geq -80$ MeV. For both orders of the phase transition, the EoS is considerably softened due to kaon condensation.

If a mixed phase is formed, there is a new degree of freedom to maximize the pressure: the redistribution of charge between the two phases [56]. There are three possible solutions for the charge density: i) the pure nucleon phase with $q_K = 0$, ii) the pure kaon condensed phase with $q_N = 0$, and iii) the mixed phase with

$$q_{\text{total}} = (1-\chi)q_N(\mu_B, \mu_e) + \chi q_K(\mu_B, \mu_e) = 0, \qquad (36)$$

where $\chi$ is the volume fraction of the two phases. The total global charge is still zero, while the two phases of the mixed phase can have very large local electric (opposite) charge densities.

The charge density in the mixed phase as a function of the volume fraction is plotted in Fig. 11. The nucleon phase starts with zero charge density. Its charge density is getting positive in the mixed phase, as it is energetically favoured to have about equal amounts of protons and neutrons in matter. The positive charge of the nucleon phase is compensate by the kaon condensed phase. The latter

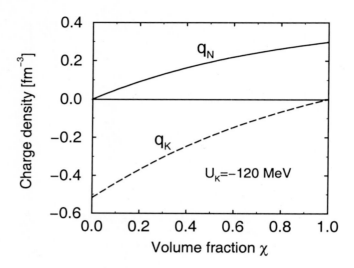

**Fig. 11.** Charge densities of the two phases in the mixed phase (taken from [14])

phase starts at large negative charge density and stops at zero charge density at the end of the mixed phase. For larger density, the pure neutrally charged kaon condensed phase prevails. The charge is distributed between the two phases in the mixed phase, and geometrical (charged) structures appear. These structures are similar to those discussed for the liquid-gas phase transition in the neutron star crust [58] and for the phase transition to deconfined matter [56].

In the selfconsistent approach used here for nucleons and kaons, the crucial ingredient for calculating the geometric structures, the surface tension between the two phases, can be calculated within the model [59,60]. The resulting sizes of the geometric structures are summarized in Fig. 12. First, bubbles of the kaon condensed phase appear. Then at larger density, here lower radii, the kaon condensed phase forms rods, then slabs. In the core, the situation reverses and nucleonic slabs form which are immersed in the kaon condensed phase. If one increases the density even further, nucleonic rods, then drops will form ending finally in a pure kaon condensed phase. For the EoS used in Fig. 12, the maximum density reached inside the neutron star is too low to achieve these phases. The size of the structures is around 10 fm and the separation 20–30 fm, not far from the size of heavy nuclei of say 7 fm. Compared to the size of the neutron star, the mixed phase structures are microscopic and will effect transport and cooling phenomena inside the neutron star (see e.g. [59,61]).

One might wonder, why there exists a phase boundary for the nucleons, as they are present in both phases. In the relativistic mean-field model, the mass of the nucleons is shifted in dense matter due to the interaction with the scalar field. In the calculation for the mixed phase, it turns out, that Gibbs conditions are

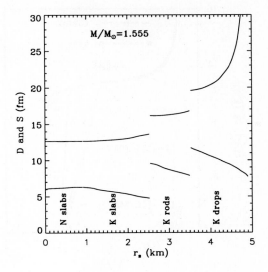

**Fig. 12.** Sizes of the geometrical structures appearing in the mixed phase (from [60])

satisfied for different field configurations, i.e. for different values of the effective nucleon masses in the two phases. The difference of the effective nucleon masses is about a factor two in the mixed phase. Hence, the nucleons in the two phases are indeed distinguishable.

The appearance of the kaon condensed phase has certain impacts also on the global properties of the neutron star. As kaon condensation softens the EoS, the maximum mass of neutron stars will be reduced (see e.g. [9]). Figure 13 shows the mass-radius relation for a neutron star with a kaon condensed phase for different strengths of the kaon-nucleon interactions. For moderate attraction for the $K^-$, the maximum mass of a neutron stars drops while the minimum radius increases slightly. For larger attraction, a considerable fraction of the neutron star is in the mixed phase and the maximum mass as well as the minimum radius decreases. For the largest attraction studied here, the minimum radius changes drastically from values of 12 km without $K^-$ condensation to 8 km with $K^-$ condensation. The maximum mass of the neutron star is lowered from 1.8 $M_\odot$ to 1.4 $M_\odot$. Note, that there are no instabilities for the Gibbs construction of the mixed phase. The Maxwell construction, shown by dashed lines, is mechanically unstable for some intermediate ranges of the radius.

### 3.3 Effects of hyperonization on kaon condensation

Hyperons may appear around twice normal nuclear density in beta-stable matter [57]. In the last few years, this picture has gained support by various model calculations. Hyperons (either the $\Lambda$ or the $\Sigma^-$) are present in neutron star matter

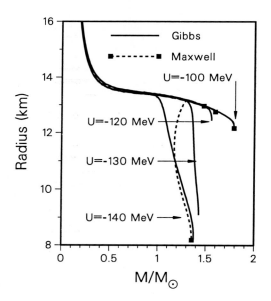

**Fig. 13.** The mass-radius relation of a neutron star with kaon condensation for different values of the $K^-$ optical potential at $\rho_0$ (taken from [14])

at $2\rho_0$ within effective nonrelativistic potential models [62], the Quark-Meson Coupling Model [63], extended Relativistic Mean-Field approaches [11,12], Relativistic Hartree-Fock [64], Brueckner-Hartree-Fock [65,66], and chiral effective Lagrangians [67]. Whether the $\Lambda$ or the $\Sigma^-$ appears first, depends sensitively on the chosen isospin interaction of the $\Sigma$ hyperons. In any case, these two hyperons appear around $(2\text{--}4)\rho_0$ in the model calculations listed above, which is before kaon condensation sets in.

In most of the modern EoS, the interaction between the baryons is mediated by meson exchange. The nucleon parameters are fitted to properties of nuclei or nuclear matter. Some of the hyperon coupling constants, say the ones to the scalar field, are fixed by the optical potential as extracted from hypernuclear data [68] (see also our discussion in the previous section). The quark model (SU(6) symmetry) can be used to constrain the hyperon coupling constants to the vector fields. The latter choice is often relaxed. While the $\Lambda$ coupling constants, and to some extend the ones for the $\Xi$, can be constrained by hypernuclear data, the ones for the $\Sigma$ hyperons can not. Studies of $\Sigma^-$ atoms indicate a repulsive potential for the $\Sigma$ in nuclei [69]. Another point of uncertainty is related with the interaction between hyperons themselves. Apart from the $\Lambda\Lambda$ interaction, nothing is known about the hyperon-hyperon interaction. Only a few calculations have addressed this issue for neutron star matter so far [12,62,66].

A representative hyperon population as a function of density is plotted in Fig. 14. The three panels shown corresponds to three different choices of the hyperon vector coupling constants. In the upper panel, the ratio of all hyperon

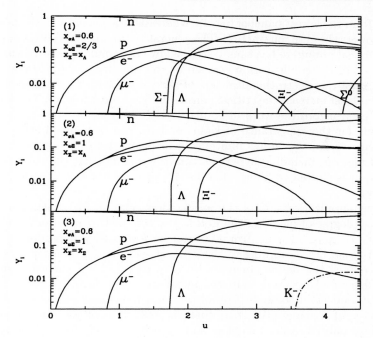

**Fig. 14.** The particle fractions in neutron star matter versus the relative baryon density $u = \rho/\rho_0$. The three panels correspond to different choices for the hyperon coupling constants (from [11])

vector coupling constants to the one of the nucleon is set to about $x_{\omega Y} = 2/3$. This is the ratio predicted by the quark model or SU(6) symmetry for the coupling constant to the $\omega$ meson for the $\Lambda$ and the $\Sigma$; for the $\Xi$ it would be, of course, only 1/3. The $\Sigma^-$ and the $\Lambda$ appear at a little bit less than $2\rho_0$. The heavier $\Xi$ hyperon is present in matter at $3.3\rho_0$. If one arbitrarily increases the vector coupling constant of the $\Sigma$ meson to be equal to the one for nucleons, $x_{\omega\Sigma} = 1$, the $\Sigma$ hyperons do not appear (see middle panel). The critical density for the onset of the $\Xi^-$ is then shifted to $2.2\rho_0$, so that the $\Xi^-$ takes over to some extent the role of the $\Sigma^-$. If the corresponding ratio for the $\Xi$ is also increased to $x_{\omega\Xi} = 1$, also the $\Xi$ population vanishes (lower panel). The critical density for the $\Lambda$ is unchanged. Only for the latter case, a kaon condensed phase emerges at $3.6\rho_0$ in these model calculations.

Note, that the electron fraction decreases, once hyperons are in the system [57]. This means, that the electron chemical potential does not only saturate but is substantially lowered when hyperons are present in matter. The obvious reason is, that the negative charge needed to cancel the positive charge of the protons is carried now by the negatively charged hyperons instead. Another reason is, that any appearance of a new degree of freedom in matter lowers the overall Fermi momenta of nucleons and leptons, be it charged or not. The latter effect is apparent from the lowest panel of Fig. 14, where just the presence of the neutral $\Lambda$ hyperon lowers the electron fraction. The general feature, that hyperons lower the electrochemical potential, has been restressed by Glendenning most recently [70]. As the onset of $K^-$ condensation is given by the equality of the effective energy of the $K^-$ and the electrochemical potential, it is evident, that the presence of hyperons at least increase the critical density for kaon condensation. Glendenning discussed even before the work of Kaplan and Nelson [5] the destructive effect of hyperons for the appearance of kaon condensation in neutron stars [57]. In more recent works, the in-medium effects for kaons were incorporated in the model calculations including hyperons, and it was found that the onset for kaon condensation is shifted to higher density [11] or does not take place at all [12] (see below).

The hyperonization effects the mass-radius relation for neutron stars. As for any new degree of freedom, the EoS is also softened by hyperons. The maximum possible mass of a neutron star can be lowered by hyperons by about $(0.4–0.7)M_\odot$ [68,11]. According to the work of [11], kaon condensation without hyperons reduce the maximum mass by only $(0.1–0.2)M_\odot$ and the combined effect of kaons and hyperons shift the maximum mass down by about $0.5M_\odot$. So, the main effect in the reduction of the maximum mass stems from the hyperon degree of freedom. We note, that the reduction of the maximum mass due to kaons only is rather model dependent. Other estimates find changes up to $\Delta M = 0.4M_\odot$ due to kaon condensation [9,14], but hyperons are ignored in these latter calculations.

There can be an additional hindrance for kaon condensation which is related to the hyperon-hyperon interaction. The vector meson $\phi$ controls the interaction between hyperons at large hyperon densities [12]. The inclusion of the $\phi$ meson models also the kaon-hyperon interaction. As the $\phi$ meson couples solely to strange quarks in SU(6) symmetry, it is repulsive for hyperons and the $K^-$ and attractive for the $K^+$. For neutron star matter with hyperons, the $\phi$ meson exchange can saturate the effective energy of the $K^-$, so that kaon condensation can not happen at all. This effect is depicted in Fig. 15 as a function of density. The effective energy of the $K^-$ drops at low density while the one for the $K^+$ increases. At $2\rho_0$, when hyperons are getting populated, the effective energy of the $K^-$ saturates as the $K^-$ feels the repulsive contribution from the hyperons. On the contrary, the $K^+$ energy starts to drop at this density as the $K^+$-hyperon interaction is attractive. The lower curves show the electrochemical potential which has to be crossed by the effective $K^-$ energy for the $K^-$ to be present. As the electrochemical potential as well as the $K^-$ effective energy saturate at large density due to the presence of hyperons, that crossing does not happen and kaon condensation is prevented in this scenario.

## 4 Summary and Outlook

We have discussed the elementary kaon-nucleon interaction as derived from scattering data and the in-medium changes of the $K^-$ as deduced from kaonic atoms.

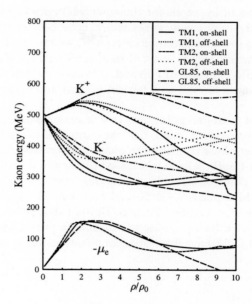

**Fig. 15.** The effective energy of kaons as well as the electrochemical potential versus the baryon density for neutron star matter (taken from [12])

Various coupled channel calculations using chiral effective interactions or the boson exchange model with higher order partial waves and their results for the $K^-$ properties in dense matter have been reviewed. The extracted range of the optical potential has been used to study the phase transition to kaon condensation in neutron star matter. We outlined the effects of a first order phase transition due to the presence of structures in the mixed phase. Consequences for the global properties of neutron stars, i.e. a reduced maximum mass and minimum radius, have been addressed. Finally, we examined the rôle of hyperons on the onset of kaon condensation in neutron stars.

It is clear from our review, that the discussion about kaon condensation is far from being complete at present. The topic is a challenge both for experimentalists as well as for theorists, demonstrating a growing interplay between traditional nuclear physics, heavy-ion physics, and astrophysics. For example, the importance of p-wave interactions also for cold, dense neutron star matter has been stressed, but further work is needed in that direction. Especially, the precise value of the optical potential of the $K^-$ is a crucial ingrediente for the neutron star matter calculation and needs to be pinned down more precisely, be it by the experimental study of deep lying levels in kaonic atoms, subthreshold kaon production in heavy-ion collisions at GSI, or by the mass-radius measurement of neutron stars.

The inclusion of hyperons for the equation of state is certainly a necessary one, but there are only a few works which have been devoted to this issue. In particular, the hyperon-hyperon interaction as well as the kaon-hyperon interaction can be important for the onset of kaon condensation and could be addressed by e.g. a chiral SU(3) symmetric model. A consistent calculation, incorporating realistic kaon-baryon as well as a baryon-baryon interactions on the same basis is still missing.

Last but not least, we point out, that kaon condensation can have impacts on other facets of neutron stars, like the evolution of proto-neutron stars and the deconfinement phase transition to quark matter. Concerning the latter point, the appearance of both, a kaon condensed phase and a quark matter phase, will lead to a triple point of strongly interacting matter inside a neutron star. At this point, the three phases, the normal hadronic, the kaon condensed and the quark matter phase, are in equilibrium. Beyond that point, the kaon condensed phase and the quark matter phase are forming a mixed phase. The presence of this triple point might have interesting implications for transport phenomena in neutron stars.

**Acknowledgments**

We thank all colleagues that have collaborated with us in obtaining the results presented here, especially M.B. Christiansen, M. Effenberger, N.K. Glendenning, S. Hirenzaki, V. Koch, T.T.S. Kuo, I.N. Mishustin, Y. Okumura, E. Oset, A. Polls, H. Toki, L. Tolos. AR acknowledges support by the DGICYT project PB98-1247, the Generalitat de Catalunya Grant 1998SGR-11 and the EU contract FMRX-CT98-0169. JSB thanks RIKEN, BNL and the U.S. Department of Energy for providing the facilities essential for the completion of this work. JW thanks the Gesellschaft für Schwerionenforschung (GSI), Darmstadt, and the Bundesministerium für Bildung und Forschung (BMBF) for their support.

# References

1. T. Schäfer and E.V. Shuryak, Rev. Mod. Phys. **70**, 323 (1998).
2. J. Gasser and H. Leutwyler, Ann. Phys. (NY) **158**, 142 (1984); Nucl. Phys. **B250**, 465 (1985).
3. N. Kaiser, P.B. Siegel and W. Weise, Nucl. Phys. **A594**, 325 (1995).
4. N. Kaiser, T. Waas and W. Weise, Nucl. Phys. **A612**, 297 (1997).
5. D.B. Kaplan and A.E. Nelson, Phys. Lett. B **175**, 57 (1986), ibid **179**, 409 (Erratum).
6. G.E. Brown, K. Kubodera, M. Rho, and V. Thorsson, Phys. Lett. B **291**, 355 (1992).
7. V. Thorsson, M. Prakash, and J.M. Lattimer, Nucl. Phys. **A572**, 693 (1994).
8. H. Fujii, T. Maruyama, T. Muto, and T. Tatsumi, Nucl. Phys. **A597**, 645 (1996).
9. G.Q. Li, C.-H. Lee, and G.E. Brown, Phys. Rev. Lett. **79**, 5214 (1997).
10. P.J. Ellis, R. Knorren, and M. Prakash, Phys. Lett. B **349**, 11 (1995).
11. R. Knorren, M. Prakash, and P.J. Ellis, Phys. Rev. C **52**, 3470 (1995).

12. J. Schaffner and I.N. Mishustin, Phys. Rev. C **53**, 1416 (1996).
13. N.K. Glendenning and J. Schaffner-Bielich, Phys. Rev. Lett. **81**, 4564 (1998).
14. N.K. Glendenning and J. Schaffner-Bielich, Phys. Rev. C **60**, 025803 (1999).
15. T. Tatsumi and M. Yasuhira, Phys. Lett. B **441**, 9 (1998); T. Tatsumi and M. Yasuhira, Nucl. Phys. **A653**, 133 (1999).
16. E. Friedman, A. Gal, and C.J. Batty, Nucl. Phys. **A579**, 518 (1994).
17. R. Barth et al., Phys. Rev. Lett. **78**, 4007 (1997); F. Laue et al., Phys. Rev. Lett. **82**, 1640 (1999).
18. W. Cassing, E.L. Bratkovskaya, U. Mosel, S. Teis, and A. Sibirtsev, Nucl. Phys. **A614**, 415 (1997); E.L. Bratkovskaya, W. Cassing, and U. Mosel, Nucl. Phys. **A622**, 593 (1997).
19. G.Q. Li, C.M. Ko, and X.S. Fang, Phys. Lett. B **329**, 149 (1994).
20. G.Q. Li and C.M. Ko, Phys. Rev. C **54**, 2159 (1996).
21. G.Q. Li, C.-H. Lee, and G.E. Brown, Nucl. Phys. **A625**, 372 (1997).
22. J. Schaffner-Bielich, V. Koch, and M. Effenberg, Nucl. Phys. **A669**, 153 (2000).
23. A. Sibirtsev and W. Cassing, Nucl. Phys. **A641**, 476 (1998).
24. C.-H. Lee, G.E. Brown, and M. Rho, Phys. Lett. B **335**, 266 (1994); C.-H. Lee, G.E. Brown, D.P. Min, and M. Rho, Nucl. Phys. **A585**, 401 (1995).
25. C.-H. Lee, Phys. Rep. **275**, 255 (1996).
26. G. Mao, P. Papazoglou, S. Hofmann, S. Schramm, H. Stöcker, and W. Greiner, Phys. Rev. C **59**, 3381 (1999).
27. J. Schaffner-Bielich, I.N. Mishustin, and J. Bondorf, Nucl. Phys. **A625**, 325 (1997).
28. K. Tsushima, K. Saito, A.W. Thomas, and S.V. Wright, Phys. Lett. B **429**, 239 (1998).
29. M. Alberg, E.M. Henley, and L. Wilets, Ann. Phys. (N.Y.) **96**, 43 (1976).
30. L.R. Staronski and S. Wycech, J. Phys. **G13**, 1361 (1987).
31. V. Koch, Phys. Lett. B **337**, 7 (1994).
32. T. Waas, N. Kaiser, and W. Weise, Phys. Lett. B **365**, 12 (1996).
33. T. Waas and W. Weise, Nucl. Phys. **A625**, 287 (1997).
34. M. Lutz, Phys. Lett. B **426**, 12 (1998).
35. A. Ramos and E. Oset, Nucl. Phys. **A671**, 481 (2000).
36. E. Oset and A. Ramos, Nucl. Phys. **A635**, 99 (1998).
37. J. A. Oller and E. Oset, Nucl. Phys. **A620**, 438 (1997); erratum Nucl. Phys. **A624**, 407 (1999).
38. J. A. Oller, E. Oset, and J. R. Peláez, Phys. Rev. Lett. **80**, 3452 (1998); Phys. Rev. D **59**, 074001 (1999); erratum Phys. Rev. D **60**, 099906 (1999).
39. J.A. Oller, E. Oset, and A. Ramos, Prog. Part. Nucl. Phys. **45**, 157 (2000).
40. L. Tolós, A. Ramos, A. Polls, and T.T.S. Kuo, nucl-th/0007042
41. A. Müller-Groeling, K. Holinde, and J. Speth, Nucl. Phys. **A513**, 557 (1990).
42. H. Bandō, T. Motoba, and J. Žofka, Int. J. Mod. Phys. **A5**, 4021 (1990).
43. C.J. Batty et al., Phys. Lett. **74B**, 27 (1978); C.J. Batty et al., Phys. Lett. **87B**, 324 (1979).
44. E. Oset, P. Fernández de Córdoba, L.L. Salcedo, and R. Brockmann, Phys. Rep. **188**, 79 (1990).
45. C.J. Batty, E. Friedman, and A. Gal, Phys. Lett. B **335**, 273 (1994).
46. J. Caro, N. Kaiser, S. Wetzel, and W. Weise, Nucl. Phys. **A672**, 249 (2000).
47. A. Ramos, E. Oset, and L.L. Salcedo, Phys. Rev. C **50**, 2314 (1994).
48. S. Hirenzaki, Y. Okumura, H. Toki, E. Oset, and A. Ramos, Phys. Rev. C **61**, 055205 (2000).
49. C. J. Batty, E. Friedman, and A. Gal, Phys. Rep. **287**, 385 (1997).

50. E. Friedman, A. Gal, J. Mareš, and A. Cieplý, Phys. Rev. C **60**, 024314 (1999).
51. C.J. Batty, Nucl. Phys. **A372**, 418 (1981).
52. A. Baca, C. García-Recio, and J. Nieves, Nucl. Phys. **A673**, 335 (2000).
53. E. Friedman and A. Gal, Phys. Lett. B **459**, 43 (1999).
54. E. Friedman and A. Gal, Nucl. Phys. **A658**, 345 (1999).
55. T. Kishimoto, Phys. Rev. Lett. **83**, 4701 (1999).
56. N.K. Glendenning, Phys. Rev. D **46**, 1274 (1992).
57. N.K. Glendenning, Astrophys. J. **293**, 470 (1985).
58. D.G. Ravenhall, C.J. Pethick, and J.R. Wilson, Phys. Rev. Lett. **50**, 2066 (1983).
59. S. Reddy, G. Bertsch, and M. Prakash, Phys. Lett. B **475**, 1 (2000); T. Norsen and S. Reddy, nucl-th/0010075 (2000).
60. M.B. Christiansen, N.K. Glendenning, and J. Schaffner-Bielich, Phys. Rev. C **62**, 025804 (2000).
61. J.A. Pons, S. Reddy, P.J. Ellis, M. Prakash and J.M. Lattimer, Phys. Rev. C **62**, 035803 (2000).
62. S. Balberg and A. Gal, Nucl. Phys. **A625**, 435 (1997).
63. S. Pal, M. Hanauske, I. Zakout, H. Stöcker, and W. Greiner, Phys. Rev. C **60**, 015802 (1999).
64. H. Huber, F. Weber, M.K. Weigel, and C. Schaab, Int. J. Mod. Phys. **E7**, 301 (1998).
65. M. Baldo, G.F. Burgio, and H.J. Schulze, Phys. Rev. C **61**, 055801 (2000).
66. I. Vidana, A. Polls, A. Ramos, L. Engvik, and M. Hjorth-Jensen, Phys. Rev. C **62**, 035801 (2000).
67. M. Hanauske, D. Zschiesche, S. Pal, S. Schramm, H. Stöcker, and W. Greiner, Astrophys. J. **537**, 50320 (2000).
68. N.K. Glendenning and S.A. Moszkowski, Phys. Rev. Lett. **67**, 2414 (1991).
69. J. Mares, E. Friedman, A. Gal, and B.K. Jennings, Nucl. Phys. **A594**, 311 (1995).
70. N.K. Glendenning, nucl-th/0009082 (2000).

# Phases of QCD at High Baryon Density

Thomas Schäfer[1,2] and Edward Shuryak[1]

[1] Department of Physics and Astronomy, State University of New York, Stony Brook, NY 11794-3800
[2] Riken-BNL Research Center, Brookhaven National Laboratory, Upton, NY 11973

**Abstract.** We review recent work on the phase structure of QCD at very high baryon density. We introduce the phenomenon of color superconductivity and discuss how the quark masses and chemical potentials determine the structure of the superfluid quark phase. We comment on the possibility of kaon condensation at very high baryon density and study the competition between superfluid, density wave, and chiral crystal phases at intermediate density.

## 1 Color Superconductivity

In the interior of compact stars matter is compressed to densities several times larger than the density of ordinary matter. Unlike the situation in relativistic heavy ion collisions, these conditions are maintained for essentially infinite periods of time. Also, compared to QCD scales, matter inside a compact star is quite cold. At low density quarks are confined, chiral symmetry is broken, and baryonic matter is described in terms of neutrons and protons as well as their excitations. At very large density, on the other hand, we expect that baryonic matter is described more effectively in terms of quarks rather than hadrons. As we shall see, these quarks can form new condensates and the phase structure of dense quark matter is quite rich.

At very high density the natural degrees of freedom are quark excitations and holes in the vicinity of the Fermi surface. Since the Fermi momentum is large, asymptotic freedom implies that the interaction between quasi-particles is weak. In QCD, because of the presence of unscreened long range gauge forces, this is not quite true. Nevertheless, we believe that this fact does not essentially modify the argument. We know from the theory of superconductivity the Fermi surface is unstable in the presence of even an arbitrarily weak attractive interaction. At very large density, the attraction is provided by one-gluon exchange between quarks in a color anti-symmetric $\bar{3}$ state. High density quark matter is therefore expected to be a color superconductor [1–4].

Color superconductivity is described by a pair condensate of the form

$$\phi = \langle \psi^T C \Gamma_D \lambda_C \tau_F \psi \rangle. \tag{1}$$

Here, $C$ is the charge conjugation matrix, and $\Gamma_D, \lambda_C, \tau_F$ are Dirac, color, and flavor matrices. Except in the case of only two colors, the order parameter cannot be a color singlet. Color superconductivity is therefore characterized by the breakdown of color gauge invariance. As usual, this statement has to be interpreted with care because local gauge invariance cannot really be broken. Nevertheless, we can study gauge invariant consequences of a quark pair condensate, in particular the formation of a gap in the excitation spectrum.

In addition to that, color superconductivity can lead to the breakdown of global symmetries. We shall see that in some cases there is a gauge invariant order parameter for the $U(1)$ of baryon number. This corresponds to true superfluidity and the appearance of a massless phonon. We shall also find that for $N_f > 2$ color superconductivity leads to chiral symmetry breaking and that quark matter may support a kaon condensate. Finally, as we move to stronger coupling we find that other forms of order can compete with color superconductivity and that quark matter may exist in the form of chiral density waves or chiral crystals.

## 2 Phase Structure in Weak Coupling

### 2.1 QCD with two flavors

In this section we shall discuss how to use weak coupling methods in order to explore the phases of dense quark matter. We begin with what is usually considered to be the simplest case, quark matter with two degenerate flavors, up and down. Renormalization group arguments suggest [5–7], and explicit calculations show [8,9], that whenever possible quark pairs condense in an $s$-wave. This means that the spin wave function of the pair is anti-symmetric. Since the color wave function is also anti-symmetric, the Pauli principle requires the flavor wave function to be anti-symmetric, too. This essentially determines the structure of the order parameter [10,11]

$$\phi^a = \langle \epsilon^{abc} \psi^b C \gamma_5 \tau_2 \psi^c \rangle. \tag{2}$$

This order parameter breaks the color $SU(3) \to SU(2)$ and leads to a gap for up and down quarks with two out of the three colors. Chiral and isospin symmetry remain unbroken.

We can calculate the magnitude of the gap and the condensation energy using weak coupling methods. In weak coupling the gap is determined by ladder diagrams with the one gluon exchange interaction. These diagrams can be summed using the gap equation [12–16]

$$\Delta(p_0) = \frac{g^2}{12\pi^2} \int dq_0 \int d\cos\theta \left( \frac{\frac{3}{2} - \frac{1}{2}\cos\theta}{1 - \cos\theta + G/(2\mu^2)} + \frac{\frac{1}{2} + \frac{1}{2}\cos\theta}{1 - \cos\theta + F/(2\mu^2)} \right) \frac{\Delta(q_0)}{\sqrt{q_0^2 + \Delta(q_0)^2}}. \tag{3}$$

Here, $\Delta(p_0)$ is the frequency dependent gap, $g$ is the QCD coupling constant and $G$ and $F$ are the self energies of magnetic and electric gluons. This gap equation is very similar to the BCS gap equations that describe nuclear superfluids. The main difference is that because the gluon is massless, the gap equation contains a collinear divergence for $\cos\theta \to 1$. In a dense medium the collinear divergence is regularized by the gluon self energy. For $q \to 0$ and to leading order in perturbation theory we have

$$F = 2m^2, \qquad G = \frac{\pi}{2}m^2\frac{q_0}{|\boldsymbol{q}|}, \qquad (4)$$

with $m^2 = N_f g^2 \mu^2/(4\pi^2)$. In the electric part, $m_D^2 = 2m^2$ is the familiar Debye screening mass. In the magnetic part, there is no screening of static modes, but non-static modes are modes are dynamically screened due to Landau damping.

We can now perform the angular integral and find

$$\Delta(p_0) = \frac{g^2}{18\pi^2}\int dq_0 \log\left(\frac{b\mu}{|p_0-q_0|}\right)\frac{\Delta(q_0)}{\sqrt{q_0^2+\Delta(q_0)^2}}, \qquad (5)$$

with $b = 256\pi^4(2/N_f)^{5/2}g^{-5}$. This result shows why it was important to keep the frequency dependence of the gap. Because the collinear divergence is regulated by dynamic screening, the gap equation depends on $p_0$ even if the frequency is small. We can also see that the gap scales as $\exp(-c/g)$. The collinear divergence leads to a gap equation with a double-log behavior. Qualitatively

$$1 \sim \frac{g^2}{18\pi^2}\left[\log\left(\frac{\mu}{\Delta}\right)\right]^2, \qquad (6)$$

from which we conclude that $\Delta \sim \exp(-c/g)$. The approximation (6) is not sufficiently accurate to determine the correct value of the constant $c$. A more detailed analysis shows that the gap on the Fermi surface is given by

$$\Delta_0 \simeq 512\pi^4(2/N_f)^{5/2}\mu g^{-5}\exp\left(-\frac{3\pi^2}{\sqrt{2}g}\right). \qquad (7)$$

We should emphasize that, strictly speaking, this result contains only an estimate of the pre-exponent. It was recently argued that wave function renormalization and quasi-particle damping give $O(1)$ contributions to the pre-exponent which substantially reduce the gap [16].

For chemical potentials $\mu < 1$ GeV, the coupling constant is not small and the applicability of perturbation theory is in doubt. If we ignore this problem and extrapolate the perturbative calculation to densities $\rho \simeq 5\rho_0$ we find gaps $\Delta \simeq 100$ MeV. This result may indeed be more reliable than the calculation on which it is based. In particular, we note that similar results have been obtained using realistic interactions which reproduce the chiral condensate at zero baryon density [10,11].

We can also determine the condensation energy. In weak coupling the grand potential can be calculated from [4]

$$\Omega = \frac{1}{2}\int\frac{d^4q}{(2\pi)^4}\left\{-\mathrm{tr}\left[S(q)\Sigma(q)\right] + \mathrm{tr}\left[S_0^{-1}(q)S(q)\right]\right\}, \qquad (8)$$

where $S(q)$ and $\Sigma(q)$ are the Nambu-Gorkov propagator and proper self energy. Using the propagator in the superconducting phase we find

$$\epsilon = 4 \frac{\mu^2}{4\pi^2} \Delta_0^2 \log\left(\frac{\Delta_0}{\mu}\right), \qquad (9)$$

where the factor $4 = 2N_f$ comes from the number of condensed species. Using $\Delta_0 \simeq 100$ MeV and $\mu \simeq 500$ MeV we find $\epsilon \simeq -50$ MeV/fm$^3$. This shows that the condensation energy is only a small $O(\Delta^2/\mu^2)$ correction to the total energy density of the quark phase. We note that the result for the condensation energy agrees with BCS theory. The same is true for the critical temperature which is given by $T_c = 0.56\Delta_0$.

## 2.2 QCD with three flavors: Color-Flavor-Locking

If quark matter is formed at densities several times nuclear matter density we expect the quark chemical potential to be larger than the strange quark mass. We therefore have to determine the structure of the superfluid order parameter for three quark flavors. We begin with the idealized situation of three degenerate flavors. From the arguments given in the last section we expect the order parameter to be color and flavor anti-symmetric matrix of the form

$$\phi_{ij}^{ab} = \langle \psi_i^a C \gamma_5 \psi_j^b \rangle. \qquad (10)$$

In order to determine the precise structure of this matrix we have to extremize grand canonical potential. We find [17,18]

$$\Delta_{ij}^{ab} = \Delta(\delta_i^a \delta_j^b - \delta_i^b \delta_j^a), \qquad (11)$$

which describes the color-flavor locked phase proposed in [19]. Both color and flavor symmetry are completely broken. There are eight combinations of color and flavor symmetries that generate unbroken global symmetries. The symmetry breaking pattern is

$$SU(3)_L \times SU(3)_R \times U(1)_V \to SU(3)_V. \qquad (12)$$

This is exactly the same symmetry breaking that QCD exhibits at low density. The spectrum of excitations in the color-flavor-locked (CFL) phase also looks remarkably like the spectrum of QCD at low density [20]. The excitations can be classified according to their quantum numbers under the unbroken $SU(3)$, and by their electric charge. The modified charge operator that generates a true symmetry of the CFL phase is given by a linear combination of the original charge operator $Q_{em}$ and the color hypercharge operator $Q = \text{diag}(-2/3, -2/3, 1/3)$. Also, baryon number is only broken modulo $2/3$, which means that one can still distinguish baryons from mesons. We find that the CFL phase contains an octet of Goldstone bosons associated with chiral symmetry breaking, an octet of vector mesons, an octet and a singlet of baryons, and a singlet Goldstone boson related to superfluidity. All of these states have integer charges.

With the exception of the $U(1)$ Goldstone boson, these states exactly match the quantum numbers of the lowest lying multiplets in QCD at low density. In addition to that, the presence of the $U(1)$ Goldstone boson can also be understood. The $U(1)$ order parameter is $\langle (uds)(uds) \rangle$. This order parameter has the quantum numbers of a $0^+$ $\Lambda\Lambda$ pair condensate. In $N_f = 3$ QCD, this is the most symmetric two nucleon channel, and a very likely candidate for superfluidity in nuclear matter at low to moderate density. We conclude that in QCD with three degenerate light flavors, there is no fundamental difference between the high and low density phases. This implies that a low density hyper-nuclear phase and the high density quark phase might be continuously connected, without an intervening phase transition.

## 2.3 QCD with one flavor: Color-Spin-Locking

In this section we shall study superfluid quark matter with only one quark flavor. As we will discuss in more detail in the next section, our results are relevant to quark matter with two or three flavors. This is the case when the Fermi surfaces of the individual flavors are too far apart in order to allow for pairing between different species.

If two quarks with the same flavor are in a color anti-symmetric wave function then there combined spin and spatial wave function cannot be antisymmetric. This means that pairing between identical flavors has to involve total angular momentum one or greater. The simplest order parameters are of the form

$$\Phi_1 = \langle \epsilon^{3ab} \psi^a C \gamma \psi^b \rangle, \qquad \Phi_2 = \langle \epsilon^{3ab} \psi^a C \hat{q} \psi^b \rangle. \tag{13}$$

The corresponding gaps can be determined using the methods introduced in section 2.1. We find $\Delta(\Phi_{1,2}) = \exp(-3c_{1,2})\Delta_0$ with $c_1 = -1.5$ and $c_2 = -2$ [8,9]. While the natural scale of the s-wave gap is $\Delta = 100$ MeV, the p-wave gap is expected to be less than 1 MeV.

The spin one order parameter (13) is a color-spin matrix. This opens the possibility that color and spin degrees become entangled, similar to the color-flavor locked phase (11) or the B-phase of liquid $^3$He. The corresponding order parameter is

$$\Phi_{CSL} = \delta_i^a \langle \epsilon^{abc} \psi^b C \left( \cos(\beta)\hat{q}_i + \sin(\beta)\gamma_i \right) \psi^c \rangle, \tag{14}$$

where the angle $\beta$ determines the mixing between the two types of condensates shown in (13). In reference [9] we showed that the color-spin locked phase (14) is favored over the "polar" phase (13).

In the color-spin locked phase color and rotational invariance are broken, but a diagonal $SO(3)$ survives. As a consequence, the gap is isotropic. There are no gapless modes except if $\beta$ takes on special values. The parameter $\beta$ is sensitive to the quark mass and to higher order corrections. In the non-relativistic limit we find $\beta = \pi/4$ whereas in the ultra-relativistic limit $\beta = \pi/2$.

## 3 The Role of the Strange Quark Mass and the Electron Chemical Potential

So far, we have only considered the case of two or three degenerate quark flavors. In the real world, the strange quark is significantly heavier than the up and down quarks. Also, in the interior of a neutron star, electrons are present and the chemical potentials for up and down quarks are not equal.

The role of the strange quark mass in the high density phase was studied in [21,22]. The main effect is a purely kinematic phenomenon that is easily explained. The Fermi surface for the strange quarks is shifted by $\delta p_F = \mu - (\mu^2 - m_s^2)^{1/2} \simeq m_s^2/(2\mu)$ with respect to the Fermi surface of the light quarks. The condensate involves pairing between quarks of different flavors at opposite points on the Fermi surface. But if the Fermi surfaces are shifted, then the pairs do not have total momentum zero, and they cannot mix with pairs at others points on the Fermi surface. If the system is superfluid then the Fermi surface is smeared out over a range $\Delta$. This means that pairing between strange and light quarks can take place as long as the mismatch between the Fermi momenta is smaller than the gap,

$$\Delta > \frac{m_s^2}{2\mu}. \tag{15}$$

This conclusion is supported by a more detailed analysis [21,22]. Since flavor symmetry is broken, we allow the $\langle ud \rangle$ and $\langle us \rangle = \langle ds \rangle$ components of the CFL condensate to be different. The $N_f = 2$ phase corresponds to $\langle us \rangle = \langle ds \rangle = 0$. We find that there is a first order phase transition from the CFL to the $N_f = 2$ phase, and that the critical strange quark mass is in rough agreement with the estimate (15).

For densities $\rho \simeq (5-10)\rho_0$ the critical strange quark mass is close to physical mass of the strange quark. It is therefore hard to predict with certainty whether superconducting strange quark matter in the interior of a neutron star is in the color-flavor locked phase. Neutron star observations may help to answer the question. In the CFL phase all quarks have large gaps, whereas in the unlocked phase up and down quarks of the first two colors have large gaps, strange quarks have a small gap, and the up and down quarks of the remaining color have tiny gaps, or may not be gapped at all.

The effects of a non-zero electron chemical potential was studied in [23]. If the electron chemical potential exceeds the gap in the up-down sector then up and down quarks cannot pair. In this case, the up and down quarks pair separately, and with much smaller gaps, in the one flavor phase discussed in section 14. Between the phases with $\langle ud \rangle$ and $\langle uu \rangle$, $\langle dd \rangle$ pairing there is a small window of electron chemical potentials where inhomogeneous superconductivity takes place [24].

## 4 Kaon Condensation

The low energy properties of dense quark matter are determined by collective modes. In the color-flavor locked phase these modes are the phonon and the pseudoscalar Goldstone bosons, the pions, kaons and etas. Some time ago, it was suggested that pions [25–27] or kaons [28,29] might form Bose condensates in dense baryonic matter. The proposed critical densities were close to nuclear matter density in the case of pion condensation, and several times nuclear matter density in the case of kaon condensation. Since mesonic modes persist in the color-flavor-locked phase of quark matter, we can now revisit the issue of Goldstone boson condensation [30]. In particular, we shall be able to use rigorous weak coupling methods in order to address the possibility of Bose condensation in dense matter.

Our starting point is the effective lagrangian for the pseudoscalar Goldstone in dense matter [31,32]

$$\mathcal{L}_{eff} = \frac{f_\pi^2}{4}\mathrm{Tr}\left[\partial_0\Sigma\partial_0\Sigma^\dagger + v_\pi^2\partial_i\Sigma\partial_i\Sigma^\dagger\right] - c\left[\det(M)\mathrm{Tr}(M^{-1}\Sigma) + h.c.\right]. \quad (16)$$

Here, $\Sigma \in SU(3)$ is the Goldstone boson field, $v_\pi$ is the velocity of the Goldstone modes and $M = \mathrm{diag}(m_u, m_d, m_s)$ is the quark mass matrix. The effective description is valid for energies and momenta below the scale set by the gap, $\omega, q \ll \Delta$. The low energy constants can be determined in weak coupling perturbation theory. The result is $v_\pi^2 = 1/3$ and [32–36]

$$f_\pi^2 = \frac{21 - 8\log(2)}{18}\frac{\mu^2}{2\pi^2}, \quad (17)$$

$$c = \frac{3\Delta^2}{2\pi^2} \cdot \frac{2}{f_\pi^2}. \quad (18)$$

We can now determine the masses of the Goldstone bosons

$$m_{\pi^\pm} = c(m_u + m_d)m_s, \qquad m_{K^\pm} = cm_d(m_u + m_s). \quad (19)$$

This result shows that the kaon is *lighter* than the pion. This can be understood from the fact that, at high density, it is more appropriate to think of the interpolating field $\Sigma$ as

$$\Sigma_{ij} \sim \epsilon_{ikl}\epsilon_{jmn}\bar{\psi}_{L,k}\bar{\psi}_{L,l}\psi_{R,m}\psi_{R,n} \quad (20)$$

rather than the more familiar $\Sigma_{ij} \sim \bar{\psi}_{L,i}\psi_{R,j}$ [31]. Using (20) we observe that the negative pion field has the flavor structure $\bar{d}\bar{s}us$ and therefore has mass proportional to $(m_u + m_d)m_s$ [32]. Putting in numerical values we find that the kaon mass is very small, $m_{K^-} \simeq 5$ MeV at $\mu = 500$ MeV and $m_{K^-} \simeq 1$ MeV at $\mu = 1000$ MeV.

There are two reasons why the pseudoscalar Goldstone bosons are anomalously light. First of all, the Goldstone boson masses in the color-flavor-locked

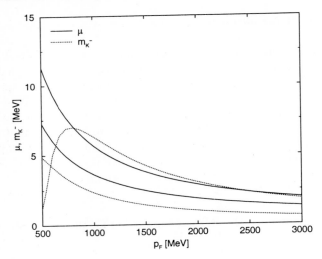

**Fig. 1.** Electron chemical potential (solid lines) and kaon mass (dashed lines) in the color-flavor-locked quark phase. The two curves for both quantities represent a simple estimate of the uncertainties due to the value of the strange quark mass and the scale setting procedure.

phase are proportional to the quark masses squared rather than linear in the quark mass, as they are at zero density. This is due to an approximate $Z_2$ chiral symmetry in the color-flavor-locked phase [19]. In addition to that, the Goldstone boson masses are suppressed by a factor $\Delta/\mu$. This is a consequence of the fact that the Goldstone modes are collective excitations of particles and holes near the Fermi surface, whereas the quark mass term connects particles and anti-particles, far away from the Fermi surface.

The fact that the meson spectrum is inverted, and that the kaon mass is exceptionally small opens the possibility that in dense quark matter electrons decay into kaons, and a kaon condensate is formed. Consider a kaon condensate $\langle K^- \rangle = v_K e^{-i\mu t}$ where $\mu$ is the chemical potential for negative charge. The thermodynamic potential $\mathcal{H} - \mu Q$ for this state is given by

$$\epsilon(\rho_q, x, y, \mu) = \frac{3}{4\pi^2}\pi^{8/3}\rho_q^{4/3}\left\{x^{4/3} + y^{4/3} + (1-x-y)^{4/3}\right.$$
$$\left. + \pi^{-4/3}\rho_q^{-2/3}m_s^2(1-x-y)^{2/3}\right\}$$
$$- (\mu^2 - m_K^2)v_K^2 + O(v_K^3) + \mu\rho_q x - \frac{1}{12\pi^2}\mu^4 \quad (21)$$

Here, $\rho_q = 3\rho_B$ is the quark density, and $x = \rho_u/\rho_q$ and $y = \rho_d/\rho_q$ are the up and down quark fractions. For simplicity, we have dropped higher order terms in the strange quark mass and neglected the electron mass. In order to determine the ground state we have to make (21) stationary with respect to $x, y, \mu$ and $v_K$. Minimization with respect to $x$ and $y$ enforces $\beta$ equilibrium, while minimization with respect to $\mu$ ensures charge neutrality. Below the onset for kaon condensation we have $v_K = 0$ and there is no kaon contribution to the charge density. Neglecting $m_e$ and higher order corrections in $m_s$ we find

$$\mu \simeq \frac{m_s^2}{4p_F} = \frac{\pi^{2/3} m_s^2}{4\rho_B^{1/3}}. \tag{22}$$

In the absence of kaon condensation, the electron chemical potential will level off at the value of the electron mass for very high baryon density. The onset of kaon condensation is determined by the condition $\mu = m_K$. At this point it becomes favorable to convert electrons into negatively charged kaons.

Results for the electron chemical potential and the kaon mass as a function of the light quark Fermi momentum are shown in Fig. 1. In order to assess some of the uncertainties we have varied the quark masses in the range $m_u = (3-5)$ MeV, $m_d = (6-8)$ MeV, and $m_s = (120-150)$ MeV. We have used the one loop result for the running coupling constant at two different scales $q = \mu$ and $q = \mu/2$. The scale parameter was set to $\Lambda_{QCD} = 238$ MeV, corresponding to $\alpha_s(m_\tau) = 0.35$. An important constraint is provided by the condition $m_s < \sqrt{2p_F\Delta}$ discussed in the previous section. We have checked that this condition is always satisfied for $p_F > 500$ MeV. Figure 1 shows that there is significant uncertainty in the relative magnitude of the chemical potential and the kaon mass. Nevertheless, the band of kaon mass predictions lies systematically below the predicted chemical potentials. We therefore conclude that kaon condensation appears likely even for moderate Fermi momenta $p_F \simeq 500$ MeV. For very large baryon density $\mu \to m_e$ while $m_K \to 0$ and kaon condensation seems inevitable.

## 5  Chiral Waves and Chiral Crystals

It is very important for the structure of compact stars to determine whether the quark matter core is in liquid or solid form. In the previous sections we discussed the case of weak coupling. In this case, particle-particle pairing is the only instability that needs to be considered. Nevertheless, in strong coupling, or if superconductivity is suppressed, other forms of pairing may take place. Obvious candidates are the formation of larger clusters or particle-hole pairing.

Particle-hole pairing is characterized by an order parameter of the form

$$\langle \bar{\psi}(x)\psi(y) \rangle = \exp(i\boldsymbol{Q} \cdot (\boldsymbol{x} + \boldsymbol{y})) \Sigma(x-y), \tag{23}$$

where $\boldsymbol{Q}$ is an arbitrary vector. This state describes a chiral density wave. It was first suggested in [37] as the ground state of QCD at large chemical potential and large $N_c$. This suggestion was based on the fact that particle-particle pairing, and superconductivity, is suppressed for large $N_c$ whereas particle-hole pairing is not. Particle-hole pairing, on the other hand, uses only a small part of the Fermi surface and does not take place in weak coupling. In the case of the one-gluon exchange interaction these issues were studied in [38]. The main conclusion is that, in weak coupling, the chiral density wave instability requires very large $N_c \gg 3$.

At moderate densities, and using realistic interactions, this is not necessarily the case. In particular, we know that at zero density the particle-anti-particle interaction is stronger, by a factor $N_c - 2$, than the particle-particle interaction. In a Nambu-Jona-Lasinio type description this interaction exceeds the critical value required for chiral symmetry breaking to take place. For this reason we have recently studied the competition between the particle-particle and particle-hole instabilities using non-perturbative, instanton generated, forces [39]. Our results are not only relevant to flavor symmetric quark matter at moderate densities, but also for the important case when there is a substantial difference between the chemical potentials for up and down quarks. As discussed in section 3 this disfavors $ud$-pairing, but it does not inhibit $uu^{-1}$ and $dd^{-1}$ particle-hole pairing.

Our results show that at low density the chiral density wave state is practically degenerate with the BCS solution. Given the uncertainties that affect the calculation this implies that both states have to be considered as realistic possibilities for the behavior of quark matter near the phase transition.

## 5.1 BCS Pairing

In order to study competing instabilities we use the standard Nambu-Gorkov formalism, in which the propagator is written as a matrix in the space of all possible pair condensates. The BCS channel is described by the $2 \times 2$ matrix

$$\hat{G}_{BCS} = \begin{pmatrix} \langle c_{k\uparrow} c^\dagger_{k\uparrow} \rangle & \langle c_{k\uparrow} c_{-k\downarrow} \rangle \\ \langle c^\dagger_{-k\downarrow} c^\dagger_{k\uparrow} \rangle & \langle c^\dagger_{-k\downarrow} c_{-k\downarrow} \rangle \end{pmatrix} \equiv \begin{pmatrix} G(k_0, \mathbf{k}, \Delta) & \bar{F}(k_0, \mathbf{k}, \Delta) \\ F(k_0, \mathbf{k}, \Delta) & \bar{G}(k_0, -\mathbf{k}, \Delta) \end{pmatrix}. \quad (24)$$

The propagator has the form

$$\hat{G}_{BCS} = \frac{1}{G_0^{-1} \bar{G}_0^{-1} - \Delta \bar{\Delta}} \begin{pmatrix} \bar{G}_0^{-1} & -\Delta \\ -\bar{\Delta} & G_0^{-1} \end{pmatrix}, \quad (25)$$

where

$$G_0 = \frac{1}{k_0 - \epsilon_k + i\delta_{\epsilon_k}}, \quad \bar{G}_0 = \frac{1}{k_0 + \epsilon_k + i\delta_{\epsilon_k}} \quad (26)$$

are the free particle propagator and its conjugate at finite chemical potential. Here, $\epsilon_k = \omega_k - \mu_q$ and $\delta_{\epsilon_k} = \text{sgn}(\epsilon_k)\delta$ determines the pole position. From this equation we can read off the diagonal and off-diagonal components of the Gorkov propagator. The off-diagonal, anomalous, propagator is

$$F(k_0, \mathbf{k}, \Delta) = \frac{-\Delta}{(k_0 - \epsilon_k + i\delta_{\epsilon_k})(k_0 + \epsilon_k + i\delta_{\epsilon_k}) - \Delta^2}. \quad (27)$$

The anomalous self energy $\Delta$ is determined by the gap equation

$$\Delta = (-i)\, \alpha_{pp} \int \frac{d^4p}{(2\pi)^4}\, F(p_0, \mathbf{p}, \Delta). \quad (28)$$

Here $\alpha_{pp}$ is the effective coupling in the particle-particle channel.

## 5.2 Chiral Density Wave

Using the same formalism we can also address pairing in the particle-hole channel at finite total pair momentum $Q$. In the mean-field approximation, the full Greens function in the presence of a single density wave takes the form

$$\hat{G}_{Ovh} = \begin{pmatrix} \langle c_{k\uparrow} c_{k\uparrow}^\dagger \rangle & \langle c_{k\uparrow} c_{k+Q\downarrow}^\dagger \rangle \\ \langle c_{k+Q\downarrow} c_{k\uparrow}^\dagger \rangle & \langle c_{k+Q\downarrow} c_{k+Q\downarrow}^\dagger \rangle \end{pmatrix} \equiv \begin{pmatrix} G(k_0, \boldsymbol{k}, \boldsymbol{Q}, \sigma) & \bar{S}(k_0, \boldsymbol{k}, \boldsymbol{Q}, \sigma) \\ S(k_0, \boldsymbol{k}, \boldsymbol{Q}, \sigma) & G(k_0, \boldsymbol{k}+\boldsymbol{Q}, \boldsymbol{Q}, \sigma) \end{pmatrix}$$
$$= \left[ \hat{G}_0^{-1} - \hat{\sigma} \right]^{-1} \tag{29}$$

Again, we can read off the anomalous part of the Greens function

$$S(k_0, \boldsymbol{k}, \boldsymbol{Q}, \sigma) = \frac{-\sigma}{(k_0 - \epsilon_k + i\delta_{\epsilon_k})(k_0 - \epsilon_{k+Q} + i\delta_{\epsilon_{k+Q}}) - \sigma^2}, \tag{30}$$

and the anomalous self energy $\sigma$ is determined by a self-consistency, or gap, equation

$$\sigma = (-i)\alpha_{ph} \int \frac{d^4 p}{(2\pi)^4} \, S(p_0, \boldsymbol{p}, \boldsymbol{Q}, \sigma; \mu_q) \, . \tag{31}$$

Notice that the energy contour integration receives non-vanishing contributions only if

$$\epsilon_p \, \epsilon_{p+Q} - \sigma^2 < 0 \, , \tag{32}$$

which ensures that the two poles in $p_0$ are in different (upper/lower) half-planes. This means that one particle (above the Fermi surface) and one hole (below the Fermi surface) participate in the interaction.

The formation of a condensate carrying nonzero total momentum $\boldsymbol{Q}$ is associated with nontrivial spatial structures. In the simplest case of particle-hole pairs with total momentum $Q$ this is a density wave of wave length $\lambda = 2\pi/Q$. In three dimensions, however, we can have several density waves characterized by different momenta $\boldsymbol{Q}$. In this case, the resulting spatial structure is a crystal. In general, the p-h pairing gap can be written as

$$\sigma(\boldsymbol{r}) = \sum_j \sum_{n=-\infty}^{+\infty} \sigma_{j,n} e^{in\boldsymbol{Q}_j \cdot \boldsymbol{r}} \, , \tag{33}$$

where the $\boldsymbol{Q}_j$ correspond to the (finite) number of fundamental waves, and the summation over $|n| > 1$ accounts for higher harmonics in the Fourier series. The matrix propagator formalism allows for the treatment of more than one density wave through a straightforward expansion of the basis states according to

$$\hat{G} = \begin{pmatrix} \langle c_{k\uparrow} c_{k\uparrow}^\dagger \rangle & \langle c_{k\uparrow} c_{k+Q_x\downarrow}^\dagger \rangle & \langle c_{k\uparrow} c_{k+Q_y\downarrow}^\dagger \rangle & \cdots \\ \langle c_{k+Q_x\downarrow} c_{k\uparrow}^\dagger \rangle & \langle c_{k+Q_x\downarrow} c_{k+Q_x\downarrow}^\dagger \rangle & \langle c_{k+Q_x\downarrow} c_{k+Q_y\downarrow}^\dagger \rangle & \cdots \\ \langle c_{k+Q_y\downarrow} c_{k\uparrow}^\dagger \rangle & \langle c_{k+Q_y\downarrow} c_{k+Q_x\downarrow}^\dagger \rangle & \langle c_{k+Q_y\downarrow} c_{k+Q_y\downarrow}^\dagger \rangle & \cdots \\ \vdots & \vdots & \vdots & \ddots \end{pmatrix} . \tag{34}$$

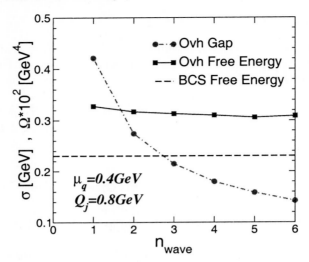

**Fig. 2.** Dependence of the free energy (upper full line) and $p$-$h$ pairing gap (dashed-dotted) on the number of waves ('patches') with fixed magnitude of the three-momentum $|Q_j| = 0.8$ GeV. The full line shows the value of the BCS ground state free energy. The results correspond to an instanton calculation with $\mu_q = 0.4$ GeV and $N/V = 1$ fm$^{-4}$.

The possibility of simultaneous BCS pairing can be incorporated by extending the Gorkov propagator to include both particle-hole and particle-particle components. In the following we will consider up to $n_w = 6$ waves in three orthogonal directions with $Q_x = Q_y = Q_z$ and $n = \pm 1$, characterizing a cubic crystal.

Note that in the propagators $G_0$ we do not include the contribution of anti-particles. This should be a reasonable approximation in the quark matter phase at sufficiently large $\mu_q$, when the standard particle-anti-particle chiral condensate has disappeared. At the same time, since our analysis is based on non-perturbative forces, the range of applicability is limited from above. Taken together, we estimate the range of validity for our calculations to be roughly given by $0.4$ GeV$\gtrsim \mu_q \gtrsim 0.6$ GeV. This coincides with the regime where, for the physical current strange quark mass of $m_s \simeq 0.14$ GeV, the two-flavor superconductor might prevail over the color-flavor locked (CFL) state so that our restriction to $N_f = 2$ is supported.

Solutions of the gap equations correspond to extrema (minima) in the energy density with respect to the gap $\sigma$. However, solutions may exist for several values of the wave vector $Q$. To determine the minimum in this quantity, one has to take into account the explicit form of the free energy density. In the mean-field approximation,

$$V_3 \, \Omega(\mu_q, Q, \sigma) = \int d^3x \left( \frac{\sigma^2(x)}{2\lambda} + \langle q^\dagger \left( i\alpha \cdot \nabla - 2\sigma(x) \right) q \rangle \right) , \quad (35)$$

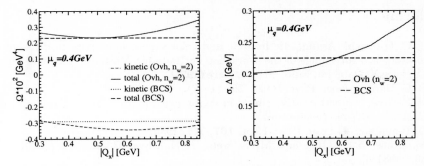

**Fig. 3.** Left panel: wave-vector dependence of the density wave free energy for one standing wave (full line: $\Omega_{tot}^{Ovh}$, short-dashed line: $\Omega_{kin}^{Ovh}$) in comparison to the BCS solution (long-dashed line: $\Omega_{tot}^{BCS}$, dotted line: $\Omega_{kin}^{BCS}$) at $\mu_q = 0.4$ GeV. Right panel: wave-vector dependence of the density wave pairing gap (full line) compared to the BCS gap (long-dashed line).

where $V_3$ is the 3-volume. The first contribution removes the double counting from the fermionic contribution in the mean-field treatment.

We have studied the coupled gap equations numerically. We do not find any solutions with simultaneous particle-particle and particle-hole condensates. This reduces the problem to the question whether the BCS or the density wave state is thermodynamically favored. The BCS solution $\Delta = 0.225$ GeV is unique and has free energy of $\Omega_{BCS}(\mu_q = 0.4 \text{ GeV}) = 2.3 \cdot 10^{-3}$ GeV$^4$. Here, we have neglected an irrelevant overall constant that does not affect the comparison with the density wave state.

The situation is more complicated in the case of particle-hole pairing. Let us start with the 'canonical' case where the momentum of the chiral density wave is fixed at twice the Fermi momentum, $Q = 2p_F$. In fig. 2 the resulting minimized free energy is displayed as a function of the number of included waves. The density wave solutions are not far above the BCS groundstate, with a slight energy gain for an increased number of waves.

However, one can further economize the energy of the chiral density wave state by exploiting the freedom associated with the wave vector $Q$ (or, equivalently, the periodicity of the lattice). For $Q > 2p_F$ the free energy rapidly increases. On the other hand, for $Q < 2p_F$ more favorable configurations are found. To correctly assess them one has to include the waves in pairs $|k \pm Q_j|$ of standing waves ($n_w = 2, 4, 6, \ldots$) to ensure that the occupied states in the Fermi sea are saturated within the first Brillouin Zone. The lowest-lying state we could find at $\mu_q = 0.4$ GeV occurs for one standing wave with $Q_{min} \simeq 0.5$ GeV and $\sigma \simeq 0.21$ GeV with a free energy $\Omega \simeq 2.3 \cdot 10^{-3}$ GeV$^4$, practically degenerate with the BCS solution. This density wave has a wavelength $\lambda \simeq 2.5$ fm. The minimum in the wave vector is in fact rather shallow, as seen from the explicit momentum dependence of the free energy displayed in fig. 3.

# References

1. S. C. Frautschi, Asymptotic freedom and color superconductivity in dense quark matter, in: Proceedings of the Workshop on Hadronic Matter at Extreme Energy Density, N. Cabibbo, Editor, Erice, Italy (1978).
2. B. C. Barrois, Nucl. Phys. **B129**, 390 (1977).
3. F. Barrois, Nonperturbative effects in dense quark matter, Ph.D. thesis, Caltech, UMI 79-04847-mc (microfiche).
4. D. Bailin and A. Love, Phys. Rept. **107**, 325 (1984).
5. N. Evans, S. D. Hsu and M. Schwetz, Nucl. Phys. **B551**, 275 (1999) [hep-ph/9808444].
6. N. Evans, S. D. Hsu and M. Schwetz, Phys. Lett. **B449**, 281 (1999) [hep-ph/9810514].
7. T. Schäfer and F. Wilczek, Phys. Lett. **B450**, 325 (1999) [hep-ph/9810509].
8. W. E. Brown, J. T. Liu and H. Ren, Phys. Rev. **D62**, 054016 (2000) [hep-ph/9912409].
9. T. Schäfer, Phys. Rev. **D62**, 094007 (2000) [hep-ph/0006034].
10. M. Alford, K. Rajagopal and F. Wilczek, Phys.Lett.B422,247,(1998).
11. R. Rapp, T. Schäfer, E. V. Shuryak and M. Velkovsky, Phys. Rev. Lett. **81**, 53 (1998) [hep-ph/9711396].
12. D. T. Son, Phys. Rev. **D59**, 094019 (1999) [hep-ph/9812287].
13. T. Schäfer and F. Wilczek, Phys. Rev. **D60**, 114033 (1999) [hep-ph/9906512].
14. R. D. Pisarski and D. H. Rischke, Phys. Rev. **D61**, 074017 (2000) [nucl-th/9910056].
15. D. K. Hong, V. A. Miransky, I. A. Shovkovy and L. C. Wijewardhana, Phys. Rev. **D61**, 056001 (2000) [hep-ph/9906478].
16. W. E. Brown, J. T. Liu and H. Ren, Phys. Rev. **D61**, 114012 (2000) [hep-ph/9908248].
17. T. Schäfer, Nucl. Phys. **B575**, 269 (2000) [hep-ph/9909574].
18. N. Evans, J. Hormuzdiar, S. D. Hsu and M. Schwetz, Nucl. Phys. **B581**, 391 (2000) [hep-ph/9910313].
19. M. Alford, K. Rajagopal and F. Wilczek, Nucl. Phys. **B537**, 443 (1999) [hep-ph/9804403].
20. T. Schäfer and F. Wilczek, Phys. Rev. Lett. **82**, 3956 (1999) [hep-ph/9811473].
21. M. Alford, J. Berges and K. Rajagopal, Nucl. Phys. **B558**, 219 (1999) [hep-ph/9903502].
22. T. Schäfer and F. Wilczek, Phys. Rev. **D60**, 074014 (1999) [hep-ph/9903503].
23. P. Bedaque, hep-ph/9910247.
24. M. Alford, J. Bowers and K. Rajagopal, Phys.Rev.D63,074016,(2001).
25. A. B. Migdal, Sov. Phys. JETP **36**, 1052 (1973).
26. R. F. Sawyer, Phys. Rev. Lett. **29**, 382 (1972).
27. D. J. Scalapino, Phys. Rev. Lett. **29**, 392 (1972).
28. D. B. Kaplan and A. E. Nelson, Phys. Lett. **B175**, 57 (1986).
29. G. E. Brown, K. Kubodera, and M. Rho, Phys. Lett. **B175**, 57 (1987).
30. T. Schäfer, nucl-th/0007021.
31. R. Casalbuoni and D. Gatto, Phys. Lett. **B464**, 111 (1999) [hep-ph/9908227].
32. D. T. Son and M. Stephanov, Phys. Rev. **D61**, 074012 (2000) [hep-ph/9910491], erratum: hep-ph/0004095.
33. M. Rho, A. Wirzba, and I. Zahed, Phys. Lett. **B473**, 126 (2000) [hep-ph/9910550].
34. D. K. Hong, T. Lee, and D. Min, Phys. Lett. **B477**, 137 (2000) [hep-ph/9912531].

35. C. Manuel and M. H. Tytgat, Phys. Lett. **B479**, 190 (2000) [hep-ph/0001095].
36. S. R. Beane, P. F. Bedaque, and M. J. Savage, Phys. Lett. **B483**, 131 (2000) [hep-ph/0002209].
37. D. V. Deryagin, D. Yu. Grigoriev, and V. A. Rubakov, Int. J. Mod. Phys. **A7**, 659 (1992).
38. E. Shuster and D. T. Son, Nucl. Phys. **B573**, 434 (2000) [hep-ph/9905448].
39. R. Rapp, E. Shuryak and I. Zahed, Phys.Rev.D63, 034008 (2001).

# Diquarks in Dense Matter

Marten B. Hecht, Craig D. Roberts, Sebastian M. Schmidt

Physics Division, Argonne National Laboratory, Argonne, Illinois, 60439-4843, USA

**Abstract.** We describe aspects of the role that diquark correlations play in understanding baryon structure and interactions. The significance of diquarks in that application motivates a study of the possibility that dense hadronic matter may exhibit diquark condensation; i.e., quark-quark pairing promoted by a quark chemical potential. A Gorkov-Nambu-like gap equation is introduced for QCD and analysed for 2-colour QCD (QC$_2$D) and, in two qualitatively different truncations, for QCD itself. Among other interesting features, we illustrate that QC$_2$D with massive fermions undergoes a second-order transition to a superfluid phase when the chemical potential exceeds $m_\pi/2$. In the QCD application we illustrate that the $\sigma := -\langle\bar{q}q\rangle^{1/3} \neq 0$ phase, which determines the properties of the mass spectrum at zero temperature and chemical potential, is unstable with respect to the superfluid phase when the chemical potential exceeds $\approx 2\sigma$, and that at this point the diquark gap is large, $\approx \sigma/2$. The superfluid phase survives to temperatures greater than that expected in the core of compact stars.

## 1 Diquarks

A diquark is a bosonic quark-quark correlation, which is necessarily coloured in all but 2-colour QCD (QC$_2$D). Therefore, in the presence of colour-confinement, diquarks cannot be directly observed in a $N_c \geq 3$ colour gauge theory's spectrum. Nevertheless evidence is accumulating that suggests confined diquark correlations play an important role in hadronic spectroscopy and interactions.

The first discussion of diquark correlations in literature addressing the strong interaction is almost coincident with that of quarks themselves [1,2]. It was quickly realised that both Lorentz scalar and vector diquarks, at least, are important for baryon spectroscopy [3] and, from a consideration of baryon magnetic moments [4], that the diquark correlations are not pointlike. This latter point is still often overlooked, although with decreasing frequency and now certainly without the imputation that it is a realistic simplification.

The motivation for considering diquarks in the constituent-quark model is that treating baryons directly as a three-body problem poses significant challenges in anything other than a mean-field approach. The task is much simplified if two of the constituents can be replaced by a single degree of freedom. However, an obvious question is whether there is any sense in which that replacement is more than just an expedient; i.e., a sense in which it captures some important aspect of QCD's dynamics? The answer is "yes" and we now turn to explaining that.

A significant step toward a description of baryons in quantum field theory can be identified in the realisation [5] that a large class of field theoretical models of the strong interaction admit the construction of a meson-diquark auxiliary-field effective action and thereby a description of baryons as loosely-bound quark-diquark composites. This is the class of theories with a chiral symmetry preserving four-fermion interaction, which includes, e.g., the Nambu–Jona-Lasinio model [6] and the Global Color Model [7], that have been widely used in analysing low energy strong interaction phenomena.

The picture of a baryon as loosely bound quark-diquark composite can also be reached via a direct analysis of the bound state contributions to the three quark scattering matrix. The associated Schwinger function (Euclidean Green function) is just that quantity whose large Euclidean-time behaviour yields a baryon's mass in numerical simulations of lattice-QCD. Considering the colour structure of this Schwinger function, we focus on the Clebsch-Gordon series for quarks in the fundamental representation of $SU_c(3)$:

$$3_c \otimes 3_c \otimes 3_c = (\bar{3}_c \oplus 6_c) \otimes 3_c = 1_c \oplus 8'_c \oplus 8_c \oplus 10_c, \qquad (1)$$

from which it is clear that a *colour singlet* 3-quark contribution is only possible when two of the quarks are combined to transform according to the *antitriplet*, $\bar{3}_c$, representation. This is the representation under which antiquarks transform.

Single gluon exchange is repulsive in the $6_c$ channel but attractive in the $\bar{3}_c$ channel. It is this feature that underpins the existence of the meson-diquark bosonisation referred to above. One way to see that is to realise that the auxiliary field effective action obtained for any element of the class of four-fermion interaction models provides a Lagrangian realisation of the rainbow-ladder truncation of the Dyson-Schwinger equations (DSEs) [8]. The rainbow-ladder truncation has been widely and successfully employed in the study of meson spectroscopy and interactions, see, e.g., Refs. [9–12], and nonpointlike colour-antitriplet diquark bound states exist in this truncation of the quark-quark Bethe-Salpeter equation (BSE) [13]. Hence they provide a real degree of freedom to be used in the bosonisation.

At first sight the existence of colour-antitriplet diquark bound states in these models, and in the rainbow-ladder truncation, appears to be a problem because such states are not observed in the QCD spectrum. However, as demonstrated in Refs. [14,15], this apparent lack of confinement is primarily an artefact of the rainbow-ladder truncation. Higher order terms in the quark-quark scattering kernel, the crossed-box and vertex corrections, whose analogue in the quark-antiquark channel do not much affect many of the colour singlet meson channels, act to ensure that the quark-quark scattering matrix does not exhibit the singularities that correspond to asymptotic (unconfined) diquark bound states.

Nevertheless, such studies with improved kernels, which do not produce diquark bound states, do support a physical interpretation of the "spurious" rainbow-ladder diquark masses. Denoting the mass in a given diquark channel (scalar, pseudovector, etc.) by $m_{qq}$, then $\ell_{qq} := 1/m_{qq}$ represents the range over which a true diquark correlation in this channel can persist *inside* a baryon. In this sense they are "pseudo-particle" masses that can be used to estimate which $\bar{3}_c$ diquark correlations should dominate the bound state contribution to the three quark scattering matrix, and hence which should be retained in deriving and solving a Poincaré covariant homogeneous Fadde'ev equation for baryons.

The simple Goldstone-theorem-preserving rainbow-ladder kernel of Ref. [16] can be used to illustrate this point. The model yields the following calculated diquark masses (isospin symmetry is assumed):

$$\begin{array}{c|cccccccc} (qq)_{J^P} & (ud)_{0^+} & (us)_{0^+} & (uu)_{1^+} & (us)_{1^+} & (ss)_{1^+} & (uu)_{1^-} & (us)_{1^-} & (ss)_{1^-} \\ \hline m_{qq}\,(\text{GeV}) & 0.74 & 0.88 & 0.95 & 1.05 & 1.13 & 1.47 & 1.53 & 1.64 \end{array} \quad (2)$$

and the results are relevant because the mass ordering is characteristic and model-independent, and lattice estimates, where available [17], agree with the masses tabulated here. Equation (2) suggests that an accurate study of the nucleon should retain the scalar and pseudovector correlations: $(ud)_{0^+}$, $(uu)_{1^+}$, $(ud)_{1^+}$, $(dd)_{1^+}$, because for these diquarks $m_{qq} \lesssim m_N$, where $m_N$ is the nucleon mass, but may neglect other correlations. Furthermore, it is obvious from the angular momentum Clebsch-Gordon series: $\frac{1}{2} \otimes 0 = \frac{1}{2}$ and $\frac{1}{2} \otimes 1 = \frac{1}{2} \oplus \frac{3}{2}$, that decuplet baryons are inaccessible without pseudovector diquark correlations. It is interesting to note that $m_{(ud)_{0^+}}/m_{(uu)_{1^+}} = 0.78$ cf. $0.76 = m_N/m_\Delta$ and hence one might anticipate that the presence of diquark correlations in baryons can provide a straightforward explanation of the $N$-$\Delta$ mass-splitting and other like effects. These ideas were first enunciated in Refs. [18,19] and Ref. [20] provides a convincing demonstration of their efficacy.

Explicit calculations; e.g., Ref. [12], show that retaining only a scalar diquark correlation in the kernel of the nucleon's Fadde'ev equation provides insufficient binding to obtain the experimental nucleon mass: the best calculated value is typically $\sim 40\%$ too large. However, with the addition of a pseudovector diquark it is easy to simultaneously obtain [12,21] the experimental masses of the nucleon and $\Delta$. Such calculations plainly verify the intuition that follows from simple mass-counting: the pseudovector diquarks are an important but subdominant element of the nucleon's Fadde'ev amplitude (cf. the scalar diquark) whilst being the sole component of the $\Delta$.

The presence of diquark correlations in baryons also affects the predictions for scattering observables, which may therefore provide a means for experimentally verifying the ideas described above. For example, their presence provides a simple explanation of the neutron's nonzero electric form factor [12]: charge separation arising from a heavy $(ud)$ diquark with electric charge $\frac{1}{3}$ holding on to a relatively light, electric charge $(-\frac{1}{3})$ $d$-quark. And also a prediction for the ratio of the proton's valence-quark distributions: $d/u := d_v(x \to 1)/u_v(x \to 1)$, which can be measured in deep inelastic scattering [22]. In this case, diquark correlations with differing masses in the nucleon's Fadde'ev amplitude are an immediate indication of the breaking of $SU(6)$ symmetry, hence $d/u \neq 1/2$. Furthermore, if it were true that $m_{(qq)_{J^P}} \gg m_{(ud)_{0^+}}$, for all $J^P \neq 0^+$, then $d/u = 0$. However, as we have seen, in reality the $1^+$ diquark is an important subdominant piece of the nucleon's Fadde'ev amplitude so that a realistic picture of diquarks in the nucleon implies $0 < d/u < \frac{1}{2}$, with the actual value being a sensitive measure of the proton's pseudovector diquark fraction.

## 2 Superfluidity in Quark Matter

We have outlined above the role and nature of diquark correlations in hadronic physics at zero temperature and density, and emphasised that diquarks are an idea as old as that of quarks themselves. Another phenomenon suggested immediately by the meson-diquark auxiliary-field effective action is that of diquark condensation; i.e., quark-quark Cooper pairing, which was first explored in this context using a simple version of the Nambu–Jona-Lasinio model [23]. A chemical potential promotes Cooper pairing in fermion systems and the possibility that such diquark pairing is exhibited in quark matter is also an old idea, early explorations of which employed [24] the rainbow-ladder truncation of the quark DSE (QCD gap equation). That interest in this possibility has been renewed is evident in a number of contributions to this volume [25–29]. A quark-quark Cooper pair is a composite boson with both electric and colour charge, and hence superfluidity in quark matter entails superconductivity and colour superconductivity. However, the last feature makes it difficult to identify an order parameter that can characterise a transition to the superfluid phase: the Cooper pair is gauge dependent and an order parameter is ideally describable by a gauge-invariant operator. This particularly inhibits an analysis of the phenomenon using lattice-QCD.

### 2.1 Gap Equation

Studies of the gap equation that suppress the possibility of diquark condensation show that cold, sparse two-flavour QCD exhibits a nonzero quark-antiquark condensate: $\langle \bar{q}q \rangle \neq 0$. If it were otherwise then the $\pi$-meson would be almost as massive as the $\rho$-meson, which would yield a very different observable world. The quark condensate is undermined by increasing $\mu$ and $T$, and there is a large domain in the physical (upper-right) quadrant of the $(\mu, T)$-plane for which $\langle \bar{q}q \rangle = 0$: for the purpose of exemplification, that domain can crudely be characterised as the set (see, e.g., Refs. [30–32]):

$$\{(\mu, T) : \mu^2/\mu_c^2 + T^2/T_c^2 > 1, \ \mu, T > 0 \, ; \mu_c \sim 0.3\text{--}0.4\,\text{GeV}, T_c \sim 0.15\,\text{GeV}\}. \tag{3}$$

Increasing temperature also opposes Cooper pairing. However, since increasing $\mu$ promotes it, there may be a (large-$\mu$,low-$T$)-subdomain in which quark matter exists in a superfluid phase. That domain, if it exists, is unlikely to be accessible at the Relativistic Heavy Ion Collider, because it operates in the high temperature regime, but may be realised in the core of compact astrophysical objects, which could undergo a transition to superfluid quark matter as they cool. Possible signals accompanying such a transition are considered in Refs. [25,27–29].

It was observed in Ref. [33] that a direct means of determining whether a $SU_c(N)$ gauge theory supports scalar diquark condensation is to study the gap equation satisfied by

$$\mathcal{D}(p,\mu) := \mathcal{S}(p,\mu)^{-1} = \begin{pmatrix} D(p,\mu) & \Delta^i(p,\mu)\,\gamma_5 \lambda^i_\wedge \\ -\Delta^i(p,-\mu)\,\gamma_5 \lambda^i_\wedge & CD(-p,\mu)^t C^\dagger \end{pmatrix}. \quad (4)$$

Here $T = 0$, for illustrative simplicity and because temperature can only act to destabilise a condensate, and, with $\omega_{[\mu]} = p_4 + i\mu$,

$$D(p,\mu) = i\boldsymbol{\gamma}\cdot\boldsymbol{p}\, A(\boldsymbol{p}^2, \omega_{[\mu]}^2) + B(\boldsymbol{p}^2, \omega_{[\mu]}^2) + i\gamma_4\, \omega_{[\mu]}\, C(\boldsymbol{p}^2, \omega_{[\mu]}^2); \quad (5)$$

i.e., the inverse of the dressed-quark propagator in the absence of diquark pairing. (NB. For $\mu = 0$, $A$, $B$ and $C$ are real functions.) It is one of the fundamental features of DSE studies that the existence of a nonzero quark condensate: $\langle\bar{q}q\rangle \neq 0$, is signalled in the solution of the gap equation by $B(\boldsymbol{p}^2, \omega_{[\mu]}^2) \neq 0$ [9].

In Eq. (4), $\{\lambda^i_\wedge, i = 1\ldots n_c^\wedge, n_c^\wedge = N_c(N_c-1)/2\}$ are the antisymmetric generators of $SU_c(N_c)$ and $C = \gamma_2\gamma_4$ is the charge conjugation matrix,

$$C\gamma_\mu^t C^\dagger = -\gamma_\mu; \quad [C,\gamma_5] = 0, \quad (6)$$

where $X^t$ denotes the matrix transpose of $X$. The key new feature here is that diquark condensation is characterised by $\Delta^i(p,\mu) \neq 0$, for at least one $i$. That is clear if one considers the quark piece of the QCD Lagrangian density: $L[\bar{q}, q]$. It is a scalar and hence $L[\bar{q}, q]^t = L[\bar{q}, q]$. Therefore $L[\bar{q}, q] \propto L[\bar{q}, q] + L[\bar{q}, q]^t$, and it is a simple exercise to show that this sum, and hence the action, can be re-expressed in terms of a $2 \times 2$ diagonal matrix using the bispinor fields

$$Q(x) := \begin{pmatrix} q(x) \\ \underline{q}(x) := \tau_f^2 C \bar{q}^t \end{pmatrix}, \quad \bar{Q}(x) := \left(\bar{q}(x)\ \ \underline{\bar{q}}(x) := q^t C \tau_f^2\right), \quad (7)$$

where $\{\tau_f^i : i = 1, 2, 3\}$ are Pauli matrices that act on the isospin index.[1] It is plain upon inspection that a nonzero entry: $d(x)\gamma_5$, in row-2–column-1 of this action-matrix would act as a source for $q^t \tau_f^2 C \gamma_5 q$; i.e., as a scalar diquark source.

It is plain now that the explicit $2 \times 2$ matrix structure of $\mathcal{D}(p,\mu)$ in Eq. (4) exhibits a quark bispinor index that is made with reference to $Q(x)$, $\bar{Q}(x)$. This approach; i.e., employing a "matrix propagator" with "anomalous" off-diagonal elements, simply exploits the Gorkov-Nambu treatment of superconductivity in fermionic systems, which is explained in textbooks, e.g., Ref. [34]. It makes possible a well-ordered treatment and makes unnecessary a truncated bosonisation, which in all but the simplest models is a procedure difficult to improve systematically.

The bispinor gap equation can be written in the form

$$\mathcal{D}(p,\mu) = \mathcal{D}_0(p,\mu) + \begin{pmatrix} \Sigma_{11}(p,\mu) & \Sigma_{12}(p,\mu) \\ \gamma_4\, \Sigma_{12}(-p,\mu) & \gamma_4\, C\Sigma_{11}(-p,\mu)^t C^\dagger \end{pmatrix}, \quad (8)$$

---

[1] We only consider theories with two light-flavours. Additional possibilities open if this restriction is lifted [25,26].

where the second term on the right-hand-side is just the bispinor self energy. Here, in the absence of a scalar diquark source term,

$$\mathcal{D}_0(p,\mu) = (i\gamma \cdot p + m)\tau_Q^0 - \mu \tau_Q^3, \qquad (9)$$

with $m$ the current-quark mass, and the additional Pauli matrices: $\{\tau_Q^\alpha, \alpha = 0,1,2,3\}$, act on the bispinor indices. As we will see, the structure of $\Sigma_{ij}(p,\mu)$ specifies the theory and, in practice, also the approximation or truncation of it.

## 3  Two Colours

Two colour QCD (QC$_2$D) provides an important and instructive example. In this case $\Delta^i \lambda_\Lambda^i = \Delta \tau_c^2$ in Eq. (4), with $\frac{1}{2}\tau_c$ the generators of $SU_c(2)$, and it is useful to employ a modified bispinor

$$Q_2(x) := \begin{pmatrix} q(x) \\ q_2 := \tau_c^2 \underline{q}(x) \end{pmatrix}, \quad \bar{Q}_2(x) := \begin{pmatrix} \bar{q}(x) & \bar{q}_2(x) := \underline{\bar{q}}(x)\,\tau_c^2 \end{pmatrix}. \qquad (10)$$

Embedding the additional factor of $\tau_c^2$ in this way makes it possible to write the Lagrangian's fermion–gauge-boson interaction term as

$$\bar{Q}_2(x) \frac{i}{2} g \gamma_\mu \tau_c^k \tau_Q^0 \, Q_2(x) \, A_\mu^k(x) \qquad (11)$$

because $SU_c(2)$ is pseudoreal; i.e., $\tau_c^2 (-\boldsymbol{\tau}_c)^{\text{t}} \tau_c^2 = \boldsymbol{\tau}_c$, and the fundamental and conjugate representations are equivalent; i.e., fermions and antifermions are practically indistinguishable. (That the interaction term takes this form is easily seen using $L[\bar{q},q]^{\text{t}} = L[\bar{q},q]$.)

Using the pseudoreality of $SU_c(2)$ it can be shown that, for $\mu = 0$ and in the chiral limit, $m = 0$, the general solution of the bispinor gap equation is [33]

$$\mathcal{D}(p) = i\gamma \cdot p\, A(p^2) + \mathcal{V}(-\boldsymbol{\pi})\, \mathcal{M}(p^2), \quad \mathcal{V}(\boldsymbol{\pi}) = \exp\left\{i\gamma_5 \sum_{\ell=1}^{5} T^\ell \pi^\ell\right\} = \mathcal{V}(-\boldsymbol{\pi})^{-1}, \qquad (12)$$

where $\pi^{\ell=1,\ldots,5}$ are arbitrary constants and

$$\{T^{1,2,3} = \tau_Q^3 \otimes \boldsymbol{\tau}_f,\ T^4 = \tau_Q^1 \otimes \tau_f^0,\ T^5 = \tau_Q^2 \otimes \tau_f^0\},\ \{T^i, T^j\} = 2\delta^{ij}, \qquad (13)$$

so that $\mathcal{D}^{-1}$ is

$$\mathcal{S}(p) = \frac{-i\gamma \cdot p A(p^2) + \mathcal{V}(\boldsymbol{\pi})\mathcal{M}(p^2)}{p^2 A^2(p^2) + \mathcal{M}^2(p^2)} := -i\gamma \cdot p\, \sigma_V(p^2) + \mathcal{V}(\boldsymbol{\pi})\, \sigma_S(p^2). \qquad (14)$$

[To illustrate this, note that inserting $\boldsymbol{\pi} = (0,0,0,0,-\frac{1}{4}\pi)$ produces an inverse bispinor propagator with the simple form in Eq. (4).]

That the gap equation is satisfied for any constants $\pi^\ell$ signals a vacuum degeneracy – it corresponds to a multidimensional "Mexican hat" structure of

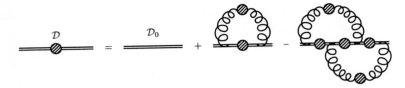

**Fig. 1.** Illustration of the vertex-corrected gap equation, which is the next-to-leading-order in the systematic, symmetry-preserving truncation scheme of Ref. [14]. Retaining only the first two diagrams on the right-hand-side yields the dressed-rainbow truncation. Each bispinor quark-gluon vertex is bare but the shaded circles mark quark and gluon 2-point functions that are dressed. The corresponding truncation in the relevant Bethe-Salpeter equations ensures the absence of diquark bound states in the strong interaction spectrum. (Adapted from Ref. [33].)

the theory's effective potential, as noted in a related context in Ref. [35]. Consequently, if the interaction supports a mass gap, then that gap describes a five-parameter continuum of degenerate condensates:

$$\langle \bar{Q}_2 \mathcal{V}(\pi) Q_2 \rangle \neq 0 \,, \tag{15}$$

and there are 5 associated Goldstone bosons: 3 pions, a diquark and an antidiquark. (Diquarks are the "baryons" of $QC_2D$.) In the construction of Eq. (12) one has a simple elucidation of a necessary consequence of the Pauli-Gürsey symmetry of $QC_2D$; i.e., the practical equivalence of particles and antiparticles.

For $m \neq 0$, the gap equation requires [33] $\text{tr}_{FQ}[T^i \mathcal{V}] = 0$, so that in this case only $\langle \bar{Q}_2 Q_2 \rangle \neq 0$ and now the spectrum contains five degenerate but massive pseudo-Goldstone bosons. This illustrates that a nonzero current-quark mass promotes a quark condensate and opposes diquark condensation.

For $\mu \neq 0$ the general solution of the gap equation has the form

$$\mathcal{D}(p,\mu) = \begin{pmatrix} D(p,\mu) & \gamma_5 \Delta(p,\mu) \\ -\gamma_5 \Delta^*(p,\mu) & CD(-p,\mu)C^\dagger \end{pmatrix} . \tag{16}$$

In the *absence* of a diquark condensate; i.e., for $\Delta \equiv 0$,

$$[U_B(\alpha), \mathcal{D}(p,\mu)] = 0 \,, \quad U_B(\alpha) := e^{i\alpha \tau_Q^3 \otimes \tau_f^0} \,; \tag{17}$$

i.e., baryon number is conserved in $QC_2D$. This makes plain that the existence of a diquark condensate dynamically breaks this symmetry.

To proceed we choose to be explicit and employ the dressed-rainbow truncation of the gap equation, see Fig. 1, with a model for the Landau gauge dressed-gluon propagator:

$$g^2 D_{\mu\nu}(k) = \left(\delta_{\mu\nu} - \frac{k_\mu k_\nu}{k^2}\right) \mathcal{F}_2(k^2) \,, \quad \mathcal{F}_2(k^2) = \frac{64}{9}\pi^4 \hat{\eta}^2 \delta^4(k) \,. \tag{18}$$

This form for $g^2 D_{\mu\nu}(k)$ was introduced [36] for the modelling of confinement in QCD but it is also appropriate here because the string tension in QC$_2$D is nonzero, and that is represented implicitly in Eq. (18) via the mass-scale $\hat{\eta}$.

Using Eq. (18) we obtain an algebraic dressed-rainbow gap equation that, for $p^2 = |\boldsymbol{p}|^2 + p_4^2 = 0$, reads:

$$A - 1 = \frac{1}{2}\hat{\eta}^2 K \left\{ A\left(B^{*2} - C^{*2}\mu^2\right) + A^* |\Delta|^2 \right\}, \tag{19}$$

$$(C-1)\mu = \frac{\mu}{2}\hat{\eta}^2 K \left\{ C\left(B^{*2} - C^{*2}\mu^2\right) - C^* |\Delta|^2 \right\}, \tag{20}$$

$$B - m = \hat{\eta}^2 K \left\{ B\left(B^{*2} - C^{*2}\mu^2\right) + B^* |\Delta|^2 \right\}, \tag{21}$$

$$\Delta = \hat{\eta}^2 K \left\{ \Delta\left(|B|^2 + |C|^2\mu^2\right) + \Delta |\Delta|^2 \right\}, \tag{22}$$

with $K^{-1} = |B^2 - C^2\mu^2|^2 + 2|\Delta|^2(|B|^2 + |C|^2\mu^2) + |\Delta|^4$. These equations possess a $B \leftrightarrow \Delta$ symmetry when $(m, \mu) = 0$, which is a straightforward illustration of the vacuum degeneracy described above using the matrix $\mathcal{V}(\boldsymbol{\pi})$. (Recall that for $\mu = 0$, $A$, $B$ and $C$ are real functions.) They also exemplify the general result that $\Delta$ is real for all $\mu$. Another exemplary result follows from a linearisation in $\mu^2$: $\mu \neq 0$ acts to promote a nonzero value of $\Delta$ but oppose a nonzero value of $B$; i.e., a nonzero chemical potential plainly acts to promote Cooper pairing at the expense of $\langle \bar{q}q \rangle$.

For $(m, \mu) = 0$ the solution of the dressed-rainbow gap equation obtained using Eq. (18) is:

$$A(p^2) = C(p^2) = \begin{cases} 2, & p^2 < \frac{\hat{\eta}^2}{4} \\ \frac{1}{2}\left(1 + \sqrt{1 + \frac{2\hat{\eta}^2}{p^2}}\right), & \text{otherwise}, \end{cases} \tag{23}$$

$$\mathcal{M}^2(p^2) := B^2(p^2) + \Delta^2(p^2) = \begin{cases} \hat{\eta}^2 - 4p^2, & p^2 < \frac{\hat{\eta}^2}{4} \\ 0, & \text{otherwise}. \end{cases} \tag{24}$$

As we have already mentioned, the dynamically generated mass function, $\mathcal{M}(p^2)$, is tied to the existence of quark and/or diquark condensates, which can be illustrated by noting that $(B = 0, \Delta \neq 0)$ corresponds to $\boldsymbol{\pi} = (0, 0, 0, 0, \frac{1}{2}\pi)$ in Eq. (15); i.e., $\langle \bar{Q}_2 i\gamma_5 \tau_Q^2 Q_2 \rangle \neq 0$, while $(B \neq 0, \Delta = 0)$ corresponds to $\boldsymbol{\pi} = (0, 0, 0, 0, 0)$; i.e., $\langle \bar{Q}_2 Q_2 \rangle \neq 0$.

The usual chiral, $SU_A(2)$, transformations are realised via

$$\mathcal{D}(p, \mu) \to V(\boldsymbol{\pi}) \mathcal{D}(p, \mu) V(\boldsymbol{\pi}), \quad V(\boldsymbol{\pi}) := e^{i\gamma_5 \boldsymbol{\pi} \cdot \boldsymbol{T}}, \quad \boldsymbol{\pi} = (\pi^1, \pi^2, \pi^3), \tag{25}$$

and therefore, since the anticommutator $\{\boldsymbol{T}, T^{4,5}\} = 0$, a diquark condensate does not dynamically break chiral symmetry. On the other hand, since $[\mathbf{1}, \boldsymbol{T}] = 0$, a quark condensate does dynamically break chiral symmetry.

In addition, and of particular importance, is the feature that in *combination* with the momentum-dependent vector self energy the momentum-dependence of $\mathcal{M}(p^2)$ ensures that the dressed-quark propagator does not have a Lehmann representation and hence can be interpreted as describing a confined quark [8–10].

The interplay between the scalar and vector self energies is the key to this realisation of confinement. The qualitative features of this simple model's dressed-quark propagator have been confirmed in recent lattice-QCD simulations [37] and the agreement between those simulations and more sophisticated DSE studies is semi-quantitative [11].

In the steepest descent (or stationary phase) approximation the contribution of dressed-quarks to the thermodynamic pressure is

$$p_\Sigma(\mu, T) = \frac{1}{2\beta V} \left\{ \text{TrLn}\left[\beta S^{-1}\right] - \frac{1}{2}\text{Tr}\left[\Sigma\, S\right]\right\}, \qquad (26)$$

where $\beta = 1/T$, and "Tr" and "Ln" are extensions of "tr" and "ln" to matrix-valued functions.

The MIT Bag Model pictures the quarks in a baryon as occupying a spatial volume from which the nontrivial quark-condensed vacuum (scalar-field) has been expelled. Therefore, as observed in Refs. [35], the bag constant can be identified as the pressure difference between the $\langle \bar{q}q \rangle \neq 0$ vacuum, the so-called Nambu-Goldstone phase in which chiral symmetry is dynamically broken, and the chirally symmetric no-condensate alternative, which is called the Wigner-Weyl vacuum. That difference is given by

$$\mathcal{B}_B(\mu) := p_\Sigma(\mu, \mathcal{S}[B, \Delta = 0]) - p_\Sigma(\mu, \mathcal{S}[B = 0, \Delta = 0]), \qquad (27)$$

and it is, of course, $\mu$-dependent because the vacuum *evolves* with changing $\mu$. $\mathcal{B}_B$ also evolves with temperature and this necessary $(\mu, T)$-dependence of the bag constant can have an important effect on quark star properties; e.g., reducing the maximum supportable mass of a quark matter star, as discussed in Ref. [38].

If we define, by analogy,

$$\mathcal{B}_\Delta(\mu) := p_\Sigma(\mu, \mathcal{S}[B = 0, \Delta]) - p_\Sigma(\mu, \mathcal{S}[B = 0, \Delta = 0]), \qquad (28)$$

then the relative stability of the quark- and diquark-condensed phases is measured by the pressure difference

$$\delta p(\mu) := \mathcal{B}_\Delta(\mu) - \mathcal{B}_B(\mu). \qquad (29)$$

For $\delta p(\mu) > 0$ the diquark condensed phase is favoured.

At $(m = 0, \mu = 0)$, $\delta p = 0$, with

$$\mathcal{B}_B(0) = \mathcal{B}_\Delta(0) = (0.092\,\hat{\eta})^4. \qquad (30)$$

This equality is a manifestation of the vacuum degeneracy identified above in connection with the matrix $\mathcal{V}(\pi)$. However,

$$\text{with } m = 0\,, \delta p > 0 \text{ for all } \mu > 0\,, \qquad (31)$$

which means that the Wigner-Weyl vacuum is unstable with respect to diquark condensation for all $\mu > 0$ [33] and that the superfluid phase is favoured over the Nambu-Goldstone phase.

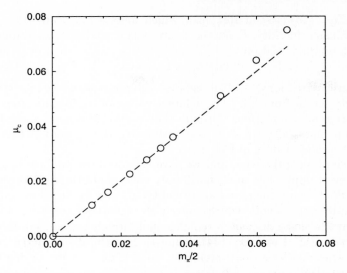

**Fig. 2.** Evolution of the critical chemical potential for diquark condensation as the current-quark mass is increased. The coordinate measures the magnitude of the current-quark mass through the mass of the theory's lightest excitation [a pseudo-Goldstone mode, as described after Eq. (15)]. $\mu_c$ and $m_\pi$ are measured in units of the model's mass scale, $\hat{\eta}$ in Eq. (18): for $m = 0$, the vector meson mass is $\frac{1}{\sqrt{2}}\hat{\eta}$.

Now, although the action for the $\mu \neq 0$ theory is invariant under

$$Q_2 \to U_B(\alpha)\, Q_2\,, \quad \bar{Q}_2 \to \bar{Q}_2\, U_B(-\alpha)\,, \tag{32}$$

which is associated with baryon number conservation, the diquark condensate breaks this symmetry:

$$\langle \bar{Q}_2 i\gamma_5 \tau_Q^2 Q_2 \rangle \to \cos(2\alpha)\, \langle \bar{Q}_2 i\gamma_5 \tau_Q^2 Q_2 \rangle - \sin(2\alpha)\, \langle \bar{Q}_2 i\gamma_5 \tau_Q^1 Q_2 \rangle\,; \tag{33}$$

i.e., it is a ground state that is not invariant under the transformation. Hence, for $(m = 0, \mu \neq 0)$, only one Goldstone mode remains. These symmetry breaking patterns and the concomitant numbers of Goldstone modes in QC$_2$D are also described in Ref. [39].

For $m \neq 0$ and small values of $\mu$ the gap equation only admits a solution with $\Delta \equiv 0$; i.e., diquark condensation is blocked because the current-quark mass is a source of the quark condensate [see, e.g., Eq. (21) and the comments after Eq. (15)]. However, with increasing $\mu$, the theory undergoes a transition to a phase in which the diquark condensate is nonzero. We identify the transition as second order because the diquark condensate falls continuously to zero as $\mu \to \mu_c^+$, where $\mu_c$ is the critical chemical potential. In Fig. 2 we plot the critical chemical potential as a function of $m_\pi/2$, where, to sidestep solving the Bethe-Salpeter equation, $m_\pi$ was obtained using a Gell-Mann–Oakes–Renner-like mass formula, Eqs. (16)-(18) in Ref. [31], which follows [40] from the axial-vector Ward-Takahashi identity. From the figure it is clear that this simple model of QC$_2$D exhibits the relation

$$\mu_c = \tfrac{1}{2} m_\pi , \qquad (34)$$

which is anticipated for QCD-like theories with pseudoreal fermions [41]. We note that the deviation from Eq. (34) at larger values of $m_\pi$ results from neglecting $O(m^2)$-corrections in the mass formula. This omission leads to an underestimate of the pion mass [42], which is responsible for the upward deflection of the calculated results evident in Fig. 2.

In exemplifying these features we have employed the rainbow-ladder truncation. However, improving on that will only yield quantitative changes of $\lesssim 20\%$ in the results because the pseudoreality of QC$_2$D and the equal dimension of the colour and bispinor spaces, which underly the theory's Pauli-Gürsey symmetry, ensure that the entire discussion remains qualitatively unchanged. In particular, the results of Fig. 2 and Eq. (34), being tied to chiral symmetry, remain unchanged because at least one systematic, chiral symmetry preserving truncation scheme exists [14].

## 4 Three Colours

The exploration of superfluidity in true QCD encounters two differences: the dimension of the colour space is greater than that of the bispinor space and the fundamental and conjugate representations of the gauge group are not equivalent. The latter is of obvious importance because it entails that the quark-quark and quark-antiquark scattering matrices are qualitatively different.

$n_c^\Lambda = 3$ in QCD and hence in canvassing superfluidity it is necessary to choose a direction for the condensate in colour space;[2] e.g., $\Delta^i \lambda_\Lambda^i = \Delta \lambda^2$ in Eq. (4), so that

$$\mathcal{D}(p,\mu) = \left( \begin{array}{c|c} D_\|(p,\mu)P_\| + D_\perp(p,\mu)P_\perp & \Delta(p,\mu)\gamma_5 \lambda^2 \\ \hline -\Delta(p,-\mu)\gamma_5 \lambda^2 & CD_\|(-p,\mu)C^\dagger P_\| + CD_\perp(p,\mu)C^\dagger P_\perp \end{array} \right) , \qquad (35)$$

where $P_\| = (\lambda^2)^2$, $P_\perp + P_\| = \mathrm{diag}(1,1,1)$, and $D_\|$, $D_\perp$ are defined via obvious generalisations of Eqs. (4), (5). In Eq. (35) the evident, demarcated block structure makes explicit the bispinor index: each block is a $3 \times 3$ colour matrix and the subscripts: $\|$, $\perp$, indicate whether or not the subspace is accessible via $\lambda_2$.

The bispinors associated with this representation are given in Eqs. (7) and in this case the Lagrangian's quark-gluon interaction term is

$$\bar{Q}(x) ig \Gamma_\mu^a Q(x) A_\mu^a(x) , \quad \Gamma_\mu^a = \left( \begin{array}{c|c} \tfrac{1}{2}\gamma_\mu \lambda^a & 0 \\ \hline 0 & -\tfrac{1}{2}\gamma_\mu (\lambda^a)^{\mathrm{t}} \end{array} \right) . \qquad (36)$$

---

[2] It is this selection of a direction in colour space that opens the possibility for colour-flavour locked diquark condensation in a theory with three effectively-massless quarks; i.e., current-quark masses $\ll \mu$ [25,26].

It is instructive to compare this with Eq. (11): with three colours the interaction term is not proportional to the identity matrix in the bispinor space, $\tau_Q^0$. This makes plain the inequivalence of the fundamental and conjugate fermion representations of $SU_c(3)$, which entails that quark-antiquark scattering is different from quark-quark scattering.

It is straightforward to derive the gap equation at arbitrary order in the truncation scheme of Ref. [14] and it is important to note that because

$$D_\parallel(p,\mu) P_\parallel + D_\perp(p,\mu) P_\perp = \lambda^0 \left\{ \tfrac{2}{3} D_\parallel(p,\mu) + \tfrac{1}{3} D_\perp(p,\mu) \right\} + \tfrac{1}{\sqrt{3}} \lambda^8 \left\{ D_\parallel(p,\mu) - D_\perp(p,\mu) \right\} \tag{37}$$

the interaction: $\Gamma_\mu^a \mathcal{S}(p,\mu) \Gamma_\nu^a$, necessarily couples the $\parallel$- and $\perp$-components. Reference [33] explored the possibility of diquark condensation in QCD using both the rainbow and vertex-corrected gap equation, illustrated in Fig. 1, with

$$g^2 D_{\mu\nu}(k) = \left( \delta_{\mu\nu} - \frac{k_\mu k_\nu}{k^2} \right) \mathcal{F}(k^2), \quad \mathcal{F}(k^2) = 4\pi^4 \eta^2 \delta^4(k). \tag{38}$$

For $(m,\mu) = 0$ the rainbow-ladder truncation yields

$$m_\omega^2 = m_\rho^2 = \tfrac{1}{2} \eta^2, \quad \langle \bar{q}q \rangle^0 = (0.11\,\eta)^3, \quad \mathcal{B}_B(\mu=0) = (0.10\,\eta)^4, \tag{39}$$

and momentum-dependent vector self energies, which lead to an interaction between the $\parallel$- and $\perp$-components of $\mathcal{D}$ that blocks diquark condensation. This is in spite of the fact that $\lambda^a \lambda^2 (-\lambda^a)^t = \tfrac{1}{2} \lambda^a \lambda^a$, which entails that the rainbow-truncation quark-quark scattering kernel is purely attractive and strong enough to produce diquark bound states [13]. (Remember that in the colour singlet meson channel the rainbow-ladder truncation gives the colour coupling $\lambda^a \lambda^a$; i.e., an interaction with the same sign but twice as strong.)

For $\mu \neq 0$ and in the *absence* of diquark condensation this model and truncation exhibits [30] coincident, first order chiral symmetry restoring and deconfining transitions at

$$\mu_{c,\,\text{rainbow}}^{B,\Delta=0} = 0.28\,\eta = 0.3\,\text{GeV}, \tag{40}$$

with $\eta = 1.06\,\text{GeV}$ fixed by fitting the $m \neq 0$ vector meson mass [14].

For $(m = 0, \mu \neq 0)$, however, the rainbow-truncation gap equation admits a solution with $\Delta(p,\mu) \neq 0$ and $B(p,\mu) \equiv 0$. The pressure difference, $\delta p(\mu)$ in Eq. (29), is again the way to determine whether the stable ground state is the Nambu-Goldstone or superfluid phase. With increasing $\mu$, $\mathcal{B}_B(\mu)$ decreases, very slowly at first, and $\mathcal{B}_\Delta(\mu)$ increases rapidly from zero. That evolution continues until

$$\mu_{c,\,\text{rainbow}}^{B=0,\Delta} = 0.25\,\eta = 0.89\,\mu_{c,\,\text{rainbow}}^{B,\Delta=0}, \tag{41}$$

where $\mathcal{B}_\Delta(\mu)$ becomes greater-than $\mathcal{B}_B(\mu)$. This signals a first order transition to the superfluid ground state and at the boundary

$$\langle \bar{Q} i \gamma_5 \tau_Q^2 \lambda^2 Q \rangle_{\mu = \mu_{c,\,\text{rainbow}}^{B=0,\Delta}} = (0.65)^3 \langle \bar{Q}Q \rangle_{\mu=0}. \tag{42}$$

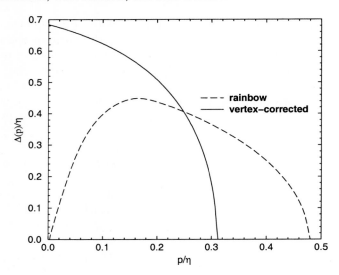

**Fig. 3.** Dashed line: $\Delta(z, \mu_c^{B,\Delta})$ obtained in rainbow truncation with the QCD model defined via Eq. (38), plotted for $\alpha = 0$ as a function of $p$, where $z = p(0, 0, \sin\alpha, i\mu + \cos\alpha)$. As $\mu$ increases, the peak position shifts to larger values of $p$ and the peak height increases. Solid line: $\Delta(z, \mu = 0)$ obtained as the solution of Eq. (43), the vertex-corrected gap equation, also with $\alpha = 0$. (Adapted from Ref. [33].)

Since $\mathcal{B}_\Delta(\mu) > 0$ for all $\mu > 0$ there is no intermediate domain of $\mu$ in which all condensates vanish.

The solution of the rainbow gap equation: $\Delta(p, \mu_c^{B,\Delta})$, which is real and characterises the diquark gap, is plotted in Fig. 3. It vanishes at $p^2 = 0$ as a consequence of the $\|$-$\perp$ coupling that blocked diquark condensation at $\mu = 0$, and also at large $p^2$, which is a manifestation of this simple model's version of asymptotic freedom.

The chemical potential[3] at which the switch to the superfluid ground state occurs, Eq. (41), is consistent with other estimates made using models comparable to the rainbow-truncation class [25,26,44–46], as is the large magnitude of the gap at this point [25,26,44,45].

A question that now arises is: How sensitive is this phenomenon to the nature of the quark-quark interaction? As we discussed in connection with Eq. (2), the inhomogeneous dressed-ladder BSE exhibits particle-like singularities in the $0^+$ diquark channels and such states do not exist in the strong interaction spec-

---

[3] We note that in a two flavour free-quark gas at $\mu = 0.3\,\text{GeV}$ the baryon number density is $1.5\,\rho_0$, where $\rho_0 = 0.16\,\text{fm}^{-3}$. In the same system at $\mu = 0.55\,\text{GeV}$ the baryon number density is $> 10\,\rho_0$. For comparison, the central core density expected in a $1.4\,M_\odot$ neutron star is 3.6-4.1 $\rho_0$ [43]. Arguments valid at "asymptotically large" quark chemical potential are therefore unlikely to be relevant to experimentally or observationally accessible systems.

trum. Does diquark condensation persist when a truncation of the gap equation is employed that does not correspond to a BSE whose solutions exhibit diquark bound states? The vertex corrected gap equation,

$$\mathcal{D}(p,\mu) = \mathcal{D}_0(p,\mu) \tag{43}$$
$$+ \tfrac{3}{16}\eta^2\,\Gamma^a_\rho\,\mathcal{S}(p,\mu)\,\Gamma^a_\rho - \tfrac{9}{256}\eta^4\,\Gamma^a_\rho\,\mathcal{S}(p,\mu)\,\Gamma^b_\sigma\,\mathcal{S}(p,\mu)\,\Gamma^a_\rho\,\mathcal{S}(p,\mu)\,\Gamma^b_\sigma\,,$$

which is depicted in Fig. 1, is just such a truncation and it was also studied in Ref. [33].

In this case there is a $\Delta \neq 0$ solution even for $\mu = 0$, which is illustrated in Fig. 3, and using the interaction of Eq. (38)

$$m_\rho^2 = (1.1)\,m_\rho^{2\,\text{ladder}}, \quad \langle\bar{Q}Q\rangle = (1.0)^3\,\langle\bar{Q}Q\rangle^{\text{rainbow}}, \quad \mathcal{B}_B = (1.1)^4\,\mathcal{B}_B^{\text{rainbow}}, \tag{44}$$

where the rainbow-ladder results are given in Eqs. (39), and

$$\langle\bar{Q}i\gamma_5\tau_Q^2\lambda^2 Q\rangle = (0.48)^3\,\langle\bar{Q}Q\rangle, \quad \mathcal{B}_\Delta = (0.42)^4\,\mathcal{B}_B\,. \tag{45}$$

The last result shows, unsurprisingly, that the Nambu-Goldstone phase is favoured at $\mu = 0$. *Precluding* diquark condensation, the solution of the vertex-corrected gap equation exhibits coincident, first order chiral symmetry restoring and deconfinement transitions at

$$\mu_c^{B,\Delta=0} = 0.77\,\mu_{c,\,\text{rainbow}}^{B,\Delta=0}\,. \tag{46}$$

Admitting diquark condensation, however, the $\mu$-dependence of the bag constants again shows there is a first order transition to the superfluid phase, here at

$$\mu_c^{B=0,\Delta} = 0.63\,\mu_c^{B,\Delta=0}\,, \text{ with } \langle\bar{Q}i\gamma_5\tau_Q^2\lambda^2 Q\rangle_{\mu=0.63\,\mu_c^{B,\Delta=0}} = (0.51)^3\,\langle\bar{Q}Q\rangle_{\mu=0}\,. \tag{47}$$

(NB. This discussion is still for $m = 0$. We saw at the end of Sec. 3 what effects to anticipate at $m \neq 0$.) Thus the material step of employing a truncation that eliminates diquark bound states leads only to small quantitative changes in the quantities characterising the still extant superfluid phase; e.g., reductions in the magnitude of both the critical chemical potential for the transition to superfluid quark matter and the gap. Hence scalar diquark condensation appears to be a robust phenomenon. One caveat to bear in mind, however, is that the gap equation studies conducted hitherto do not obviate the question of whether the diquark condensed phase is stable with-respect-to dinucleon condensation [47], which requires further attention.

Heating causes the diquark condensate to evaporate. Existing studies suggest that it will disappear for $T \gtrsim 60$–$100$ MeV [25,26,44,46]. However, such temperatures are high relative to that anticipated inside dense astrophysical objects, which may indeed therefore provide an environment for detecting quark matter superfluidity.

## 5 Summary

The idea that diquark correlations play an important role in strong interaction physics is an old one. However, modern computational resources and theoretical techniques make possible a more thorough and quantitative exploration of the merits of this idea and its realisation in QCD. These advances are in part responsible for the contemporary resurgence of interest in all aspects of diquark-related phenomena.

Herein we have attempted to provide a qualitative understanding of the nature of diquark condensation using exemplary, algebraic models, and focusing on two flavour theories for simplicity.

The gap equation is a primary tool in all studies of pairing. Using the special case of 2-colour QCD, $QC_2D$, we illustrated via an analysis of the gap equation how a nonzero chemical potential promotes Cooper pairing and how that pairing can overwhelm a source for quark-antiquark condensation, such as a fermion current-mass. As we saw, the pseudoreality of $SU(N_c = 2)$ entails that $QC_2D$ has a number of special symmetry properties, which dramatically affect the spectrum.

Turning to QCD itself, we saw that one can expect a nonzero quark condensate at zero chemical potential: $\sigma := -\langle \bar{q}q \rangle^{1/3} \neq 0$, to give way to a diquark condensate when the chemical potential exceeds $\approx 2\sigma$, and at this point the diquark gap is $\approx \sigma/2$. The diquark condensate melts when the temperature exceeds $\sim 60-100$ MeV; i.e., one-third to one-half of the chiral symmetry restoring temperature in two-flavour QCD. These features are model-independent in the sense that the many, disparate models applied recently to the problem yield results in semi-quantitative agreement.

### Acknowledgments

This work was supported by the US Department of Energy, Nuclear Physics Division, under contract no. W-31-109-ENG-38 and the National Science Foundation under grant no. INT-9603385. S.M.S. is grateful for financial support from the A.v. Humboldt foundation.

## References

1. M. Ida and R. Kobayashi, Prog. Theor. Phys. **36** (1966) 846; D.B. Lichtenberg and L.J. Tassie, Phys. Rev. **155** (1967) 1601.
2. D.B. Lichtenberg, "Why Is It Necessary To Consider Diquarks," in Proc. of the Workshop on Diquarks, edited by Mauro Anselmino and Enrico Predazzi (World Scientific, Singapore, 1989) pp. 1-12.
3. D.B. Lichtenberg, L.J. Tassie and P.J. Keleman, Phys. Rev. **167** (1968) 1535.
4. J. Carroll, D.B. Lichtenberg and J. Franklin, Phys. Rev. **174** (1968) 1681.
5. R.T. Cahill, J. Praschifka and C.J. Burden, Austral. J. Phys. **42** (1989) 161; R.T. Cahill, *ibid* 171.

6. S.P. Klevansky, Rev. Mod. Phys. **64** (1992) 649; G. Ripka and M. Jaminon, Annals Phys. **218** (1992) 51.
7. P.C. Tandy, Prog. Part. Nucl. Phys. **39** (1997) 117; R.T. Cahill and S.M. Gunner, Fizika **B 7** (1998) 171.
8. C.D. Roberts and A.G. Williams, Prog. Part. Nucl. Phys. **33** (1994) 477.
9. C.D. Roberts and S.M. Schmidt, Prog. Part. Nucl. Phys. **45** (2000) S1.
10. R. Alkofer and L. v. Smekal, "The infrared behavior of QCD Green's functions: Confinement, dynamical symmetry breaking, and hadrons as relativistic bound states," hep-ph/0007355.
11. P. Maris, "Continuum QCD and light mesons," nucl-th/0009064.
12. M.B. Hecht, C.D. Roberts and S.M. Schmidt, "Contemporary applications of Dyson-Schwinger equations," nucl-th/0010024.
13. R.T. Cahill, C.D. Roberts and J. Praschifka, Phys. Rev. **D 36** (1987) 2804.
14. A. Bender, C.D. Roberts and L. v. Smekal, Phys. Lett. **B 380** (1996) 7.
15. G. Hellstern, R. Alkofer and H. Reinhardt, Nucl. Phys. **A 625** (1997) 697.
16. C.J. Burden, L. Qian, C.D. Roberts, P.C. Tandy and M.J. Thomson, Phys. Rev. **C 55** (1997) 2649.
17. M. Hess, F. Karsch, E. Laermann and I. Wetzorke, Phys. Rev. **D 58** (1998) 111502.
18. R.T. Cahill, C.D. Roberts and J. Praschifka, Austral. J. Phys. **42** (1989) 129.
19. C.J. Burden, R.T. Cahill and J. Praschifka, Austral. J. Phys. **42** (1989) 147.
20. M. Oettel, G. Hellstern, R. Alkofer and H. Reinhardt, Phys. Rev. **C 58** (1998) 2459.
21. M. Oettel, R. Alkofer and L. v Smekal, Eur. Phys. J. **A 8** (2000) 553.
22. G.G. Petratos, I.R. Afnan, F. Bissey, J. Gomez, A.T. Katramatou, W. Melnitchouk and A.W. Thomas, "Measurement of the $F_2^n/F_2^p$ and $d/u$ Ratios in Deep Inelastic Electron Scattering off $^3$H and $^3$He," nucl-ex/0010011.
23. D. Kahana and U. Vogl, Phys. Lett. **B 244** (1990) 10.
24. D. Bailin and A. Love, Phys. Rept. **107** (1984) 325; and references therein.
25. M. Alford, J. A. Bowers and K. Rajagopal, "Color superconductivity in compact stars," hep-ph/0009357, this volume.
26. T. Schäfer and E. Shuryak, "Phases of QCD at high baryon density," nucl-th/0010049, this volume.
27. D. Blaschke, H. Grigorian and G. Poghosyan, "Phase Diagram for Spinning and Accreting Neutron Stars," this volume.
28. M. Prakash, J. M. Lattimer, J. A. Pons, A. W. Steiner, "Evolution of a Neutron Star From its Birth to Old Age," this volume.
29. D. N. Voskresensky, "Medium effects in Neutrino Cooling of Neutron Stars," this volume.
30. D. Blaschke, C.D. Roberts and S.M. Schmidt, Phys. Lett. **B 425** (1998) 232.
31. A. Bender, G.I. Poulis, C.D. Roberts, S.M. Schmidt and A.W. Thomas, Phys. Lett. **B 431** (1998) 263.
32. A. Höll, P. Maris and C.D. Roberts, Phys. Rev. **C 59** (1999) 1751.
33. J.C.R. Bloch, C.D. Roberts and S.M. Schmidt, Phys. Rev. **C 60** (1999) 065208.
34. R.D. Mattuck: *A Guide to Feynman Diagrams in the Many-Body Problem* (McGraw-Hill, New York 1976).
35. R.T. Cahill and C.D. Roberts, Phys. Rev. **D 32** (1985) 2419.
36. H.J. Munczek and A.M. Nemirovsky, Phys. Rev. **D 28** (1983) 181.
37. J.I. Skullerud and A.G. Williams, Phys.Rev.D63,054508, (2001).
38. D. Blaschke, H. Grigorian, G. Poghosyan, C.D. Roberts and S.M. Schmidt, Phys. Lett. **B 450** (1999) 207.

39. J.B. Kogut, M.A. Stephanov and D. Toublan, Phys. Lett. **B 464** (1999) 183.
40. M.R. Frank and C.D. Roberts, Phys. Rev. **C 53** (1996) 390.
41. S. Hands and S. Morrison, "Diquark condensation in dense matter: A Lattice perspective," in Proc. of the International Workshop on Understanding Deconfinement in QCD, edited by D. Blaschke, F. Karsch and C.D. Roberts (World Scientific, Singapore, 2000) pp. 31-42; J.B. Kogut, M.A. Stephanov, D. Toublan, J.J. Verbaarschot and A. Zhitnitsky, Nucl. Phys. **B 582** (2000) 477; R. Aloisio, V. Azcoiti, G. Di Carlo, A. Galante and A. F. Grillo, "Fermion condensates in two colours finite density QCD at strong coupling," hep-lat/0009034; R. Aloisio, V. Azcoiti, G. Di Carlo, A. Galante and A.F. Grillo, "Probability Distribution Function of the Diquark Condensate in Two Colours QCD," hep-lat/0011079.
42. P. Maris and C.D. Roberts, Phys. Rev. **C 56** (1997) 3369.
43. R. B. Wiringa, V. Fiks and A. Fabrocini, Phys. Rev. **C 38** (1988) 1010.
44. D. Blaschke and C.D. Roberts, Nucl. Phys. **A 642** (1998) 197c.
45. G.W. Carter and D. Diakonov, "Chiral symmetry breaking and color superconductivity in the Instanton picture," in Proc. of the International Workshop on Understanding Deconfinement in QCD, edited by D. Blaschke, F. Karsch and C.D. Roberts (World Scientific, Singapore, 2000) pp. 239-250.
46. J. Berges, Nucl. Phys. **A 642** (1998) 51c.
47. S. Pepin, M. C. Birse, J.A. McGovern and N.R. Walet, Phys. Rev. **C 61** (2000) 055209.

# Color Superconductivity in Compact Stars*

Mark Alford[1], Jeffrey A. Bowers[2], and Krishna Rajagopal[2]

[1] Dept. of Physics and Astronomy, University of Glasgow, G12 8QQ, U.K.
[2] Center for Theoretical Physics, Massachusetts Institute of Technology, Cambridge, MA, USA 02139

**Abstract.** After a brief review of the phenomena expected in cold dense quark matter, color superconductivity and color-flavor locking, we sketch some implications of recent developments in our understanding of cold dense quark matter for the physics of compact stars. We give a more detailed summary of our recent work on crystalline color superconductivity and the consequent realization that (some) pulsar glitches may originate in quark matter.

## 1 Color Superconductivity and Color-Flavor Locking

Because QCD is asymptotically free, its high temperature and high baryon density phases are more simply and more appropriately described in terms of quarks and gluons as degrees of freedom, rather than hadrons. The chiral symmetry breaking condensate which characterizes the vacuum melts away. At high temperatures, in the resulting quark-gluon plasma phase all of the symmetries of the QCD Lagrangian are unbroken and the excitations have the quantum numbers of quarks and gluons. At high densities, on the other hand, quarks form Cooper pairs and new condensates develop. The formation of such superconducting phases [1–6] requires only weak attractive interactions; these phases may nevertheless break chiral symmetry [5] and have excitations with the same quantum numbers as those in a confined phase [5,7–9]. These cold dense quark matter phases may arise in the cores of neutron stars; understanding this region of the QCD phase diagram requires an interplay between QCD and neutron star phenomenology.

The relevant degrees of freedom in cold dense quark matter are those which involve quarks with momenta near the Fermi surface. At high density, where the quark number chemical potential $\mu$ (and hence the quark Fermi momentum) is

---

* We acknowledge helpful discussions with P. Bedaque, J. Berges, D. Blaschke, I. Bombaci, G. Carter, D. Chakrabarty, J. Madsen, C. Nayak, M. Prakash, D. Psaltis, S. Reddy, M. Ruderman, S.-J. Rey, T. Schäfer, A. Sedrakian, E. Shuryak, E. Shuster, D. Son, M. Stephanov, I. Wasserman, F. Weber and F. Wilczek. KR thanks the organizers of the ECT Workshop on Neutron Star Interiors for providing a stimulating environment within which many of the helpful discussions acknowledged above took place. This work is supported in part by the DOE under cooperative research agreement #DF-FC02-94ER40818. The work of JB was supported in part by an NDSEG Fellowship; that of KR was supported in part by a DOE OJI Award and by the A. P. Sloan Foundation.

large, the QCD gauge coupling $g(\mu)$ is small. However, because of the infinite degeneracy among pairs of quarks with equal and opposite momenta at the Fermi surface, even an arbitrarily weak attraction between quarks renders the Fermi surface unstable to the formation of a condensate of quark Cooper pairs. Creating a pair costs no free energy at the Fermi surface and the attractive interaction results in a free energy benefit. Pairs of quarks cannot be color singlets, and in QCD with two flavors of massless quarks, they form in the (attractive) color $\bar{\mathbf{3}}$ channel in which the quarks in a Cooper pair are color-antisymmetric [1–4]. The resulting condensate creates a gap $\Delta$ at the Fermi surfaces of quarks with two out of three colors, but quarks of the third color remain gapless. Five gluons get a Meissner mass by the Anderson-Higgs mechanism [10]; a $SU(2)_{\text{color}}$ subgroup remains unbroken. The Cooper pairs are flavor singlets and no flavor symmetries are broken. There is also an unbroken global symmetry which plays the role of $U(1)_B$. Thus, no global symmetries are broken in this 2SC phase. There must therefore be a phase transition between the hadronic and 2SC phases at which chiral symmetry is restored. This phase transition is first order [3,11–13] since it involves a competition between chiral condensation and diquark condensation [11,13].

In QCD with three flavors of massless quarks, the Cooper pairs *cannot* be flavor singlets, and both color and flavor symmetries are necessarily broken. The symmetries of the phase which results have been analyzed in [5,7]. The attractive channel favored by one-gluon exchange exhibits "color-flavor locking." A condensate of the form

$$\langle \psi_L^{\alpha a} \psi_L^{\beta b} \rangle \propto \Delta \epsilon^{\alpha \beta A} \epsilon^{abA} \tag{1}$$

involving left-handed quarks alone, with $\alpha$, $\beta$ color indices and $a$, $b$ flavor indices, locks $SU(3)_L$ flavor rotations to $SU(3)_{\text{color}}$: the condensate is not symmetric under either alone, but is symmetric under the simultaneous $SU(3)_{L+\text{color}}$ rotations.[1] A condensate involving right-handed quarks alone locks $SU(3)_R$ flavor rotations to $SU(3)_{\text{color}}$. Because color is vectorial, the combined effect of the $LL$ and $RR$ condensates is to lock $SU(3)_L$ to $SU(3)_R$, breaking chiral symmetry.[2] Thus, in quark matter with three massless quarks, the $SU(3)_{\text{color}} \times SU(3)_L \times SU(3)_R \times U(1)_B$ symmetry is broken down to the global diagonal $SU(3)_{\text{color}+L+R}$ group. A gauged $U(1)$ subgroup of the original symmetry group — a linear combination of one color generator and electromagnetism, which lives within $SU(3)_L \times SU(3)_R$ — also remains unbroken. All nine quarks have a gap. All eight gluons get a mass [5,15]. There are nine massless Nambu-Goldstone bosons. All the quarks, all the massive vector bosons, and all the

---

[1] It turns out [5] that condensation in the color $\bar{\mathbf{3}}$ channel induces a condensate in the color $\mathbf{6}$ channel because this breaks no further symmetries [8]. The resulting condensates can be written in terms of $\kappa_1$ and $\kappa_2$ where $\langle \psi_L^{\alpha a} \psi_L^{\beta b} \rangle \sim \kappa_1 \delta^{\alpha a} \delta^{\beta b} + \kappa_2 \delta^{\alpha b} \delta^{\beta a}$. Here, the Kronecker $\delta$'s lock color and flavor rotations. The pure color $\bar{\mathbf{3}}$ condensate (1) has $\kappa_2 = -\kappa_1$.

[2] Once chiral symmetry is broken by color-flavor locking, there is no symmetry argument precluding the existence of an ordinary chiral condensate. Indeed, instanton effects do induce a nonzero $\langle \bar{q}q \rangle$ [5], but this is a small effect [14].

Nambu-Goldstone bosons have integer charges under the unbroken gauged $U(1)$ symmetry, which therefore plays the role of electromagnetism. The CFL phase therefore has the same symmetries (and many similar non-universal features) as baryonic matter with a condensate of Cooper pairs of baryons [7]. This raises the possibility that quark matter and baryonic matter may be continuously connected [7].

Nature chooses two light quarks and one middle-weight strange quark, rather than three degenerate quarks. A nonzero $m_s$ weakens condensates which involve pairing between light and strange quarks. The CFL phase requires $\langle us \rangle$ and $\langle ds \rangle$ condensates; such condensates which pair quarks with differing Fermi momenta can only exist if the resulting gaps are larger than of order $m_s^2/2\mu$, the difference between $u$ and $s$ Fermi momenta in the absence of pairing. This means that upon increasing $m_s$ at fixed $\mu$, one must find a first-order unlocking transition [8,9]: for larger $m_s$ only $u$ and $d$ quarks pair and the 2SC phase is obtained. For any $m_s \neq \infty$, the CFL phase is the ground state at arbitrarily high density [8]. For large values of $m_s$, there is a 2SC interlude: as a function of increasing $\mu$, one finds a first order phase transition at which hadronic matter is replaced by quark matter in the 2SC phase and a subsequent first order phase transition at a higher $\mu$ above which CFL quark matter takes over. For smaller values of $m_s$, the possibility of quark-hadron continuity [7] arises.[3]

Much effort has gone into estimating the magnitude of the gaps in the 2SC and CFL phases, and the consequent critical temperature above which quark matter ceases to be superconducting. It would be ideal if this task were within the scope of lattice gauge theory. Unfortunately, lattice methods relying on importance sampling have to this point been rendered exponentially impractical at nonzero baryon density by the complex action at nonzero $\mu$. There are more sophisticated algorithms which have allowed the simulation of theories which are simpler than QCD but which have as severe a fermion sign problem as that in QCD at nonzero chemical potential [16]. This bodes well for the future.[4] Given the current absence of suitable lattice methods, the magnitude of the gaps in

---

[3] Note that even if the strange and light quarks are not degenerate, the CFL phase may be continuous with a baryonic phase which is dense enough that the Fermi momenta of all the nucleons and hyperons are comparable; there must, however, be phase transition(s) between this hypernuclear phase and ordinary nuclear matter [8].

[4] Note that quark pairing can be studied on the lattice in some models with four-fermion interactions and in two-color QCD [17]. The $N_c = 2$ case has also been studied analytically in Refs. [4,18]; pairing in this theory is simpler to analyze because quark Cooper pairs are color singlets. The $N_c \to \infty$ limit of QCD is often one in which hard problems become tractable. However, the ground state of $N_c = \infty$ QCD is a chiral density wave, not a color superconductor [19]. At asymptotically high densities color superconductivity persists up to $N_c$'s of order thousands [20,21] before being supplanted by the phase described in Ref. [19]. At any finite $N_c$, color superconductivity occurs at arbitrarily weak coupling whereas the chiral density wave does not. For $N_c = 3$, color superconductivity is still favored over the chiral density wave (although not by much) even if the interaction is so strong that the color superconductivity gap is $\sim \mu/2$ [22].

quark matter at large but accessible $\mu$ has been estimated using two broad strategies. The first class of estimates are done within the context of models whose parameters are chosen to reproduce zero density physics [3,4,11,5,8,9,13,14,23–25,22]. The second strategy for estimating gaps and critical temperatures is to use $\mu = \infty$ physics as a guide. At asymptotically large $\mu$, models with short-range interactions are bound to fail because the dominant interaction is due to the long-range magnetic interaction coming from single-gluon exchange [12,26]. The collinear infrared divergence in small angle scattering via one-gluon exchange (which is regulated by dynamical screening [26]) results in a gap which is parametrically larger at $\mu \to \infty$ than it would be for any point-like four-fermion interaction [26]. Weak coupling estimates of the gap [26–38] are valid at asymptotically high densities, with chemical potentials $\mu \gg 10^8$ MeV [37]. Neither class of methods can be trusted quantitatively for quark number chemical potentials $\mu \sim 400 - 500$ MeV, as appropriate for the quark matter which may occur in the cores of neutron stars. It is nevertheless satisfying that two very different approaches, one using zero density phenomenology to normalize models, the other using weak-coupling methods valid at asymptotically high density, yield predictions for the gaps and critical temperatures at accessible densities which are in good agreement: the gaps at the Fermi surface are of order tens to 100 MeV, with critical temperatures about half as large.

$T_c \sim 50$ MeV is much larger relative to the Fermi momentum than in low temperature superconductivity in metals. This reflects the fact that color superconductivity is induced by an attraction due to the primary, strong, interaction in the theory, rather than having to rely on much weaker secondary interactions, as in phonon mediated superconductivity in metals. Quark matter is a high-$T_c$ superconductor by any reasonable definition. Its $T_c$ is nevertheless low enough that it is unlikely the phenomenon can be realized in heavy ion collisions.

## 2 Color Superconductivity in Compact Stars

Our current understanding of the color superconducting state of quark matter leads us to believe that it may occur naturally in compact stars. The critical temperature $T_c$ below which quark matter is a color superconductor is high enough that any quark matter which occurs within neutron stars that are more than a few seconds old is in a color superconducting state. In the absence of lattice simulations, present theoretical methods are not accurate enough to determine whether neutron star cores are made of hadronic matter or quark matter. They also cannot determine whether any quark matter which arises will be in the CFL or 2SC phase: the difference between the $u$, $d$ and $s$ Fermi momenta will be a few tens of MeV which is comparable to estimates of the gap $\Delta$; the CFL phase occurs when $\Delta$ is large compared to all differences between Fermi momenta. Just as the higher temperature regions of the QCD phase diagram are being mapped out in heavy ion collisions, we need to learn how to use neutron star phenomena to determine whether they feature cores made of 2SC quark matter, CFL quark matter or hadronic matter, thus teaching us about the high density region

## 2.1 Equation of State

Much of the work on the consequences of quark matter within a compact star has focussed on the effects of quark matter on the equation of state, and hence on the radius of the star. As a Fermi surface phenomenon, color superconductivity has little effect on the equation of state: the pressure is an integral over the whole Fermi volume. Color superconductivity modifies the equation of state at the $\sim (\Delta/\mu)^2$ level, typically by a few percent [3]. Such small effects can be neglected in present calculations, and for this reason we will not attempt to survey the many ways in which observations of neutron stars are being used to constrain the equation of state [39].

We will describe one current idea, however. As a neutron star in a low mass X-ray binary (LMXB) is spun up by accretion from its companion, it becomes more oblate and its central density decreases. If it contains a quark matter core, the volume fraction occupied by this core decreases, the star expands, and its moment of inertia increases. This raises the possibility [40] of a period during the spin-up history of an LMXB when the neutron star is gaining angular momentum via accretion, but is gaining sufficient moment of inertia that its angular frequency is hardly increasing. In their modelling of this effect, Glendenning and Weber [40] discover that LMXB's should spend a significant fraction of their history with a frequency of around 200 Hz, while their quark cores are being spun out of existence, before eventually spinning up to higher frequencies. This may explain the observation that LMXB frequencies are clustered around 250-350 Hz [41], which is otherwise puzzling in that it is thought that LMXB's provide the link between canonical pulsars and millisecond pulsars, which have frequencies as large as 600 Hz [42]. It will be interesting to see how robust the result of Ref. [40] is to changes in model assumptions and also how its predictions fare when compared to those of other explanations which posit upper bounds on LMXB frequencies [43], rather than a most probable frequency range with no associated upper bound [40]. We note here that because Glendenning and Weber's effect depends only on the equation of state and not on other properties of quark matter, the fact that the quark matter must in fact be a color superconductor will not affect the results in any significant way. If Glendenning and Weber's explanation for the observed clustering of LMXB frequencies proves robust, it would imply that pulsars with lower rotational frequencies feature quark matter cores.

## 2.2 Cooling by Neutrino Emission

We turn now to neutron star phenomena which *are* affected by Fermi surface physics. For the first $10^{5-6}$ years of its life, the cooling of a neutron star is governed by the balance between heat capacity and the loss of heat by neutrino emission. How are these quantities affected by the presence of a quark matter core? This has been addressed recently in Refs. [44,45], following earlier work in Ref. [46]. Both the specific heat $C_V$ and the neutrino emission rate $L_\nu$ are dominated by physics within $T$ of the Fermi surface. If, as in the CFL phase, all quarks have a gap $\Delta \gg T$ then the contribution of quark quasiparticles to $C_V$ and $L_\nu$ is suppressed by $\sim \exp(-\Delta/T)$. There may be other contributions to $L_\nu$ [44], but these are also very small. The specific heat is dominated by that of the electrons, although it may also receive a small contribution from the CFL phase Goldstone bosons. Although further work is required, it is already clear that both $C_V$ and $L_\nu$ are much smaller than in the nuclear matter outside the quark matter core. This means that the total heat capacity and the total neutrino emission rate (and hence the cooling rate) of a neutron star with a CFL core will be determined completely by the nuclear matter outside the core. The quark matter core is "inert": with its small heat capacity and emission rate it has little influence on the temperature of the star as a whole. As the rest of the star emits neutrinos and cools, the core cools by conduction, because the electrons keep it in good thermal contact with the rest of the star. These qualitative expectations are nicely borne out in the calculations presented by Page et al. [45].

The analysis of the cooling history of a neutron star with a quark matter core in the 2SC phase is more complicated. The red and green up and down quarks pair with a gap many orders of magnitude larger than the temperature, which is of order 10 keV, and are therefore inert as described above. Any strange quarks present will form a $\langle ss \rangle$ condensate with angular momentum $J = 1$ which locks to color in such a way that rotational invariance is not broken [47]. The resulting gap has been estimated to be of order hundreds of keV [47], although applying results of Ref. [48] suggests a somewhat smaller gap, around 10 keV. The blue up and down quarks also pair, forming a $J = 1$ condensate which breaks rotational invariance [3]. The related gap was estimated to be a few keV [3], but this estimate was not robust and should be revisited in light of more recent developments given its importance in the following. The critical temperature $T_c$ above which no condensate forms is of order the zero-temperature gap $\Delta$. ($T_c = 0.57\Delta$ for $J = 0$ condensates [27].) Therefore, if there are quarks for which $\Delta \sim T$ or smaller, these quarks do not pair at temperature $T$. Such quark quasiparticles will radiate neutrinos rapidly (via direct URCA reactions like $d \to u + e + \bar{\nu}$, $u \to d + e^+ + \nu$, etc.) and the quark matter core will cool rapidly and determine the cooling history of the star as a whole [46,45]. The star will cool rapidly until its interior temperature is $T < T_c \sim \Delta$, at which time the quark matter core will become inert and the further cooling history will be dominated by neutrino emission from the nuclear matter fraction of the star. If future data were to show that neutron stars first cool rapidly (direct URCA) and then cool more slowly, such data would allow an estimate of the smallest quark matter gap. We are unlikely to be so lucky. The simple observation of rapid cooling would *not* be an unambiguous discovery of quark matter with small gaps; there are other circumstances in which the direct URCA processes occur. However, if as data on neutron star temperatures improves in coming years the standard cooling scenario proves correct, indicating the absence of the direct

URCA processes, this *would* rule out the presence of quark matter with gaps in the 10 keV range or smaller. The presence of a quark matter core in which *all* gaps are $\gg T$ can never be revealed by an analysis of the cooling history.

### 2.3 Supernova Neutrinos

We now turn from neutrino emission from a neutron star which is many years old to that from the protoneutron star during the first seconds of a supernova. Carter and Reddy [49] have pointed out that when this protoneutron star is at its maximum temperature of order 30-50 MeV, it may have a quark matter core which is too hot for color superconductivity. As such a protoneutron star core cools over the next few seconds, this quark matter will cool through $T_c$, entering the color superconducting regime of the QCD phase diagram. For $T \sim T_c$, the specific heat rises and the cooling slows. Then, as $T$ drops further and $\Delta$ increases to become greater than $T$, the specific heat drops rapidly. Furthermore, as the number density of quark quasiparticles becomes suppressed by $\exp(-\Delta/T)$, the neutrino transport mean free path rapidly becomes very long [49]. This means that all the neutrinos previously trapped in the now color superconducting core are able to escape in a sudden burst. If a terrestrial neutrino detector sees thousands of neutrinos from a future supernova, Carter and Reddy's results suggest that there may be a signature of the transition to color superconductivity present in the time distribution of these neutrinos. Neutrinos from the core of the protoneutron star will lose energy as they scatter on their way out, but because they will be the last to reach the surface of last scattering, they will be the final neutrinos received at the earth. If they are released from the quark matter core in a sudden burst, they may therefore result in a bump at late times in the temporal distribution of the detected neutrinos. More detailed study remains to be done in order to understand how Carter and Reddy's signature, dramatic when the neutrinos escape from the core, is processed as the neutrinos traverse the rest of the protoneutron star and reach their surface of last scattering.

### 2.4 R-mode Instabilities

Another arena in which color superconductivity comes into play is the physics of r-mode instabilities. A neutron star whose angular rotation frequency $\Omega$ is large enough is unstable to the growth of r-mode oscillations which radiate away angular momentum via gravitational waves, reducing $\Omega$. What does "large enough" mean? The answer depends on the damping mechanisms which act to prevent the growth of the relevant modes. Both shear viscosity and bulk viscosity act to damp the r-modes, preventing them from going unstable. The bulk viscosity and the quark contribution to the shear viscosity both become exponentially small in quark matter with $\Delta > T$ and as a result, as Madsen [50] has shown, a compact star made *entirely* of quark matter with gaps $\Delta = 1$ MeV or greater is unstable if its spin frequency is greater than tens to 100 Hz. Many compact stars spin faster than this, and Madsen therefore argues that compact stars cannot be strange quark stars unless some quarks remain ungapped. Alas, this powerful argument becomes much less powerful in the context of a neutron star with a quark matter core. First, the r-mode oscillations have a wave form whose amplitude is largest at large radius, outside the core. Second, in an ordinary neutron star there is a new source of damping: friction at the boundary between the crust and the neutron superfluid "mantle" keeps the r-modes stable regardless of the properties of a quark matter core [51,50].

## 2.5 Magnetic Field Evolution

Next, we turn to the physics of magnetic fields within color superconducting neutron star cores [52,53]. The interior of a conventional neutron star is a superfluid (because of neutron-neutron pairing) and is an electromagnetic superconductor (because of proton-proton pairing). Ordinary magnetic fields penetrate it only in the cores of magnetic flux tubes. A color superconductor behaves differently. At first glance, it seems that because a diquark Cooper pair has nonzero electric charge, a diquark condensate must exhibit the standard Meissner effect, expelling ordinary magnetic fields or restricting them to flux tubes within whose cores the condensate vanishes. This is not the case [53]. In both the 2SC and CFL phase a linear combination of the $U(1)$ gauge transformation of ordinary electromagnetism and one (the eighth) color gauge transformation remains unbroken even in the presence of the condensate. This means that the ordinary photon $A_\mu$ and the eighth gluon $G_\mu^8$ are replaced by new linear combinations

$$A_\mu^{\tilde{Q}} = \cos\alpha_0\, A_\mu + \sin\alpha_0\, G_\mu^8$$
$$A_\mu^X = -\sin\alpha_0\, A_\mu + \cos\alpha_0\, G_\mu^8 \qquad (2)$$

where $A_\mu^{\tilde{Q}}$ is massless and $A_\mu^X$ is massive. That is, $B_{\tilde{Q}}$ satisfies the ordinary Maxwell equations while $B_X$ experiences a Meissner effect. The mixing angle $\alpha_0$ is the analogue of the Weinberg angle in electroweak theory, in which the presence of the Higgs condensate causes the $A_\mu^Y$ and the third $SU(2)_W$ gauge boson to mix to form the photon, $A_\mu$, and the massive $Z$ boson. $\sin(\alpha_0)$ is proportional to $e/g$ and turns out to be about $1/20$ in the 2SC phase and $1/40$ in the CFL phase [53]. This means that the $\tilde{Q}$-photon which propagates in color superconducting quark matter is mostly photon with only a small gluon admixture. If a color superconducting neutron star core is subjected to an ordinary magnetic field, it will either expel the $X$ component of the flux or restrict it to flux tubes, but it can (and does [53]) admit the great majority of the flux in the form of a $B_{\tilde{Q}}$ magnetic field satisfying Maxwell's equations. The decay in time of this "free field" (i.e. not in flux tubes) is limited by the $\tilde{Q}$-conductivity of the quark matter. A color superconductor is not a $\tilde{Q}$-superconductor — that is the whole point — but it turns out to be a very good $\tilde{Q}$-conductor due to the presence of electrons: the $B_{\tilde{Q}}$ magnetic field decays only on a time scale which is much longer than the age of the universe [53]. This means that a quark matter core within a neutron star serves as an "anchor" for the magnetic field: whereas in ordinary nuclear matter the magnetic flux tubes can be dragged outward by the neutron superfluid vortices as the star spins down [54], the magnetic flux within the color superconducting core simply cannot decay. Even though this distinction is a qualitative one, it will be difficult to confront it with data since what is observed is the total dipole moment of the neutron star. A color superconducting core anchors those magnetic flux lines which pass through the core, while in a neutron star with no quark matter core the entire internal magnetic field can decay over time. In both cases, however, the total dipole moment can change since the magnetic flux lines which do not pass through the core can move.

## 3 Crystalline Color Superconductivity and Glitches in Quark Matter

The final consequence of color superconductivity we wish to discuss is the possibility that (some) glitches may originate within quark matter regions of a compact star [48]. In any context in which color superconductivity arises in nature, it is likely to involve pairing between species of quarks with differing chemical potentials. If the chemical potential difference is small enough, BCS pairing occurs as we have been discussing. If the Fermi surfaces are too far apart, no pairing between the species is possible. The transition between the BCS and unpaired states as the splitting between Fermi momenta increases has been studied in electron superconductors [55], nuclear superfluids [56] and QCD superconductors [8,9,57], assuming that no other state intervenes. However, there is good reason to think that another state can occur. This is the "LOFF" state, first explored by Larkin and Ovchinnikov [58] and Fulde and Ferrell [59] in the context of electron superconductivity in the presence of magnetic impurities. They found that near the unpairing transition, it is favorable to form a state in which the Cooper pairs have nonzero momentum. This is favored because it gives rise to a region of phase space where each of the two quarks in a pair can be close to its Fermi surface, and such pairs can be created at low cost in free energy. Condensates of this sort spontaneously break translational and rotational invariance, leading to gaps which vary periodically in a crystalline pattern. If in some shell within the quark matter core of a neutron star (or within a strange quark star) the quark number densities are such that crystalline color superconductivity arises, rotational vortices may be pinned in this shell, making it a locus for glitch phenomena.

In Ref. [48], we have explored the range of parameters for which crystalline color superconductivity occurs in the QCD phase diagram, upon making various simplifying assumptions. We focus primarily on a toy model in which the quarks interact via a four-fermion interaction with the quantum numbers of single gluon exchange. Also, we only consider pairing between $u$ and $d$ quarks, with $\mu_d = \bar{\mu} + \delta\mu$ and $\mu_u = \bar{\mu} - \delta\mu$, whereas we expect a LOFF state wherever the difference between the Fermi momenta of any two quark flavors is near an unpairing transition, including, for example, near the unlocking phase transition between the 2SC and CFL phases.

In the LOFF state, each Cooper pair carries momentum $2\mathbf{q}$ with $|\mathbf{q}| \approx 1.2\delta\mu$. The condensate and gap parameter vary in space with wavelength $\pi/|\mathbf{q}|$. In Ref. [48], we simplify the calculation by assuming that the condensate varies in space like a plane wave, leaving the determination of the crystal structure of the QCD LOFF phase to future work. We give an ansatz for the LOFF wave function, and by variation obtain a gap equation which allows us to solve for the gap parameter $\Delta_A$, the free energy and the values of the diquark condensates which characterize the LOFF state at a given $\delta\mu$ and $|\mathbf{q}|$. We then vary $|\mathbf{q}|$, to find the preferred (lowest free energy) LOFF state at a given $\delta\mu$, and compare the free energy of the LOFF state to that of the BCS state with which it competes. We show results for one choice of parameters[5] in Fig. 1(a). The LOFF state is characterized by a gap parameter $\Delta_A$ and a diquark condensate, but not by an energy gap in the dispersion relation: we obtain the quasiparticle dispersion relations [48] and find that they vary with the direction of the momentum, yielding gaps that vary from zero up to a maximum of $\Delta_A$. The condensate is dominated by the regions in momentum space in which a quark pair with total momentum $2\mathbf{q}$ has both members of the pair within $\sim \Delta_A$ of their respective Fermi surfaces.

Because it violates rotational invariance by involving Cooper pairs whose momenta are not antiparallel, the quark matter LOFF state necessarily features condensates in both the $J=0$ and $J=1$ channels. (Cooper pairs in the symmetric $J=1$ channel are antisymmetric in color but symmetric in flavor, and are impossible in the original LOFF context of pairing between electrons, which have neither color nor flavor.) Both $J=0$ and $J=1$ condensates are present even if there is no interaction in the $J=1$ channel, as is the case when we use a four-fermion interaction with the quantum numbers of Lorentz-invariant single gluon exchange. Because there is no interaction in the $J=1$ channel, the $J=1$ condensate does not affect the quasiparticle dispersion relations; that is, the $J=1$ gap parameter vanishes.

The LOFF state is favored for values of $\delta\mu$ which satisfy $\delta\mu_1 < \delta\mu < \delta\mu_2$ as shown in Fig. 1(b), with $\delta\mu_1/\Delta_0 = 0.707$ and $\delta\mu_2/\Delta_0 = 0.754$ in the weak coupling limit in which $\Delta_0 \ll \mu$. (For $\delta\mu < \delta\mu_1$, we have the 2SC phase with gap $\Delta_0$.) At weak coupling, the LOFF gap parameter decreases from $0.23\Delta_0$ at $\delta\mu = \delta\mu_1$ (where there is a first order BCS-LOFF phase transition) to zero at $\delta\mu = \delta\mu_2$ (where there is a second order LOFF-normal transition). Except for very close to $\delta\mu_2$, the critical temperature above which the LOFF state melts will be much higher than typical neutron star temperatures. At stronger coupling the LOFF gap parameter decreases relative to $\Delta_0$ and the window of $\delta\mu/\Delta_0$ within which the LOFF state is favored shrinks, as seen in Fig. 1(b). The single gluon

---

[5] Our model Hamiltonian has two parameters, the four-fermion coupling $G$ and a cutoff $\Lambda$. We often use the value of $\Delta_0$, the BCS gap obtained at $\delta\mu = 0$, to describe the strength of the interaction: small $\Delta_0$ corresponds to small $G$. When we wish to study the dependence on the cutoff, we vary $\Lambda$ while at the same time varying the coupling $G$ such that $\Delta_0$ is kept fixed. We expect that the relation between other physical quantities and $\Delta_0$ will be reasonably insensitive to variation of $\Lambda$.

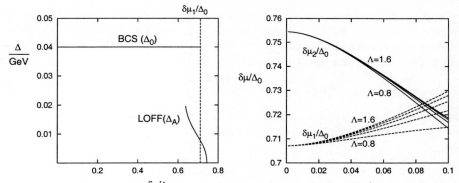

**Fig. 1.** (a) LOFF and BCS gap parameters as a function of $\delta\mu$, with coupling chosen so that $\Delta_0 = 40$ MeV. At each $\delta\mu$ we have varied $|\mathbf{q}|$ to find the LOFF state with the lowest free energy. The vertical dashed line marks $\delta\mu = \delta\mu_1$, the value of $\delta\mu$ above which the LOFF state has lower free energy than BCS. (b) The interval of $\delta\mu$ within which the LOFF state occurs as a function of the coupling, parametrized by the BCS gap $\Delta_0$ shown in GeV. Here and in (a), the average quark chemical potential $\bar{\mu}$ has been set to 0.4 GeV, corresponding to a baryon density of about 4 to 5 times that in nuclear matter. A crude estimate [48] suggests that in quark matter at this density, $\delta\mu \sim 15 - 30$ MeV depending on the value of the density-dependent effective strange quark mass. Below the solid line, there is a LOFF state. Below the dashed line, the BCS state is favored. The different lines of each type correspond to different cutoffs on the four-fermion interaction: $\Lambda = 0.8$ GeV to 1.6 GeV. $\delta\mu_1/\Delta_0$ and $\delta\mu_2/\Delta_0$ show little cutoff-dependence, and the cutoff-dependence disappears completely as $\Delta_0, \delta\mu \to 0$. $\Lambda = 1$ GeV in (a).

exchange interaction used in Fig. 1 is neither attractive nor repulsive in the $J = 1$ channel: the width of the LOFF window grows if the interaction is modified to include an attraction in this channel [48].

Near the second-order critical point $\delta\mu_2$, we can describe the phase transition with a Ginzburg-Landau effective potential. The order parameter for the LOFF-to-normal phase transition is

$$\Phi(\mathbf{r}) = -\frac{1}{2}\langle \epsilon_{ab}\epsilon_{\alpha\beta 3}\psi^{a\alpha}(\mathbf{r})C\gamma_5\psi^{b\beta}(\mathbf{r})\rangle \tag{3}$$

so that in the normal phase $\Phi(\mathbf{r}) = 0$, while in the LOFF phase $\Phi(\mathbf{r}) = \Gamma_A e^{i2\mathbf{q}\cdot\mathbf{r}}$. (The gap parameter is related to the order parameter by $\Delta_A = G\Gamma_A$.) Expressing the order parameter in terms of its Fourier modes $\tilde{\Phi}(\mathbf{k})$, we write the LOFF free energy (relative to the normal state) as

$$F(\{\tilde{\Phi}(\mathbf{k})\}) = \sum_{\mathbf{k}}\left(C_2(k^2)|\tilde{\Phi}(\mathbf{k})|^2 + C_4(k^2)|\tilde{\Phi}(\mathbf{k})|^4 + \mathcal{O}(|\tilde{\Phi}|^6)\right). \tag{4}$$

For $\delta\mu > \delta\mu_2$, all of the $C_2(k^2)$ are positive and the normal state is stable. Just below the critical point, all of the modes $\tilde{\Phi}(\mathbf{k})$ are stable except those on the sphere $|\mathbf{k}| = 2q_2$, where $q_2$ is the value of $|\mathbf{q}|$ at $\delta\mu_2$ (so that $q_2 \simeq 1.2\delta\mu_2 \simeq 0.9\Delta_0$

at weak coupling). In general, many modes on this sphere can become nonzero, giving a condensate with a complex crystal structure. We consider the simplest case of a plane wave condensate where only the one mode $\tilde{\Phi}(\mathbf{k} = 2\mathbf{q}_2) = \Gamma_A$ is nonvanishing. Dropping all other modes, we have

$$F(\Gamma_A) = a(\delta\mu - \delta\mu_2)(\Gamma_A)^2 + b(\Gamma_A)^4 \tag{5}$$

where $a$ and $b$ are positive constants. Finding the minimum-energy solution for $\delta\mu < \delta\mu_2$, we obtain simple power-law relations for the condensate and the free energy:

$$\Gamma_A(\delta\mu) = K_\Gamma(\delta\mu_2 - \delta\mu)^{1/2}, \qquad F(\delta\mu) = -K_F(\delta\mu_2 - \delta\mu)^2. \tag{6}$$

These expressions agree well with the numerical results obtained by solving the gap equation [48]. The Ginzburg-Landau method does not specify the proportionality factors $K_\Gamma$ and $K_F$, but analytical expressions for these coefficients can be obtained in the weak coupling limit by explicitly solving the gap equation [60,48], yielding

$$\begin{aligned} GK_\Gamma &= 2\sqrt{\delta\mu_2}\sqrt{(q_2/\delta\mu_2)^2 - 1} \simeq 1.15\sqrt{\Delta_0} \\ K_F &= (4\bar{\mu}^2/\pi^2)((q_2/\delta\mu_2)^2 - 1) \simeq 0.178\bar{\mu}^2. \end{aligned} \tag{7}$$

Notice that because $(\delta\mu_2 - \delta\mu_1)/\delta\mu_2$ is small, the power-law relations (6) are a good model of the system throughout the entire LOFF interval $\delta\mu_1 < \delta\mu < \delta\mu_2$ where the LOFF phase is favored over the BCS phase. The Ginzburg-Landau expression (5) gives the free energy of the LOFF phase near $\delta\mu_2$, but it cannot be used to determine the location $\delta\mu_1$ of the first-order phase transition where the LOFF window terminates. (Locating the first-order point requires a comparison of LOFF and BCS free energies.)

The quark matter which may be present within a compact star will be in the crystalline color superconductor (LOFF) state if $\delta\mu/\Delta_0$ is in the requisite range. For a reasonable value of $\delta\mu$, say 25 MeV, this occurs if the gap $\Delta_0$ which characterizes the uniform color superconductor present at smaller values of $\delta\mu$ is about 40 MeV. This is in the middle of the range of present estimates. Both $\delta\mu$ and $\Delta_0$ vary as a function of density and hence as a function of radius in a compact star. Although it is too early to make quantitative predictions, the numbers are such that crystalline color superconducting quark matter may very well occur in a range of radii within a compact star. It is therefore worthwhile to consider the consequences.

Many pulsars have been observed to glitch. Glitches are sudden jumps in rotation frequency $\Omega$ which may be as large as $\Delta\Omega/\Omega \sim 10^{-6}$, but may also be several orders of magnitude smaller. The frequency of observed glitches is statistically consistent with the hypothesis that all radio pulsars experience glitches [61]. Glitches are thought to originate from interactions between the rigid neutron star crust, typically somewhat more than a kilometer thick, and rotational vortices in a neutron superfluid. The inner kilometer of crust consists of a crystal lattice of nuclei immersed in a neutron superfluid [62]. Because the pulsar is spinning, the neutron superfluid (both within the inner crust and deeper inside the star) is threaded with a regular array of rotational vortices. As the pulsar's spin gradually slows, these vortices must gradually move outwards since the rotation frequency of a superfluid is proportional to the density of vortices. Deep within the star, the vortices are free to move outwards. In the crust, however, the vortices are pinned by their interaction with the nuclear lattice. Models [63] differ in important respects as to how the stress associated with pinned vortices is released in a glitch: for example, the vortices may break and rearrange the crust, or a cluster of vortices may suddenly overcome the pinning force and move macroscopically outward, with the sudden decrease in the angular momentum of the superfluid within the crust resulting in a sudden increase in angular momentum of the rigid crust itself and hence a glitch. All the models agree that the fundamental requirements are the presence of rotational vortices in a superfluid and the presence of a rigid structure which impedes the motion of vortices and which encompasses enough of the volume of the pulsar to contribute significantly to the total moment of inertia.

Although it is premature to draw quantitative conclusions, it is interesting to speculate that some glitches may originate deep within a pulsar which features a quark matter core, in a region of that core which is in a LOFF crystalline color superconductor phase. A three flavor analysis is required to estimate over what range of densities LOFF phases may arise, as either $\langle ud \rangle$, $\langle us \rangle$ or $\langle ds \rangle$ condensates approach their unpairing transitions. Comparison to existing models which describe how $p_F^u$, $p_F^d$ and $p_F^s$ vary within a quark matter core in a neutron star [64] would then permit an estimate of how much the LOFF region contributes to the moment of inertia of the pulsar. Furthermore, a three flavor analysis is required to determine whether the LOFF phase is a superfluid. If the only pairing is between $u$ and $d$ quarks, this 2SC phase is not a superfluid [3,8], whereas if all three quarks pair in some way, a superfluid *is* obtained [5,8]. Henceforth, we suppose that the LOFF phase is a superfluid, which means that if it occurs within a pulsar it will be threaded by an array of rotational vortices. It is reasonable to expect that these vortices will be pinned in a LOFF crystal, in which the diquark condensate varies periodically in space. Indeed, one of the suggestions for how to look for a LOFF phase in terrestrial electron superconductors relies on the fact that the pinning of magnetic flux tubes (which, like the rotational vortices of interest to us, have normal cores) is expected to be much stronger in a LOFF phase than in a uniform BCS superconductor [65].

A real calculation of the pinning force experienced by a vortex in a crystalline color superconductor must await the determination of the crystal structure of the LOFF phase. We can, however, attempt an order of magnitude estimate along the same lines as that done by Anderson and Itoh [66] for neutron vortices in the inner crust of a neutron star. In that context, this estimate has since been made quantitative [67,68,63]. For one specific choice of parameters [48], the LOFF phase is favored over the normal phase by a free energy $F_{\text{LOFF}} \sim 5 \times (10 \text{ MeV})^4$ and the spacing between nodes in the LOFF crystal is $b = \pi/(2|\mathbf{q}|) \sim 9$ fm. The thickness of a rotational vortex is given by the correlation length $\xi \sim 1/\Delta \sim 25$ fm. The pinning energy is the difference between the energy of a section of vortex of length $b$ which is centered on a node of the LOFF crystal vs. one which is centered on a maximum of the LOFF crystal. It is of order $E_p \sim F_{\text{LOFF}}\, b^3 \sim 4$ MeV. The resulting pinning force per unit length of vortex is of order $f_p \sim E_p/b^2 \sim (4 \text{ MeV})/(80 \text{ fm}^2)$. A complete calculation will be challenging because $b < \xi$, and is likely to yield an $f_p$ which is somewhat less than that we have obtained by dimensional analysis. Note that our estimate of $f_p$ is quite uncertain both because it is only based on dimensional analysis and because the values of $\Delta$, $b$ and $F_{\text{LOFF}}$ are uncertain. (We have a good understanding of all the ratios $\Delta/\Delta_0$, $\delta\mu/\Delta_0$, $q/\Delta_0$ and consequently $b\Delta_0$ in the LOFF phase. It is of course the value of the BCS gap $\Delta_0$ which is uncertain.) It is premature to compare our crude result to the results of serious calculations of the pinning of crustal neutron vortices as in Refs. [67,68,63]. It is nevertheless remarkable that they prove to be similar: the pinning energy of neutron vortices in the inner crust is $E_p \approx 1 - 3$ MeV and the pinning force per unit length is $f_p \approx (1 - 3 \text{ MeV})/(200 - 400 \text{ fm}^2)$.

The reader may be concerned that a glitch deep within the quark matter core of a neutron star may not be observable: the vortices within the crystalline color superconductor region suddenly unpin and leap outward; this loss of angular momentum is compensated by a gain in angular momentum of the layer outside the LOFF region; how quickly, then, does this increase in angular momentum manifest itself at the *surface* of the star as a glitch? The important point here is that the rotation of any superfluid region within which the vortices are able to move freely is coupled to the rotation of the outer crust on very short time scales [69]. This rapid coupling, due to electron scattering off vortices and the fact that the electron fluid penetrates throughout the star, is usually invoked to explain that the core nucleon superfluid speeds up quickly after a crustal glitch: the only long relaxation time is that of the vortices within the inner crust [69]. Here, we invoke it to explain that the outer crust speeds up rapidly after a LOFF glitch has accelerated the quark matter at the base of the nucleon superfluid. After a glitch in the LOFF region, the only long relaxation times are those of the vortices in the LOFF region and in the inner crust.

A quantitative theory of glitches originating within quark matter in a LOFF phase must await further calculations, in particular a three flavor analysis and the determination of the crystal structure of the QCD LOFF phase. However, our rough estimate of the pinning force on rotational vortices in a LOFF region suggests that this force may be comparable to that on vortices in the inner crust of a conventional neutron star. Perhaps, therefore, glitches occurring in a region of crystalline color superconducting quark matter may yield similar phenomenology to those occurring in the inner crust. This is surely strong motivation for further investigation.

Perhaps the most interesting consequence of these speculations arises in the context of compact stars made entirely of strange quark matter . The work of Witten [70] and Farhi and Jaffe [71] raised the possibility that strange quark matter may be stable relative to nuclear matter even at zero pressure. If this is the case it raises the question whether observed compact stars—pulsars, for example—are strange quark stars [72,73] rather than neutron stars. A conventional neutron star may feature a core made of strange quark matter, as we have been discussing above.[6] Strange quark stars, on the other hand, are made (almost) entirely of quark matter with either no hadronic matter content at all or with a thin crust, of order one hundred meters thick, which contains no neutron superfluid [73,74]. The nuclei in this thin crust are supported above the quark matter by electrostatic forces; these forces cannot support a neutron fluid. Because of the absence of superfluid neutrons, and because of the thinness of the crust, no successful models of glitches in the crust of a strange quark star have been proposed. Since pulsars are observed to glitch, the apparent lack of a glitch mechanism for strange quark stars has been the strongest argument that pulsars cannot be strange quark stars [75–77]. This conclusion must now be revisited.

Madsen's conclusion [50] that a strange quark star is prone to r-mode instability due to the absence of damping must also be revisited, since the relevant oscillations may be damped within or at the boundary of a crystalline color superconductor region.

The quark matter in a strange quark star, should one exist, would be a color superconductor. Depending on the mass of the star, the quark number densities increase by a factor of about two to ten in going from the surface to the center [73]. This means that the chemical potential differences among the three quarks will vary also, and there could be a range of radii within which the quark matter is in a crystalline color superconductor phase. This raises the possibility of glitches in strange quark stars. Because the variation in density with radius is gradual, if a shell of LOFF quark matter exists it need not be particularly thin. And, we have seen, the pinning forces may be comparable in magnitude to those in the inner crust of a conventional neutron star. It has recently been suggested (for reasons unrelated to our considerations) that certain accreting compact stars may be strange quark stars [78], although the evidence is far from unambiguous [79]. In contrast, it has been thought that, because they glitch, conventional radio pulsars cannot be strange quark stars. Our work questions this assertion by raising the possibility that glitches may originate within a layer of quark matter which is in a crystalline color superconducting state.

There has been much recent progress in our understanding of how the presence of color superconducting quark matter in a compact star would affect five different phenomena: cooling by neutrino emission, the pattern of the arrival times of supernova neutrinos, the evolution of neutron star magnetic fields, r-mode instabilities and glitches. Nevertheless, much theoretical work remains to be done before we can make sharp proposals for which astrophysical observations can teach us whether compact stars contain quark matter, and if so whether it is in the 2SC or CFL phase.

---

[6] Note that a convincing discovery of a quark matter core within an otherwise hadronic neutron star would demonstrate conclusively that strange quark matter is *not* stable at zero pressure, thus ruling out the existence of strange quark stars. It is not possible for neutron stars with quark matter cores and strange quark stars to both be stable.

# References

1. B. Barrois, Nucl. Phys. **B129**, 390 (1977); S. Frautschi, Proceedings of workshop on hadronic matter at extreme density, Erice 1978; B. Barrois, "Nonperturbative effects in dense quark matter", Cal Tech PhD thesis, UMI 79-04847-mc (1979).
2. D. Bailin and A. Love, Phys. Rept. **107**, 325 (1984), and references therein.
3. M. Alford, K. Rajagopal and F. Wilczek, Phys. Lett. **B422**, 247 (1998) [hep-ph/9711395].
4. R. Rapp, T. Schäfer, E. V. Shuryak and M. Velkovsky, Phys. Rev. Lett. **81**, 53 (1998).
5. M. Alford, K. Rajagopal and F. Wilczek, Nucl. Phys. **B537**, 443 (1999) [hep-ph/9804403].
6. For reviews, see F. Wilczek, Nucl. Phys. A663-664 (200) 257; T. Schäfer, nucl-th/9911017; S. Hsu, hep-ph/0003140; M. Alford, hep-ph/0003185; D. Rischke and R. Pisarski, nucl-th/0004016; K. Rajagopal, hep-ph/0009058.
7. T. Schäfer and F. Wilczek, Phys. Rev. Lett. **82**, 3956 (1999) [hep-ph/9811473].
8. M. Alford, J. Berges and K. Rajagopal, Nucl. Phys. **B558**, 219 (1999) [hep-ph/9903502].
9. T. Schäfer and F. Wilczek, Phys. Rev. **D60**, 074014 (1999) [hep-ph/9903503].
10. D. H. Rischke, Phys. Rev. **D62**, 034007 (2000) [nucl-th/0001040]; G. Carter and D. Diakonov, Nucl. Phys. **B582**, 571 (2000) [hep-ph/0001318].
11. J. Berges and K. Rajagopal, Nucl. Phys. **B538**, 215 (1999) [hep-ph/9804233].
12. R. D. Pisarski and D. H. Rischke, Phys. Rev. Lett. **83**, 37 (1999) [nucl-th/9811104].
13. G. W. Carter and D. Diakonov, Phys. Rev. **D60**, 016004 (1999) [hep-ph/9812445].
14. R. Rapp, T. Schäfer, E. V. Shuryak and M. Velkovsky, Ann. Phys. **280**, 35 (2000).
15. D. H. Rischke, Phys. Rev. **D62**, 054017 (2000) [nucl-th/0003063].
16. S. Chandrasekharan and U. Wiese, Phys. Rev. Lett. **83**, 3116 (1999) [cond-mat/9902128].
17. UKQCD Collaboration, Phys. Rev. **D59** (1999) 116002; S. Hands, J. B. Kogut, M. Lombardo and S. E. Morrison, Nucl. Phys. **B558**, 327 (1999) [hep-lat/9902034]; S. Hands, I. Montvay, S. Morrison, M. Oevers, L. Scorzato and J. Skullerud, Eur. Phys. J. **C 17**, 285 (2000) [hep-lat/0006018].
18. J. B. Kogut, M. A. Stephanov and D. Toublan, Phys. Lett. **B464**, 183 (1999) [hep-ph/9906346]; J. B. Kogut, M. A. Stephanov, D. Toublan, J. J. Verbaarschot and A. Zhitnitsky, Nucl. Phys. **B582**, 477 (2000) [hep-ph/0001171].
19. D. V. Deryagin, D. Yu. Grigoriev and V. A. Rubakov, Int. J. Mod. Phys. **A7**, 659 (1992).
20. E. Shuster and D. T. Son, Nucl. Phys. **B573**, 434 (2000) [hep-ph/9905448].
21. B. Park, M. Rho, A. Wirzba and I. Zahed, Phys. Rev. **D62**, 034015 (2000) [hep-ph/9910347].
22. R. Rapp, E. Shuryak and I. Zahed, Phys. Rev. D63, 034008 (2001).
23. N. Evans, S. D. H. Hsu and M. Schwetz, Nucl. Phys. **B551**, 275 (1999) [hep-ph/9808444]; Phys. Lett. **B449**, 281 (1999) [hep-ph/9810514].
24. T. Schäfer and F. Wilczek, Phys. Lett. **B450**, 325 (1999) [hep-ph/9810509].
25. B. Vanderheyden and A. D. Jackson, Phys. Rev. **D 62**, 094010 (2000) Phys.Rev.D62, 094010 (2000).
26. D. T. Son, Phys. Rev. **D59**, 094019 (1999) [hep-ph/9812287].

27. R. D. Pisarski and D. H. Rischke, Phys. Rev. **D60**, 094013 (1999) [nucl-th/9903023]; Phys. Rev. **D61**, 051501 (2000) [nucl-th/9907041]; Phys. Rev. **D61**, 074017 (2000) [nucl-th/9910056];
28. D. K. Hong, Phys. Lett. **B473**, 118 (2000) [hep-ph/9812510]; Nucl. Phys. **B582**, 451 (2000) [hep-ph/9905523].
29. D. K. Hong, V. A. Miransky, I. A. Shovkovy and L. C. Wijewardhana, Phys. Rev. **D61**, 056001 (2000), erratum *ibid.* **D62**, 059903 (2000) [hep-ph/9906478].
30. T. Schäfer and F. Wilczek, Phys. Rev. **D60**, 114033 (1999) [hep-ph/9906512].
31. W. E. Brown, J. T. Liu and H. Ren, Phys. Rev. **D61**, 114012 (2000) [hep-ph/9908248]; Phys. Rev. **D62**, 054016 (2000) [hep-ph/9912409]; Phys. Rev. **D62**, 054013 (2000) [hep-ph/0003199]..
32. S. D. Hsu and M. Schwetz, Nucl. Phys. **B572**, 211 (2000) [hep-ph/9908310].
33. I. A. Shovkovy and L. C. Wijewardhana, Phys. Lett. **B470**, 189 (1999) [hep-ph/9910225].
34. T. Schäfer, Nucl. Phys. **B575**, 269 (2000) [hep-ph/9909574].
35. N. Evans, J. Hormuzdiar, S. D. Hsu and M. Schwetz, Nucl. Phys. **B581**, 391 (2000) [hep-ph/9910313].
36. S. R. Beane, P. F. Bedaque and M. J. Savage, nucl-th/0004013.
37. K. Rajagopal and E. Shuster, Phys. Rev. **D 62**, 085007 (2000)
38. C. Manuel, Phys. Rev. **D 62**, 076009 (2000) [hep-ph/0005040].
39. For a review, see H. Heiselberg and M. Hjorth-Jensen, Phys. Rept. **328**, 237 (2000).
40. N. K. Glendenning and F. Weber, astro-ph/0003426. See also G. Poghosyan, H. Grigorian, D. Blaschke, ApJ 551, (2001) L73.
41. M. van der Klis, Ann. Rev. Astronomy and Astrophysics **38**, 717 (2000)
42. R. Wijnands and M. van der Klis, Nature **394**, 344 (1998); D. Chakrabarty and E. Morgan, Nature **394**, 346 (1998).
43. L. Bildsten, Astrophys. J. **501** L89 [astro-ph/9804325]; A. Andersson, D. I. Jones, K. D. Kokkotas and N. Stergioulas, Astrophys. J. **534**, L75 (2000) [astro-ph/0003426].
44. D. Blaschke, T. Klähn and D. N. Voskresensky, Astrophys. J. **533**, 406 (2000) [astro-ph/9908334]; D. Blaschke, H. Grigorian and D. N. Voskresensky, Astron. & Astrophys. **368** (2001) 561
45. D. Page, M. Prakash, J. M. Lattimer and A. Steiner, Phys. Rev. Lett. **85**, 2048 (2000).
46. C. Schaab *et al*, Astrophys. J. Lett **480** (1997) L111 and references therein.
47. T. Schäfer, Phys. Rev. **D 62**, 094007 (2000) [hep-ph/0006034].
48. M. Alford, J. Bowers and K. Rajagopal, Phys. Rev. **D63**, 074016 (2001).
49. G. W. Carter and S. Reddy, Phys. Rev. **D62**, 103002 (2001).
50. J. Madsen, Phys. Rev. Lett. **85**, 10 (2000) [astro-ph/9912418].
51. L. Bildsten and G. Ushomirsky, Astrophys. J. **529**, L75 (2000) [astro-ph/9911155]; N. Andersson *et al.* in [43]; L. Lindblom, B. J. Owen and G. Ushomirsky, astro-ph/0006242.
52. D. Blaschke, D. M. Sedrakian and K. M. Shahabasian, Astron. and Astrophys. **350**, L47 (1999) [astro-ph/9904395].
53. M. Alford, J. Berges and K. Rajagopal, Nucl. Phys. **B571**, 269 (2000) [hep-ph/9910254].
54. For reviews, see J. Sauls, in Timing Neutron Stars, J. Ögleman and E. P. J. van den Heuvel, eds., (Kluwer, Dordrecht: 1989) 457; and D. Bhattacharya and G. Srinivasan, in X-Ray Binaries, W. H. G. Lewin, J. van Paradijs, and E. P. J. van den Heuvel eds., (Cambridge University Press, 1995) 495.

55. A. M. Clogston, Phys. Rev. Lett. **9**, 266 (1962); B. S. Chandrasekhar, App. Phys. Lett. **1**, 7 (1962).
56. A. Sedrakian and U. Lombardo, Phys. Rev. Lett. **84**, 602 (2000).
57. P. F. Bedaque, hep-ph/9910247.
58. A. I. Larkin and Yu. N. Ovchinnikov, Zh. Eksp. Teor. Fiz. **47**, 1136 (1964); translation: Sov. Phys. JETP **20**, 762 (1965).
59. P. Fulde and R. A. Ferrell, Phys. Rev. **135**, A550 (1964).
60. S. Takada and T. Izuyama, Prog. Theor. Phys. **41**, 635 (1969).
61. M. A. Alpar and C. Ho, Mon. Not. R. Astron. Soc. **204**, 655 (1983). For a recent review, see A.G. Lyne in *Pulsars: Problems and Progress*, S. Johnston, M. A. Walker and M. Bailes, eds., 73 (ASP, 1996).
62. J. Negele and D. Vautherin, Nucl. Phys. **A207**, 298 (1973).
63. For reviews, see D. Pines and A. Alpar, Nature **316**, 27 (1985); D. Pines, in *Neutron Stars: Theory and Observation*, J. Ventura and D. Pines, eds., 57 (Kluwer, 1991); M. A. Alpar, in *The Lives of Neutron Stars*, M. A. Alpar et al., eds., 185 (Kluwer, 1995). For more recent developments and references to further work, see M. Ruderman, Astrophys. J. **382**, 587 (1991); R. I. Epstein and G. Baym, Astrophys. J. **387**, 276 (1992); M. A. Alpar, H. F. Chau, K. S. Cheng and D. Pines, Astrophys. J. **409**, 345 (1993); B. Link and R. I. Epstein, Astrophys. J. **457**, 844 (1996); M. Ruderman, T. Zhu, and K. Chen, Astrophys. J. **492**, 267 (1998); A. Sedrakian and J. M. Cordes, Mon. Not. R. Astron. Soc. **307**, 365 (1999).
64. N. K. Glendenning, Phys. Rev. **D46**, 1274 (1992); N. K. Glendenning, Compact Stars (Springer-Verlag, 1997); F. Weber, J. Phys. G. Nucl. Part. Phys. **25**, R195 (1999).
65. R. Modler et al., Phys. Rev. Lett. **76**, 1292 (1996).
66. P. W. Anderson and N. Itoh, Nature **256**, 25 (1975).
67. M. A. Alpar, Astrophys. J. **213**, 527 (1977).
68. M. A. Alpar, P. W. Anderson, D. Pines and J. Shaham, Astrophys. J. **278**, 791 (1984).
69. M. A. Alpar, S. A. Langer and J. A. Sauls, Astrophys. J. **282**, 533 (1984).
70. E. Witten, Phys. Rev. **D30**, 272 (1984).
71. E. Farhi and R. L. Jaffe, Phys. Rev. **D30**, 2379 (1984).
72. P. Haensel, J. L. Zdunik and R. Schaeffer, Astron. Astrophys. **160**, 121 (1986).
73. C. Alcock, E. Farhi and A. Olinto, Phys. Rev. Lett. **57**, 2088 (1986); Astrophys. J. **310**, 261 (1986).
74. N. K. Glendenning and F. Weber, Astrophys. J. **400**, 647 (1992).
75. A. Alpar, Phys. Rev. Lett. **58**, 2152 (1987).
76. J. Madsen, Phys. Rev. Lett. **61**, 2909 (1988).
77. R. R. Caldwell and J. L. Friedman, Phys. Lett. **B264**, 143 (1991).
78. X.-D. Li, I. Bombaci, M. Dey, J. Dey, E. P. J. van den Heuvel, Phys. Rev. Lett. **83**, 3776 (1999); X.-D. Li, S. Ray, J. Dey, M. Dey, I. Bombaci, Astrophys. J. **527**, L51 (1999); B. Datta, A. V. Thampan, I. Bombaci, Astron.&Astrophys. 355, L19, (2000). ; I. Bombaci, astro-ph/0002524.
79. D. Psaltis and D. Chakrabarty, Astrophys. J. **521**, 332 (1999); D. Chakrabarty, Phys. World **13**, No. 2, 26 (2000).

# Strange Quark Stars: Structural Properties and Possible Signatures for Their Existence

Ignazio Bombaci

Dipartimento di Fisica "E. Fermi", Universitá di Pisa, and INFN Sezione di Pisa,
via Buonarroti, 2, I-56127 Pisa, Italy
Email: bombaci@mail.df.unipi.it

**Abstract.** We give a brief introduction to the physics of strange quark matter, and explore the possibility this novel deconfined phase of matter might be absolutely stable. Strange quark stars represent one of the most intriguing consequences of such a possibility. We study the structural properties of this hypothetical new class of stellar compact objects, both for non-rotating and rapidly rotating configurations in general relativity. Next, using recent observational data for the X-ray burster 4U 1820–30, the newly discovered millisecond X-ray pulsar SAX J1808.4-3658, and for the atoll source 4U 1728–34, we argue that the compact stars in these X-ray sources are likely strange star candidates. Finally, we study the conversion of a neutron star to a strange quark star. We show that the total amount of energy liberated in the conversion process is in the range $(1-4) \times 10^{53}$ erg, in agreement with the energy required to power gamma-ray bursts at cosmological distances.

## 1 Introduction

The core of a neutron star is one of the best candidates in the Universe where a deconfined phase of quark matter (QM) could be found. This possibility was realized by several researchers [1–6] soon after the introduction of quarks as the fundamental building blocks of hadrons.

Even more intriguing than the existence of a quark core in a neutron star, is the possible existence of a new family of compact stars consisting completely of a deconfined mixture of *up* (*u*), *down* (*d*), and *strange* (*s*) quarks, together with an appropriate number of electrons to guarantee electrical neutrality. Such compact stars have been referred to in the literature, as *strange quark stars* or shortly *strange stars* (SS), and their constituent matter as *strange quark matter* (SQM). The investigation of such a possibility is relevant not only for astrophysics, but for high energy physics too. In fact, the search for a deconfined phase of quark matter is one of the main goals in heavy ion physics. Experiments at Brookhaven National Lab's Relativistic Heavy Ion Collider (RHIC) and at CERN's Large Hadron Collider (LHC), will hopefully clarify this issue in the near future.

The possible existence of SS is a direct consequence of the so called *strange matter hypothesis* [7–9]. According to this hypothesis, SQM could be the true ground state of matter. In other words, the energy per baryon of SQM (at the baryon density where the pressure is equal to zero) is supposed to be less than the lowest energy per baryon found in nuclei, which is about 930 MeV for $^{56}$Fe.

If the strange matter hypothesis is true, then a nucleus with $A$ nucleons, could in principle lower its energy by converting to a *strangelet* (a drop of SQM). However, this process requires a very high-order simultaneous weak interactions to convert about a number $A$ of $u$ and $d$ quarks of the nucleus into strange quarks. The probability for such a process is proportional to $G_F^{2A}$, with $G_F$ the Fermi constant, and assuming a number $A$ of simultaneous weak processes. Thus, for a large enough baryon number ($A > A_{min} \sim 5$), this probability is extremely low, and the mean life time for an atomic nucleus to decay to a strangelet is much higher than the age of the Universe. In addition, finite size effects (surface and shell effects) place a lower limit ($A_{min} \sim 10\text{--}10^3$, depending on the assumed model parameters) on the baryon number of a stable strangelet even if bulk SQM is stable [10–12]. On the other hand, a step by step production of $s$ quarks, at different times, will produce hyperons in the nucleus, *i.e.* a system (hypernucleus) with a higher energy per baryon with respect to the original nucleus. Thus, according to the strange matter hypothesis, the ordinary state of matter, in which quarks are confined within hadrons, is a metastable state.

The success of traditional nuclear physics, in explaining an astonishing amount of experimental data, provides a clear indication that quarks in a nucleus are confined within protons and neutrons. Thus, the energy per baryon for a droplet of $u,d$ quark matter (nonstrange quark matter) must be higher than the energy per baryon of a nucleus with the same baryon number.

These stability conditions in turn may be used to constrain the parameters entering in models for the equation of state (EOS) of SQM [10]. Our present understanding of the properties of ultra-dense hadronic matter, does not allow us to exclude or to accept *a priory* the validity of the strange matter hypothesis. Thus *strange stars may exist in the Universe*.

In the present chapter, we will not consider the so called *hybrid stars*, *i.e.* neutron stars with a quark matter core, or with a region where a mixed phase of hadronic and quark matter is present. These hybrid stars are thoroughly discussed in the chapters by D. Blaschke *et al.* [13] and by N. Glendenning and F. Weber [14] in the present volume.

Also, in this work, we will consider only *bare* strange stars, *i.e.* we neglect the possible presence of a crust of normal (confined) matter above the deconfined quark matter core [15]. For stars with $M \sim 1 \, M_\odot$, the thickness of this crust is on the order of 10–100 m, therefore the presence of a crust will not affect the considerations on the radius of strange star candidates we will make in section 5. However, this crust might be relevant to model glitches in strange stars [16], and it represents the largest part of the stellar radius in the case of the so called *strange dwarfs* [17].

Strange stars are the natural site for a color superconducting state of quark matter (see the chapter by Alford, Bowers and Rajagopal [18] in the present volume, and ref. [19] for a detailed introduction to this subject). Particularly, there could be a region inside a strange star where quark matter is in a crystalline ("LOFF" [20]) superconducting phase [18,19]. This raises the possibility to successfully model pulsar glitches with strange stars [18,19].

To close this section, we want to give a simple argument to explain why *charmed* (c), *bottom* (b), and *top* (t) quark flavors are not expected in neutron star cores or in quark stars. The reason is that c, b, t quarks are much more massive than u, d, s quarks. Suppose for the moment u, d, s quarks to be massless and non–interacting, and consider u,d,s quark matter (SQM) in equilibrium with respect to weak interactions. Under these hypotheses the number densities for the three quark flavor species are equal, *i.e.* $n_B = n_u = n_d = n_s$ (see discussion below). Then the creation of the lightest massive quark, *e.g.* through the weak process $s \to c + e^- + \bar{\nu}_e$, requires the Fermi energy of these massless quarks should be at least equal to the rest mass of charmed quark:

$$E_{Fq} = \hbar c \, k_{Fq} = \hbar c \left(3\pi^2 n_q\right)^{1/3} = \hbar c \left(3\pi^2 n_B\right)^{1/3} \geq m_c = 1.3 \, GeV \qquad (1)$$

which implies $n_B \geq 9.7$ fm$^{-3}$, *i.e* a baryon number density at least equal to about 60 times the normal saturation density of nuclear matter, far above the central density expected for neutron stars or strange stars.

## 2 The equation of state for strange quark matter

From a basic point of view the equation of state for SQM should be calculated solving the equations of Quantum Chromo-Dynamics (QCD) at finite density. As we know, such a fundamental approach is presently not doable [21], and the usual way to circumvent this difficulty is to make use of simple phenomenological models, which incorporate from the beginning some of the fundamental characteristic of QCD. Here, we discuss two phenomenological models for the EOS of strange quark matter. One is a model [22,23,10] which is related to the MIT bag model [24] for hadrons. The other is a recent model proposed by Dey *et al.* [25].

The concentrations of different quark flavors and leptons in SQM is determined by the requirement of electric charge neutrality

$$\frac{2}{3}n_u - \frac{1}{3}n_d - \frac{1}{3}n_s - n_e = 0 \qquad (2)$$

and equilibrium with respect to the weak processes:

$$u + e^- \to d + \nu_e \qquad (3)$$

$$u + e^- \to s + \nu_e \qquad (4)$$

$$d \to u + e^- + \bar{\nu}_e \qquad (5)$$

$$s \to u + e^- + \bar{\nu}_e \qquad (6)$$

$$s + u \to d + u. \qquad (7)$$

The latter equilibrium conditions can be written in terms of the corresponding chemical potentials $\mu_i$,

$$\mu_d = \mu_s \equiv \mu \qquad (8)$$

$$\mu = \mu_u + \mu_e, \tag{9}$$

with $\mu_{\nu_e} = \mu_{\bar{\nu}_e} = 0$ (neutrino–free matter). Charge neutral matter in equilibrium with respect to the weak interactions will be referred to as $\beta$–stable matter.

In the case of massless free quarks, the charge neutrality and $\beta$–equilibrium conditions imply:

$$n_u = n_d = n_s \,, \qquad n_e = 0 \,, \tag{10}$$

*i.e.* for massless free quarks, $\beta$–stable SQM is composed by an equal number of $u$, $d$, $s$ quarks with no electrons. As a consequence of this property of bulk SQM, strangelets are expected to have a very low charge to mass ratio $Z/A$ ($Ze$ being the net electric charge). This is in general true also in the more realistic case of finite mass and interacting quarks [10]. Strangelets, having $Z/A \ll 1$, are not destabilized by Coulomb effects. Thus they can grow in baryon number (and in size) and form ultra-dense objects of macroscopic size. The property of low $Z/A$ provides the basis for current experimental searches for strangelets in heavy ion collisions. For heavy "ordinary" nuclei, to a large extend the balance between the nuclear symmetry energy and the Coulomb energy, makes $Z/A \sim 0.4$. Thus the strong Coulomb repulsion makes very heavy nuclear systems unstable with respect to fission. This sets the upper limit ($A_{max} \sim 240$) on the baryon number of stable atomic nuclei.

All the thermodynamical properties of SQM can be deduced from the Gibbs grand canonical potential. The grand canonical potential per unit volume can be written

$$\Omega = \Omega^{(0)} + \Omega_{int} \tag{11}$$

where $\Omega^{(0)}$ is the contribution of a non–interacting $u,d,s$ Fermi gas, and $\Omega_{int}$ is the contribution coming from the interaction among quarks.

We write the grand potential density $\Omega_{int}$ as a sum of two pieces, which characterize the two different regimes of strong interactions:

$$\Omega_{int} \simeq \Omega_{short} + \Omega_{long} \tag{12}$$

The short range contribution $\Omega_{short}$ can be calculated using perturbative QCD. The long range contribution $\Omega_{long}$, is very hard to evaluate because of the difficulties involved in solving nonperturbative QCD. A very promising approach to deal with the nonperturbative regime of strong interactions is to solve QCD equations on a discrete lattice of space–time [26,27,21]. In the first model for SQM, we consider in this work, $\Omega_{long}$ is approximated by a phenomenological term,

$$\Omega_{long} \simeq B, \tag{13}$$

which represents the difference between the energy density of "perturbative vacuum" and true QCD vacuum. The parameter $B$ accounts in a crude phenomenological way of nonperturbative aspects of QCD. $B$ is related to the "bag constant" which in the MIT bag model for hadrons gives the confinement of quarks within the hadronic bag. This is a rough approximation, which is expected to be reasonable at very high density, but which is not appropriate in the density region where quarks clusterize to form hadrons, *i.e.* in the region where the phase transition between hadronic and quark matter takes place.

In the following we assume massless $u$ and $d$ quarks and finite mass $s$ quarks:

$$m_u = m_d = 0, \qquad m_s \neq 0. \tag{14}$$

The free Fermi gas contribution to $\Omega$, at zero temperature, is:

$$\Omega^{(0)} = \Omega_u^{(0)} + \Omega_d^{(0)} + \Omega_s^{(0)} \tag{15}$$

$$\Omega_q^{(0)} = -\frac{1}{(\hbar c)^3} \frac{1}{4\pi^2} \mu_q^4, \qquad (q = u, d) \tag{16}$$

$$\Omega_s^{(0)} = -\frac{1}{(\hbar c)^3} \frac{1}{4\pi^2} \left\{ \mu_s \mu_s^* \left( \mu_s^2 - \frac{5}{2} m_s^2 \right) + \frac{3}{2} m_s^4 \ln\left(\frac{\mu_s + \mu_s^*}{m_s}\right) \right\} \tag{17}$$

where $\mu_u, \mu_d, \mu_s$ are the quark chemical potentials, and

$$\mu_s^* \equiv \left(\mu_s^2 - m_s^2\right)^{1/2} = \hbar c \, k_{F_s} \tag{18}$$

The perturbative expansion of $\Omega_{short}$ up to linear terms in the QCD structure constant $\alpha_c$ gives [22,23,10]

$$\Omega_{short} \simeq \Omega^{(1)} = \Omega_u^{(1)} + \Omega_d^{(1)} + \Omega_s^{(1)}, \tag{19}$$

$$\Omega_q^{(1)} = \frac{1}{(\hbar c)^3} \frac{1}{4\pi^2} \frac{2\alpha_c}{\pi} \mu_q^4, \qquad (q = u, d) \tag{20}$$

$$\Omega_s^{(1)} = \frac{1}{(\hbar c)^3} \frac{1}{4\pi^2} \frac{2\alpha_c}{\pi} \left\{ 3\left[\mu_s \mu_s^* - m_s^2 \ln\left(\frac{\mu_s + \mu_s^*}{m_s}\right)\right]^2 - 2\mu_s^{*4} \right.$$

$$\left. - 3m_s^4 \ln^2\left(\frac{m_s}{\mu_s}\right) + 6 \ln\left(\frac{\rho_{ren}}{\mu_s}\right)\left[\mu_s \mu_s^* m_s^2 - m_s^4 \ln\left(\frac{\mu_s + \mu_s^*}{m_s}\right)\right] \right\} \tag{21}$$

where $\rho_{ren}$ is the so called *renormalization point* (see ref. [10]). In the case of massless $u$ and $d$ quarks a standard choice [10] is $\rho_{ren} = 313$ MeV.

To summarize, the equation of state of SQM (at zero temperature) in the approximation for the grand potential density we are considering is:

$$P(\mu_u, \mu_d, \mu_s) = -\Omega \simeq -\Omega^{(0)} - \Omega^{(1)} - B \tag{22}$$

$$\rho(\mu_u, \mu_d, \mu_s) \simeq \frac{1}{c^2} \left\{ \Omega^{(0)} + \Omega^{(1)} + \sum_{f=u,d,s} \mu_f n_f + B \right\} \tag{23}$$

**Table 1.** Parameters for the MIT bag model equations of state, and corresponding ground state properties of SQM. $B$ (in MeV/fm$^3$) is the bag constant, $m_s$ (MeV) the strange quark mass, and $\alpha_c$ the QCD structure constant. $(E/A)_{gs}$ (in MeV) is the energy per baryon, $\rho_{gs}$ ($\times 10^{14}$ g/cm$^3$) the mass density, and $n_{gs}$ (fm$^{-3}$), give the properties of SQM at the point where the pressure is equal to zero.

| EOS | $B$ | $m_s$ | $\alpha_c$ | $(E/A)_{gs}$ | $\rho_{gs}$ | $n_{gs}$ |
|---|---|---|---|---|---|---|
| B60$_0$ | 60 | 0 | 0.00 | 837 | 4.28 | 0.287 |
| B60$_{200}$ | 60 | 200 | 0.00 | 908 | 4.78 | 0.295 |
| B90$_0$ | 90 | 0 | 0.00 | 926 | 6.42 | 0.388 |

**Table 2.** Ground state properties of SQM within the model by Dey et al. [25]. The two equations of state SS1 and SS2 differ for the choice of the parameter $\nu$ entering in the expression of the in-medium quark masses. The value of the current quark masses are: $m_u = 4$ MeV, $m_d = 7$ MeV, and $m_s = 150$ MeV. The ground state properties are given in the same units as in the previous table.

| EOS | $\nu$ | $(E/A)_{gs}$ | $\rho_{gs}$ | $n_{gs}$ |
|---|---|---|---|---|
| SS1 | 0.333 | 888 | 12.3 | 0.779 |
| SS2 | 0.286 | 926 | 14.1 | 0.858 |

The number densities for each flavor can be calculated using the thermodynamical relation:

$$n_f = -\left(\frac{\partial \Omega_f}{\partial \mu_f}\right)_{TV}, \qquad (f = u, d, s) \qquad (24)$$

and the total baryon number density is

$$n_B = \frac{1}{3}(n_u + n_d + n_s). \qquad (25)$$

In the particular case of massless quarks, with gluon exchange interactions to the first order in the QCD structure constant $\alpha_c$, the EOS of $\beta$–stable SQM can be written in the following simple form:

$$\varepsilon = K n_B^{4/3} + B, \qquad P = \frac{1}{3} K n_B^{4/3} - B, \qquad K \equiv \frac{9}{4} \pi^{2/3} \left(1 + \frac{2\alpha_c}{3\pi}\right) \hbar c \qquad (26)$$

where $\varepsilon$ is the energy density. Eliminating the baryon number density $n_B$ one gets:

$$P = \frac{1}{3}(\varepsilon - 4B). \qquad (27)$$

Within this approximation, the density of zero pressure SQM is $\rho_{gs} = 4B/c^2$, which is the value of the surface density of a bare strange star.

In summary, within this model for the equation of state of SQM, there are three phenomenological parameters, namely: $B$, $m_s$, and $\alpha_c$. It is possible to determine ranges in the values of these parameters in which SQM is stable, and nonstrange quark matter is not [10]. For example, in the case of non–interacting quarks ($\alpha_c = 0$) one has $B \simeq 57$–91 MeV/fm$^3$ for $m_s = 0$, and $B \simeq 57$–75 MeV/fm$^3$ for $m_s = 150$ MeV. Using a popular terminology, in the following, we will refer to this model for the equation of state as the "*MIT bag model EOS for strange quark matter*".

The schematic model outlined above becomes less and less trustworthy going from very high density region (asymptotic freedom regime) to lower densities, where confinement (hadrons formation) takes place. Recently, Dey *et al.* [25] derived a new equation of state for SQM using a "dynamical" density-dependent approach to confinement. The EOS of ref. [25] has asymptotic freedom built in, shows confinement at zero baryon density, deconfinement at high density. In this model, the quark interaction is described by a colour-Debye-screened inter-quark vector potential originating from gluon exchange, and by a density-dependent scalar potential which restores chiral symmetry at high density (in the limit of massless quarks). The density-dependent scalar potential arises from the density dependence of the in-medium effective quark masses $M_q$, which, in this model, are taken to depend upon the baryon number density according to

$$M_q = m_q + 310 \cdot sech\left(\nu \frac{n_B}{n_0}\right) \quad \text{(MeV)}, \qquad (28)$$

where $n_0 = 0.16$ fm$^{-3}$ is the normal nuclear matter density, $q(= u, d, s)$ is the flavor index, and $\nu$ is a parameter. The effective quark mass $M_q(n_B)$ goes from its constituent masses at zero density, to its current mass $m_q$ as $n_B$ goes to infinity. Here we consider two different parameterizations of the EOS of ref. [25], which correspond to a different choice for the parameter $\nu$. The equation of state SS1 (SS2) corresponds to $\nu = 0.333$ ($\nu = 0.286$). These two models for the EOS give absolutely stable SQM according to the strange matter hypothesis (see Tab. 2).

Medium dependent mechanisms for confinement and their consequences for the EOS of quark matter, have been explored by many authors using different QCD motivated phenomenological models. Here, due to the lack of space, we can not discuss all these different attempts, and we refer the interested reader to the original literature [28–37].

## 3 Structural properties of strange quark stars

The structural properties of non-rotating strange stars are obtained by integrating the Tolman–Oppenheimer–Volkoff (TOV) equations [38,39],

$$\frac{dP(r)}{dr} = -G\frac{m(r)\rho(r)}{r^2}\left\{\left[1 + \frac{P(r)}{c^2\rho(r)}\right]\left[1 + \frac{4\pi r^3 P(r)}{c^2 m(r)}\right]\left[1 - \frac{2Gm(r)}{c^2 r}\right]^{-1}\right\} \qquad (29)$$

$$\frac{dm(r)}{dr} = 4\pi r^2 \rho(r) \qquad (30)$$

where $G$ is the universal constant of gravitation, $c$ the speed of light, $m(r)$ is the gravitational mass enclosed within a "radius" $r$ [determined by (proper surface area) $= 4\pi r^2$)]. The pressure $P(r)$ and the mass density $\rho(r)$ specify the equation of state. The TOV equations can be integrated numerically, with the following boundary conditions:

$$m(0) = 0, \qquad P(R) = P_{surf}. \tag{31}$$

The first condition means that the density and the pressure are finite at the center of the star. The second boundary condition is the definition of the surface of the star, which specifies the *radius* $R$ of the compact star through the surface area $4\pi R^2$. In the case of a bare strange star $P_{surf} = 0$.

The total mass

$$M \equiv M_G = m(R) = \int_0^R 4\pi r^2 \rho(r) dr \tag{32}$$

is the *gravitational mass* of the compact star measured by a distant observer in keplerian orbit around the star. The volume of a spherical layer of the star in the Schwarzschild metric (proper volume) is

$$dV = 4\pi\, e^{\lambda(r)/2}\, r^2 dr = 4\pi \left[1 - \frac{2Gm(r)}{c^2 r}\right]^{-1/2} r^2 dr \ . \tag{33}$$

**Table 3.** Properties of the maximum mass configuration obtained from different equations of state of Strange Quark Matter. $M_G$ is the gravitational (maximum) mass in unit of the solar mass $M_\odot$, R is the corresponding radius, $\rho_c$ the central density, $n_c$ the central number density ($n_0 = 0.16$ fm$^{-3}$), $P_c$ the central pressure, $M_B$ the baryonic mass.

| EOS | $M_G/M_\odot$ | R(km) | $\rho_c$(g/cm$^3$) | $n_c/n_0$ | $P_c$(dyne/cm$^2$) | $M_B/M_\odot$ |
|---|---|---|---|---|---|---|
| B60$_0$ | 1.964 | 10.71 | $2.06 \times 10^{15}$ | 6.94 | $0.49 \times 10^{36}$ | 2.625 |
| B60$_{200}$ | 1.751 | 9.83 | $2.44 \times 10^{15}$ | 7.63 | $0.54 \times 10^{36}$ | 2.141 |
| B90$_0$ | 1.603 | 8.75 | $3.09 \times 10^{15}$ | 9.41 | $0.73 \times 10^{36}$ | 1.937 |
| SS1 | 1.438 | 7.09 | $4.65 \times 10^{15}$ | 14.49 | $1.40 \times 10^{36}$ | 1.880 |
| SS2 | 1.324 | 6.53 | $5.60 \times 10^{15}$ | 16.34 | $1.64 \times 10^{36}$ | 1.658 |

The *total number of baryons* $N_B$ in the star is given by

$$N_B = \int n\, dV = \int_0^R 4\pi r^2 n_B(r) \left[1 - \frac{2Gm(r)}{c^2 r}\right]^{-1/2} dr \ , \tag{34}$$

and the *baryonic mass* (or *rest mass*) of the star is

$$M_B = m_u N_B, \tag{35}$$

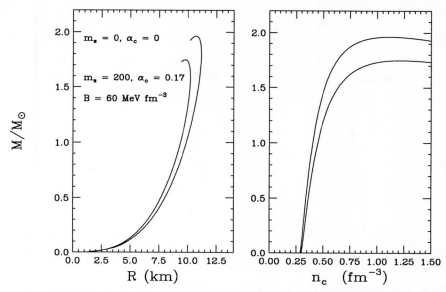

**Fig. 1.** Gravitational mass versus radius (left panel), and versus central number density (right panel), for MIT bag model strange stars. Stable equilibrium configurations are those on the positive slope branch of the $M_G(n_c)$ curve.

where $m_u$ is an average baryonic mass unit. $M_B$ is the rest mass of $N_B$ baryons dispersed at infinity, which form the compact star. Clearly, due to color confinement, in the case of a strange star, we can not have the dispersion at infinity of isolated quarks, but only of baryons. The so called *proper mass* of the star is given by

$$M_P = \int \rho \, dV = \int_0^R 4\pi r^2 \rho(r) \left[1 - \frac{2Gm(r)}{c^2 r}\right]^{-1/2} dr \, . \tag{36}$$

$M_P$ is equal to the sum of the mass elements on the whole volume of the star, it includes the contributions of rest mass and internal energy (kinetic and interactions (other than gravitation)) of the constituents of the star. This particular role of the gravitational interaction which does not enter in the expression of the energy density (EOS), and which contribute to the gravitational mass (total energy) of the star through the general relativistic field equations is due to the long-range nature of the gravitational interaction. Since $e^{\lambda(r)/2} \geq 1$, it follows that $M_P \geq M_G$.

The difference

$$E_G = (M_G - M_P)c^2 = c^2 \int_0^R 4\pi r^2 dr \rho(r) \left[1 - e^{\lambda(r)/2}\right] \leq 0 \tag{37}$$

is the *gravitational energy* of the star. Its opposite $BE_G = -E_G$ is called the *gravitational binding energy*. It is the gravitational energy released moving the infinitesimal mass elements $\rho dV$ from infinity to form the star. In the Newtonian

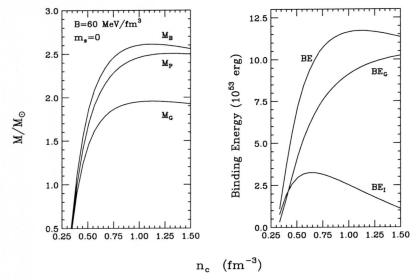

**Fig. 2.** Left panel: gravitational, baryonic, and proper masses of strange stars versus the central number density. Right panel: total ($BE$), gravitational ($BE_G$), and internal ($BE_I$) binding energies versus the central number density. Results are relative to the bag model EOS with massless non–interacting quarks ($\alpha_c = 0$).

limit, from eq.(37), one recovers the well known relation for the (Newtonian) gravitational energy

$$E_G^{Newt} = -G \int_0^R 4\pi r^2 dr \frac{m(r)\rho(r)}{r} \ . \tag{38}$$

The *internal energy* of the star is

$$E_I = (M_P - M_B)c^2 = \int_0^R \varepsilon'(r) dV \ , \tag{39}$$

where $\varepsilon'(r)$ is the internal energy density, *i.e.* the total energy density apart from the rest-mass energy density. The *internal binding energy* is thus $BE_I = -E_I$.

The *total binding energy* is given by

$$BE = BE_G + BE_I = (M_B - M_G)c^2, \tag{40}$$

it is the total energy released during the formation of a static compact star configuration from a rarefied gas of $N_B$ baryons. The stability of a such configuration requires $BE > 0$.

Therefore, the gravitational mass of the compact star, represents the total energy ($E = M_G c^2$) of the star, including both the rest mass energy $M_B c^2$ of its constituents dispersed at infinity, and the mass–energy contribution coming from

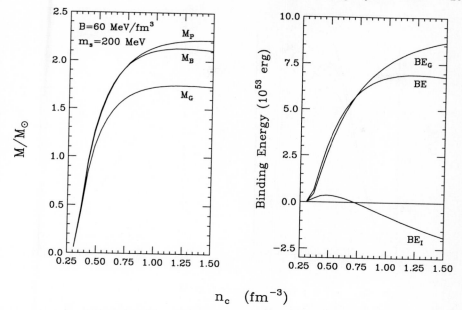

**Fig. 3.** Same as in figure 2, but for $m_s = 200$ MeV.

the microscopic motion and the interactions (including gravitation) between the star's constituents:

$$M_G = M_B + \frac{1}{c^2}E_I + \frac{1}{c^2}E_G = M_P + \frac{1}{c^2}E_G \qquad (41)$$

Since the work of Haensel et al. [40] and Alcock et al. [15], non-rotating strange star configurations have been calculated by many authors. In Fig. 1, we show the calculated gravitational mass of strange stars as a function of the radius (left panel) and as a function of the central number density (right panel) using the MIT bag model EOS. Stable equilibrium configurations are those on the positive slope branch of the $M_G(n_c)$ curve. The higher curve is the result for massless non–interacting gas ($m_s = 0$, $\alpha_c = 0$) and $B = 60$ MeV/fm$^3$. To illustrate how these properties depend on the value of the strange quark mass and on the QCD structure constant, we report, in the same figure, the results obtained for $m_s = 200$ MeV, $\alpha_c = 0.17$. The overall effect of finite $m_s$ and $\alpha_c$ is to reduce the maximum mass and the radius of the strange star. The properties of the maximum mass configuration, using different EOSs, are summarized in Tab. 3. The larger value of the maximum mass for the SS1 model, with respect to the SS2 model, can be traced back to role of the parameter $\nu$ in eq. (28) for the effective quark mass $M_q$. In fact, a larger value of $\nu$ (SS1 model) gives a faster decrease of $M_q$ with density, producing a stiffer EOS.

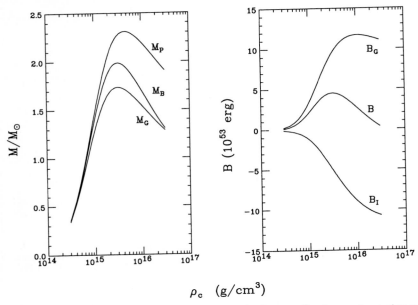

**Fig. 4.** Masses (left panel) and binding energies (right panel) of a neutron star versus central density $\rho_c$ assuming a generic EOS. Results are relative to the BPAL22 EOS. Configurations on the decreasing branch of the function $M_G(\rho_c)$ are unstable.

A transparent analysis of the properties of strange stars can be made in the case of the simple EOS

$$P = \frac{1}{3}(\rho c^2 - 4B) \qquad (42)$$

valid for massless quarks. There is a striking qualitative difference in the mass–radius (MR) relation of strange stars with respect to that of neutron stars. For strange stars with "small" ($M_G \ll M_{max}$) gravitational mass, $M_G$ is proportional to $R^3$ (see Fig. 1). In contrast, neutron stars have radii that decrease with increasing mass (see Fig. 7). Another related consequence of an EOS of the form (42) is that "low" mass strange stars are bound by the strong interaction, contrary to the case of neutron stars which are bound by gravity. In fact, for "low" values of the central density the internal binding energy of a strange star is positive (see Figs 2 and 3, and compare with the corresponding Fig. 4 for the neutron stars case).

As we know, there is a minimum mass for a neutron star ($M_{min} \sim 0.1\, M_\odot$). In the case of a strange star, there is essentially no minimum mass. As $\rho_c \to \rho_{surf}$ (surface density), a strange star (or better a lump of SQM for very low baryon number) is a self-bound system, until the baryon number becomes so low that finite size effects destabilize it.

A strange star has a very sharp boundary. In fact, the density drops abruptly from $\rho_{surf} \sim 4 - 10 \times 10^{14}$ g/cm$^3$ to zero on a length scale typical of the strong interaction, *i.e.* the thickness of the "quark surface" is of the order of $10^{-13}$cm.

This is of the same order of the thickness of the surface of an atomic nucleus. In the case of the simple EOS (26) the surface density is:

$$\rho_{surf} = \frac{4B}{c^2} \qquad n_{surf} = \left(\frac{3B}{K}\right)^{3/4}. \qquad (43)$$

## 4 Rapidly rotating strange stars

The possible existence of strange stars in binary stellar systems, implies that these compact stars may possess rapid rotation rates (see ref. [41] and references therein). Particularly, the two SS candidates in SAX J1808.4–3658 and 4U 1728–34 are millisecond pulsars having spin periods $P = 2.49$ ms and $P = 2.75$ ms respectively. This makes the incorporation of general relativistic effects of rotation imperative for a satisfactory treatment of the problem.

General relativity predicts the existence of a limiting stable circular orbit for the motion of a test particle around a compact star. This orbit is called the *innermost stable circular orbit* (ISCO) or the marginally stable orbit. For material particles within the radius of such orbit, no keplerian orbit is possible and the particles will undergo free fall under gravity. For a non-rotating compact star the radius ($R_{ISCO}$) of the ISCO is equal to three times the Schwartzschild radius ($R_s$) of the compact star

$$R_{ISCO} = 3R_s = \frac{6GM}{c^2} \simeq 8.86 \frac{M}{M_\odot} \text{ km}. \qquad (44)$$

$R_{ISCO}$ can be calculated for equilibrium sequences of rapidly rotating strange stars in a general relativistic space–time in the same way as for neutron stars [42].

Most of the calculations on the rotational properties of SS, reported so far, have relied on the slow rotation approximation [43,16]. This approximation loses its validity as the star's spin frequency approaches the mass shedding limit. Rapidly rotating SS sequences have been recently computed by the authors of ref. [44–46]. However, all the calculations mentioned above make use of the equation of state for SQM related to the MIT bag model for hadrons. Within this EOS model, as we will show in the following pages, the calculated strange star radii are seen to be incompatible with the mass–radius relation [47] for SAX J1808.4-3658, and only marginally compatible (see Fig. 8) with that for 4U 1728-34 (ref.[48]). In this section, we present some recent calculations, reported in ref. [49], for the equilibrium sequences of rapidly rotating SS in general relativity using both the MIT bag model EOS, and the new EOS by Dey *et al.* [25].

To calculate the structure of rapidly rotating SS, we use the methodology described in detail in Datta *et al.* [42]. For completeness, we briefly describe the method here. For a general axisymmetric and stationary space–time, assuming a perfect fluid configuration, the Einstein field equations reduce to ordinary integrals (using Green's function approach). These integrals may be self consistently (numerically and iteratively) solved to yield the value of metric coefficients in all space. Using these metric coefficients, one may then compute the structure parameters, angular momentum and moment of inertia corresponding to initially assumed central density and polar to equatorial radius ratio. These may then be used (as described in ref.[50]) to calculate parameters connected with stable circular orbits (like the ISCO and the keplerian angular velocities) around the configuration in question.

The sequences that we present are: constant rest (baryonic) mass sequences, constant angular velocity sequences, constant central density sequences and constant angular momentum sequences. We also calculate the radius of the ISCO and its dependence on the spin rate of the strange star, which will be relevant for modeling X–ray burst sources involving strange stars.

The equilibrium sequences of rotating SS depend on two parameters: the central density ($\rho_c$) and the rotation rate ($\Omega$). For purpose of illustration, we choose three limits in this parameter space. These are: (i) the static or non–rotating limit, (ii) the limit at which instability to quasi-radial mode sets in and (iii) the centrifugal mass shed limit. The last limit corresponds to the maximum $\Omega$ for which centrifugal forces are able to balance the inward gravitational force.

The result of the calculations of ref. [49] for the EOS SS1 is displayed in Fig. 5. In panel (a) of this figure, we show the functional dependence of the gravitational mass with $\rho_c$. In these set of figures, the bold solid curve represents the non–rotating or static limit, and the dotted curve the centrifugal mass shed limit. The thin solid curves are the constant baryonic mass evolutionary sequences. The evolutionary sequences above the maximum stable non–rotating mass configuration are the supramassive evolutionary sequences, and those that lie below this limit are the normal evolutionary sequences. The maximum mass sequence for

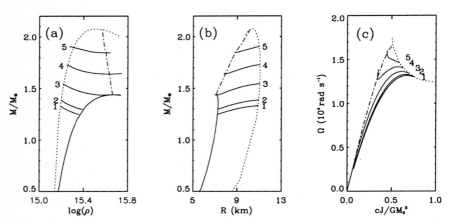

**Fig. 5.** Structure parameters for rotating strange stars corresponding to EOS SS1. The bold-solid line represents the non–rotating limit, the dotted line the mass–shed limit and the almost vertical dot–dashed line is the instability limit to quasi–radial mode perturbations. The thin solid lines (labelled 1, 2 ...) represent constant baryonic mass sequences: 1: 1.59 $M_\odot$, 2: 1.66 $M_\odot$, 3: 1.88 $M_\odot$, 4: 2.14 $M_\odot$, 5: 2.41 $M_\odot$.

this EOS corresponds to $M_B = 2.2\ M_\odot$. The dot–dashed line (slanted towards left) represents instability to quasi–radial perturbations. In the central panel of the same figure, we give a plot of $M$ as a function of the equatorial radius $R$. For the millisecond pulsar PSR 1937 +21 with an assumed mass value of $1.4\ M_\odot$, this corresponds to a radius of 7.3 km.

In panel (c) of Fig. 5, we display the plot of $\Omega$ as a function of the specific angular momentum $\tilde{j} = cJ/GM_B^2$ (where $J$ is the angular momentum of the configuration). Unlike for neutron stars (e.g. [51,42]), the $\Omega$–$\tilde{j}$ curve does not show a turn–over to lower $\tilde{j}$ values for SS. This is due to the effect of the long–range (non–perturbative) interaction in QCD, which is responsible for quark confinement in hadrons, and makes low mass SS self-bound objects. $\Omega$ for the mass shed limit appears to asymptotically tend to a non–zero value for rapidly rotating low mass stars. A further ramification of this result is that the ratio of the rotational energy to the total gravitational energy $(T/W)$ becomes greater than 0.21 (as also reported in ref. [44]) thus probably making the configurations susceptible to triaxial instabilities.

In Table 4 we display the values of the structure parameters for the maximum mass non–rotating strange star models. Table 5 and Table 6 display the maximum mass rotating and maximum angular momentum models for the EOS models under consideration. While for EOS SS1, the maximum mass rotating model and the maximum angular momentum models are the same, for EOS

**Table 4.** Structure parameters for the non–rotating maximum mass configurations. Listed, are the central density ($\rho_c$) in units of g cm$^{-3}$, the gravitational mass ($M$) in solar units, the equatorial radius ($R$) in km, the baryonic mass ($M_B$) in solar units, the radius of the marginally stable orbit ($R_{\rm ISCO}$) in km and the moment of inertia ($I$) in units of $10^{45}$ g cm$^2$.

| EOS | $\rho_c$ | $M$ | $R$ | $M_B$ | $R_{\rm ISCO}$ | $I$ |
|---|---|---|---|---|---|---|
| SS1 | $4.65 \times 10^{15}$ | 1.438 | 7.093 | 1.880 | 12.740 | 0.733 |
| SS2 | $5.60 \times 10^{15}$ | 1.324 | 6.533 | 1.658 | 11.730 | 0.576 |
| B90_0 | $3.09 \times 10^{15}$ | 1.603 | 8.745 | 1.937 | 14.202 | 1.146 |

**Table 5.** Structure parameters for the maximally rotating ($\Omega = \Omega_{\rm ms}$) maximum mass configuration. In addition to the quantities listed in the previous table, we display, the rotation rate ($\Omega$) in $10^4$ rad s$^{-1}$, the ratio of the rotational to the total gravitational energy $(T/W)$ and the specific angular momentum ($\tilde{j} = cJ/GM_B^2$).

| EOS | $\rho_c$ | $\Omega$ | $I$ | $M$ | $T/W$ | $R$ | $\tilde{j}$ | $R_{\rm ISCO}$ | $M_B$ |
|---|---|---|---|---|---|---|---|---|---|
| SS1 | $3.10 \times 10^{15}$ | 1.613 | 2.072 | 2.077 | 0.219 | 10.404 | 0.524 | 11.656 | 2.694 |
| SS2 | $3.60 \times 10^{15}$ | 1.738 | 1.613 | 1.904 | 0.218 | 9.612 | 0.570 | 10.758 | 2.366 |
| B90_0 | $1.90 \times 10^{15}$ | 1.190 | 3.369 | 2.272 | 0.232 | 13.213 | 0.633 | 14.612 | 2.683 |

**Table 6.** Structure parameters for the maximum angular momentum configuration.

| EOS | $\rho_c$ | $\Omega$ | $I$ | $M$ | $T/W$ | $R$ | $\tilde{j}$ | $R_{ISCO}$ | $M_B$ |
|---|---|---|---|---|---|---|---|---|---|
| SS1 | $3.10 \times 10^{15}$ | 1.613 | 2.072 | 2.077 | 0.219 | 10.404 | 0.524 | 11.656 | 2.694 |
| SS2 | $3.40 \times 10^{15}$ | 1.719 | 1.633 | 1.899 | 0.220 | 9.693 | 0.575 | 10.837 | 2.355 |
| B90_0 | $1.70 \times 10^{15}$ | 1.161 | 3.456 | 2.254 | 0.239 | 13.447 | 0.650 | 14.864 | 2.650 |

**Table 7.** Structure parameters for the constant angular velocity sequence for EOS SS1. This sequence corresponds to the rotation rate of the pulsar PSR 1937 +21, having $\Omega = 4.03 \times 10^3$ rad s$^{-1}$ or period $P = 1.556$ ms.

| $\rho_c$ | $I$ | $M$ | $T/W$ | $R$ | $\tilde{j}$ | $R_{ISCO}$ | $M_B$ |
|---|---|---|---|---|---|---|---|
| $1.70 \times 10^{15}$ | 0.353 | 0.852 | 0.013 | 6.663 | 0.153 | 6.663 | 1.027 |
| $1.80 \times 10^{15}$ | 0.433 | 0.963 | 0.012 | 6.884 | 0.143 | 7.664 | 1.178 |
| $1.90 \times 10^{15}$ | 0.502 | 1.052 | 0.019 | 7.036 | 0.136 | 8.378 | 1.301 |
| $2.40 \times 10^{15}$ | 0.701 | 1.297 | 0.010 | 7.326 | 0.116 | 10.355 | 1.660 |
| $2.60 \times 10^{15}$ | 0.737 | 1.346 | 0.010 | 7.344 | 0.112 | 10.741 | 1.735 |
| $4.60 \times 10^{15}$ | 0.769 | 1.458 | 0.008 | 7.139 | 0.096 | 11.732 | 1.914 |
| $5.65 \times 10^{15}$ | 0.734 | 1.449 | 0.007 | 7.007 | 0.093 | 11.703 | 1.899 |

SS2, the two models are slightly different, with the maximum angular momentum model coming earlier (with respect to $\rho_c$) than the maximum mass rotating configuration.

In Table 7 we list the values of the various parameters for the constant $\Omega$ sequences for EOS SS1. The first entry in this table corresponds to $\rho_c$ for which $R_{ISCO} = R$. For higher values of $\rho_c$, $R_{ISCO} > R$; for large values of $\rho_c$, the boundary layer (separation between the surface of the SS and its innermost stable orbit) can be substantial ($\sim 5$ km for the maximum value of the listed $\rho_c$).

## 5 Strange star candidates

To distinguish whether a compact star is a neutron star or a strange star, one has to find a clear observational signature. As we saw in section 3, the most striking qualitative difference between neutron stars and strange stars is in their mass–radius (MR) relation. In the following we report about recent studies where it has been claimed that some compact objects associated with discrete X-ray sources are possibly strange stars.

## 5.1 4U 1820–30

We begin with the X-ray burst source 4U 1820–30. X-ray bursts are sudden emission of X-rays from discrete cosmic sources (referred to as *X-ray bursters*). They were discovered [53] in 1975 by the ANS satellite in the globular cluster NG6624. The bursts have very short rise times ($\leq 1$ s) and decay times in the range 3–100 s. The recurrence interval between bursts is generally in the range 1 hour–1 day. The X-ray energy emitted in each burst is typically $\sim 10^{39}$ erg and the peak burst luminosity $\sim 10^{38}$ erg/s. Most bursters are also sources of persistent X-ray emission with average steady-state luminosity $\sim 10^{37}$ erg/s. Bursts with the above features are known as Type-I X-ray bursts, to be distinguished from Type-II bursts [54], which have recurrence intervals between bursts on time scales of seconds to minutes when the source is active. Presently only two Type-II bursters are known: the source MXB 1730-335 (which is also known as the *Rapid Burster*) and the source GRO 1744-28 (see ref. [55]). According to the currently accepted models [56–58], Type-I X-ray bursts originate from thermonuclear flashes in matter that accumulates on the surface of an accreting compact object (usually assumed to be a neutron star) in a low-mass X-ray binary. The thermonuclear flashes model is able to account very successfully most of the observed properties of the X-ray bursts.

The X-ray burst source 4U 1820–30 is located in the globular cluster NGC 6624, and its distance from earth is 6.4±0.6 kpc. Using the burst spectra collected by EXOSAT [59], Haberl and Titarchuk [60] were able to extract a semiempirical mass–radius relation for the underlying compact object. This MR relation is shown in fig. 6 by the trapezium-like region labeled 4U 1820–30. In the same figure, we show [61] the theoretical MR relation as calculated for "conventional" neutron stars. Results in left panel of Fig. 6 are relative to phenomenological models for the EOS. The dashed line refers to Skyrme SkM* nuclear interaction. Continuous lines refer to the so called SL EOS, based on a generalized Skyrme interaction (see ref. [62,63]). The three models SL12, SL22 and SL32 differ for the value of the nuclear incompressibility $K_0$, equal to 120, 180, and 240 MeV respectively. In the right panel of Fig. 6, show the MR relation as calculated from other models for asymmetric nuclear matter EOS. References to these models are given in the figure caption. In general, for a fixed value of the neutron star mass, a soft EOS is expected to give a smaller radius with respect to a stiff EOS. Therefore, as a limiting case, we considered the SL12 and BPAL12 EOSs [62,63], which give $K_0 = 120$ MeV. The value 120 MeV for the incompressibility is unrealistically small when compared with the value (180–240 MeV) extracted from monopole nuclear oscillations [64,65] and nuclear systematic [66]. However, SL12 and BPAL12 EOSs are still consistent with the measured neutron star masses, which set a lower limit of about 1.44 $M_\odot$, for the theoretical value of the limiting mass of a neutron star. All neutron star models considered in Fig. 6 are excluded by the observational data from 4U 1820–30.

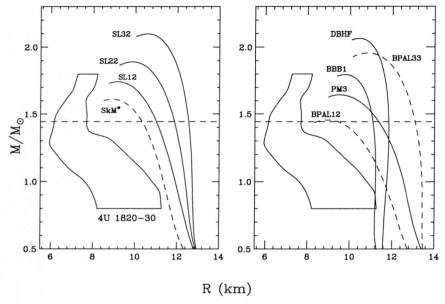

**Fig. 6.** Theoretical mass–radius (MR) relations (curves) for "conventional" neutron stars are compared with the semiempirical MR relation extracted in ref. [60] for the X-ray burster 4U 1820–30 (closed region of the MR plane labeled 4U 1820–30). MR curve PM3 has been obtained using the EOS of ref. [67], and the curve DBHF with the EOS given in ref. [68]. The dashed horizontal line represents the gravitational mass of the pulsar PSR1916+13.

## 5.2 SAX J1808.4–3658

The transient X-ray burst source SAX J1808.4–3658 was discovered in September 1996 by the BeppoSAX satellite. Two bright type-I X-ray bursts were detected, each lasting less than 30 seconds. Analysis of the bursts in SAX J1808.4–3658 indicates that it is 4 kpc distant and has a peak X-ray luminosity of $6 \times 10^{36}$ erg/s in its bright state, and a X-ray luminosity lower than $10^{35}$ erg/s in quiescence [69]. The object is nearly certainly the same as the transient X-ray source detected with the Proportional Counter Array on board the Rossi X-ray Timing Explorer (RXTE). Coherent pulsations at a period of 2.49 milliseconds were discovered [70]. The star's surface dipolar magnetic moment was derived to be less than $10^{26}$ G cm$^3$ from detection of X-ray pulsations at a luminosity of $10^{36}$ erg/s [70], consistent with the weak fields expected for type-I X-ray bursters and millisecond radio pulsars [41]. The binary nature of SAX J1808.4–3658 was firmly established with the detection of a 2 hour orbital period [71] as well as with the optical identification of the companion star. SAX J1808.4–3658 is the first pulsar to show both coherent pulsations in its persistent emission and X-ray bursts, and by far the fastest-rotating, lowest-field accretion-driven pulsar known. It presents direct evidence for the evolutionary link between low-mass X-ray binaries (LMXBs) and millisecond radio pulsars [41].

A mass–radius relation for the compact star in SAX J1808.4–3658 has been recently obtained by Li et al. [47] (see also ref. [72,73]) using the following two requirements. (*i*) Detection of X-ray pulsations requires that the inner radius $R_0$ of the accretion disk should be larger than the stellar radius $R$. In other words, the stellar magnetic field must be strong enough to disrupt the disk flow above the stellar surface. (*ii*) The radius $R_0$ must be less than the so called co-rotation radius $R_c$, *i.e.* the stellar magnetic field must be weak enough that accretion is not centrifugally inhibited: $R_0 \lesssim R_c = [GMP^2/(4\pi^2)]^{1/3}$. Here $G$ is the gravitation constant, $M$ is the mass of the star, and $P$ is the pulse period. The inner disk radius $R_0$ is generally evaluated in terms of the Alfvén radius $R_A$, at which the magnetic and material stresses balance [41]: $R_0 = \xi R_A = \xi[B^2 R^6/\dot{M}(2GM)^{1/2}]^{2/7}$, where $B$ and $\dot{M}$ are respectively the surface magnetic field and the mass accretion rate of the pulsar, and $\xi$ is a parameter of order of unity almost independent [74] of $\dot{M}$. Since X-ray pulsations in SAX J1808.4–3658 were detected over a wide range of mass accretion rate (say, from $\dot{M}_{\min}$ to $\dot{M}_{\max}$), the two conditions (*i*) and (*ii*) give $R \lesssim R_0(\dot{M}_{\max}) < R_0(\dot{M}_{\min}) \lesssim R_c$. Next, the authors of ref. [47] assume that the mass accretion rate $\dot{M}$ is proportional to the X-ray flux $F$ observed with RXTE. This is guaranteed by the fact that the X-ray spectrum of SAX J1808.4–3658 was remarkably stable and there was only slight increase in the pulse amplitude when the X-ray luminosity varied by a factor of $\sim 100$ during the 1998 April/May outburst [75,76,73]. Therefore, Li et al. [47]

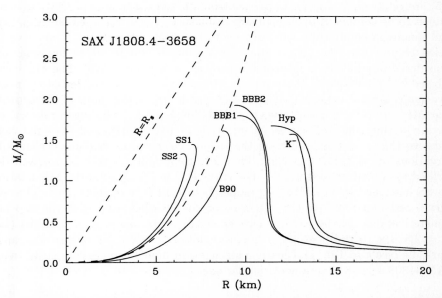

**Fig. 7.** Comparison of the MR relation of SAX J1808.4–3658 determined from RXTE observations with theoretical models of neutron stars and of SS. The solid curves represents theoretical MR relations for neutron stars and strange stars.

get the following upper limit of the stellar radius: $R < (F_{min}/F_{max})^{2/7} R_c$, or

$$R < 27.5 \left(\frac{F_{min}}{F_{max}}\right)^{2/7} \left(\frac{P}{2.49 \text{ ms}}\right)^{2/3} \left(\frac{M}{M_\odot}\right)^{1/3} \text{ km}, \qquad (45)$$

where $F_{\max}$ and $F_{\min}$ denote the X-ray fluxes measured during X-ray high- and low-state, respectively, $M_\odot$ is the solar mass. Note that in writing inequality (45) it is assumed that the pulsar's magnetic field is basically dipolar. Arguments to support this hypothesis are given in ref. [47], whereas a study of the influence on the MR relation for SAX J1808.4–3658 of a quadrupole magnetic moment, and of a *non-standard* disk–magnetosphere interaction model is reported in ref. [73].

Given the range of X-ray flux at which coherent pulsations were detected, inequality (45) defines a limiting curve in the MR plane for SAX J1808.4–3658, as plotted in the dashed curve in Fig. 7. The authors of ref. [47] adopted the flux ratio $F_{\max}/F_{\min} \simeq 100$ from the measured X-ray fluxes with the RXTE during the 1998 April/May outburst [76,73]. The dashed line $R = R_s \equiv 2GM/c^2$ represents the Schwarzschild radius - the lower limit of the stellar radius to prevent the star collapsing into a black hole. Thus the allowed range of the mass and radius of SAX J1808.4–3658 is the region confined by these two dashed curves in Fig. 7.

In the same figure, we report the theoretical MR relations (solid curves) for neutron stars given by some recent realistic models for the EOS of dense matter (see ref. [47] for references to the EOS models). Models BBB1 and BBB2 [95] are relative to "conventional" neutron stars (*i.e.* the core of the star is assumed to be composed by an uncharged mixture of neutrons, protons, electrons and muons in equilibrium with respect to the weak interaction). The curve labeled Hyp depicts the MR relation for a neutron star in which hyperons are considered in addition to nucleons as hadronic constituents [63]. The MR curve labeled $K^-$ is relative to neutron stars with a Bose-Einstein condensate of negative kaons in their cores [63]. It is clearly seen in Fig. 7 that none of the neutron star MR curves is consistent with SAX J1808.4–3658. Including rotational effects will shift the $MR$ curves to up-right in Fig. 1 [42], and does not help improve the consistency between the theoretical neutron star models and observations of SAX J1808.4–3658. Therefore SAX J1808.4–3658 is not well described by a neutron star model. The curve B90 in Fig. 7 gives the MR relation for SS described by the MIT bag model EOS with $B = 90$ MeV/fm$^3$. The two curves SS1 and SS2 give the MR relation for SS calculated with the two parameterizations for the EOS of Dey *et al.*[25]. Rotation at $P = 2.49$ ms (*i.e.* $\Omega = 2.523 \times 10^3$ rad/s $\ll \Omega_{ms}$) has negligible effects on the radius of strange stars [49] (see also Tab. 7). Clearly a strange star model is more compatible with SAX J1808.4–3658 than a neutron star one.

## 5.3  4U 1728–34

Recently, Li *et al.* [48] investigated possible signatures for the existence of SS in connection with the newly discovered phenomenon of kilohertz quasi–periodic oscillations (kHz QPOs) in the X-ray flux from LMXB (for a review see ref. [77]).

Initially, kHz QPO data from various sources were interpreted assuming a simple *beat-frequency model* (see *e.g.* ref.[78]). In many cases, two simultaneous kHz QPO peaks ("twin peaks") are observed. The QPO frequencies vary and are strongly correlated with source flux. In the beat-frequency model the highest observed QPO frequency $\nu_u$ is interpreted as the keplerian orbital frequency $\nu_K$ at the inner edge of the accretion disk. The frequency $\nu_l$ of the lower QPO peak is instead interpreted as the beat frequency between $\nu_K$ and the neutron star spin frequency $\nu_0$, which within this model is equal to the separation frequency $\Delta\nu \equiv \nu_u - \nu_l$ of the two peaks. Thus $\Delta\nu$ is predicted to be constant. Nevertheless, novel observations for different kHz QPO sources have challenged this simple beat-frequency model. The most striking case is the source 4U 1728−34, where it was found that $\Delta\nu$ decreases significantly, from 349.3±1.7 Hz to 278.7±11.6 Hz, as the frequency of the lower kHz QPO increases [79]. Furthermore, in the spectra observed by the RXTE for 4U 1728−34, Ford and van der Klis [80] found low-frequency Lorentian oscillations with frequencies between 10 and 50 Hz. These frequencies as well as the break frequency ($\nu_{break}$) of the power spectrum density for the same source were shown [80] to be correlated with $\nu_u$ and $\nu_l$.

A different model was recently developed by Titarchuk and Osherovich [81] who proposed a unified classification of kHz QPOs and the related observed low frequency phenomena. In this model, kHz QPOs are modeled as keplerian oscillations under the influence of the Coriolis force in a rotating frame of reference (magnetosphere). The frequency $\nu_l$ of the lower kHz QPO peak is the

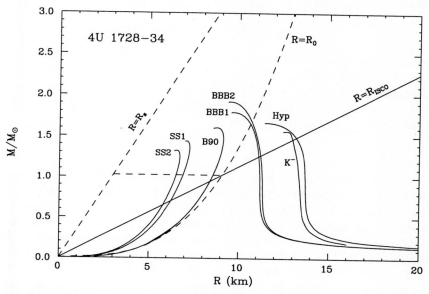

**Fig. 8.** Comparison of the MR relation of 4U 1728−34 determined from RXTE observations with theoretical models of neutron stars and of strange stars. The range of mass and radius of 4U 1728−34 is allowed in the region outlined by the dashed curve $R = R_0$, the horizontal dashed line, and the dashed line $R = R_s$.

keplerian frequency at the outer edge of a viscous transition layer between the keplerian disk and the surface of the compact star. The frequency $\nu_u$ is a hybrid frequency related to the rotational frequency $\nu_m$ of the star's magnetosphere by: $\nu_u^2 = \nu_K^2 + (2\nu_m)^2$. The observed low Lorentzian frequency in 4U 1728–34 is suggested to be associated with radial oscillations in the viscous transition layer of the disk, whereas the observed break frequency is determined by the characteristic diffusion time of the inward motion of the matter in the accretion flow [81]. Predictions of this model regarding relations between the QPO frequencies mentioned above compare favorably with recent observations for 4U 1728–34, Sco X-1, 4U 1608–52, and 4U 1702–429.

The presence of the break frequency and the correlated Lorentzian frequency suggests the introduction of a new scale in the phenomenon. One attractive feature of the model of ref.[81] is the introduction of such a scale in the model through the Reynolds number for the accretion flow. The best fit for the observed data was obtained by Titarchuk and Osherovich [81] when

$$a_k = (M/M_\odot)(R_0/3R_s)^{3/2}(\nu_0/364\,\mathrm{Hz}) = 1.03, \qquad (46)$$

where $M$ is the stellar mass, $R_0$ is the inner edge of the accretion disk, $R_s$ is the Schwarzschild radius, and $\nu_0$ is the spin frequency of the star. Given the 364 Hz spin frequency [82] of 4U 1728–34, the inner disk radius can be derived from the previous equation. Since the innermost radius of the disk must be larger than the radius $R$ of the star itself, this leads to a mass-dependent upper bound on the stellar radius (plotted by the dashed curve in Fig. 8)

$$R \leq R_0 \simeq 8.86\, a_k^{2/3} (M/M_\odot)^{1/3}\,\mathrm{km}. \qquad (47)$$

A second constraint on the mass and radius of 4U 1728–34 results from the requirement that the inner radius $R_0$ of the disk must be larger than the radius of innermost stable circular orbit $R_{ISCO}$ around the star. To make our discussion more transparent, neglect for a moment the rotation of the compact star. For a non-rotating star $R_{ISCO} = 3R_s$, then the second condition gives:

$$R_0 \geq 3R_s = 8.86\, (M/M_\odot)\,\mathrm{km}. \qquad (48)$$

Therefore, the allowed range of the mass and radius for 4U 1728–34 is the region in the lower left corner of the MR plane confined by the dashed curve ($R = R_0$), by the horizontal dashed line, and by the Schwarzschild radius (dashed line $R = R_s$). In the same figure, we compare with the theoretical MR relations for non-rotating neutron stars and strange stars, for the same models for the EOS considered in Fig. 7. It is clear that a strange star model is more compatible with 4U 1728–34 than a neutron star one. Including the effect of rotation ($\nu_0$ =364 Hz) in the calculation of $R_{ISCO}$ and in the theoretical MR relations, does not change the previous conclusion [48,49]. These results strongly suggest that the compact star in 4U 1728–34 might be a strange star.

# 6 Conversion of neutron stars to strange stars as the central engine of gamma-ray bursts

There is now compelling evidence to suggest that a substantial fraction of all gamma-ray bursts (GRBs) occur at cosmological distances (red shift $z \sim 1-3$). In particular, the measured red shift $z = 3.42$ for GRB 971214 (ref. [83]), and $z \sim 1.6$ for GRB 990123 (ref. [84]) implies an energy release of $3 \times 10^{53}$ erg and $3.4 \times 10^{54}$ erg respectively, in the $\gamma$-rays alone, assuming isotropic emission. The latter energy estimate could be substantially reduced if the energy emission is not isotropic, but displays a jet-like geometry [85,84]. Models in which the burst is produced by a narrow jet are able to explain the complex temporal structure observed in many GRBs [86]. In any case, a cosmological origin of GRBs leads to the conclusion of a huge energy output. Depending on the degree of burst beaming and on the efficiency of $\gamma$-ray production, the central engine powering these extraordinary events should be capable of releasing a total energy of a few $10^{53}$ erg.

Many cosmological models for GRBs have been proposed. Among the most popular is the merging of two neutron stars (or a neutron star and a black hole) in a binary system [87]. Recent results [88] within this model, indicate that, even under the most favorable conditions, the energy provided by $\nu\bar{\nu}$ annihilation during the merger is too small by at least an order of magnitude, and more probably two or three orders of magnitude, to power typical GRBs at cosmological distances. An alternative model is the so-called "failed supernova" [89], or "hypernova" model [87].

In the following, we consider the conversion of a neutron star (NS) to a strange star (hereafter NS→SS conversion) as a possible central engine for GRBs. In particular, we focus on the energetics of the NS→SS conversion, and not on the mechanism by which $\gamma$-rays are produced. Previous estimate of the total energy $E^{conv}$ released in the NS→SS conversion [90,91] or in the conversion of a neutron star to hybrid star [92] gave $E^{conv} \sim 10^{52}$ erg, too low to power GRBs at cosmological distances. These calculations did not include the various details of the neutron star and strange star structural properties, which go into the binding energy release considerations. Here we report recent accurate and systematic calculations [93] of the total energy released in the NS→SS conversion using different models for the equation of state of neutron star matter (NSM) and strange quark matter. As shown by the authors of ref. [93], the total amount of energy liberated in the conversion is in the range $E^{conv} = (1-4) \times 10^{53}$ erg, in agreement with the energy required to power gamma-ray burst sources at cosmological distances.

Originally, the idea that GRBs could be powered by the conversion of a neutron star to a strange star was proposed by Alcock *et al.* [15] (see also ref. [90]), and recently reconsidered by other authors [91]. A similar model has been discussed by Ma and Xie [92] for the conversion of a neutron star to a hybrid star.

A number of different mechanisms have been proposed for the NS→SS conversion. All of them are based on the formation of a "seed" of SQM inside the neutron star. For example: (i) a seed of SQM enters in a NS and converts it to a SS [90]. These seeds of SQM, according to Witten [9], are relics of the primordial quark–hadron phase transition microseconds after the Big Bang. (ii) A seed of SQM forms in the core of a neutron star as a result of a phase transition from neutron star matter to deconfined strange quark matter (NSM→SQM phase transition). This could possibly happen when a neutron star is a member of a binary stellar system. The NS accretes matter from the companion star. The central density of the NS increases and it may overcome the critical density for the NSM→SQM phase transition. The NS is then converted to a SS. In the case of accretion induced conversion in a binary stellar system, the conversion rate has been estimated [91] to be in the range $(3-30)\times 10^{-10}$ conversions per day per galaxy. This rate is consistent with the observed GRBs rate.

However, once there is a seed of SQM inside a neutron star, it is possible to calculate the rate of growth [90,94]. The SQM front absorbs neutrons, protons, and hyperons (if present), liberating their constituent quarks. Weak equilibrium is then re-established by the weak interactions. As shown by Horvath and Benvenuto [94], the conversion of the whole star will then occur in a very short time (detonation mode), in the range 1 ms – 1 s, which is in agreement with the typical observed duration of GRBs. A detailed simulation of the conversion process is still lacking, and only rough estimates of the total energy liberated in the conversion have been made.

As we show below, following ref.[93], the dominant contribution to $E^{conv}$ arises from the internal energy released in the conversion, *i.e.* in the NSM→SQM phase transition. Moreover, the gravitational mass of the star will change during the conversion process, even under the assumption that the total number of baryons in the star is conserved.

The total energy released in the NS→SS conversion is given by the difference between the total binding energy of the strange star BE(SS) and the total binding energy of the neutron star BE(NS)

$$E^{conv} = BE(SS) - BE(NS). \tag{49}$$

Here we assume that the baryonic mass $M_B$ of the compact object is conserved in the conversion process, *i.e.* $M_B(SS) = M_B(NS) \equiv M_B$. Then $E^{conv}$ is given in terms of the difference between the gravitational mass of the NS and SS: $E^{conv} = [M_G(NS) - M_G(SS)]c^2$.

In general the total binding energy for a compact object can be written $BE = BE_I + BE_G = (M_B - M_P)c^2 + (M_P - M_G)c^2$, where $BE_I$ and $BE_G$ denote the internal and gravitational binding energies respectively, and $M_P$ is the proper mass of the compact object. The total conversion energy can then be written as the sum of two contributions

$$E^{conv} = E_I^{conv} + E_G^{conv} \tag{50}$$

related to the internal and gravitational energy changes in the conversion. These two contributions are given by:

$$E_I^{conv} = BE_I(SS) - BE_I(NS) = [M_P(NS) - M_P(SS)]c^2, \tag{51}$$

$$E_G^{conv} = BE_G(SS) - BE_G(NS) = [M_P(SS) - M_G(SS) - M_P(NS) + M_G(NS)]c^2, \tag{52}$$

and these can be evaluated solving the structural equations for non-rotating compact objects. To highlight the dependence of $E^{conv}$ upon the present uncertainties in the microphysics, the authors of ref. [93] employed different models for the EOS of both NSM and SQM.

Recently, a microscopic EOS of dense stellar matter has been calculated by Baldo et al. [95], and used to compute the structure of static [95] as well as rapidly rotating neutron stars [42]. In this model for the EOS, the neutron star core is composed of asymmetric nuclear matter in equilibrium, with respect to the weak interactions, with electrons and muons ($\beta$-stable matter). In particular, we consider their EOS based on the Argonne $v14$ nucleon-nucleon interaction implemented by nuclear three-body forces (hereafter BBB1 EOS).

At the high densities expected in the core of a neutron star, additional baryonic states besides the neutron and the proton may be present, including the hyperons $\Lambda$, $\Sigma$, $\Xi$, $\Omega$, and the isospin 3/2 nucleon resonance $\Delta$. The equation of state of this hyperonic matter is traditionally investigated in the framework of Lagrangian field theory in the mean field approximation [96,97,63]. According to this model, the onset for hyperon formation in $\beta$-stable–charged neutral dense matter is about 2–3 times the normal nuclear matter density ($n_0 = 0.17$ fm$^{-3}$). The latter result has been confirmed by recent microscopic calculations based on the Brueckner-Hartree-Fock theory [98]. The appearance of hyperons, in general, gives a softening of the EOS with respect to the pure nucleonic case. In the present work we considered one of the EOS for hyperonic matter given in ref. [63]. For strange quark matter, we consider the simple EOS based on the MIT bag model for hadrons and the EOS by Dey et al. [25].

**Table 8.** Conversion to strange star of a neutron star with $M_G \sim 1.4\ M_\odot$, for different EOSs for NSM and SQM. $M_B$ is the baryonic mass (which is conserved in the conversion process), $M_G(NS)$ is the neutron star gravitational mass, and $M_G(SS)$ is the gravitational mass of the corresponding strange star. All masses are in unit of the solar mass $M_\odot = 1.989 \times 10^{33}$ g. $E_G^{conv}$, $E_I^{conv}$, and $E^{conv}$ are respectively the gravitational, internal, and total conversion energy, in unit of $10^{53}$ erg.

| NSM→SQM | $M_B$ | $M_G(NS)$ | $M_G(SS)$ | $E_G^{conv}$ | $E_I^{conv}$ | $E^{conv}$ |
|---|---|---|---|---|---|---|
| BBB1→ $B60_0$ | 1.574 | 1.409 | 1.254 | -1.436 | 4.215 | 2.779 |
| BBB1→ $B60_{200}$ | 1.574 | 1.409 | 1.340 | -0.677 | 1.920 | 1.243 |
| BBB1→ $B90_0$ | 1.573 | 1.409 | 1.343 | -0.057 | 1.241 | 1.184 |
| BBB1→SS1 | 1.558 | 1.397 | 1.235 | 0.580 | 2.308 | 2.888 |
| BBB1→SS2 | 1.566 | 1.403 | 1.268 | 1.604 | 0.800 | 2.404 |
| Hyp→ $B60_0$ | 1.530 | 1.401 | 1.223 | -0.617 | 3.802 | 3.185 |
| Hyp→SS1 | 1.530 | 1.401 | 1.217 | 1.291 | 2.002 | 3.293 |

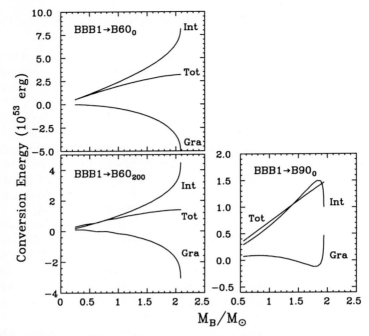

**Fig. 9.** The total energy liberated in the conversion of a neutron star to a strange star and the partial contributions from internal energy $E_I^{conv}$ (curves labeled "Int") and from gravitational energy $E_G^{conv}$ (curves labeled "Gra") as a function of $M_B$. See text for details on the equations of state for neutron star matter and strange quark matter

To begin with, we fix as a "standard" EOS for neutron star matter the BBB1 EOS [95], to explore how the energy budget in the NS→SS conversion depends on the details of the EOS for strange quark matter. First we consider the $B60_0$ equation of state. The NS→SS conversion based on this couple of EOSs, will be referred to as the BBB1→ $B60_0$ conversion model. Similar notation will be employed according to the EOS of NSM and SQM. The total conversion energy, together with the partial contributions, is shown in the upper panel of Fig. 9. As we can see, for $M_B$ larger than $\sim 1~M_\odot$ (i.e. values of the baryonic mass compatible with the measured neutron star gravitational masses) the total energy released in the NS→SS conversion is in the range $(1$–$3)\times 10^{53}$ erg, one order of magnitude larger than previous estimates [90–92]. Moreover, contrary to a common expectation, the gravitational conversion energy $E_G^{conv}$ is negative for this couple of EOSs. To make a more quantitative analysis, we consider a neutron star with a baryonic mass $M_B = 1.574 M_\odot$ (see tab. 8), which has a gravitational mass $M_G = 1.409 M_\odot$, a radius $R(NS) = 11.0$ km, and a gravitational binding energy $BE_G(NS) = 4.497 \times 10^{53}$ erg. After conversion, the corresponding strange star has $M_G = 1.254 M_\odot$, $R(SS) = 10.5$ km, and $BE_G(SS) = 3.061 \times 10^{53}$ erg. The NS→SS conversion is energetically possible in this case, thanks to the large amount of (internal) energy liberated in the NSM→SQM phase transition.

Similar qualitative results for the total conversion energy are obtained for other choices of the two EOSs, but as we show below the magnitude of the two partial contributions are strongly dependent on the underlying EOS for NSM and SQM. The total conversion energy for the BBB1$\to$ $B60_{200}$ model is plotted in the lower left panel of Fig. 9. Comparing with the previous case, we notice that the strange quark mass produces a large modification of the conversion energy, which is reduced by a factor between 2–3 with respect to the case $m_s = 0$. The bag constant $B$ has also a sizeable influence on the conversion energy. Increasing the value of $B$ reduces $E^{conv}$ and strongly modifies $E_G^{conv}$. This can be seen comparing the results for the BBB1$\to$ $B60_0$ conversion model with those in the lower right panel of Fig. 1 for the BBB1$\to$ $B90_0$ model. These results are a consequence of the sizeable effects of the strange quark mass and of the bag constant mainly on the internal binding energy $BE_I(SS)$ for strange stars (compare for example the results depicted in figures 2 and 3). In fact, all strange stars configurations within the $B60_0$ EOS are self–bound objects (*i.e.* $BE_I(SS) > 0$). Strange star configurations within the $B90_0$ ($B60_{200}$) EOS are self–bound objects up to $M_G \sim 0.8~M_\odot$ ($M_G \sim 1.6~M_\odot$), to compare with the corresponding maximum gravitational mass $M_{max} = 1.60~M_\odot$ ($M_{max} = 1.75~M_\odot$).

The results depicted in the two upper panels of Fig. 10 have been obtained using the EOS of ref. [25] for SQM, for two different choices of the parameter $\nu$. The parameter $\nu$ has a strong influence on the internal binding energy of the strange star. In fact, we found that strange stars within the SS2 (SS1) EOS are self–bound objects up to $M_G \sim 0.7~M_\odot$ ($M_G \sim 1.4~M_\odot$), to compare with the maximum gravitational mass $M_{max} = 1.33~M_\odot$ ($M_{max} = 1.44~M_\odot$). This effect is the main source for the differences in the calculated conversion energies for the two conversion models BBB1$\to$SS1 and BBB1$\to$SS2.

The next step in our study is to consider a different neutron star matter EOS, which allows for the presence of hyperons in the neutron star core. We consider one of the EOS (hereafter Hyp) for hyperonic matter given in ref. [63]. In the two lower panels of Fig. 10 we plot the total conversion energy, together with the partial contributions, for the Hyp$\to$ $B60_0$ and for the Hyp$\to$SS1 conversion models. These results are in qualitative agreement with those reported in the previous figures.

In table 8, we report the conversion energy, together with the partial contributions for the conversion of a neutron star with a gravitational mass $M_G(NS) \sim 1.4 M_\odot$ and for various conversion models.

# 7 Search for strangelets in cosmic rays and in heavy ion collision experiments

Circumstantial evidence for strange stars could be supplied by the experimental detection of stable or metastable strangelets.

Lumps of SQM might be present in cosmic radiation, or might be formed when cosmic rays penetrate the earth atmosphere. There have been several re-

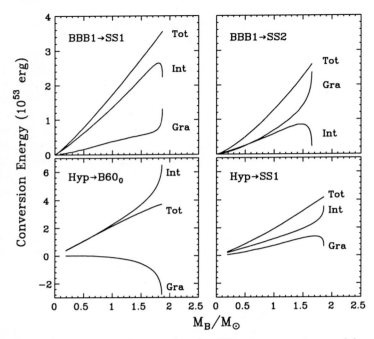

**Fig. 10.** Same as in figure 9, but for different conversion models

ports of events with $A \sim 350$–$500$ and $Z \sim 10$–$20$ in cosmic ray events [99–103]. Due to the very low charge to mass ratio, these so called *exotic cosmic ray events* have been interpreted as a signal for the existence of strangelets. Nevertheless, due to various experimental uncertainties and to other possible theoretical interpretations [104,105], it is still premature to conclude these exotic events are really due to strangelets.

Collisions between heavy atomic nuclei provide a promising possible way to produce and to prove the existence of strangeletes in the laboratory. Many of these experiments have been performed at the AGS accelerator at the Brookhaven National Laboratory (BNL) [106–109], or at the Super Proton Synchrotron (SPS) at CERN [110,111]. For example, the recent experiment E864 [109] at BNL–AGS made use of a beam of gold ions at 11.5 GeV per nucleon, with platinum or lead targets.

Several mechanisms have been proposed for strangelets production in nucleus–nucleus collisions. In the so called coalescence model [112] a conglomerate of baryons is produced in the collision, and subsequently may fuse to form a SQM drop. In another model strangelets are generated following Quark Gluon Plasma (QGP) production [113,114].

To date all these nucleus–nucleus collision experiments have not found any signal for the existence of strangelets. Thus, presently these experiments are only able to set upper limits for strangelets production.

Strangelets search will be carried on in the next generation of ultra-relativistic heavy ion colliders, *i.e.* RHIC at BNL and LHC at CERN, although these machines were mainly planned to detect the formation of QGP. Particularly, a specific detector system for strangelets search will be CASTOR [115] at CERN–LHC as a part of the ALICE experiment.

A very unpleasant consequence of the strange matter hypothesis could be the possible formation of stable negatively charged strangelets during heavy ion collisions at RHIC or at LHC. In fact, it has been pointed out [116] that these "dangerous" negatively charged strangelets may trigger the disruption of our planet. Luckily, there are various theoretical as well as experimental arguments [116,117] to rule out this "Disaster Scenario".

## 8 Final remarks

The main result of the present work (*i.e.* the likely existence of strange stars) is based on the analysis of observational data for X-ray burster 4U 1820–30, the X-ray sources SAX J1808.4–3658 and 4U 1728–34. The interpretation of these data is done using *standard* models for the accretion mechanism, which is responsible for the observed phenomena. The present uncertainties in our knowledge of the accretion mechanism, and the disk–magnetosphere interaction, do not allow us to definitely rule out the possibility of a neutron star for the X-ray sources we discussed. For example, making *a priori* the *conservative* assumption that the compact object in SAX J1808.4–3658 is a neutron star, and using a MR relation similar to our eq.(45), Psaltis and Chakrabarty [73] try to constrain disk–magnetosphere interaction models or to infer the presence of a quadrupole magnetic moment in the compact star.

The X-ray binary systems discussed in the present work, are not the only LMXBs which could harbour a strange star. Recent studies have shown that the compact objects associated with the bursting X-ray pulsar GRO J1744-28 [118] and the X-ray pulsar Her X-1 (ref.[25]) are likely strange star candidates. For each of these X-ray sources (strange star candidates) the conservative assumption of a neutron star as the central accretor would require some particular (possibly *ad hoc*) assumption about the nature of the plasma accretion flow and/or the structure of the stellar magnetic field. On the other hand, the possibility of a strange star gives a simple and unifying picture for all the systems mentioned above.

### Acknowledgments

Most of the ideas and results in this lecture stem from a fruitful collaboration, during the last four years, with B. Datta, J. Dey, M. Dey, E.P.J. van den Heuvel, X.D. Li, S. Ray, B.C. Samanta, and A. Thampan. It is a pleasure to acknowledge stimulating discussions with David Blaschke, Hovik Grigorian, Sergei B. Popov, Krishna Rajagopal, and Julien L. Zdunik during the workshop Neutron Star Interiors, held at the ECT* in Trento during the summer of year 2000.

# References

1. D. Ivanenko, D.F. Kurdgelaidze: Lett. Nuovo Cimento **2**, 13 (1969)
2. N. Itoh: Prog. Theor. Phys. **44**, 291 (1970)
3. F. Iachello, W.D. Langer, A. Lande: Nucl. Phys. A **219**, 612 (1974)
4. J.C. Collins, M.J. Perry: Phys. Rev. Lett. **34**, 1353 (1975)
5. G. Baym, S.A. Chin: Phys. Lett. B **62**, 241 (1976)
6. B.D. Keister, L.S. Kisslinger: Phys. Lett. B **64**, 117 (1976)
7. A. R. Bodmer: Phys. Rev. D **4**, 1601 (1971)
8. H. Terazawa: INS Rep. **336** (Univ. Tokyo, INS) (1979); J. Phys. Soc. Jpn. **58**, 3555 (1989); **58**, 4388 (1989); **59**, 1199 (1990)
9. E. Witten: Phys. Rev. D **30**, 272 (1984)
10. E. Farhi, R. L. Jaffe: Phys. Rev. D **30**, 2379 (1984)
11. E.P. Gilson, R. L. Jaffe: Phys. Rev. Lett. **71**, 332 (1993)
12. M.G. Mustafa, A. Ansari: Phys. Rev. D **53**, 5136 (1996)
13. D. Blaschke, H. Grigorian, G. Poghosyan: 'Phase Diagram for Spinning and Accreting Neutron Stars', this volume
14. N. Glendenning, F. Weber: 'Signal of Quark Deconfinement in Millisecond Pulsars and Reconfinement in Accreting X-ray Neutron Stars', this volume
15. C. Alcock, E. Farhi, A. Olinto: Astrophys. J. **310**, 261 (1986)
16. N.K. Glendenning, F. Weber: Astrophys. J. **400**, 647 (1992)
17. N.K. Glendenning, Ch. Kettner, F. Weber: Phys. Rev. Lett. **74**, 3519 (1995); Astrophys. J. **450**, 253 (1995)
18. M.Alford, J.A. Bowers, K. Rajagopal: 'Color superconductivity in compact stars', this volume (hep-ph/0009357)
19. K. Rajagopal, F. Wilczek: 'The condensed matter physics of QCD'. In: At the frontiers of particle physics/Handbook of QCD. M. Shifman ed. (World Scientific), (hep-ph/0011333)
20. A.I. Larkin, Yu.N. Ovchinnikov: Zh. Eksp. Teor. Fiz. **47**, 1136 (1964), translation: Sov. Phys. JETP **20**, 762 (1965); P. Fulde, R.A. Ferrell: Phys. Rev. **135**, A550 (1964)
21. I.M. Barbour, S.E. Morrison, E.G. Klepfish, J. Kogut, M.-P. Lombardo: Nucl. Phys. Proc. Suppl. **60A**, 220 (1998).
22. G. Baym, S.A. Chin: Nucl. Phys. A **262**, 527 (1976)
23. B. Freedman, L. McLerran: Phys. Rev. D **17**, 1109 (1978)
24. A. Chodos, *et al.* : Phys. Rev. D **9**, 3471 (1974)
25. M. Dey, I. Bombaci, J. Dey, S. Ray, B.C. Samanta, Phys. Lett. B **438**, 123 (1998); erratum, Phys. Lett. B **467**, 303 (1999)
26. H. Rothe: *Lattice Gauge Theories: An Introduction.* (World Scientific, 1992).
27. I. Montvay, G. Münster: *Quantum Fields on a Lattice.* (Cambridge University Press, 1994).
28. G.N. Fowler, S. Raha, R.M. Weiner: Z. Phys. C **9**, 271 (1981)
29. P.A.M. Guichon: Phys. Lett. B **200**, 235 (1988)
30. S. Chakrabarty, S. Raha, B. Sinha: Phys. Lett. B **229**, 112 (1989)
31. S. Chakrabarty: Phys. Rev. D **43**, 627 (1991)
32. O.G. Benvenuto, G. Lugones: Phys. Rev. D **51**, 1989 (1995)
33. A. Drago, M. Fiolhais, U. Tambini: Nucl. Phys. A **588**, 801 (1995)
34. A. Drago, U. Tambini, M. Hjorth-Jensen: Phys. Lett. B **380**, 13 (1996)
35. G.X. Peng, P.Z. Ning, H.Q. Chiang: Phys. Rev. C **56**, 491 (1997)
36. P.K. Panda, A.Mishra, J.M. Eisenberg, W. Greiner: Phys. Rev. C **56**, 3134 (1997)

37. M. Buballa, M. Oertel: Phys. Lett. B **457**, 261 (1999)
38. R.C. Tolman: Proc. Nat. Acad. Sci. USA **20**, 3 (1934)
39. J. Oppenheimer, G. Volkoff: Phys. Rev. **55**, 374, (1939)
40. P. Haensel, J.L. Zdunik, R. Schaeffer: Astron. Astrophys. **160**, 121 (1986)
41. D. Bhattacharya, E. P. J. van den Heuvel: Phys. Rep. **203**, 1 (1991).
42. B. Datta, A.V. Thampan, I. Bombaci: Astron. Astrophys. **334**, 943 (1998)
43. M. Colpi, J.C. Miller: Astrophys. J. **388**, 513 (1992)
44. E. Gourgoulhon, et al.: Astron. Astrophys. **349**, 851 (1999)
45. N. Stergioulas, W. Kluźniak, T. Bulik: Astron. Astrophys. **352**, L116 (1999)
46. J.L. Zdunik, P. Haensel, D. Gondek–Rosinska, E. Gourgoulhon: Astron. Astrophys. **356**, 612 (2000)
47. X.-D. Li, I. Bombaci, M. Dey, J. Dey, E.P.J. van den Heuvel: Phys. Rev. Lett. **83**, 3776 (1999)
48. X.-D. Li, S. Ray, J. Dey, M. Dey, I. Bombaci: Astrophys. J. **527**, L51 (1999)
49. I. Bombaci, A.V. Thampan, B. Datta: Astrophys. Jour. **541**, L71, (2000)
50. A.V. Thampan, B. Datta: MNRAS **297**, 570 (1998)
51. G.B. Cook, S.L. Shapiro, and S.A. Teukolsky, Astrophys. J. **424**, 823 (1994)
52. D.C. Backer, S. Kulkarni, C. Heiles, M.M. Davis, W.M. Goss: Nature **300**, 615 (1982)
53. J. Grindlay, H. Gursky, H. Schnopper, *et al.*: Astrophys. J. **205**, L127 (1976)
54. J.A. Hoffman, H.L. Marshall, and W.H.G. Lewin: Nature **271**, 630 (1978)
55. W.H.G. Lewin, R.E. Rutledge, J.M. Kommers, J. van Paradijs, C. Kouveliotou: Astrophys. J. **462**, L39 (1996)
56. P.C. Joss: Nature **270**, 310 (1977)
57. W.H.G. Lewin, J. van Paradijs, R.E. Taam: Space Sci. Rev. **62**, 223 (1993)
58. L. Titarchuk: Astrophys. J. **429**, 340 (1994)
59. F. Haberl, L. Stella, N.E. White, W.C. Priedhorsky, M. Gottwald: Astrophys. J. **314**, 266, (1987)
60. F. Haberl, L. Titarchuk: Astron. Astrophys. **299**, 414 (1995)
61. I. Bombaci: Phys. Rev. C **55**, 1587 (1997)
62. I. Bombaci: 'An equation of state for asymmetric nuclear matter and the structure of neutron stars'. In: Perspectives on Theoretical Nuclear Physics, Proceedings of the conference Problems in Theoretical Nuclear Physics, Cortona, Italy, 12–14 October 1995, ed. by I. Bombaci et al. (ETS Pisa 1996) pp.223–237
63. M. Prakash, I. Bombaci, M. Prakash, P.J. Ellis, R. Knorren J.M. Lattimer: Phys. Rep. **280**, 1 (1997)
64. J.P. Blaizot: Phys. Rep. **64**, 171, (1980)
65. M. Farine, J.M. Pearson, F. Tondeur: Nucl. Phys. A **615**, 135, (1997)
66. W.D. Myers, W.J. Swiatecky: Nucl. Phys. A **601**, 141 (1996)
67. T. Muto, T. Tatsumi: Phys. Lett. B **238**, 165 (1992); T. Maruyama, H. Fujii, T. Muto, T. Tatsumi: Phys. Lett. B **337**, 19 (1994); T. Tatsumi: Prog. Theor. Phys. Suppl. **120**, 111 (1995)
68. G.Q. Li, R. Machleidt, R. Brockmann: Phys. Rev. C **45**, 2782, (1992)
69. J.J.M. in't Zand, et al.: Astron. Astrophys. **331**, L25 (1998)
70. R. Wijnands, M. van der Klis: Nature **394**, 344 (1998)
71. D. Chakrabarty, E.H. Morgan: Nature **394**, 346 (1998)
72. L. Burderi, A.R. King: Astrophys. J. **505**, L135 (1998)
73. D. Psaltis, D. Chakrabarty: Astrophys. J. **521**, 332 (1999)
74. X.-D. Li: Astrophys. J. **476**, 278 (1997)
75. M. Gilfanov, M. Revnivtsev, R. Sunyaev, E. Churazov: Astron. Astrophys. **338**, L83 (1998)

76. W. Cui, E.H. Morgan, L. Titarchuk: Astrophys. J. **504**, L27 (1998)
77. M. van der Klis: Ann. Rev. Astr. Astrophys. **38**, 717 (2000). [astro-ph/0001167]
78. P. Kaaret, E.C. Ford: Science **276**, 1386 (1997)
79. M. Méndez, M. van der Klis: Astrophys. J. **517**, L51 (1999)
80. E. Ford, M. van der Klis: Astrophys. J. **506**, L39 (1998)
81. L. Titarchuk, V. Osherovich: Astrophys. J. **518**, L95 (1999); V. Osherovich, L. Titarchuk: Astrophys. J. **522**, L113 (1999)
82. T.E.Strohmayer, J.H. Swank, W. Zhang: Astrophys. J. **503**, L147 (1998)
83. S.R. Kulkarni, et al.: Nature **393**, 35 (1998)
84. S.R. Kulkarni, et al.: Nature **398**, 389 (1999)
85. A. Dar: Astrophys. J. **500**, L93 (1998)
86. R. Sari, T. Piran: Astrophys. J. **485**, 270 (1997); R. Sari, T. Piran, J.P. Halperen: astro-ph/9903339.
87. B. Paczynski: Astrophys. J. **494**, L45 (1998)
88. H.-Th. Janka, M. Ruffert: Astron. Astrophys. **307**, L33 (1996)
89. S.E. Woosley: Astrophys. J. **405**, 273 (1993)
90. A. Olinto: Phys. Lett. B **192**, 71 (1987)
91. K.S.Cheng, Z.G. Dai: Phys. Rev. Lett. **77**, 1210 (1996)
92. F. Ma, B. Xie: Astrophys. J. **462**, L63 (1996)
93. I. Bombaci, and B. Datta, Astrophys. J. 530, L69 (2000).
94. J.E. Horvath, O.G. Benvenuto: Phys. Lett. B **213**, 516 (1988)
95. M. Baldo, I. Bombaci, G.F. Burgio: Astron. Astrophys. **328**, 274 (1997)
96. N.K. Glendenning: Astrophys. J. **293**, 470 (1985)
97. J. Schaffner, I.N. Mishustin: Phys. Rev. C **53**, 1416 (1996)
98. M. Baldo, G.F. Burgio, H.-J. Schulze: Phys. Rev. C **58**, 3688 (1998)
99. P.B. Price et al.: Phys. Rev. Lett. **35**, 487 (1975)
100. T. Saito, Y. Hatano, Y. Fukuda, H. Oda: Phys. Rev. Lett. **65**, 2094 (1990)
101. M. Kasuya et al.: Phys. Rev. D **47**, 2153 (1993)
102. M. Ichimura et al.: Il Nuovo Cim. A **106**, 843 (1993)
103. J.N. Capdevielle: Il Nuovo Cim. **19C**, 623 (1996)
104. H. Terazawa: J. Phys. Soc. Jpn. **62**, 1415 (1993)
105. H. Terazawa: 'Are color-balled nuclei found in cosmic rays?', KEK preprint 99-146 (IPNS, HEARO, November, 1999)
106. J. Barrette et al.: Phys. Lett. B **252**, 550 (1990)
107. A. Aoki et al.: Phys. Rev. Lett. **69**, 2345 (1992)
108. D. Beavis et al.: Phys. Rev. Lett. **75**, 3078 (1995); A. Rusek et al.: Phys. Rev. C **54**, R15 (1996)
109. T.A. Armstrong et al.: nucl-ex/0010017
110. K. Borer et al.: Phys. Rev. Lett. **72**, 1415 (1994)
111. A. Aoki et al.: Phys. Rev. Lett. **76**, 3907 (1996)
112. A.J. Baltz, B.C. Dover, S.H. Kahana, Y. Pang, T.J. Schlagel, E. Schnedermann: Phys. Lett. B **325**, 7 (1994)
113. C. Greiner, P. Koch, H. Stocker: Phys. Rev. Lett. **58**, 1825 (1987); C. Greiner, H. Stocker: Phys. Rev. D **44**, 3517 (1991)
114. J. Kapusta, A.P. Vischer, R. Venugopalan: Phys. Rev. C **51**, 901 (1995); J. Kapusta, A.P. Vischer: Phys. Rev. C **52**, 2725 (1995)
115. A.L.S.Angelis et al.: Nucl. Phys. B (Proc. Suppl.) **75**, 203 (1999)
116. A. Dar, A. De Rújula, U. Heinz: Phys. Lett. B **470**, 142 (1999); R.L. Jaffe, W. Busza, J. Sandweiss, F. Wilczek: Rev. Mod. Phys. (in press)
117. J. Madsen: Phys. Rev. Lett. **85**, 4687 (2000)
118. K.S. Cheng, Z.G. Dai, D.M. Wai, T. Lu: Science **280**, 407 (1998)

# Phase Diagram for Spinning and Accreting Neutron Stars

David Blaschke[1], Hovik Grigorian[2], and Gevorg Poghosyan[2]

[1] Department of Physics, University of Rostock, D-18051 Rostock, Germany
[2] Department of Physics, Yerevan State University,
Alex Manoogian 1, 375025 Yerevan, Armenia

**Abstract.** Neutron star configurations are considered as thermodynamical systems for which a phase diagram in the angular velocity ($\Omega$) - baryon number ($N$) plane is obtained with a dividing line $N_{\rm crit}(\Omega)$ for quark core configurations. Trajectories of neutron star evolution in this diagram are studied for different scenarios defined by the external torque acting on the star due to radiation and/or mass accretion. They show a characteristic change in the rotational kinematics when the star enters the quark core regime. For isolated pulsars the braking index signal for deconfinement has been studied in its dependence on the mass of the star. Model calculations of the spin evolution of accreting low-mass X-ray binaries in the phase diagram have been performed for different values of the initial magnetic field, its decay time as well as initial mass and mass accretion rate. Population clustering of these objects at the line $N_{\rm crit}(\Omega)$ in the phase diagram is suggested as an observable signal for the deconfinement phase transition if it exists for spinnning and accreting neutron stars.

## 1 Neutron stars as thermodynamical systems

Quantum Chromodynamics (QCD) as the fundamental theory for strongly interacting matter predicts a deconfined state of quarks and gluons under conditions of sufficiently high temperatures and/or densities which occur, e.g., in heavy-ion collisions, a few microseconds after the Big Bang or in the cores of pulsars. The unambiguous detection of the phase transition from hadronic to quark matter (or vice-versa) has been a challenge to particle and astrophysics over the past two decades [1,2]. While the diagnostics of a phase transition in experiments with heavy-ion beams faces the problems of strong nonequilibrium and finite size, the dense matter in a compact star forms a macroscopic system in thermal and chemical equilibrium for which signals of a phase transition shall be more pronounced.

Such signals have been suggested in the form of characteristic changes of observables such as the surface temperature [3], brightness [4], pulse timing [5] and rotational mode instabilities [6] during the evolution of the compact object. In particular the pulse timing signal has attracted much interest since it is due to changes in the kinematics of rotation. Thus it could be used not only to detect the occurrence but also to determine the size of the quark core from the magnitude of the braking index deviation from the magnetic dipole value [7]. Besides of the isolated pulsars, one can consider also the accreting compact stars in low-mass X-ray binaries (LMXBs) as objects from which we can expect signals of a deconfinement transition in their interior [7–9]. The observation of quasiperiodic brightness oscillations (QPOs) [10] for some LMXBs has lead to very stringent constraints for masses and radii [11] which according to [12,13] could even favour strange quark matter interiors over hadronic ones for these objects. Due to the mass accretion flow these systems are candidates for the formation of the most massive compact stars from which we expect to observe signals of the transition to either quark core stars, to a third family of stars [14] or to black holes. Each compact star configuratrion can be identified with a thermodynamical system characterized by the total baryon number (or mass), temperature, spin frequency and magnetic field as thermodynamical variables.

Since the evolutionary processes for the compact objects accompanying the structural changes are slow enough we will consider here the case of rigid rotation only and restrict ourselves to the degenerate systems at $T = 0$. The magnetic and thermal evolution of the neutron stars we will consider as decoupled from the mechanical evolution.

In this approximation we can introduce a classification of isolated and accreting compact stars in the plane of their angular frequency $\Omega$ and mass (baryon number $N$) which we will call *phase diagram*, see Fig. 1.

Each point in this phase diagram corresponds to a mechanical state of a neutron star. Mechanical equilibrium of thermal pressure with gravitational and centrifugal forces leads to a stationary distribution of matter inside the configuration. This distribution is determined by the central density and rotational frequency. Thermal and chemical equilibrium are described by an equation of state which determines the structure and composition of the compact star configuration.

The requirements of stability restrict the region of stability of the quasistationary rotating star configurations. From the right hand side of the phase diagram there is a line which separates a region of stars which are collapsed to black holes (BH) from the stable ones. At high baryon numbers beyond this line gravitational forces dominate over pressure and centrifugal forces of matter.

From the top of the diagram the region of stable star configurations is separated by the Keplerian frequencies $\Omega_K(N)$ from that where mass shedding under the centrifugal forces does not allow stationary rotating objects.

Inside the region of stationary rotators one can distinguish two types of stationary rotating stars: those with quark matter cores (QCSs) and hadronic stars (HS) [15]. The region of the QCSs is expected to be located at the bottom right of the phase diagram in the region of the most massive and slowly rotating compact stars. The critical line $N_{\mathrm{crit}}(\Omega)$ which separates the QCS region from that of HSs is correlated with the local maxima of the moment of inertia with respect to changes of the baryon number at given angular velocity $\Omega$ due to the change of the internal structure of the compact object at the deconfinement transition .

Using the analogy with the phase diagram of a conventional thermodynamical system we can consider trajectories in this diagram as quasistationary evolutionary processes.

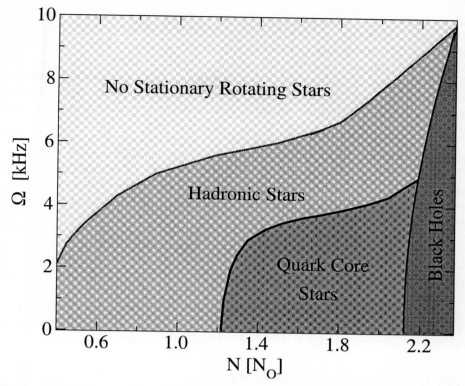

**Fig. 1.** Sketch of a phase diagram for spinning neutron stars with deconfinement phase transition.

Since these processes for neutron stars have characteristic time scales ($\approx 10^6 \div 10^8$ yr) much longer than those of typical observers it is almost impossible to trace these trajectories directly. A possible strategy for measuring evolutionary tracks will be to perform a statistics of the population of the different regions in the phase diagram.

Our aim is to investigate the conditions under which the passage of the phase border leads to observable significant clustering in the populations.

We will provide criteria under which a particular astrophysical scenario with spin evolution could be qualified to signal a deconfinement transition.

## 2 Phase diagram for a stationary rotating star model

### 2.1 Equation of state with deconfinement transition

Since our focus is on the elucidation of qualitative features of signals from the deconfinement transition in the pulsar timing we will use a generic form of an equation of state (EoS) with such a transition [7] which is not excluded by the mass and radius constraints derived from QPOs. In our case as well as in most of the approaches to quark deconfinement in neutron star matter a standard two-phase description of the equation of state is applied where the hadronic phase and the quark matter phase are modelled separately. The ambiguity in the choice of the bag constant for the description of the quark matter phase can be removed by a derivation of this quantity [16,17] from a dynamical confining approach [18]. The resulting EoS is obtained by imposing Gibbs' conditions for phase equilibrium with the constraints of globally conserved baryon number and electric charge [19–21].

Since nuclear forces are not fundamental, our knowledge about the equation of state for nuclear and neutron star matter at high densities is not very precise. There is not yet a description of the equation of state for strongly interacting matter on the fundamental level in terms of quark and gluon degrees of freedom where nucleons and mesons appear as composite structures. It is one of the goals of nuclear astrophysics and of neutron star physics to use pulsars and other compact objects as laboratories for studies of the nuclear forces and the phases of nuclear matter [20,21]. The existence of several models for the equation of state for dense stellar matter allows a variability in the phase structure of neutron stars [22,23], therefore the phase diagram (Fig. 1) remains robust only qualitativly.

For the detailed introduction of the phase diagram and a quantitative analysis using thermodynamical methods, we will employ a particular EoS model [7] which is characterized by a relatively hard hadronic matter part.

## 2.2 Configurations of rotating stars

In our model calculations we assume quasistationary evolution with negligible convection and without differential rotation which is justified when both the mass load onto the star and the transfer of the angular momentum are sufficiently slow processes.

For our treatment of rotation within general relativity we employ a perturbation expansion with respect to the ratio of the rotational and gravitational energies for the homogeneous Newtonian spherical rotator with the mass density $\rho(0)$ equal to the central density, $E_{\rm rot}/E_{\rm grav} = (\Omega/\bar\Omega)^2$, where $\bar\Omega^2 = 4\pi G\rho(0)$. This ratio is a small parameter, less than one up to the mass shedding limit [7].

The general form of the expansion allows us to describe the metric coefficients and the distributions of pressure, energy density and hydrodynamical enthalpy in the following form

$$X(r,\theta;\Omega) = X^{(0)}(r) + (\Omega/\bar\Omega)^2 X^{(2)}(r,\theta) + O(\Omega^4) , \qquad (1)$$

where $X$ stands for one of the above mentioned quantities [7]. The series expansion allows one to transform the Einstein equations into a coupled set of differential equations for the coefficient functions defined in (1), which can be solved by recursion. The static solutions, obeying the Oppenheimer-Volkoff equations, are contained in this expansion for the case $\Omega = 0$ when only the functions with superscript (0) remain. The other terms are corrections due to the rotation. We truncate higher order terms $\sim O(\Omega^4)$ in this expansion and neglect the change of the frame dragging frequency, which appears at $O(\Omega^3)$. For a more detailed description of the method and analytic results in the integral representation of the moment of inertia we refer to [7] and to works of Hartle and Thorne[24,25], Sedrakian and Chubarian [26,27].

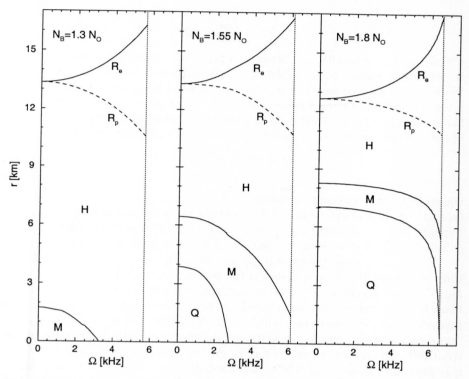

**Fig. 2.** Phase structure of rotating hybrid stars in equatorial direction in dependence of the angular velocity $\Omega$ for stars with different total baryon number: $N_B/N_\odot = 1.3, 1.55, 1.8$.

In Fig. 2 we show the critical regions of the phase transition in the inner structure of the star configuration as well as the equatorial and polar radii in the plane of angular velocity $\Omega$ versus distance from the center of the star . It is obvious that with the increase of the angular velocity the star is deformating its shape. The maximal excentricities of the configurations with $N_B = 1.3\ N_\odot$, $N_B = 1.55\ N_\odot$ and $N_B = 1.8\ N_\odot$ are $\epsilon(\Omega_{\max}) = 0.7603$, $\epsilon(\Omega_{\max}) = 0.7655$ and $\epsilon(\Omega_{\max}) = 0.7659$, respectively. Due to the changes of the central density the quark core could disappear above a critical angular velocity.

It is the aim of the present paper to investigate the conditions for a verification of the existence of the critical line $N_{\mathrm{crit}}(\Omega)$ by observation. We will show evidence that in principle such a measurement is possible since this deconfinement transition line corresponds to a maximum of the moment of inertia, which is the key quantity for the rotational behavior of compact objects, see Fig. 3.

In the case of rigid rotation the moment of inertia is defined by

$$I(\Omega, N) = J(\Omega, N)/\Omega ,\qquad(2)$$

where the angular momentum $J(\Omega, N)$ of the star can be expressed in invariant form as

$$J(\Omega, N) = \int T^t_\phi \sqrt{-g}dV ,\qquad(3)$$

with $T^t_\phi$ being the nondiagonal element of the energy momentum tensor, $\sqrt{-g}dV$ the invariant volume and $g = \det ||g_{\mu\nu}||$ the determinant of the metric tensor. We assume that the superdense compact object rotates stationary as a rigid body, so that for a given time-interval both the angular velocity as well as the baryon number can be considered as global parameters of the theory. The result of our calculations for the moment of inertia (2) can be cast into the form

$$I = I^{(0)} + \sum_\alpha \Delta I_\alpha,\qquad(4)$$

where $I^{(0)}$ is the moment of inertia of the static configuration with the same central density and $\Delta I_\alpha$ stands for contributions to the moment of inertia from different rotational effects which are labeled by $\alpha$: matter redistribution, shape deformation, and changes in the centrifugal forces and the gravitational field [7].

In Fig. 3 we show the resulting phase diagram for compact star configurations which exhibits four regions: (i) the region above the maximum frequency $\Omega_K(N)$ where no stationary rotating configurations are found, (ii) the region of black holes for baryon numbers exceeding the maximum value $N_{\max}(\Omega)$, and the region of stable compact stars which is subdivided by the critical line $N_{\mathrm{crit}}(\Omega)$ into a region of (iii) quark core stars and another one of (iv) hadronic stars, respectively. The numerical values for the critical lines are model dependent. For this particular model EoS due to the hardness of the hadronic branch (linear Walecka model [20]) there is a maximum value of the baryonic mass on the critical line $N_{\mathrm{crit}}(\Omega_k) = 1.8 N_\odot$, such that for stars more massive than that one all stable rotating configurations have to have a quark core. This property can be seen from the dependence of the phase structure of the star on angular velocity in Fig. 2. For the whole interval of possible frequencies in the case of $N = 1.8 N_\odot$ the quark core radius remains approximately unchanged: $R_{core} \sim 7$ km.

## 3  Evolution scenarios with phase transitions

We want to explain why the occurence of a sharply peaked maximum for the moment of inertia in the $\Omega - N$ plane entails observational consequences for the angular velocity evolution of rotating compact objects. The basic formula which governs the rotational dynamics is

$$\dot{\Omega} = \frac{K(N, \Omega)}{I(N, \Omega) + \Omega \left(\partial I(N, \Omega)/\partial \Omega\right)_N} ,\qquad(5)$$

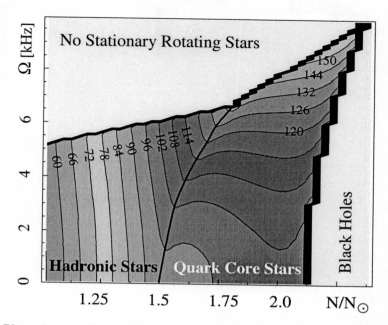

**Fig. 3.** Phase diagram for configurations of rotating compact objects in the plane of angular velocity $\Omega$ and mass (baryon number $N$). Contour lines show the values of the moment of inertia in $M_\odot \text{km}^2$. The line $N_{\text{crit}}(\Omega)$ which separates hadronic from quark core stars corresponds the set of configurations with a central density equal to the critical density for the occurence of a pure quark matter phase.

where $K = K_{\text{int}} + K_{\text{ext}}$ is the net torque acting on the star due to internal and external forces. The internal torque is given by $K_{\text{int}}(N,\Omega) = -\Omega \dot{N} \left(\partial I(N,\Omega)/\partial N\right)_\Omega$, the external one can be subdivided into an accretion and a radiation term $K_{\text{ext}} = K_{\text{acc}} + K_{\text{rad}}$. The first one is due to all processes which change the baryon number, $K_{\text{acc}} = \dot{N} \, dJ/dN$ and the second one contains all processes which do not. For the example of magnetic dipole and/or gravitational wave radiation it can be described by a power law $K_{\text{rad}} = \beta \Omega^n$, see [28,29].

### 3.1 Spin-down scenario for isolated pulsars

The simple case of the spindown evolution of isolated (non-accreting, $\dot{N} = 0$) pulsars due to magnetic dipole radiation would be described by vertical lines in Fig. 1. These objects can undergo a deconfinement transition if the baryon number lies within the interval $N_{\min} < N < N_{\max}$, where for our model EoS the endpoints of $N_{\text{crit}}(\Omega)$ are $N_{\min} = 1.49 \, N_\odot$ and $N_{\max} = 1.78 \, N_\odot$. As it has been shown in [7], the braking index $n(\Omega)$ changes its value from $n(\Omega) > 3$ in the region (iii) to $n(\Omega) < 3$ in (iv). This is the braking index signal for a deconfinement transition introduced by [5].

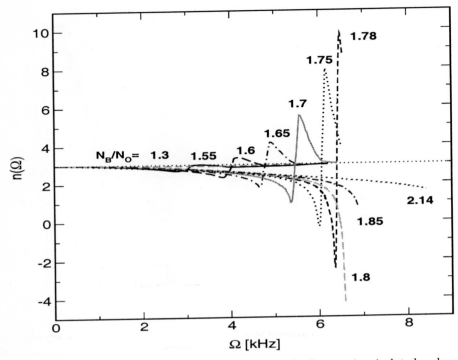

**Fig. 4.** Braking index due to dipole radiation from fastly rotating isolated pulsars as a function of the angular velocity. The minima of $n(\Omega)$ indicate the appearance/disappearance of quark matter cores.

In Fig. 4 we display the result for the braking index $n(\Omega)$ for a set of configurations with fixed total baryon numbers ranging from $N_B = 1.55\, N_\odot$ up to $N_B = 1.9\, N_\odot$, the region where during the spin-down evolution a quark matter core could occur for our model EOS. We observe that only for configurations within the interval of total baryon numbers $1.4 \leq N_B/N_\odot \leq 1.9$ a quark matter core occurs during the spin-down as a consequence of the increasing central density, see also Fig. 2, and the braking index shows variations. The critical angular velocity $\Omega_{\rm crit}(N_B)$ for the appearance of a quark matter core can be found from the minimum of the braking index $n(\Omega)$. As can be seen from Fig. 4, all configurations with a quark matter core have braking indices $n(\Omega) < 3$ and braking indices significantly larger than 3 can be considered as precursors of the deconfinement transition. The magnitude of the jump in $n(\Omega)$ during the transition to the quark core regime is a measure for the size of the quark core. It would be even sufficient to observe the maximum of the braking index $n_{\rm max}$ in order to infer not only the onset of deconfinement ($\Omega_{\rm max}$) but also the size of the quark core to be developed during further spin-down from the maximum deviation $\delta n = n_{\rm max} - 3$ of the braking index. For the model EOS we used a significant enhancement of the braking index does only occur for pulsars with periods $P < 1.5$ ms (corresponding to $\Omega > 4$ kHz) which have not yet been observed in nature. Thus the signal seems to be a weak one for most of the possible candidate pulsars. However, this statement is model dependent since, e.g., for the model EOS used by [5], which includes the strangeness degree of freedom, a more dramatic signal at lower spin frequencies has been reported. Therefore, a more complete investigation of the braking index for a set of realistic EOS should be performed.

### 3.2 Scenarios with mass accretion

All other possible trajectories correspond to processes with variable baryon number (accretion). In the phase of hadronic stars, $\dot{\Omega}$ first decreases as long as the moment of inertia monotonously increases with $N$. When passing the critical line $N_{\mathrm{crit}}(\Omega)$ for the deconfinement transition, the moment of inertia starts decreasing and the internal torque term $K_{\mathrm{int}}$ changes sign. This leads to a narrow dip for $\dot{\Omega}(N)$ in the vicinity of this line. As a result, the phase diagram gets populated for $N \lesssim N_{\mathrm{crit}}(\Omega)$ and depopulated for $N \gtrsim N_{\mathrm{crit}}(\Omega)$ up to the second maximum of $I(N, \Omega)$ close to the black-hole line $N_{\mathrm{max}}(\Omega)$. The resulting population clustering of compact stars at the deconfinement transition line is suggested to emerge as a signal for the occurence of stars with quark matter cores. In contrast to this scenario, in the case without a deconfinement transition, the moment of inertia could at best saturate before the transition to the black hole region and consequently $\dot{\Omega}$ would also saturate. This would entail a smooth population of the phase diagram without a pronounced structure.

The clearest scenario could be the evolution along lines of constant $\Omega$ in the phase diagram. These trajectories are associated with processes where the external and internal torques are balanced. A situation like this has been described, e.g. by [30] for accreting binaries emitting gravitational waves.

In the following we would like to explore which influence the magnitude of the external torque $K_{\mathrm{ext}}$ has on the pronouncedness of the quark matter signal. In Fig. 5 we show evolutionary tracks (dotted) of configurations in the phase diagram of Fig. 1 for different parameter values of the accretion torque and different initial values $J_0$ of the angular momentum.

As we have discussed above, the narrow dip for $\dot{\Omega}$ as a quark core signal occurs when configurations cross the critical line during a spin-down phase. We can quantify this criterion by introducing a minimal frequency $\Omega_{\mathrm{min}}$ above which spin-down occurs. It can be found as a solution of the equation for the torque balance at the phase border

$$dJ/dN = K_{\mathrm{int}}(N_{\mathrm{crit}}(\Omega_{\mathrm{min}}), \Omega_{\mathrm{min}})/\dot{N} \, . \tag{6}$$

The dependence of $\Omega_{\mathrm{min}}$ on $dJ/dN$ shown in Fig. 6 can be used to sample accreting compact objects from the region in which the suggested quark matter signal should be most pronounced before making a population statistics.

The ideal candidates for such a search program are LMXBs for which the discovery of strong and remarkably coherent high-frequency QPOs with the Rossi

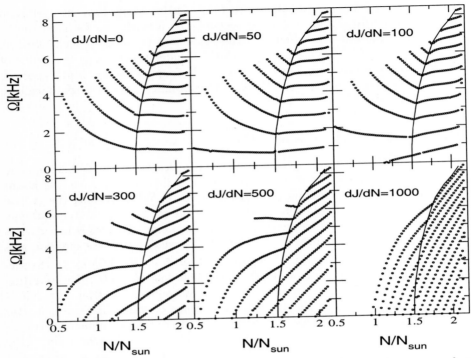

**Fig. 5.** Phase diagram for compact stars in the angular velocity - baryon number plane with a dividing line for quark core configurations. Trajectories of spin evolution are given for different parameter values of the accretion torque $dJ/dN$ in units of $[M_\odot$ km$^2$ kHz/$N_\odot]$ and different initial values $J_0$ of the angular momentum. The curves in the upper left panel correspond to $J_0[M_\odot$ km$^2$ kHz$] = 300, 400, \ldots, 1400$ from bottom to top.

X-ray Timing Explorer has provided new information about the masses and rotation frequencies of the central compact object [10,11]. As a strategy for the quark matter search in compact stars one should perform a population statistics among those LMXBs exhibiting the QPO phenomenon which have a small $dJ/dN$ and a sufficiently large angular velocity, see Fig. 6. If, e.g., the recently discussed period clustering for Atoll- and Z-sources [8,30] will correspond to objects in a narrow region of masses, this could be interpreted as a signal for the deconfinement transition to be associated with a fragment of the critical line in the phase diagram for rotating compact stars [9].

## 4 Signal for deconfinement in LMXBs

### 4.1 Spin-up trajectories for accretor

We consider the spin evolution of a compact star under mass accretion from a low-mass companion star as a sequence of stationary states of configurations

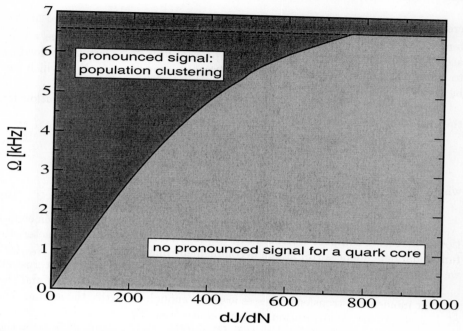

**Fig. 6.** Dividing line $\Omega_{\min}$ for the deconfinement signal in binary systems with spin-up. At a given rate of change of the total angular momentum $dJ/dN$ in units of [M$_\odot$ km$^2$ kHz/N$_\odot$] all configurations with spin frequency $\Omega_{\min} < \Omega < \Omega_{\max}$ have quark matter cores and the population clustering signal is most pronounced. The dashed line is the maximum spin frequency $\Omega_{\max} \leq \Omega_K(N)$ for which quark core configurations exist.

(points) in the phase diagram spanned by $\Omega$ and $N$. The process is governed by the change in angular momentum of the star

$$\frac{d}{dt}(I(N,\Omega)\,\Omega) = K_{\text{ext}}, \tag{7}$$

where

$$K_{\text{ext}} = \sqrt{GM\dot{M}^2 r_0} - N_{\text{out}} \tag{8}$$

is the external torque due to both the specific angular momentum transferred by the accreting plasma and the magnetic plus viscous stress given by $N_{\text{out}} = \kappa \mu^2 r_c^{-3}$, $\kappa = 1/3$ [31]. For a star with radius $R$ and magnetic field strength $B$, the magnetic moment is given by $\mu = R^3 B$. The co-rotating radius $r_c = (GM/\Omega^2)^{1/3}$ is very large ($r_c \gg r_0$) for slow rotators. The inner radius of the accretion disc is

$$r_0 \approx \begin{cases} R, & \mu < \mu_c \\ 0.52\, r_A, & \mu \geq \mu_c \end{cases}$$

where $\mu_c$ is that value of the magnetic moment of the star for which the disc would touch the star surface. The characteristic Alfvén radius for spherical accretion with the rate $\dot{M} = m\dot{N}$ is $r_A = \left(2\mu^{-4}GM\dot{M}^2\right)^{-1/7}$. Since we are interested in the case of fast rotation for which the spin-up torque due to the accreting plasma in Eq. (8) is partly compensated by $N_{\text{out}}$, eventually leading to a saturation of the spin-up, we neglect the spin-up torque in $N_{\text{out}}$ which can be important only for slow rotators [32].

From Eqs. (7), (8) one can obtain the first order differential equation for the evolution of angular velocity

$$\frac{d\Omega}{dt} = \frac{K_{\text{ext}}(N,\Omega) - K_{\text{int}}(N,\Omega)}{I(N,\Omega) + \Omega(\partial I(N,\Omega)/\partial\Omega)_N}, \qquad (9)$$

where

$$K_{\text{int}}(N,\Omega) = \Omega\dot{N}(\partial I(N,\Omega)/\partial N)_\Omega . \qquad (10)$$

Solutions of (9) are trajectories in the $\Omega - N$ plane describing the spin evolution of accreting compact stars, see Fig. 7. Since $I(N,\Omega)$ exhibits characteristic functional dependences [15] at the deconfinement phase transition line $N_{\text{crit}}(\Omega)$ we expect observable consequences in the $\dot{P} - P$ plane when this line is crossed.

In our model calculations we assume that both the mass accretion and the angular momentum transfer processes are slow enough to justify the assumption of quasistationary rigid rotation without convection. The moment of inertia of the rotating star can be defined as $I(N,\Omega) = J(N,\Omega)/\Omega$, where $J(N,\Omega)$ is the angular momentum of the star. For a more detailed description of the method and analytic results we refer to [7] and the works of [24,25], as well as [26,27].

The time dependence of the baryon number for the constant accreting rate $\dot{N}$ is given by

$$N(t) = N(t_0) + (t - t_0)\dot{N} . \qquad (11)$$

For the magnetic field of the accretors we consider the exponential decay [33]

$$B(t) = [B(0) - B_\infty]\exp(-t/\tau_B) + B_\infty . \qquad (12)$$

We solve the equation for the spin-up evolution (9) of the accreting star for decay times $10^7 \leq \tau_B[\text{yr}] \leq 10^9$ and initial magnetic fields in the range $0.2 \leq B(0)[\text{TG}] \leq 4.0$. The remnant magnetic field is chosen to be $B_\infty = 10^{-4}\text{TG}^1$ [34].

At high rotation frequency, both the angular momentum transfer from accreting matter and the influence of magnetic fields can be small enough to let the evolution of angular velocity be determined by the dependence of the moment of inertia on the baryon number, i.e. on the total mass. This case is similar to the one with negligible magnetic field considered in [7,29,35,36] where $\mu \leq \mu_c$ in Eq. (9), so that only the so called internal torque term (10) remains.

In Fig. 7 we show evolutionary tracks of accretors in phase diagrams (left panels) and show the corresponding spin evolution $\Omega(t)$ (right panels). In the lower panels, the paths for possible spin-up evolution are shown for accretor models initially having a quark matter core ($N(0) = 1.55\ N_\odot$, $\Omega(0) = 1$ Hz).

---

[1] 1 TG= $10^{12}$ G

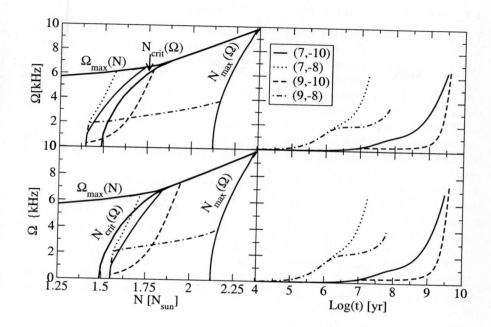

**Fig. 7.** Spin evolution of an accreting compact star for different decay times of the magnetic field and different accretion rates. Upper panels: initial configuration with $N(0) = 1.4\,N_\odot$; Lower panels: $N(0) = 1.55\,N_\odot$; $\Omega(0) = 1$ Hz in both cases. The numbers in the legend box stand for $(\log(\tau_B[\mathrm{yr}]), \log(\dot{N}[N_\odot/\mathrm{yr}]))$. For instance (9,-8) denotes $\tau_B = 10^9$ yr and $\dot{N} = 10^{-8} N_\odot/$ yr.

The upper panels show evolution of a hybrid star without a quark matter core in the initial state ($N(0) = 1.4\,N_\odot$, $\Omega(0) = 1$ Hz), containing quarks only in mixed phase. We assume a value of $\dot{N}$ corresponding to observations made on LMXBs, which are divided into Z sources with $\dot{N} \sim 10^{-8} N_\odot/\mathrm{yr}$ and A(toll) sources accreting at rates $\dot{N} \sim 10^{-10} N_\odot/\mathrm{yr}$ [10,33,37].

For the case of a small magnetic field decay time $\tau_B = 10^7$ yr (solid and dotted lines in Fig.7) the spin-up evolution of the star cannot be stopped by the magnetic braking term so that the maximal frequency consistent $\Omega_{\max}(N)$ with stationary rotation can be reached regardless whether the star did initially have a pure quark matter core or not.

For long lived magnetic fields ($\tau_B = 10^9$ yr, dashed and dot-dashed lines in Fig. 7) the spin-up evolution deviates from the monotonous behaviour of the previous case and shows a tendency to saturate. At a high accretion rate (dot-dashed lines) the mass load onto the star can be sufficient to transform it to a black hole before the maximum frequency could be reached whereas at low accretion (dashed lines) the star spins up to the Kepler frequency limit.

## 4.2 Waiting time and population clustering

The question arises whether there is any characteristic feature in the spin evolution which distinguishes trajectories that traverse the critical phase transition line from those remaining within the initial phase.

For an accretion rate as high as $\dot{N} = 10^{-8}\ N_\odot/$ yr the evolution of the spin frequency in Fig. 7 shows a plateau where the angular velocity remains within the narrow region between $2.1 \leq \Omega[\text{kHz}] \leq 2.3$ for the decay time $\tau_B = 10^9$ yr and between $0.4 \leq \Omega[\text{kHz}] \leq 0.5$ when $\tau_B = 10^7$ yr. This plateau occurs for stars evolving into the QCS region (upper panels) as well as for stars remaining within the QCS region (lower panels). This saturation of spin frequencies is mainly related to the compensation of spin-up and spin-down torques at a level determined by the strength of the magnetic field. In order to perform a more quantitative discussion of possible signals of the deconfinement phase transition we present in Fig. 8 trajectories of the spin-up evolution in the $\dot{P} - P$ plane for stars with $N(0) = 1.4\ N_\odot$ and $\Omega(0) = 1$ Hz in the initial state; the four sets of accretion rates and magnetic field decay times coincide with those in Fig. 7.

When we compare the results for the above hybrid star model (solid lines) with those of a hadronic star model (quark matter part of the hybrid model omitted; dotted lines) we observe that only in the case of high accretion rate ($\dot{N} = 10^{-8} N_\odot/$yr, e.g. for Z sources) and long-lived magnetic field ($\tau_B = 10^9$yr) there is a significant difference in the behaviour of the period derivatives. The evolution of a star with deconfinement phase transition shows a dip in the period derivative in a narrow region of spin periods. This feature corresponds to a plateau in the spin evolution which can be quantified by the *waiting time* $\tau = \left|P/\dot{P}\right| = \Omega/\dot{\Omega}$. In Fig. 9 (lower and middle panels) we present this waiting time in dependence on the rotation frequencies $\nu = \Omega/(2\pi)$ for the relevant case labeled (9,-8) in Figs. 7,8. The comparison of the trajectory for a hybrid star surviving the phase trasition during the evolution (solid line) with those of a star evolving within the hadronic and the QCS domains (dotted line and dashed lines, respectively), demonstrates that an enhancement of the waiting time in a narrow region of frequencies is a characteristic indicator for a deconfinement transition in the accreting compact star.

The position of this peak in the waiting time depends on the initial value of magnetic field, see Fig. 9. In the middle and lower panels of that Figure, we show the waiting time distribution for $B(0) = 0.75$ TG and $B(0) = 0.82$ TG, respectively. Maxima of the waiting time in a certain frequency region have the consequence that the probability to observe objects there is increased (*population clustering*). In the upper panel of this Figure the spin frequencies for observed Z sources in LMXBs with QPOs [10] are shown for comparison. In order to interpret the clustering of objects in the frequency interval $225 \leq \nu[\text{Hz}] \leq 375$ as a phenomenon related to the increase in the waiting time, we have to chose initial magnetic field values in the range $1.0 \geq B(0)[\text{TG}] \geq 0.6$ for the scenario labeled (9,-8), see also the dashed lines in Fig.8.

The results of the previous section show that the waiting time for accreting stars along their evolution trajectory is larger in a hadronic configuration than in

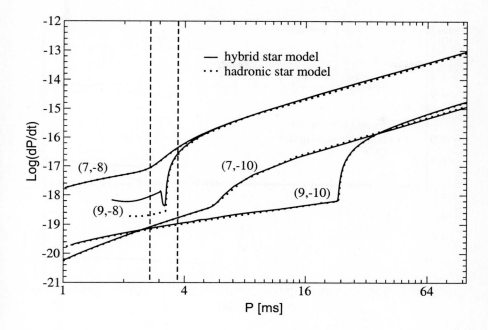

**Fig. 8.** Spin-up evolution of accreting compact stars in the $\dot{P} - P$ diagram. For labels and initial values see to Fig. 7. The region of the vertical dashed lines corresponds to the clustering of periods observed for LMXBs with QPOs. A dip (waiting point) occurs at the deconfinement transition for parameters which correspond to Z sources ($\dot{N} = 10^{-8} N_\odot/\text{yr}$) with slow magnetic field decay ($\tau_B = 10^9 \text{yr}$).

a QCS, after a time scale when the mass load onto the star becomes significant. This suggests that if a hadronic star enters the QCS region, its spin evolution gets enhanced thus depopulating the higher frequency branch of its trajectory in the $\Omega - N$ plane.

In Fig. 10 we show contours of waiting time regions in the phase diagram. The initial baryon number is $N(0) = 1.4 N_\odot$ and the initial magnetic field is taken from the interval $0.2 \leq B(0)[\text{TG}] \leq 4.0$.

The region of longest waiting times is located in a narrow branch around the phase transition border and does not depend on the evolution scenario after the passage of the border, when the depopulation occurs and the probability to find an accreting compact star is reduced. Another smaller increase of the waiting time and thus a population clustering could occur in a region where the accretor is already a QCS. For an estimate of a population statistics we show the region of evolutionary tracks when the values of initial magnetic field are within $0.6 \leq B(0)[\text{TG}] \leq 1.0$ as suggested by the observation of frequency clustering in the narrow interval $375 \geq \nu[\text{Hz}] \geq 225$, see Fig. 9.

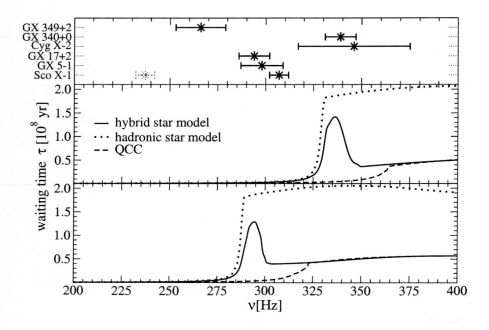

**Fig. 9.** *Upper panel:* frequency interval for observed Z source LMXBs [10]; *middle panel:* waiting times $\tau = P/\dot{P}$ for scenario (9,-8) and initial magnetic field $B(0) = 0.75$ TG; *lower panel:* same as middle panel for $B(0) = 0.82$ TG. Spin evolution of a hybrid stars (solid lines) shows a peak in the waiting time characteristic for the deconfinement transition. Hadronic stars (dotted lines) and QCSs (dashed lines) have no such structure.

As a strategy of search for QCSs we suggest to select from the LMXBs exhibiting the QPO phenomenon those accreting close to the Eddington limit [33] and to determine simultaneously the spin frequency and the mass [38] for sufficiently many of these objects. The emerging statistics of accreting compact stars should then exhibit the population clustering shown in Fig. 10 when a deconfinement transition is possible. If a structureless distribution of objects in the $\Omega - N$ plane will be observed, then no firm conclusion about quark core formation in compact stars can be made.

For the model equation of state on which the results of our present work are based, we expect a baryon number clustering rather than a frequency clustering to be a signal of the deconfinement transition in the compact stars of LMXBs. The model independent result of our study is that a population clustering in the phase diagram for accreting compact stars shall measure the critical line $N_{\rm crit}(\Omega)$ which separates hadronic stars from QCSs where the shape of this curve can discriminate between different models of the nuclear EoS at high densities.

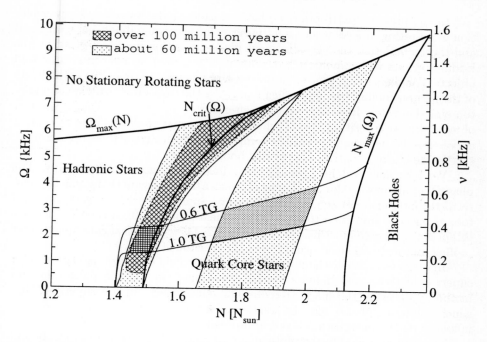

**Fig. 10.** Regions of waiting times in the phase diagram for compact hybrid stars for the (9,-8) scenario. For an estimate of a population statistics we show the region of evolutionary tracks when the interval of initial magnetic field values is restricted to $0.6 \leq B(0)[\text{TG}] \leq 1.0$. Note that the probability of finding a compact star in the phase diagram is enhanced in the vicinity of the critical line for the deconfinement phase transition $N_{\text{crit}}(\Omega)$ by at least a factor of two relative to all other regions in the phase diagram.

## 5 Summary and Outlook

On the example of the deconfinement transition from hadronic to quark matter we have demonstrated that the rotational frequency of accreting neutron stars is sensitive to changes of their inner structure. Overclustering of the population of Z-sources of LMXBs in the frequency-mass plane ($\Omega - N$ plane) is suggested as a direct measurement of the critical line for the deconfinement phase transition since it is correlated to a maximum in the moment of inertia of the compact star.

A generalization of the phase diagram method to other thermodynamical observables like thermal and electromagnetic processes is possible and will provide a systematical tool for the study of observable consequences of the deconfinement phase transition. A population statistics in the space of the appropriate thermodynamic degrees of freedom allows a direct measurement of the shape of hypersurfaces representing the phase border which is characteristic for the equation of state of superdense stellar matter.

In the present contribution we have considered the quasistationary evolution of accreting objects where the timescale for accreting about one solar mass and thus completing the deconfinement phase transition processes is set by the Eddington rate to be about $10^8$ yr. Therefore this analysis concerns rather old objects. For the cooling process, however, a characteristic timescale is the begin of the photon cooling era determined mainly by the relation of neutrino and photon emission rates. It occurs typically after $10^2 - 10^4$ yr, i.e. for much younger objects. Phase transition effects, changing the emission rates, modify cooling curves for the neutron star surface temperature typically in this time interval, see Voskresensky's contribution to this book [39].

We have focussed here on the deconfinement phase transition only. The phase diagram method for developing strategies to investigate changes in the neutron star interiors is more general and can be applied to other phase transitions too, as e.g. the transition to superfluid nuclear matter [40], a kaon condensate [41] or color superconducting quark matter [42–44] with typical consequences for cooling and magnetic field evolution of the neutron star.

In order to consider the problem of magnetic field evolution in case of color superconductivity and diquark condensation one can use the phase diagram method to classify the magneto-hydrodynamical effects in the superfluid and superconducting phases, e.g. the creation of vortices in the quark core and their influence on the postglitch relaxation processes can be discussed [45].

One can also modify the phase diagram approach to the case of self bound rotating quark stars (strange stars) [13]. The application of our method could visualize the evolution of quark stars from normal to strange stars.

The present approach opens new perspectives for the search for an understanding of the occurence and properties of superdense stellar matter in neutron star interiors.

**Acknowledgement**

This work of H.G. and G.P. has been supported by DFG under grant Nos. ARM 436 17/1/00 and ARM 436 17/7/00, respectively. D.B. and G.P. acknowledge a visiting Fellowship granted by the ECT* Trento. The authors thank the participants of the workshop "Physics of Neutron Star Interiors" at the ECT* Trento for many stimulating discussions.

# References

1. *Quark Matter '99*, edited by L. Riccati, M. Masera and E. Vercellin (Elsevier, Amsterdam, 1999).
2. *Understanding Deconfinement in QCD*, edited by D. Blaschke, F. Karsch and C.D. Roberts (World Scientific, Singapore, 2000).
3. Ch. Schaab, B. Hermann, F. Weber, and M.K. Weigel, Astrophys. J. **480**, L111 (1997).
4. A. Dar and A. DeRújula, *SGRs and AXPs - Magnetars or Young Quark Stars?*, [astro-ph/0002014].

5. N.K. Glendenning, S. Pei and F. Weber, Phys. Rev. Lett. **79**, 1603 (1997).
6. J. Madsen, Phys. Rev. Lett. **85**, 10 (2000).
7. E. Chubarian, H. Grigorian, G. Poghosyan and D. Blaschke, Astron.& Astrophys. **357**, 968 (2000).
8. N.K. Glendenning, F. Weber, "Signal of Quark Deconfinement in Millisecond Pulsars and Reconfinement in Accreting X-ray Neutron Stars", this volume.
9. G. Poghosyan, H. Grigorian, D. Blaschke, Astrophys. J. Lett. **551**, L73 (2001).
10. M. Van der Klis, Ann. Rev. Astron. Astrophys. **38**, 717 (2000).
11. M. C. Miller, F. K. Lamb, D. Psaltis, Astrophys. J. **508**, 791 (1998).
12. X.-D. Li, I. Bombaci, M. Dey, J. Dey, E. P. J. van der Heuvel, Phys. Rev. Lett. **83**, 3776 (1999).
13. I. Bombaci, "Strange Quark Stars: structural properties and possible signatures for their existence", this volume.
14. N.K. Glendenning, Ch. Kettner, Astron.& Astrophys. **353**, L9 (1999).
15. D. Blaschke, H. Grigorian, G. Poghosyan, *Conditions for deconfinement transition signals from compact star rotation*, [astro-ph/0008005].
16. D. Blaschke, H. Grigorian, G. Poghosian, C.D. Roberts, S. Schmidt, Phys. Lett. **B 450**, 207 (1999).
17. D. Blaschke, P.C. Tandy, in [2], p. 218.
18. A. Bender, D. Blaschke, Yu. Kalinovsky, C.D. Roberts, Phys. Rev. Lett. **77**, 3724 (1996).
19. N.K. Glendenning, Phys. Rev. **D 46**, 1274 (1992).
20. N.K. Glendenning, *Compact Stars*, (Springer, New York, 1997).
21. F. Weber, *Pulsars as Astrophysical Laboratories for Nuclear and Particle Physics*, (IoP Publishing, Bristol, 1999).
22. M. Baldo, F. Burgio, "Microscopic Theory of the Nuclear Equartion f State and NEutron Star Structure", this volume.
23. M. Prakash, J. M. Lattimer, J. A. Pons, A. W. Steiner, "Evolution of a Neutron Star From its Birth to Old Age," this volume.
24. J.B. Hartle, Astrophys. J. **150**, 1005 (1967).
25. J.B. Hartle, K.S. Thorne, Astrophys. J. **153**, 807 (1967).
26. D.M. Sedrakian, E.V. Chubarian, Astrofizika **4**, 239 (1968).
27. D.M. Sedrakian, E.V. Chubarian, Astrofizika **4**, 551 (1968).
28. P. Ghosh, F.K. Lamb, Astrophys. J. **234**, 296 (1979).
29. S.L. Shapiro, S.A. Teukolsky, *Black Holes, White Dwarfs, and Neutron Stars*, (Wiley, New York 1983).
30. L. Bildsten, Astrophys. J. **501**, L89 (1998).
31. V.M. Lipunov, *Astrophysics of Neutron Stars*, (Springer, Berlin, 1992).
32. P. Ghosh, F.K. Lamb, Astrophys. J. **234**, 296 (1979).
33. D. Bhattacharya and E.P.J. van den Heuvel, Phys. Rep. **203**, 1 (1991).
34. D. Page, U. Geppert, T. Zannias, Astron. & Astrophys. **360**, 1052 (2000).
35. L. Burderi, A. Possenti, M. Colpi, T. Di Salvo, N. D'Amico, Astrophys. J. **519**, 285 (1999).
36. M. Colpi, A. Possenti, S. Popov, F. Pizzolato "Spin and Magnetism in Old Neutron Stars", this volume.
37. N.K. Glendenning and F. Weber, *Possible evidence of quark matter in neutron star X-ray binaries*, [astro-ph/0003426].
38. F.K. Lamb, M.C. Miller, HEAD **32**, 2401L, (2000).
39. D.Voskresensky, "Neutrino cooling and neutron stars medium effects", this volume.
40. U. Lombardo, H.-J. Schulze, "Superfluidity in Neutron Star Matter", this volume.

41. A. Ramos, J. Schaffner-Bielich, J. Wambach, "Kaon Condensation in Neutron Stars", this volume.
42. M. Alford, J. A. Bowers, K. Rajagopal, "Color Superconductivity in Compact Stars", this volume.
43. T. Schäfer, E. Shuryak, "Phases of QCD at High Baryon Density", this volume.
44. M.B. Hecht, C.D. Roberts, S.M. Schmidt, "Diquarks and Density", this volume.
45. D. Blaschke, D. Sedrakian, "Ginzburg-Landau equations for superconducting quark matter in neutron stars", [nucl-th/0006038].

# Signal of Quark Deconfinement in Millisecond Pulsars and Reconfinement in Accreting X-ray Neutron Stars

Norman K. Glendenning[1] and Fridolin Weber[2]

[1] Nuclear Science Division, & Institute for Nuclear and Particle Astrophysics, Lawrence Berkeley National Laboratory, University of California, Berkeley, CA 94720

[2] University of Notre Dame, Department of Physics, 225 Nieuwland Science Hall, Notre Dame, IN 46556-5670

**Abstract.** Theoretically, the phase transition between the confined and deconfined phases of quarks can have a remarkable effect on the spin properties of millisecond pulsars and on the spin distribution of the population of x-ray neutron stars in low-mass binaries. In the latter class of stars, the effect has already been observed—a strong clustering in the population in a narrow band of spins. The observed clustering cannot presently be uniquely assigned to the phase transition as cause. However, there is another possible signal—not so far observed—in millisecond pulsars that we also discuss which would have the same origin, and whose discovery would tend to confirm the interpretation in terms of a phase transition in the stellar core.

## 1 Motivation

One of the great fascinations of neutron stars is the deep interior where the density is a few times larger than the density of normal nuclei. There, in matter inaccessible to us other than fleetingly in relativistic collisions, unfamiliar states may exist. The most exotic of these is, of course, quark matter—the deconfined phase of hadronic matter. It is quite plausible that ordinary canonical pulsars—those like the Crab, and more slowly rotating ones—have a quark matter core, or at least a mixed phase of quark and confined hadronic matter. If one were able to close pack nucleons to a distance such that they were touching at a radius of 0.5 fm, the density would be a mere $1/[(2r)^3 \rho_0] = 6.5$ larger than normal nuclear density. If nucleons were close packed to their rms charge radius of 0.8 fm, the density would be 1.6 times nuclear density. Of course, fermions cannot be close packed according to the exclusion principle since their localization would give them enormous uncertainty in their momentum. They would be torn apart before they could be squeezed so much. This simple, and perhaps oversimplified argument, makes it plausible that ordinary (slowly rotating pulsars) have a quark matter core essentially from birth.

By comparison, millisecond (ms) pulsars are centrifugally flattened in the equatorial plane and the density is diluted in the interior. We shall suppose that the critical phase transition density lies between the diluted density of millisecond pulsars and the high density at the center of canonical pulsars. Then as a millisecond neutron star spins down because it is radiating angular momentum in a broad band of electromagnetic frequencies as well as in a wind of particle-antiparticle pairs, or as a canonical neutron star at some stage begins accreting matter from a companion and is spun up, a change in phase of matter in the inner part of the star will occur.

Whatever the high density phase, and we assume here it is the quark deconfined one, it is softer than the normal phase. Otherwise no phase transition. Consequently, a change in the phase of matter will be accompanied by a change in the density distribution in the star. In the case of spindown of an isolated ms pulsar, the weight of the surrounding part of the star will squeeze the softer high density phase that is forming in the core. Conversely, an accreting neutron star that is being spun up, will spin out the already present quark phase. In either case, the star's moment of inertia will change, and therefore its spin rate will change to accommodate the conservation of angular momentum. Timing of pulsar and x-ray neutron star rotation is a relatively easy observation, and the effect of a phase change in the deep interior of neutron stars on timing or frequency distribution is what we study here.

The deconfinement or reconfinement of quark matter in a rotating star is a very slow process because it is governed by the rate of change of period, which is very small. This is an advantage. If the processes were fast, we would not likely witness the epoch of phase change, which is long, but nevertheless short (but not too short) compared to the timescales of spindown of ms pulsars or of spinup of accreting neutron stars in low-mass binaries. What we may see in the case of isolated ms pulsars is an occasional one that is spinning up, even though losing angular momentum to radiation. That would be a spectacular signal. What we may see in the case of accreting x-ray neutron stars in low-mass binaries is an unusual number of them falling within a small spin-frequency range—the range that corresponds to the spinout of the quark matter phase. That also is an easily observable signal. For stars of the same mass, the spontaneous spinup of isolated pulsars should occur at about the same spin frequency as the clustering in the population of neutron star accretors.

To place the canonical pulsars, ms pulsars, and x-ray accreting neutron stars in context, we refer to Fig. 1. Canonical pulsars have large surface magnetic fields, of the order of $10^{12}$ to $10^{13}$ G, and relatively long periods with an average of about 0.7 s. The ms pulsars have low fields, of the order of $10^8$ to $10^9$ G, and periods ranging from 1 to 10 ms. The x-ray stars that are accreting matter from a low-mass non-degenerate companion are believed to be the link [1–4] between the populations of canonical and ms pulsars—their path is indicated schematically in the figure. (Of course, defining ms pulsars to be those lying in the range of periods 1 to 10 ms is arbitrary.)

At present, fewer ms pulsars have been discovered than canonical pulsars. This is likely to be a selection effect. It is much more difficult to detect ms pulsars because there is radio noise in all directions (and therefore a possible signal) and because of the dispersion by the interstellar electrons (from ionized hydrogen) of any pulsed signal which might be present in the direction the telescope is

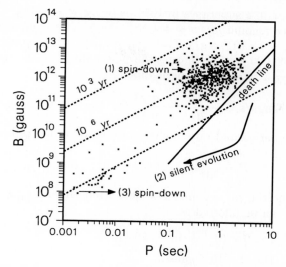

**Fig. 1.** Magnetic field of radio pulsars as a function of rotation period. Death line shows a combination of the two properties beyond which the beaming mechanism apparently fails. Canonical pulsars are designated as (1), millisecond pulsars as (3) and a schematic path for x-ray accreting neutron stars as (2). From Ref. [5].

pointing. If present the signal is weak and contains a band of radio frequencies. Because of the different time-delay of each frequency, the frequencies are placed into a number of bins. For ms pulsars, the time-delay across the range of frequencies in a pulse is greater than the interval between pulses. Detection therefore depends on a tedious analysis that corrects the time-delay in each frequency bin for an assumed density of intervening interstellar gas; this is repeated for a succession of assumed densities. If the process converges to a periodic pulse, a pulsar has been discovered. Otherwise, the radio telescopes are pointed in a new direction. Systematically covering the sky is obviously very costly and time consuming, and moreover, the present instrumentation is unable to detect pulsars with periods less than 1 millisecond.

## 2 Spontaneous Spinup of Millisecond Pulsar

We discuss here the effect that a change of phase in the core of a neutron star can have on the rotational properties of ms pulsars [6–9]. The analysis begins with the energy-loss equation for a rotating magnetized star whose magnetic axis is tilted with respect to the rotation axis. It is of historical interest to note that even before pulsars were discovered and soon identified with hypothetical neutron stars, Pacini had postulated the existence of a highly magnetized rotating neutron star inside the Crab nebula as the energy source of the nebula and inferred some of the star's properties [10]. It was already known before 1967 that the nebula, formed in a supernova in 1054, is being accelerated and illuminated by a power source amounting to $\sim 4 \times 10^{38}$ erg/s. This compares with a solar luminosity of $L_\odot \sim 4 \times 10^{33}$ erg/s. The power output of the rotating Crab pulsar equals that of 100,000 suns. Equating the power input to the nebula with the power radiated by a rotating magnetic dipole,

$$4 \times 10^{38} \text{ ergs/s} = -\frac{dE}{dt} = -\frac{d}{dt}\left(\frac{1}{2}I\Omega^2\right) = \frac{2}{3}R^6 B^2 \Omega^4 \sin^2\alpha, \tag{1}$$

and knowing the period and rate of change of period of the pulsar,

$$P \sim \frac{1}{30} \text{ s}, \qquad \dot{P} \sim 4 \times 10^{-13} \text{ s/s}, \tag{2}$$

the moment of inertia, surface magnetic field strength, and rotational energy of the Crab pulsar can be inferred as

$$\begin{aligned} I &\sim 9 \times 10^{44} \text{ g cm}^2 \sim 70 \text{ km}^3, \\ B &\sim 4 \times 10^{12} \text{ gauss}, \\ E_{\text{rot}} &\sim \tfrac{1}{2}I\Omega^2 \sim 2 \times 10^{49} \text{ ergs} \sim 10^{55} \text{ MeV}. \end{aligned} \tag{3}$$

For these estimates, we have assumed that $\sin\alpha = 1$, where $\alpha$ is the angle between magnetic and rotational axis. (Gravitational units $G = c = 1$ are used frequently. For convenient conversion formula to other units see ch. 3 of Ref. [5].)

Returning to the effect that a phase change can have on the rotational properties of a pulsar through changes in the moment of inertia induced by a phase transition, we rewrite Eq. (1) in greater generality:

$$\dot{E} = \frac{d}{dt}\left(\frac{1}{2}I\Omega^2\right) = -C\Omega^{n+1}. \tag{4}$$

Here we have written for convenience

$$C = (2/3)R^6 B^2 \sin^2\alpha. \tag{5}$$

We find the deceleration equation

$$\dot{\Omega} = -\frac{C}{I}\Omega^n \left(1 + \frac{I'\Omega}{2I}\right)^{-1}. \tag{6}$$

In work previous to ours, $I' \equiv dI/d\Omega$ was assumed to vanish. This would be a good approximation for canonical pulsars but not for millisecond pulsars.

The angular momentum of a rotating star in General Relativity can be obtained numerically as a solution of Einstein's equations or as a very complicated algebraic expression in a perturbative expansion. The moment of inertia is then obtained as $I = J/\Omega$. The complication arises in two ways. First, a rotating star sets the local inertial frames into rotation. This is referred to as frame dragging. Second, the structure of the rotating star depends on the frame dragging frequency, $\omega$, which is position dependent, and on the spacetime metric, which also depends on $\omega$. Algebraic expressions were obtained in Refs. [11,9].

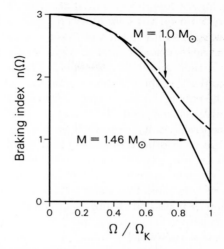

**Fig. 2.** Apparent braking index for pure dipole radiation ($n = 3$) as function of rotational frequency for stars of two different masses containing hyperons (but no first order phase transition). (The mass indicated is for slow rotation.)

The index $n$ in equation Eq. (4) equals three for magnetic dipole radiation. In principle, it can be measured in terms of the frequency $\Omega$ and its first two time derivatives. But the dependence of $I$ on $\Omega$ introduces a correction so that the dimensionless *measurable* braking index is given by

$$n(\Omega) = \frac{\Omega\ddot{\Omega}}{\dot{\Omega}^2} = n - \frac{3I'\Omega + I''\Omega^2}{2I + I'\Omega} \ . \tag{7}$$

So the measurable braking index will differ from 3 for ms pulsars, if for no other reason than that the centrifugal deformation of the star relaxes as the star spins down. Such a dependence on $\Omega$ is illustrated in Fig. 2

It is especially noteworthy that the *magnitude* of the signal—the deviation of $n(\Omega)$ from $n$—depends on $dI/d\Omega$ which is large, especially for millisecond pulsars, but the *duration* of the signal depends on $d\Omega/dt$ which is small. Therefore, the variation of the braking index over time is very slow, the time-scale being astronomical. So what one would observe over any observational era is a constant braking index $n(\Omega)$ that is less (if no phase transition) than the $n$ characteristic of the energy-loss mechanism Eq. (4) even if the radiation were a pure multipole. We will see that the effect of a phase transition on the measurable braking index can be (but not necessarily) much more dramatic. (We distinguish between the constant $n = 3$ of the energy-loss equation and the measurable quantity of Eq. (7).)

We turn now to our original postulate—that canonical pulsars have a quark matter phase in their central region but that ms pulsars which are centrifugally diluted, do not. In the slow course of spindown the central density will rise however. When the critical density is reached, stiff nuclear matter will be replaced slowly in the core by highly compressible quark matter. The overlaying layers of nuclear matter weigh down on the core and compress it. Its density rises. The star shrinks—mass is redistributed with growing concentration at the center. The by-now more massive central region gravitationally compresses the outer nuclear matter even further, amplifying the effect. The density profile for a star at three angular velocities, (1) the limiting Kepler angular velocity at which the star is stretched in the equatorial plane and its density is centrally diluted, (2) an intermediate angular velocity, and (3) a non-rotating star, are shown in Fig. 3. We see that the central density rises with decreasing angular velocity by a factor of three and the equatorial radius decreases by 30 percent. In contrast, for a model for which the phase transition did not take place, the central density would change by only a few percent [12]. The phase boundaries are shown in Fig. 4 from the highest rotational frequency to zero rotation.

The redistribution of mass and shrinkage of the star change its moment of inertia and hence the characteristics of its spin behavior. The star must spin up to conserve angular momentum which is being carried off only slowly by the weak

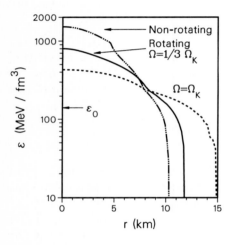

**Fig. 3.** Mass profiles as a function of equatorial radius of a star rotating at three different frequencies, as marked. At low frequency the star is very dense in its core, having a 4 km central region of highly compressible pure quark matter. At intermediate frequency, the pure quark matter phase is absent and the central 8 km is occupied by the mixed phase. At higher frequency (nearer $\Omega_K$) the star is relatively dilute in the center and centrifugally stretched. Inflections at $\epsilon = 220$ and 950 are the boundaries of the mixed phase.

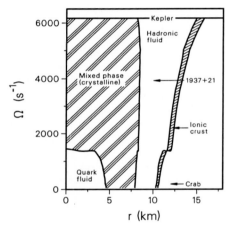

**Fig. 4.** Radial boundaries at various rotational frequencies separating (1) pure quark matter, (2) mixed phase, (3) pure hadronic phase, (4) ionic crust of neutron rich nuclei and surface of star. The pure quark phase appears only when the frequency is below $\Omega \sim 1370$ rad/s. Note the decreasing radius as the frequency falls. The frequencies of two pulsars, the Crab and PSR 1937+21 are marked for reference.

electromagnetic dipole radiation. The star behaves like an ice skater who goes into a spin with arms outstretched, is slowly spun down by friction, temporarily spins up by pulling the arms inward, after which friction takes over again.

The behavior of the moment of inertia in the vicinity of the critical region of growth of the quark matter core is shown in Fig. 5. The critical period of rotation is $P = 2\pi/\Omega \sim 4.6$ ms. Much the same phenomenon was observed in rotational nuclei during the 1970s [13,14] and had been predicted by Mottelson and Valatin [15] though of course the phase changes are different (see Fig. 6). The mechanism is evidently quite robust, but it need not occur in every model star. In the present instance, the mass of the neutron star is very close to the limiting mass of the non-rotating counterpart. This does not necessarily mean that it will be a rare event. The mass of neutron stars is bounded from below by the Chandrasekhar mass limit which is established by electron pressure and has a value of $\sim 1.4 M_\odot$. The limiting mass of neutron stars may be very little more because of the softening of the nuclear equation of state by hyperonization and quark deconfinement. This could be the reason that neutron star masses seem to lie in a very small interval [5].

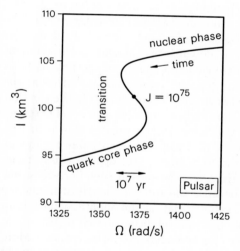
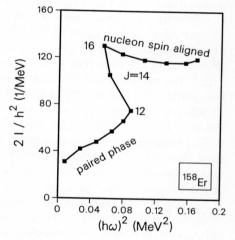

**Fig. 5.** Moment of inertia of a neutron star at angular velocities for which the central density rises from below to above critical density for the pure quark matter phase as the centrifugal force decreases. Time flows from large to small $I$. The most arresting signal of the phase change is the spontaneous spin-up that an isolated pulsar would undergo during the growth in the region of pure quark matter. (Adapted from Ref. [6].)

**Fig. 6.** Nuclear moment of inertia as a function of squared frequency for $^{158}$Er, showing backbending in the nuclear case. Quantization of spin yields the unsmooth curve compared to the one in Fig. 5.

As already pointed out, the progression in time of the growth of the quark core is very slow, being governed by the weak processes that cause the loss of angular momentum to radiation. Using the computed moment of inertia for a star of constant baryon number as shown in Fig. 5 we can integrate Eq. (6) to find the epoch of spinup to endure for $2 \times 10^7$ y. This is a small but significant fraction of the spin-down time of ms pulsars which is $\sim 10^9$y. So if ms pulsars are near their limiting mass and are approximately described by our model, about 1 in 50 *isolated* ms pulsars should be spinning up instead of spinning down. Presently, about 60 ms pulsars have been identified, and about half of them are isolated, the others being in binary systems. Because the period of ms pulsars can be measured with an accuracy that rivals atomic clocks, identification of the direction of change of period would not take long. For example, PSR1937+21 has a period (measured on 29 November 1982 at 1903 UT)

$$P = 1.5578064487275(3) \text{ ms}.$$

Its rate of change of period is a mere $\dot{P} \sim 10^{-19}$. But because of the high accuracy of the period measurement, it would take only two measurements spaced 0.3 hours apart to detect a unit change in the last significant figure, and hence to detect in which direction the period is changing.

In principal, the period and first two time derivatives can be measured. The dimensionless ratio formed from them yields the braking index $n$ of the energy-loss mechanism, corrected by a term that depends on changes in the moment of inertia. In the case that a phase change causes a spinup of the pulsar, as in Fig. 5, the measurable dimensionless quantity $n(\Omega)$ has two singularities at the frequencies at which $dI/d\Omega$ switches between $\pm\infty$. From Eq. (6) it is clear that $\dot{\Omega}$ will pass through zero and change sign at both turning points. Therefore, the measurable braking index Eq. (7) will have singularities as shown in Fig. 7. The braking index with a nominal value of 3 will make enormous excursions from that value. The braking index is shown in Fig. 8 over one decade in time, and has an anomalous value for $10^8$ y. However, the second time derivative has never been measured for a ms pulsar because the rate of change of frequency is so slow. So the one, and easily observable signal, is the spontaneous spinup of an isolated ms pulsar. Such a pulsar has not yet been observed. But then, only about 30 isolated ms pulsars have so far been detected, whereas, according to our estimate, only 1 in 50 ms pulsars of mass close to the maximum would presently be passing through the epoch in which a quark matter core grows.

We briefly describe the nuclear and quark matter phases used in this and the following sections. More details can be found in the Appendix and in Ref. [6]. The initial mass of the star in our examples is $1.42M_\odot$. Briefly, confined nuclear matter is described by a covariant Lagrangian describing the interaction of the members of the baryon octet with scalar, vector and vector-isovector mesons and solved in the meanfield approximation. Quark matter is described by the MIT bag model.

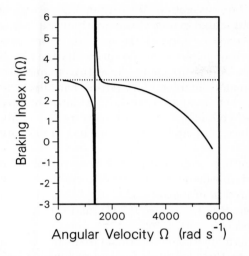 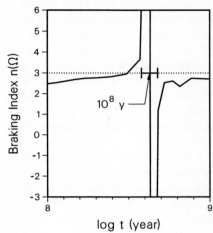

**Fig. 7.** Braking index for a pulsar that passes through a change of phase in the central region as a function of angular velocity.

**Fig. 8.** Braking index as a function of time over one decade in the critical region in case of a phase transition in the interior of the star. (From Ref. [7].)

## 3 Population Clustering in the Spin of Accreting X-ray Neutron Stars in Binaries

Some neutron stars have non-degenerate companions. Such neutron stars are radio silent because a wind from the hot surface of the companion disperses the pulsed radio signal which a rotating magnetized neutron star would otherise radiate into space. Late in the life of the neutron star, when the slowly evolving companion begins to overflow the Roche lobe, mass transfer onto the neutron star commences. The drag of the magnetic dipole torque will be eclipsed by the transfer of mass and angular momentum onto the neutron star. It has begun its evolution from an old slowly rotating neutron star with long period and high magnetic field, to a ms pulsar with low field (sometimes referred to as a "recycled pulsar"). During the long intermediate stage, when the surface and accretion ring are heated to high temperature, the star radiates x-rays.

Any asymmetry in the mass accretion pattern will cause a variability in x-ray emission. At an accretion rate of $10^{-9} M_\odot/y$ it would take only $10^8$ y to spin up the neutron star to a period of 2 ms (500 Hz). Consequently, millisecond variability in the x-ray luminosity is expected and observed.

Accreting x-ray neutron stars provide a very interesting contrast to the spin-down of isolated ms pulsars [16]. Presumably they are the link between the canonical pulsars with mean period of 0.7 sec and the ms pulsars [1–4]. If the critical deconfinement density falls within the range spanned by canonical pulsars, quark matter will already exist in them but may be "spun" out of x-ray stars as their rotational frequency increases during accretion. We can anticipate that in a certain frequency range, the changing radial extent of the quark matter phase will actually inhibit changes in frequency because of the increase in moment of inertia occasioned by the gradual disappearance of the quark matter phase. Accretors will tend to spend a greater length of time in the critical frequencies than otherwise. There will be an anomalous number of accretors that appear at or near the same frequency. This is what was found recently in data obtained with the Rossi X-ray Timing Explorer (RXTE). For an extensive review of the discoveries made in the short time since this satellite was launched (1995), see Ref. [17].

The spinup evolution of an accreting neutron star is a more complicated problem than that of the spindown of an isolated ms pulsar of constant baryon number. It is complicated by the accretion of matter ($\dot{M} \gtrsim 10^{-10} M_\odot$ yr$^{-1}$), a changing magnetic field strength (from $B \sim 10^{12}$ to $\sim 10^8$ G), and the interaction of the field with the accretion disk.

The change in moment of inertia as a function of rotational frequency caused by spinup due to accretion is similar to that described in the previous section, but in reverse [6]. However, there are additional phenomena as just mentioned. The spin-up torque of the accreting matter causes a change in the star's angular momentum according to the relation [18–20]

$$\frac{d}{dt} J \equiv \frac{d}{dt}(I\Omega) = N_A(r_{\rm m}) - N_M(r_{\rm c}). \tag{8}$$

The first term on the right-hand-side is the torque exerted on the star by a mass element rotating at the base of the accretion disk with Keplerian velocity $\omega_{\rm K}$. Denoting this distance with $r_{\rm m}$, one readily finds that $N_A(r_{\rm m})$ is given by ($G = c = 1$)

$$N_A(r_{\rm m}) = r_{\rm m}^2 \, \dot{M} \, \omega_{\rm K}$$
$$= r_{\rm m}^2 \, \dot{M} \left(\frac{M}{r_{\rm m}^3}\right)^{1/2}$$
$$= \dot{M} \, \sqrt{M r_{\rm m}} \equiv \dot{M} \, \tilde{l}(r_{\rm m}), \tag{9}$$

where $\dot{M}$ stands for the accretion rate, and $\tilde{l}(r_{\rm m})$ is the specific angular momentum of the accreting matter (angular momentum added to the star per unit mass of accreted matter). The second term on the right-hand-side of Eq. (8) stands for the magnetic plus viscous torque term ($\kappa \sim 0.1$),

$$N_M(r_{\rm c}) = \kappa \mu^2 \, r_{\rm c}^{-3}, \tag{10}$$

with $\mu \equiv R^3 B$ the star's magnetic moment. Upon substituting Eqs. (9) and (10) into (8) and writing the time derivative $d(I\Omega)/dt$ as $d(I\Omega)/dt = (dI/dt)\Omega + I(d\Omega/dt)$, the time evolution equation for the angular velocity $\Omega$ of the accreting star can be written as

$$I(t)\frac{d\Omega(t)}{dt} = \dot{M}\tilde{l}(t) - \Omega(t)\frac{dI(t)}{dt} - \kappa\mu(t)^2 \, r_{\rm c}(t)^{-3}. \tag{11}$$

The quantities $r_\mathrm{m}$ and $r_\mathrm{c}$ denote fundamental length scales of the system. The former, as mentioned above, denotes the radius of the inner edge of the accretion disk and is given by ($\xi \sim 1$)

$$r_\mathrm{m} = \xi\, r_\mathrm{A}\,. \tag{12}$$

The latter stands for the co-rotating radius defined as

$$r_\mathrm{c} = \left(M\Omega^{-2}\right)^{1/3}\,. \tag{13}$$

Accretion will be inhibited by a centrifugal barrier if the neutron star's magnetosphere rotates faster than the Kepler frequency at the magnetosphere. Hence $r_\mathrm{m} < r_\mathrm{c}$, otherwise accretion onto the star will cease. A further fundamental lengthscale is set by the Alfén radius $r_\mathrm{A}$, which enters in Eq. (12). It is the radius at which the magnetic energy density, $B^2(r)/8\pi$, equals the total kinetic energy density, $\rho(r)v^2(r)/2$, of the accreting matter. For a dipole magnetic field outside the star of magnitude

$$B(r) = \frac{\mu}{r^3}\,, \tag{14}$$

this condition reads

$$\frac{1}{8\pi}B^2(r_\mathrm{A}) = \frac{1}{2}\rho(r_\mathrm{A})v^2(r_\mathrm{A})\,, \tag{15}$$

where $v(r_\mathrm{A})$ is of the order of the Keplerian velocity, or, which is similar, the escape velocity at distance $r_\mathrm{A}$ from the neutron star,

$$v(r_\mathrm{A}) = \left(\frac{2M}{r_\mathrm{A}}\right)^{1/2}\,. \tag{16}$$

The density $\rho(r_\mathrm{A})$ in Eq. (15) can be replaced by

$$\rho(r_\mathrm{A}) = \frac{\dot M}{4\pi r_\mathrm{A}^2 v^2(r_\mathrm{A})}\,, \tag{17}$$

which follows from the equation of continuity. With the aid of Eqs. (14), (16) and (17), one obtains from (15) for the Alfén radius the following expression:

$$r_\mathrm{A} = \left(\frac{\mu^4}{2M\dot M^2}\right)^{1/7}\,. \tag{18}$$

It is instructive to write this equation as

$$r_\mathrm{A} = 7 \times 10^3\, \dot M_{-10}^{-2/7}\, \mu_{30}^{4/7} \left(\frac{M}{M_\odot}\right)^{-1/7}\ \mathrm{km}\,, \tag{19}$$

where $\dot M_{-10} \equiv \dot M/(10^{-10}\,M_\odot\,\mathrm{yr}^{-1})$ and $\mu_{30} \equiv \mu/(10^{30}\,\mathrm{G\,cm^3})$. It shows that canonical accretors ($\dot M_{-10} = 1$) with strong magnetic fields of $B \sim 10^{12}$ G have Alfén radii, and thus accretion disks, that are thousands of kilometers away from their surfaces. This is dramatically different for accretors whose magnetic fields have weakened over time to values of $\sim 10^8$ G, for instance. In this case, the Alfén radius has shrunk to $\sim 40$ km, which is just a few times the stellar radius.

We assume that the magnetic field evolves according to

$$B(t) = B(\infty) + [B(0) - B(\infty)]e^{-t/t_\mathrm{d}} \qquad (20)$$

with $t = 0$ at the start of accretion, and where $B(0) = 10^{12}$ G, $B(\infty) = 10^8$ G, and $t_\mathrm{d} = 10^6$ yr. Such a decay to an asymptotic value seems to be a feature of some treatments of the magnetic field evolution of accreting neutron stars [21]. Moreover, it expresses the fact that canonical neutron stars have high magnetic fields and ms pulsars have low fields (see Fig. 1). Beyond that, as just mentioned, the condition that accretion can occur demands that $r_\mathrm{m} < r_\mathrm{c}$ which inequality places an upper limit on the magnitude of the magnetic field of ms neutron star accretors of 2 to $6 \times 10^8$ G [3].

Frequently, it has been assumed that the moment of inertia in Eq. (11) does not respond to changes in the centrifugal force, and in that case, the above equation yields a well-known estimate of the period to which a star can be spun up [2]. The approximation is true for slow rotation. However, the response of the star to rotation becomes increasingly important as the star is spun up. Not only do changes in the distribution of matter occur but internal changes in composition occur also because of changes induced in the central density by centrifugal dilution [6]; both changes effect the moment of inertia and hence the response of the star to accretion.

The moment of inertia of ms pulsars or of neutron star accretors has to be computed in GR without making the usual assumption of slow rotation [22,23]. We use a previously obtained expression for the moment of inertia of a rotating star, good to second order in $\Omega$ [11]. The expression is too cumbersome to reproduce here. Stars that are spun up to high frequencies close to the breakup limit (Kepler frequency) undergo dramatic interior changes; the central density may change by a factor of four or so over that of a slowly rotating star if a phase change occurs during spin-up (cf. Fig. 3) [7,24].

Figure 9 shows how the moment of inertia changes for neutron stars in binary systems that are spun up by mass accretion according to Eq. (11) until $0.4 M_\odot$ has been accreted. The neutron star models are fully described in Ref. [6] and references therein and briefly in the Appendix of this article. The initial mass of the star in our examples is $1.42 M_\odot$. In one case, it is assumed that a phase transition between quark matter and confined hadronic matter occurs, and in the other that it does not. This accounts for the different initial moments of inertia, and also, as we see, the response to spinup. Three accretion rates are assumed, which range from $\dot{M}_{-10} = 1$ to $100$ (where $\dot{M}_{-10}$ is measured in units of $10^{-10} M_\odot$ per year). These rates are in accord with observations made on low-mass X-ray binaries (LMXBs) observed with the Rossi X-ray Timing Explorer [17]. The observed objects, which are divided into Z sources and A(toll) sources, appear to accrete at rates of $\dot{M}_{-10} \sim 200$ and $\dot{M}_{-10} \sim 2$, respectively. Although in a given binary, $\dot{M}$ varies on a timescale of days, we take it to be the constant average rate in our calculations.

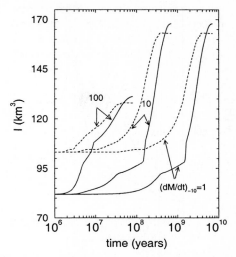

**Fig. 9.** Moment of inertia of neutron stars as a function of time with (solid curves) and without (dashed curves) quark matter core assuming $0.4M_\odot$ is accreted. Results for three *average* accretion rates are illustrated.

Figure 10 shows the spin evolution of accreting neutron stars as determined by the changing moment of inertia and the spin evolution equation, Eq. (11). Neutron stars *without* quark matter in their centers are spun up along the dashed lines to equilibrium frequencies between about 600 Hz and 850 Hz, depending on accretion rate and magnetic field. The $dI/dt$ term for these sequences manifests itself only insofar as it limits the equilibrium periods to values smaller than the Kepler frequency, $\nu_{\rm K}$. In both Figs. 9 and 10 we assume that $0.4M_\odot$ is accreted. Otherwise, the maximum frequency attained is less, the less matter is accreted.

The spin-up scenario is dramatically different for neutron stars in which a first order phase transition occurs. In this case, as known from Fig. 9, the temporal conversion of quark matter into its mixed phase of quarks and confined hadrons is accompanied by a pronounced increase of the stellar moment of inertia. This increase contributes so significantly to the torque term $N(r_{\rm c})$ in Eq. (11) that the spinup rate $d\Omega/dt$ is driven to a plateau around those frequencies at which the pure quark matter core in the center of the neutron star gives way to the mixed phase of confined hadronic matter and quark matter. The star resumes ordinary spin-up when this transition is completed. The epoch during which the spin rates reach a plateau are determined by attributes like the accretion rate, magnetic field, and its assumed decay time. The epoch lasts between $\sim 10^7$ and $10^9$ yr depending on the accretion rate at the values taken for the other factors.

We can translate the information in Fig. 10 into a frequency distribution of X-ray stars by assuming that neutron stars begin their accretion evolution at the average rate of one per million years. A different rate will only shift

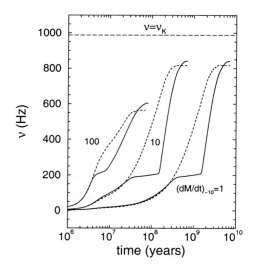

**Fig. 10.** Evolution of spin frequencies ($\nu \equiv \Omega/2\pi$) of accreting neutron stars with (solid curves) and without (dashed curves) quark deconfinement if $0.4 M_\odot$ is accreted. (If the mass of the donor star is less, then so is the maximum attainable frequency.) The spin plateau around 200 Hz signals the ongoing process of quark confinement in the stellar centers. Note that an equilibrium spin is eventually reached which is less than the Kepler frequency.

some neutron stars from one bin to an adjacent one, but will not change the basic form of the distribution. The donor masses in the binaries are believed to range between 0.1 and $0.4 M_\odot$. For lack of more precise knowledge, we assume a uniform distribution of donor masses (or mass accreted) in this range and repeat the calculation shown in Fig. 10 at intervals of $0.1 M_\odot$.

The result for the computed distribution of rotational frequency of x-ray neutron stars is shown in Fig. 11; it is striking. Spinout of the quark matter core as the neutron star spins up is signalled by a spike in the spin distribution which would be absent if there were no phase transition in our model of the neutron star. We stress that what we plot is our prediction of the *relative* frequency distribution of the *underlying* population of x-ray neutron stars—but the weight given to the spike as compared to the high frequency tail depends sensitively on the weight with which the donor masses are assigned. As already mentioned, we give equal weight to donor masses between 0.1 and $0.4 M_\odot$.

The objects above 400 Hz in Fig. 11 are actually unstable and will collapse to black holes. Donors of all masses in the range just mentioned contribute to neutron stars of spin up to 400 Hz. Neutron stars of lower initial mass than our $1.42 M_\odot$ with donors of mass at the higher end of their range will produce spins above 400 Hz. In other words, the relative population in the peak as compared to the background will be sensitive to the unknown factors (1) accretion rate (2) initial mass distribution of neutron stars in LMXBs (3) mass distribution

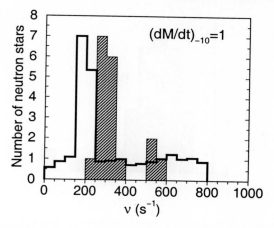

**Fig. 11.** Frequency distribution of X-ray neutron stars. Calculated distribution (open histogram) for the underlying population is normalized to the number of observed objects (18) at the peak. (The normalization causes a fractional number to appear in many bins of the calculated distribution.) Data on neutron stars in LMXBs (shaded histogram) is from Ref. [17]. See text and especially the reference for caveats to the interpretation. The spike in the calculated spin distribution corresponds to the spinout of the quark matter phase and the corresponding growth of the moment of inertia as compressible quark matter is replaced by relatively incompressible nuclear matter. Otherwise the spike would be absent.

of donor stars. However, the position of the peak in the spin distribution of x-ray neutron stars is a property of nuclear matter and independent of the above unknowns.

The calculated concentration in frequency of x-ray neutron stars is centered around 200 Hz; this is about 100 Hz lower than the observed spinup anomaly (see discussion below). This discrepancy should not be surprising in view of our total ignorance[1] of the equation of state above saturation density of nuclear matter and the necessarily crude representation of hadronic matter in the two phases in the absence of relevant solutions to the fundamental QCD theory of strong interactions. We represent the confined phase by relativistic nuclear field theory and the deconfined phase by the MIT bag model. However crude these or any other models of hadronic matter may be, the physics underlying the effect of a phase transition on spin rate is robust, although not inevitable. We have cited the example of an analogous phenomenon found in rotating nuclei in the previous section.

The data that we have plotted in Fig. 11 is gathered from Tables 2, 3, and 4 of the review article of van der Klis concerning discoveries made with the Rossi X-ray Timing Explorer, launched near the end of 1995 [17]. The interpretation of millisecond oscillations in the x-ray emission, either that found in bursts or of

---

[1] There are upcoming radioactive beam experiments from which it is hoped to gain information on the equation of state of asymmetric nuclear matter [25–27].

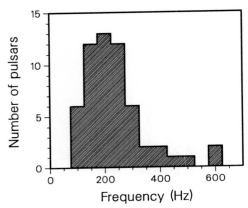

**Fig. 12.** Data on frequency distribution of millisecond pulsars ($1 \leq P < 10$ ms). Frequency bins are 50 Hz wide.

the difference between two observed quasi-periodic oscillations (QPOs) in x-ray brightness, is ambiguous in some cases. In particular, the highest frequency near 600 Hz in the "observed" data displayed in our figure, may actually be twice the rotational frequency of the star [17]. The millisecond variability in x-ray phenomena associated with accretion onto neutron stars that has been observed since the launch of the satellite was anticipated several decades ago. However, the consistency of the phenomena from one binary to another raises questions about interpretation. In this sense, the field is quite young. We refer to the above cited review article for details and references to the extensive literature.

Nevertheless, the basic feature will probably survive—a clustering in rotational frequencies of x-ray neutron stars and a higher frequency tail. Certainly there are high frequency *pulsars*. A histogram of *ms pulsar* frequencies shows a broad distribution around 200 Hz, and a tail extending to $\sim$ 600 Hz as shown in Fig. 12. So both the (sparse) data on X-ray objects and on ms pulsars seem to agree on a peak in the number of stars at moderately high rotational frequency and on an attenuation at high frequency. For ms pulsars, however, the attenuation at high frequency may be partly a selection effect due to interstellar disperion of the radio signal.

There have been other suggestions as to the cause of the spin clustering of x-ray accretors, several of which we cite (c.f. [28-30]). These works are concerned with the balance of the spinup torque by a gravitational radiation torque. Our proposal has several merits: (1) the mechanism involving a change in moment of inertia triggered by a phase transition in the stellar core due to changing density profile in the star as its spin changes is robust—it is known to occur in rotating nuclei; (2) The phase transition should occur in reverse in isolated ms pulsars and in neutron star accretors and at about the same frequency for similar mass stars; (3) The phase transition causes accretion induced spinup to stall for a long epoch, but to resume after the quark core has been expelled, thus accounting for high spin objects like very fast ms pulsars as well as the clustering in spin of the popuation of accretors.

## 4  Evolution from Canonical to Millisecond Pulsar

In the foregoing we have discussed possible signals of a phase transition in isolated ms pulsars and in accreting x-ray neutron stars in binary orbit with a low-mass companion. Spinup by mass accretion is believed to be the pathway from the relatively slowly rotating canonical pulsars formed by conservation of angular momentum in the core collapse of massive stars, and the rapidly rotating millisecond pulsars [1,2]. In this section we trace some of the possible evolutionary routes under the various physical conditions under which accretion occurs [31].

The evolutionary track between canonical and ms pulsars in the coordinates of magnetic field strength and rotational period (refer to Fig. 1) will depend on the rate of mass accretion, its duration, the centrifugal change in the moment of inertia of the star, the strength of the magnetic field and timescale of its decay, and possibly other affects. Many papers have been devoted to the decay of the magnetic field. It is an extremely complicated subject, with many physical uncertainties such as the actual location of the field, whether in the core or crust, the degree to which the crust is impregnated with impurities, crustal heating and resultant reduction in conductivity and therefore increase in ohmic field decay, screening of the magnetic field by accreted material, and so on.

The field is believed to decay only weakly due to ohmic resistance in canonical pulsars, but very significantly if in binary orbit with a low-mass non-degenerate star, when the companion fills its Roche lobe. This era can last up to $10^9$ y and cause field decay by several orders of magnitude. For a review of the literature and several evolutionary scenarios, see Refs. [20,21,32–34].

While there is no consensus concerning the magnetic field decay, observationally, we know that canonical pulsars have fields of $\sim 10^{11}$ to $10^{13} G$, while millisecond pulsars have fields that lie in the range $\sim 10^8$ to $10^9 G$. We shall rely on this observational fact, and assume that the field decays according to Eq. (20). where $B(\infty) = 10^8$ G, $B(0) = 10^{12}$ G and $t_d = 10^5$ to $10^7$ yr. Moreover, this is the general form found in some scenarios [21]. However, we shall also make a comparison with a purely exponential decay.

There are three distinct aspects to developing an evolutionary framework. One has to do with the accretion process itself, which has been developed by a number of authors in the framework of classical physics([18–20]) and which we employed in the previous section. Another has to do with the field decay, which in a complete theory will be coupled to the accretion process. The third aspect has to do with the structure of the neutron star and its response to added mass, but most especially to its response to changes in rotational frequency due to the changing centrifugal forces.

Typically, the moment of inertia has been computed in general relativity for a non-rotating star [22,23]. It is based on the Oppenheimer-Volkoff metric.

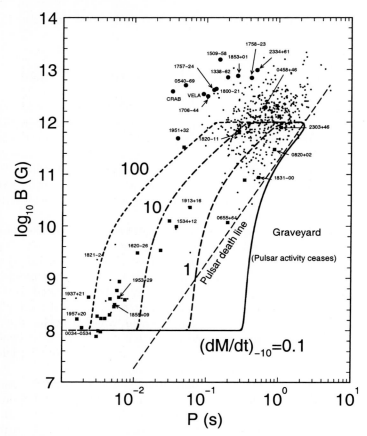

**Fig. 13.** Evolutionary tracks traced by neutron stars in the X-ray accretion stage, beginning on the death line with large $B$ field and ending as millisecond stars, for various accretion rates. Here $t_d = 10^6$ yr.

However, for the purpose of tracing the evolution of an accreting star from $\sim 1$ Hz to $400 - 600$ Hz we do not neglect the response of the shape, structure and composition of the star as it is spun up over this vast range of frequencies from essentially zero to values that approach the Kepler frequency. Nor do we neglect the dragging of local inertial frames. These features are included in our calculation of the tracks of neutron stars from canonical objects starting with large fields and very small frequencies at the "death line" to the small fields but rapid rotation of millisecond pulsars. However, the expression of the moment of inertia and a definition of the various factors that enter are too long to reproduce here. We refer instead to our derivation given in Refs. [11,9] which is computed to second order in the rotational angular momentum and found to be accurate to $\sim 10$ % when compared to numerical solutions for a rotating star [35]. In the

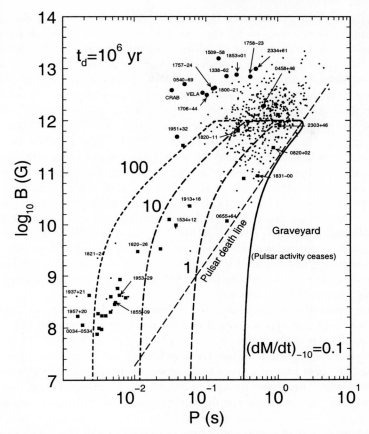

**Fig. 14.** Evolutionary tracks traced by neutron stars in the X-ray accretion stage, beginning on the death line with purely exponential decay of the $B$ field. As in Fig. 13, $t_d = 10^6$ yr.

present context, numerical solutions were obtained in Ref. [36], and semianalytic approximations were employed in Ref. [37].

The initial conditions for the evolution are arbitrary to a high degree. Canonical pulsars have a broad range of magnetic field strengths. The period of the pulsar at the time that the companion overflows its Roche lobe and accretion commences is also arbitrary. Any observed sample of x-ray accretors presumably spans a range in these variables. For concreteness, we assume that the pulsar has a field of $10^{12}$ G and that the period of the pulsar is 1 ms when accretion begins. The donor mass in the low-mass binaries are in the range 0.1 to $0.4 M_\odot$. Our sample calculations are for accretion of up to $0.4 M_\odot$.

We find essentially a continuum of evolutionary tracks in the $B - P$ plane according to the rate at which matter is accreted from the companion, and the rate at which the magnetic field decays. The evolutionary tracks essentially fill all the space in the B-P plane, starting at our assumed initial condition of an old canonical pulsar with field of $10^{12}$ G, and extending downward in field strength, broadening to fill the space on both sides of the deathline, and extending to the small periods of millisecond pulsars. All are potential tracks of some particular binary pair, since accretion rates vary by several orders of magnitude and presumably so do decay rates of the magnetic field. As a first orientation as to our results and how they relate to known pulsars as regards their magnetic field strength and their rotational period, we show the evolutionary tracks for four different accretion rates given in units of $\dot{M} = 10^{-10}$ solar masses per year in Fig. 13. The decay rate of the magnetic field is taken to have the value $t_d = 10^6$ yr in each case. The x-ray neutron star gains angular momentum and its period decreases, and over a longer timescale, the magnetic field decays. One can see already that a wide swathe of $B$ and $P$ is traced out.

In the above example, the field was assumed to decay to a finite asymptotic value of $10^8$ Gauss. A very different assumption, namely that the field decays eventually to zero, $B(t) = B(0)e^{-t/t_d}$, modifies only the results below the asymptotic value, as is seen by comparing Fig. 13 and 14. However, the conclusion concerning the origin of millisecond pulsars is quite different. For purely exponential decay, one would conclude that high frequency pulsars are created only in high accretion rate binaries.

In the remainder of the paper, we assume the field decays to an asymptotic value, since from the above comparison we see how exponential decay would modify the picture.

We show time tags on a sample track in Fig. 15 which provides some sense of time lapse. The first part of a track is traversed in short time, but the remainder ever more slowly. This shows up also in $dP/dt$ as a function of time. For each of three accretion rates we show the dependence on three field decay constants in Figs. 16, 17 and 18. Depending on decay rate of the field and accretion rate, an X-ray neutron star may spend some time on either side of the death line, but if it accreted long enough, always ends up as a candidate for a millisecond pulsar *if* the magnetic filed decays to an asymptotic value such as was assumed. However, if the field decays exponentially to zero, only high accretion rates would lead to millisecond pulsars. Of course, if accretion turns off at some time, the evolution is arrested.

In summary, we have computed the evolutionary tracks in the $B - P$ plane due to mass accretion onto neutron stars beginning at the death line with a typical field strength of $10^{12}$ Gauss, to shorter periods and low fields. According to the assumed accretion rate and decay constant for the magnetic field, the tracks indicate that the individual binaries with characteristics ranging from Z to Atoll sources will evolve along paths that cover a broad swathe in the $B - P$ plane. These include tracks of X-ray stars corresponding to low accretion rates that follow a path beyond the death line in the so-called "graveyard'. We have assumed two particular forms for the law of decay of the magnetic field. (1) The field approaches an asymptotic value of $10^8$ Gauss such as is typical of millisecond pulsars. This assumption leads to a particular form for the termination of evolutionary tracks. All accretors, no matter what the accretion

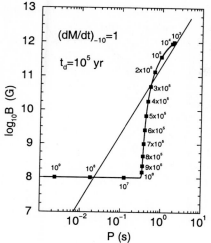 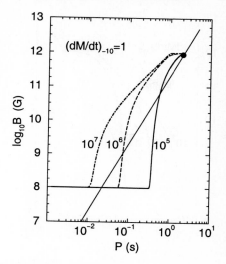

**Fig. 15.** Time tags expressed in years are shown for one of the evolutionary tracks corresponding to a decay constant for the magnetic field of $t_d = 10^5$ y.

**Fig. 16.** Evolutionary tracks for a neutron star starting at the death line and evolving by accretion to lower field and high frequency for an accretion rate 1 in units of $10^{-10}$ solar masses per year, and for three values of the magnetic field decay rate $t_d$ as marked.

rate, will end with millisecond periods, unless accretion ceases beforehand. (2) If instead, we had assumed a purely exponential decay, the tracks would not tend to an asymptotic value, but would continue to decrease in the strength of $B$. The tracks would still cover a broad swathe in the $B - P$ plane. But one would conclude that only the higher accretion rate binaries, particularly the Z-sources, could produce millisecond period neutron stars. If accretion continues for too long a time, the neutron star will be carried to very low fields and across the death line, or an overcritical mass will have been accreted, leading instead to a black hole.

## 5 Appendices

### First Order Phase Transition in Neutron Stars

We briefly recall some of the main characteristics of a first order phase transition in any substance having more than one conserved quantity, or as we will call it in the context of physics—conserved charge—such as baryon number or electric charge [38]. These are the two conserved quantities relevant to neutron stars. They have zero net electric charge and are made from baryons.

What makes a substance having more than one conserved charge different from a substance having only one, is that charges in the first case can be exchanged between two phases of the substance in equilibrium so as to minimize

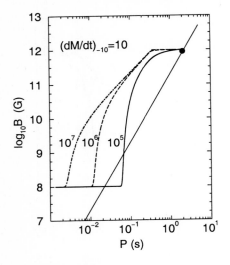 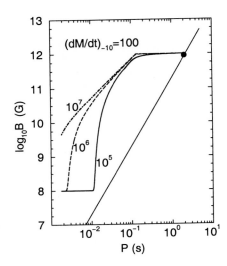

**Fig. 17.** Evolutionary tracks similar to Fig. 16 but with a different accretion rate $(dM/dt)_{-10} = 10$.

**Fig. 18.** Evolutionary tracks similar to Fig. 16 but with a different accretion rate $(dM/dt)_{-10} = 100$.

the energy. And the concentration of the charges can readjust at each proportion of the phases to minimize the energy. There is no such degree of freedom in a single-component substance, and there are $n - 1$ degrees of freedom in a substance having $n$ independent components or conserved charges.

In a single-component substance, the pressure at constant temperature remains unaltered for all proportions of the two phases as the substance is compressed. Only the proportion of the phases changes. This is not so for a substance of more than one conserved charge. The degree(s) of freedom to readjust the concentrations of the charges at each proportion of the phases causes a change in all internal properties of the two phases as their proportion changes including their common pressure under conditions of constant temperature.

The fact that the pressure varies as the proportion of phases in equilibrium in neutron star matter has an immediate consequence. The mixed phase of two phases in equilibrium will span a finite radial distance in a star. If the pressure were a constant for all proportions, then the mixed phase would be squeezed out of the star because the pressure varies monotonically, being greatest at the center and zero at the edge of the star. In early work on phase transitions in neutron stars, they were always treated as having constant pressure at the zero temperature of the star, so the mixed phase never appeared in any of those models.

The difference in properties of first order phase transitions in a one-component substance and one with several independent components or conserved charges is easily proven. First consider a one-component substance for which equilibrium of two phases, $A$ and $B$ is expressed by

$$p_A(\mu, T) = p_B(\mu, T).\tag{21}$$

At constant $T$, the solution for the chemical potential is obviously unique and independent of the proportion of the phases. So all properties of the two phases remain unaltered as long as they are in equilibrium, no matter the proportion.

Now consider a substance having two conserved charges (or independent components). For definiteness we consider the two independent conserved charges of neutron star matter, baryon number and electric charge, whose densities we denote by $\rho$ and $q$. For the corresponding chemical potentials we choose those of the neutron, $\mu_n$, and electron, $\mu_e$. The chemical potentials of all other particles can be written in terms of these independent ones. Gibbs phase equilibrium between the confined hadronic phase $C$ and the deconfined quark matter phase $D$ is now expressed as

$$p_C(\mu_n, \mu_e, T) = p_D(\mu_n, \mu_e, T).\tag{22}$$

This equation is insufficient to find the chemical potentials. At fixed $T$, it must be supplemented by another, say a statement of the conservation of one of the conserved charges. How should that statement be made? If one of the charges is the electric charge, demanding that the electric charge density should vanish identically in both phases would satisfy the condition of charge neutrality as required of a star.[2] That is in fact how charge neutrality in neutron stars was enforced for many years. However, it is overly restrictive. The net charge must vanish, but the charge density need not. Only the integrated electric charge density must vanish, $\int q(r) d^3r = 0$.[3] We refer to this as global conservation rather than local. It releases a degree of freedom that the physical system can exploit to find the minimum energy. To express this explicitly, we note that according to the preparation of the system, whether in the laboratory, or in a supernova, the concentration of the charges is fixed when the system is in a single phase. Denote the concentration by

$$c = Q/B.\tag{23}$$

However, when conditions of temperature or pressure change to bring the system into two phases equilibrium, the concentrations in each can be different

$$c_C = Q_C/B_C, \qquad c_D = Q_D/B_D\tag{24}$$

provided only that the total charges are conserved,

$$Q_C + Q_D = Q, \qquad B_C + B_D = B.\tag{25}$$

Here $Q$ and $B$ denote the total electric and baryon charge in a volume $V$. The rearrangement of charges will take place to minimize the energy of the system.

---

[2] The Coulomb force is so much stronger than the gravitational that the net charge per baryon has to be less than $10^{-36}$ which we can call zero.

[3] Familiar examples of neutral systems that have finite charge densities of opposite sign are atoms and neutrons.

The force that is responsible for exploiting this degree of freedom in neutron stars is the one responsible for the symmetry energy in nuclei and the valley of beta stability. Since neutron stars are far from symmetry, the symmetry energy is quite large; the difference in Fermi energies of neutron and proton, and the coupling of baryon isospin to the neutral rho meson are responsible.

For a uniform medium, and every sufficiently small region $V$ in a neutron star is uniform to high accuracy, the statement of global conservation is

$$\int_{V_C} q_C(r)\,d^3r + \int_{V_D} q_D(r)\,d^3r = V_C\,q_C + V_D\,q_D = Q \qquad (26)$$

where $V_C$ and $V_D$ denote the volume occupied by the two phases respectively. This can be written more conveniently as

$$(1-\chi)\,q_C(\mu_n,\mu_e) + \chi\,q_D(\mu_n,\mu_e) = Q/V \equiv \bar{q} \qquad (27)$$

where $q_C$ denotes the density of the conserved charge in the confined phase, $q_D$ in the deconfined phase, and

$$\chi = V_D/V, \quad V = V_C + V_D \qquad (28)$$

is the volume proportion of phase $D$ and $\bar{q}$ is the volume averaged electric charge (which for a star is zero). Now, equations (22) and (27) are sufficient to find $\mu_n$ and $\mu_e$. But notice that the solutions depend on the volume proportion $\chi$. Therefore, also all properties of the two phases depend on their proportion, including the common pressure. Having solved for the chemical potentials (and all field quantities specified by their equations of motion), the densities of the baryon conserved charge in the phases $C$ and $D$, are given by $\rho_C(\mu_n,\mu_e)$ and $\rho_D(\mu_n,\mu_e)$. The volume average of the baryon density is given by an equation corresponding to Eq. (27).

Let us now discuss the consequences of opening the degree of freedom embodied in Eq. (27), ie., in allowing electric charge (and strangeness) to be exchanged by the two phases in equilibrium so as to achieve the minimum energy at the corresponding baryon density. Because of the long range of the Coulomb force the Coulomb energy will be minimized when regions of like charge are small, whereas the surface interface energy will be minimized when the surface areas of the regions of the two phases is small. These are opposing tendencies, and in first order can be reconciled by minimizing their sum. A Coulomb lattice will form [38] of such a size and geometry of the rare phase immersed at spacings [39,40] in the dominant phase so as to minimize the energy.

In better approximation, the total energy, consisting of the sum of bulk energies, the surface and Coulomb energies and higher corrections such as the curvature energy,

$$E_{\text{Total}} \approx E_{\text{Bulk}} + E_{\text{Surface}} + E_{\text{Coulomb}} + E_{\text{Curvature}} + \cdots, \qquad (29)$$

should be minimized. In still better approximation, the convenient partition of the energy as above would be replaced by a lattice calculation of the total energy

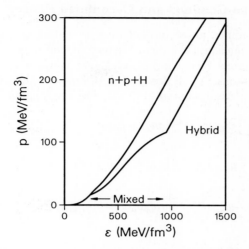

**Fig. 19.** Equation of state for a first order phase transition from neutron star matter in its confined to deconfined phases (marked hybrid). Note the monotonically increasing pressure. Comparison is made with the equation of state of neutron star matter in the confined phase with nucleons hyperons and leptons in equilibrium.

[41]. In general, the opening of the degree(s) of freedom to conserve charges globally rather than locally, or any other arbitrary way, can only lower the total energy from the value it would have were the degree of freedom closed, or in very special cases leave it unchanged. However, in the case of a neutron star, the degree of freedom allows the bulk energy in the normal phase to be lowered by decreasing the charge asymmetry of neutron star matter, so exchange of charge between the confined and deconfined phases is evidently favorable. It is unphysical to choose an arbitrary value of surface tension such that the mixed phase is energetically unfavored. The surface tension should be *calculated self-consistently* by minimizing the total energy Eq. (29), when possible [42].

However, our purpose here is not to calculate the geometric structure[4] of the mixed phase but to exhibit the equation of state for a first order deconfinement transition in neutron star matter according to the above principles. Fig. 19 shows the equation of state with the pure phases at low and high density and the mixed phase between. Note, as pointed out before that though the phase transition is first order, having as it does a mixed phase, the pressure (and all other internal properties) vary with density. The monotonic increase of pressure contrasts with early treatments of phase transitions in neutron stars prior to 1991, for which approximations rendered a constant pressure in the mixed phase [43]. For a constant pressure phase transition, gravity will squeeze out the mixed phase. Pressure is a monotonically decreasing function of distance from the center of a star, just as it is in our atmosphere.

---

[4] See Ref. [47] for a review.

## Description of the Confined and Deconfined Phases

To describe the confined phase of neutron star matter, we use a generalization [44] of relativistic nuclear field theory solved at the mean field level in which nucleons and hyperons (the baryon octet) are coupled to scalar, vector and vector-isovector mesons. A full description of how the theory can be solved for neutron star matter can be found in [5,44]. The Lagrangian is

$$\mathcal{L} = \sum_B \overline{\psi}_B (i\gamma_\mu \partial^\mu - m_B + g_{\sigma B}\sigma - g_{\omega B}\gamma_\mu \omega^\mu$$
$$- \frac{1}{2} g_{\rho B} \gamma_\mu \tau \cdot \rho^\mu)\psi_B + \frac{1}{2}(\partial_\mu \sigma \partial^\mu \sigma - m_\sigma^2 \sigma^2)$$
$$- \frac{1}{4}\omega_{\mu\nu}\omega^{\mu\nu} + \frac{1}{2}m_\omega^2 \omega_\mu \omega^\mu - \frac{1}{4}\rho_{\mu\nu}\cdot\rho^{\mu\nu} + \frac{1}{2}m_\rho^2 \rho_\mu \cdot \rho^\mu$$
$$- \frac{1}{3}bm_n(g_\sigma\sigma)^3 - \frac{1}{4}c(g_\sigma\sigma)^4$$
$$+ \sum_{e^-,\mu^-} \overline{\psi}_\lambda (i\gamma_\mu \partial^\mu - m_\lambda)\psi_\lambda. \qquad (30)$$

The sum on $B$ is over all charge states of the baryon octet. The parameters of the nuclear Lagrangian can be algebaically determined (see 2'nd ed. of [5]) so that symmetric nuclear matter has the following properties: binding energy of symmetric nuclear matter $B/A = -16.3$ MeV, saturation density $\rho = 0.153$ fm$^{-3}$, compression modulus $K = 300$ MeV, symmetry energy coefficient $a_{\text{sym}} = 32.5$ MeV, nucleon effective mass at saturation $m^\star_{\text{sat}} = 0.7m$ and ratio of hyperon to nucleon couplings $x_\sigma = 0.6$, $x_\omega = 0.653 = x_\rho$ that yield, together with the foregoing parameters, the correct $\Lambda$ binding in nuclear matter [45]).

Quark matter is treated in a version of the MIT bag model with the three light flavor quarks ($m_u = m_d = 0$, $m_s = 150$ MeV) as described in Ref. [46]. A value of the bag constant $B^{1/4} = 180$ MeV is employed.

## Acknowledgments

This work was supported by the Director, Office of Energy Research, Office of High Energy and Nuclear Physics, Division of Nuclear Physics, of the U.S. Department of Energy under Contract DE-AC03-76SF00098. F. Weber was supported by the Deutsche Forschungsgemeinschaft.

# References

1. M. A. Alpar, A. F. Cheng, M. A. Ruderman and J. Shaham, Nature **300** (1982) 728.
2. D. Bhattacharya and E. P. J. van den Heuvel, Phys. Rep., **203** (1991) 1.
3. R. Wijnands and M. van der Klis, Nature **394** (1998) 344.

4. D. Chakrabarty and E. H. Morgan, Nature **394** (1998) 346.
5. N. K. Glendenning, *COMPACT STARS* (Springer–Verlag New York, 1'st ed. 1996, 2'nd ed. 2000).
6. N. K. Glendenning, S. Pei and F. Weber, Phys. Rev. Lett. **79** (1997) 1603.
7. N. K. Glendenning, Nucl. Phys. **A638** (1998) 239c.
8. H. Heiselberg and M. Hjorth-Jensen, Phys. Rev. Lett. **80** (1998) 5485.
9. E. Chubarian, H. Grigorian, G. Poghosyan and D. Blaschke, Astron. Astrophys. **357** (2000) 968.
10. F. Pacini, Nature **216** (1967) 567.
11. N. K. Glendenning and F. Weber, Astrophys. J. **400** (1992) 647.
12. F. Weber and N. K. Glendenning, Astrophys. J. **390** (1992) 541.
13. A. Johnson, H. Ryde and S. A. Hjorth, Nucl. Phys. **A179** (1972) 753.
14. F. S. Stephens and R. S. Simon, Nucl. Phys. **A183** (1972) 257.
15. B. R. Mottelson and J. G. Valatin, Phys. Rev. Lett. **5** (1960) 511.
16. N. K. Glendenning and F. Weber, astro-ph/0003426, (2000).
17. M. van der Klis, Millisecond Oscillations in X-Ray Binaries, to appear in Ann. Rev. Astron. Astrophys, (2000).
18. R. F. Elsner and F. K. Lamb, Astrophys. J. **215** (1977) 897.
19. P. Ghosh, F. K. Lamb and C. J. Pethick, Astrophys. J. **217** (1977) 578.
20. V. M. Lupinov, *Astrophysics of Neutron Stars*, (Springer-Verlag, New York, 1992.
21. S. Konar and D. Bhattacharya, MNRAS **303** (1999) 588; op. cit. **308** (1999) 795.
22. J. B. Hartle, Astrophys. J. **150** (1967) 1005.
23. J. B. Hartle and K. S. Thorne, Astrophys. J. **153** (1968) 807.
24. F. Weber, J. Phys. G: Nucl. Part. Phys. **25** (1999) R195.
25. I. Tanihata, Prog. Part. Nucl. Phys. **35** (1995) 505.
26. P. G. Hansen, A. S. Jensen and B. Johnson, Ann. Rev. Nucl. Part. Sci. **45** (1997) 1644.
27. Bao-An Li, C. M. Ko and Zhongzhou Ren, Phys. Rev. Lett. **78** (1997) 1644.
28. L. Bildsten, Astrophys. J. **501** (1998) L89.
29. N. Anderson, D. I. Jones, K. D. Kokkotas and N. Sterigioulas, Astrophys. J. **534** (2000) L89.
30. Y. Levin, Astrophys. J. **517** (1999) 328.
31. N. K. Glendenning and F. Weber, astro-ph/0010336 (2000).
32. V. A. Urpin and U. Geppert, MNRAS **278** (1996) 471.
33. V. A. Urpin and D. Konenkov, MNRAS **284** (1997) 741.
34. V. A. Urpin, U. Geppert and D. Konenkov, MNRAS **295** (1998) 907.
35. M. Salgado, S. Bonazzola, E. Gourgoulhon and P. Haensel, Astron. Astrophys. **291** (1994) 155.
36. G. B. Cook, S. L. Shapiro and S. A. Teukolsky, Astrophys. J. Lett, **423** (1994) L117; op. cit. **424** (1994) 823.
37. L. Burderi, M. Colpi, T. Di Salvo, N. D'Amico, Astrophys. J. **519** (1999) 285.
38. N. K. Glendenning, Nucl. Phys. B (Proc. Suppl.) **24B** (1991) 110; Phys. Rev. D **46** (1992) 1274.
39. N. K. Glendenning and S. Pei, Phys. Rev. C **52** (1995) 2250.
40. N. K. Glendenning and S. Pei, in the Eugene Wigner Memorial Issue of Heavy Ion Physics (Budapest) **1** (1995) 1.
41. R. D. Williams and S. E. Koonin, Nucl. Phys. **A435** (1985) 844.
42. M. Christiansen, N. K. Glendenning and J. Schaffner-Bielich, Phys. Rev. C **62** (2000) 025804.
43. N. K. Glendenning, Phys. Rev. D **46** (1992) 1274.

44. N. K. Glendenning, Astrophys. J. **293** (1985) 470.
45. N. K. Glendenning and S. A. Moszkowski, Phys. Rev. Lett. **67** (1991) 2414.
46. E. Farhi and R. L. Jaffe, Phys. Rev. D **30** (1984) 2379.
47. N. K. Glendenning, Phys. Rep. **342** (2001) 393.

# Supernova Explosions and Neutron Star Formation

Hans-Thomas Janka, Konstantinos Kifonidis, and Markus Rampp

Max-Planck-Institut für Astrophysik, Karl-Schwarzschild-Str. 1, D-85741 Garching, Germany

**Abstract.** The current picture of the collapse and explosion of massive stars and the formation of neutron stars is reviewed. According to the favored scenario, however by no means proven and undisputed, neutrinos deposit the energy of the explosion in the stellar medium which surrounds the nascent neutron star. Observations, in particular of Supernova 1987A, suggest that mixing processes play an important role in the expanding star, and multi-dimensional simulations show that these are linked to convective instabilities in the immediate vicinity of the neutron star. Convectively enhanced energy transport inside the neutron star can have important consequences for the neutrino emission and thus the neutrino-heating mechanism. This also holds for a suppression of the neutrino interactions at nuclear densities. Multi-dimensional hydrodynamics, general relativity, and a better understanding of the neutrino interactions in neutron star matter may be crucial to resolve the problem that state-of-the-art spherical models not yield explosions even with a very accurate treatment of neutrino transport by solving the Boltzmann equation.

## 1 Introduction

Baade and Zwicky [2] were the first who speculated about a connection between supernova explosions and the origin of neutron stars. They recognized that stellar cores must become unstable because neutrons, being produced by captures of degenerate electrons on protons, define an energetically advantageous state. The gravitational binding energy liberated by the collapse of a stellar core could power a supernova explosion. More than thirty years later, Colgate and White explored this idea by performing numerical simulations [21]. Their models showed that the prompt hydrodynamical shock, which forms at the moment of core bounce, is not able to reach the outer layers of the star due to severe energy losses by photointegration of iron nuclei. Therefore they suggested that the neutrinos, which are emitted in huge numbers from the hot, collapsed core and carry away the binding energy of the forming neutron star, deposit a fraction of their energy in the stellar medium external to the shock. Later, more accurate simulations with an improved treatment of the input physics such as nuclear equation of state and neutrino transport [9–12,48,3,59] confirmed the failure of the prompt shock, but could not find the neutrino-driven explosion imagined by Colgate and White.

The discovery of weak neutral currents, whose existence has been predicted by the standard model of electro-weak interactions, had a major influence also on supernova models. It became clear that the neutrinos are strongly coupled to the stellar plasma and can escape only on timescales of seconds. This implied that the neutrino luminosities did not become large enough that the rate of energy or momentum transfer by neutrino interactions with the nuclei in the stellar plasma ahead of the supernova shock could become dynamically important. In 1982, however, Jim Wilson discovered [62] that the stalled prompt shock can be revived by neutrino heating in the dissociated postshock medium on a longer timescale. While immediately after shock breakout neutrino emission extracts energy from the shocked gas, energy transfer from neutrinos to the postshock medium is favored hundreds of milliseconds later. If the shock has expanded to a larger radius and the postshock temperature has decreased, energetic neutrinos, which stream up from deeper regions, can deposit a small fraction of their energy in a gain layer behind the shock [8]. Although the efficiency is much lower than imagined by Colgate and White — typically only a few per cent of the neutrino energy are left in the stellar medium — the neutrino heating mechanism can yield the energy of about $10^{51}$ erg for the explosion of a massive star like, e.g., Supernova 1987A.

Although Wilson's simulations showed the principle viability of this mechanism and theoretical investigations enlightened the underlying physics [4–7], later simulations revealed a strong sensitivity to the details of the post-bounce evolution of the collapsed stellar core [65,3,32,44]. General relativity, the nuclear equation of state and corresponding properties of the nascent neutron star, the treatment of the neutrino transport and neutrino-matter interactions, and the structure of the collapsing star have important influence. In particular, later models of Mayle and Wilson [63,64] produced sufficiently powerful explosions only in case of enhanced neutrino emission from the nascent neutron star due to neutron-finger instabilities. Whether such mixing processes take place, however, is not clear to date [15]. Alternatively, the hot neutron star might develop Ledoux convection [22,17] due to negative gradients of lepton number and/or entropy, a possibility which is suggested by the structure obtained in state-of-the-art spherical models for the neutrino-cooling phase of proto-neutron stars [49,47] and by two-dimensional hydrodynamical simulations [37], but is not generally accepted [43].

Supernova 1987A has provided us with evidence for large-scale mixing processes, which involve the layers in the near vicinity of the neutron star where radioactive nuclei are produced. Indeed, multi-dimensional simulations have shown the existence of hydrodynamical instabilities in the neutrino-heated region already during the first second of the explosion [28,46,31,32,20]. This convective overturn behind the shock was recognized to allow for an explosion even when spherical models do not explode. Also, it enhances the explosion energy and sets the timescale until the shock gains momentum. But still it is not clear whether it is sufficient and crucial for the neutrino-driven mechanism to be at work, or whether it is just an unavoidable side-effect [44]. The existence of these mixing processes, however, definitely means that the events leading to an explosion, the energetics of the explosion, the nucleosynthesis of radioactive elements during the first second, and the composition and distribution of the ejecta can be understood quantitatively only by multi-dimensional simulations.

Unfortunately, the 24 neutrinos from Supernova 1987A, which were recorded in the underground experiments of Kamioka, IMB and Baksan [1], did not provide enough statistical information to draw conclusions on the events which instigated the explosion of the star. Although these neutrinos confirm the basic picture of stellar core collapse and neutron star formation, they are not suitable to support the neutrino-driven mechanism. Therefore the latter may be considered as the currently most favorable explanation for the explosion, but empirical evidence cannot be put forward and numerical models do not draw a clear and unambiguous picture.

It cannot be excluded that the energy for supernova explosions of massive stars is provided by some other mechanism than neutrino heating, for example by magneto-hydrodynamical processes. However, we know that neutrinos with the expected characteristics are emitted from collapsed stellar cores, we know that these neutrinos carry away the gravitational binding energy of the nascent neutron star, we know that neutrino heating *must* occur behind the stalled shock some time after core bounce, we know that analytic studies and numerical simulations find explosions for a suitable combination of conditions, we know that the energy transfer by neutrinos can be strong enough to account for "normal" explosions with a canonical energy of $\sim 10^{51}$ erg, and we know that the timescale of shock rejuvenation by neutrino heating determines the mass cut such that accretion of the collapsed stellar core and later fallback lead to neutron star masses and supernova nucleosynthesis in rough agreement with observations [23,25,66]. Of course, the discrepant results of numerical simulations are unsatisfactory, and major problems are nagging: Why do the supposedly best and most advanced spherical models not produce explosions? Can one trust current multi-dimensional simulations with their greatly simplified and approximate treatment of neutrino transport? Is rotation in neutrino-driven explosions sufficient to explain the observed large asphericities and anisotropies in many supernovae [24]? What is the reason for the kicks by which pulsars are accelerated to average velocities of several hundred km/s presumably during the supernova explosion? What powers hyperenergetic supernovae, which seem to release up to 50 times more kinetic energy than ordinary explosions of massive stars [30]?

This paper gives an overview over the major lines of research on the standard explosion scenario of massive stars, and on neutron star formation, where progress has been achieved during the past few years or will be coming up in the near future. In Section 2 the physics of neutrino-driven explosions will be discussed in some detail, and an analytic toy model will be described which allows one a deeper understanding of the requirements of neutrino-driven explosions. In Section 3 the current status of supernova modeling in spherical symmetry will be outlined, with particular focus on results with neutrino transport by solving the Boltzmann equation. In Section 4 the relevance of convective overturn in the neutrino-heated layer will be discussed. New results of two-dimensional simulations will be presented, which take into account the nucleosynthesis during

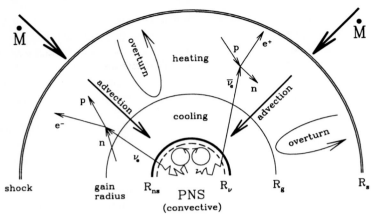

**Fig. 1.** Sketch of the post-collapse stellar core during the neutrino heating and shock revival phase. $R_\nu$ is the neutrinosphere radius, $R_{\rm ns}$ the protoneutron star radius, $R_{\rm g}$ the gain radius outside of which net neutrino heating exceeds neutrino cooling, and $R_{\rm s}$ is the shock radius. The shock expansion is impeded by mass infall at a rate $\dot{M}$, but supported by convective energy transport from the region of strongest neutrino heating into the post-shock layer. Convection inside the nascent neutron star raises the neutrino luminosities.

the explosion and follow the shock from the moment of its formation until it breaks out of the surface of the star. In Section 5, two-dimensional hydrodynamical models of the neutrino-cooling phase of newly formed neutron stars will be described. The effects of rotation and of a suppression of neutrino-nucleon interactions by nucleon correlations in the nuclear medium will be addressed. Section 6 will contain a summary and conclusions.

## 2 The Explosion Mechanism

The physics of neutrino-driven explosions is discussed, first on the level of basic considerations, then with the help of an analytic toy model, which allows one to study the competing effects that determine the destiny of the stalled supernova shock.

### 2.1 Neutrino Heating

Figure 1 displays a sketch of the neutrino cooling and heating regions outside the proto-neutron star at the center. The main processes of neutrino energy deposition are the charged-current reactions $\nu_e + n \to p + e^-$ and $\bar{\nu}_e + p \to n + e^+$ [8]. With the neutrino luminosity, $L$, the average squared neutrino energy, $\langle \epsilon^2 \rangle$, and the mean value of the cosine of the angle $\theta_\nu$ of the direction of neutrino propagation relative to the radial direction, $\langle \mu \rangle = \langle \cos \theta_\nu \rangle$, being defined as moments of the neutrino phase space distribution function $f(r, t, \mu, \epsilon)$ by integration over energies $\epsilon$ and angles $\mu$ according to

$$L = 4\pi r^2 \frac{2\pi c}{(hc)^3} \int_{-1}^{+1} d\mu \int_0^\infty d\epsilon \, \epsilon^3 \mu f \,, \tag{1}$$

$$\langle \epsilon^2 \rangle = \int_{-1}^{+1} d\mu \int_0^\infty d\epsilon \, \epsilon^5 f \cdot \left\{ \int_{-1}^{+1} d\mu \int_0^\infty d\epsilon \, \epsilon^3 f \right\}^{-1} \,, \tag{2}$$

$$\langle \mu \rangle = \int_{-1}^{+1} d\mu \int_0^\infty d\epsilon \, \epsilon^3 \mu f \cdot \left\{ \int_{-1}^{+1} d\mu \int_0^\infty d\epsilon \, \epsilon^3 f \right\}^{-1} \,, \tag{3}$$

the heating rate per nucleon ($N$) is approximately given by

$$\begin{aligned} Q_\nu^+ &\approx 110 \cdot \left( \frac{L_{\nu_e,52} \langle \epsilon^2_{\nu_e,15} \rangle}{r_7^2 \, \langle \mu \rangle_{\nu_e}} Y_n + \frac{L_{\bar{\nu}_e,52} \langle \epsilon^2_{\bar{\nu}_e,15} \rangle}{r_7^2 \, \langle \mu \rangle_{\bar{\nu}_e}} Y_p \right) \left[ \frac{\text{MeV}}{\text{s} \cdot N} \right] \\ &\approx 55 \cdot \frac{L_{\nu,52} \langle \epsilon^2_{\nu,15} \rangle}{r_7^2 \, f} \left[ \frac{\text{MeV}}{\text{s} \cdot N} \right] \,, \end{aligned} \tag{4}$$

where $Y_n$ and $Y_p$ are the number fractions of free neutrons and protons (number densities divided by baryon number density), respectively. In the second equation $Y_n + Y_p \approx 1$, and equal luminosities and spectra for $\nu_e$ and $\bar{\nu}_e$ were assumed. $L_{\nu,52}$ denotes the total luminosity of $\nu_e$ plus $\bar{\nu}_e$ in $10^{52}$ erg/s, $r_7$ the radial position in $10^7$ cm, $\langle \epsilon^2_{\nu,15} \rangle$ is measured in units of 15 MeV, and $f = \langle \mu \rangle_\nu$ is very small in the opaque regime where the neutrinos are isotropic, adopts a value of about 0.25 around the neutrinosphere, and approaches unity for radially streaming neutrinos very far out. Note that the "flux factors" $\langle \mu \rangle_\nu$ determines the neutrino energy density at a radius $r$ according to $\varepsilon_\nu(r) = L_\nu/(4\pi r^2 c \langle \mu \rangle_\nu)$.

Using this energy deposition rate, neglecting energy losses due to the re-emission of neutrinos, and assuming that the gravitational binding energy of a nucleon in the neutron star potential is (roughly) balanced by the sum of internal and nuclear recombination energies after accretion of the infalling matter through the shock, one can estimate (very approximately) the explosion energy to be of the order

$$E_{\text{exp}} \sim 10^{51} \cdot \frac{L_{\nu,52} \langle \epsilon^2_{\nu,15} \rangle}{R_{\text{g},7}^2 \, f} \left( \frac{\Delta M}{0.1\,M_\odot} \right) \left( \frac{\Delta t}{0.1\,\text{s}} \right) - E_{\text{gb}} + E_{\text{nuc}} \quad [\text{erg}] \,. \tag{5}$$

Here $\Delta M$ is the heated mass, $\Delta t$ the typical heating timescale, $E_{\text{gb}}$ the (net) total gravitational binding energy of the overlying, outward accelerated stellar layers, and $E_{\text{nuc}}$ the additional energy from explosive nucleosynthesis, which is a significant contribution of a few $10^{50}$ erg only for progenitors with main sequence masses above $20\,M_\odot$, and which roughly compensates $E_{\text{gb}}$[1].

---

[1] The latter statement is supported by the following argument (S. Woosley, personal communication): Material with a specific gravitational binding energy $\Phi_{\text{grav}}$ which is equal to or larger than the nuclear energy release per gram in Si burning, $e_{\text{nuc}} \sim 10^{18}$ erg/g, is located interior to the radius where the temperatures can become high

It is not easy to infer from Eq. (5) the dependence of the explosion energy on the neutrino luminosity. On the one hand,

$$E_{\text{exp}} \propto Q_\nu^+ V \Delta t \propto \frac{L_\nu \langle \epsilon_\nu^2 \rangle}{R_g^2} \Delta M \Delta t \propto L_\nu \Delta \tau \Delta t, \qquad (6)$$

where $V$ is the heated volume between gain radius and shock, and $\Delta \tau$ the optical depth of the heating layer. On the other hand,

$$\Delta \tau \propto \langle \epsilon_\nu^2 \rangle R_g \rho_g \propto \langle \epsilon_\nu^2 \rangle R_g T_g^3 \propto L_\nu^{1/2} \langle \epsilon_\nu^2 \rangle^{3/2}. \qquad (7)$$

Here $\rho \propto T^3$ was assumed for the relation between density and temperature in the heating layer [7], and $Q_\nu^+(R_g) = Q_\nu^-(R_g)$ was used at the gain radius, where neutrino heating is balanced by neutrino cooling. The energy loss rate $Q_\nu^-$ by neutrinos produced in capture reactions of nondegenerate electrons and positrons on nucleons, scales with $T^6$. Combining Eqs. (6) and (7) yields

$$E_{\text{exp}} \propto L_\nu^{3/2} \langle \epsilon_\nu^2 \rangle^{3/2} \Delta t \propto L_\nu^{9/4} \Delta t, \qquad (8)$$

when $\langle \epsilon_\nu^2 \rangle \propto T_\nu^2 \propto L_\nu^{1/2}$ is used for black-body like emission.

If the expansion velocity were simply proportional to $E_{\text{exp}}^{1/2}$, in which case the time $\Delta t$ for the shock to reach a given radius would be $\Delta t \propto E_{\text{exp}}^{-1/2}$, then $E_{\text{exp}} \propto L_\nu^{3/2}$ [7]. However, when shock expansion sets in, most of the energy is internal energy, but not kinetic energy, making the relation $\Delta t \propto E_{\text{exp}}^{-1/2}$ very questionable. The actual variation of $\Delta t$ with the inverse of the neutrino luminosity can be steeper.

## 2.2 Requirements for Neutrino-Driven Explosions

In order to get explosions by the delayed neutrino-heating mechanism, certain conditions need to be fulfilled. Expansion of the postshock region requires sufficiently large pressure gradients near the radius $R_{\text{cut}}$ of the developing mass cut. If one neglects self-gravity of the gas in this region and assumes the density profile to be a power law, $\rho(r) \propto r^{-n}$ (which is well justified according to numerical simulations which yield a power law index of $n \approx 3$ [5]), one gets $P(r) \propto r^{-n-1}$ for the pressure in an atmosphere near hydrostatic equilibrium. Outward acceleration is therefore maintained as long as the following condition for the "critical" internal energy density $\varepsilon$ holds:

$$\left. \frac{\varepsilon_c}{GM\rho/r} \right|_{R_{\text{cut}}} > \frac{1}{(n+1)(\gamma-1)} \cong \frac{3}{4}, \qquad (9)$$

---

enough ($T \gtrsim 5 \times 10^9$ K) for explosive nucleosynthesis of $^{56}$Ni. From $\Phi_{\text{grav}} = GM/r \gtrsim 10^{18}$ erg/g one estimates a radius of $r \lesssim 2 \times 10^8$ cm, and from $\frac{4}{3}\pi r^3 a T^4 \sim 10^{51}$ erg with $T \gtrsim 5 \times 10^9$ K one finds $r \lesssim 4 \times 10^8$ cm. This means that the energy release from explosive nucleosynthesis is easily able to account for the gravitational binding energy of the burning material.

where use was made of the relation $P = (\gamma - 1)\varepsilon$. The numerical value was obtained for $\gamma = 4/3$ and $n = 3$. This condition can be converted into a criterion for the entropy per baryon, $s$. Using the thermodynamical relation for the entropy density normalized to the baryon density $n_b$, $s = (\varepsilon + P)/(n_b T) - \sum_i \eta_i Y_i$ where $\eta_i$ ($i = n, p, e^-, e^+$) are the particle chemical potentials divided by the temperature, and assuming completely disintegrated nuclei behind the shock so that the number fractions of free protons and neutrons are $Y_p = Y_e$ and $Y_n = 1 - Y_e$, respectively, one gets

$$s_c(R_{\text{cut}}) \gtrsim 14 \left.\frac{M_1}{r_7 T}\right|_{R_{\text{cut}}} - \left.\ln\left(1.27 \cdot 10^{-3} \frac{\rho_9 Y_n}{T^{3/2}}\right)\right|_{R_{\text{cut}}} \quad [k_B/N] \; . \tag{10}$$

In this approximate expression a term with a factor $Y_e$ was dropped (its absolute value being usually less than 0.5 in the considered region), nucleons were assumed to obey Boltzmann statistics, and $T$ is measured in MeV, $M_1$ in units of $M_\odot$, $\rho_9$ in $10^9$ g/cm$^3$, and $r_7$ in $10^7$ cm. Inserting typical numbers ($T \approx 1.5$ MeV, $Y_n \approx 0.3$, $R_{\text{cut}} \approx 1.5 \cdot 10^7$ cm), one finds $s > 15\, k_B/N$, which gives an estimate of the entropy in the heating region when expansion is going to take place.

Since the entropy and energy density in the postshock later are raised by neutrino energy deposition, the conditions of Eqs. (9) and (10) imply requirements on the neutrino emission of the proto-neutron star. These can be derived by the following considerations. A stalled shock is converted into a moving one only, when the neutrino heating is strong enough to increase the pressure behind the shock by a sufficient amount. From the Rankine-Hugoniot relations at the shock, a criterion can be deduced for the heating rate per unit mass, $q_\nu$, behind the shock, which leads to a positive postshock velocity ($u_1 > 0$) [12]:

$$q_\nu > \frac{2\beta - 1}{\beta^3(\beta - 1)(\gamma - 1)} \frac{|u_0|^3}{\eta R_s} \; . \tag{11}$$

Here $\beta$ is the ratio of postshock to preshock density, $\beta = \rho_1/\rho_0$, $\gamma$ the adiabatic index of the gas (assumed to be the same in front and behind the shock), and $\eta$ defines the fraction of the shock radius $R_s$ where net heating by neutrino processes occurs: $\eta = (R_s - R_g)/R_s$. $u_0$ is the preshock velocity, which is a fraction $\alpha$ (analytical and numerical calculations show that typically $\alpha \approx 1/\sqrt{2}$) of the free fall velocity: $u_0 = \alpha\sqrt{2GM/r}$. Assuming a strong shock, one has $\beta = (\gamma + 1)/(\gamma - 1)$, which becomes $\beta = 7$ for $\gamma = 4/3$. With typical numbers, $R_s = 100$ km, $M = M_1 = 1\,M_\odot$, and $\eta \approx 0.4$, one derives for the critical luminosity of $\nu_e$ plus $\bar{\nu}_e$:

$$L_{\nu,52}\langle\epsilon^2_{\nu,15}\rangle > 4.4\, \frac{M_1^{3/2}}{R_{s,7}^{1/2}} \; . \tag{12}$$

Since this discussion was very approximate, e.g., the reemission of neutrinos was ignored and properties depending on the structure of the collapsed stellar core were absorbed into free parameters, the analysis cannot yield a quantitatively meaningful value for the threshold luminosity. However, the existence of a lower

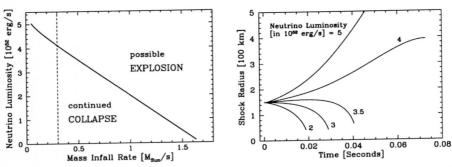

**Fig. 2.** *Left:* Phase diagram for successful explosion or continued stellar collapse in dependence upon the rate at which gas falls into the shock and upon the $\nu_e$ plus $\bar\nu_e$ luminosity of the nascent neutron star. The mass of the neutron star was assumed to be $1.25\,M_\odot$. *Right:* Shock positions as functions of time for different values of the $\nu_e$ plus $\bar\nu_e$ luminosity of the neutron star (according to the labels) and fixed value of the rate at which gas falls into the shock (marked by the dashed line in the left plot). For neutrino luminosities above the critical value (where the dashed line crosses the solid line in the left plot) explosions can occur.

bound on the neutrino luminosity as found in numerical simulations [31,32], is confirmed. Above this threshold value, neutrino heating of the gas behind the shock is strong enough to drive an expansion of the gain layer.

### 2.3 Analytic Toy Model

The discussion in the previous section was overly simplified. The behavior of the stagnant shock does not only depend on the neutrino heating in the gain region, but is also influenced by the energy loss and the settling of the cooling layer, and by the mass infall from the collapsing star star ahead of the shock. Only hydrodynamical simulations can determine the shock evolution in response to these different, partly competing effects. Analytic discussions, however, can help one understanding the significance of different effects and thus can supplement more detailed, but less transparent supercomputer calculations.

The conditions in the region of neutrino heating and the details of the heating process were analysed by Bethe and Wilson [8] and Bethe [4–7]. Burrows and Goshy [18] considered the post-bounce accretion phase of the supernova shock as a quasi steady-state situation, and thus replaced the time-dependent partial differential equations of hydrodynamics by a set of ordinary differential equations to determine the radial position of the standing shock by an eigenvalue analysis. While this approach captures an interesting aspect of the problem, it has serious weaknesses. The accretion flow between the shock and the neutron star does not need to be stationary, but neutrino-heated matter may stay in the gain region, or gas will pile up on the forming neutron star when neutrinos are unable to remove energy quickly enough for the gas to settle. In particular, when the shock accelerates outward, the transition from accretion to outflow cannot be described as a stationary situation.

Since the gas falling into the stalled shock is strongly decelerated and the postshock velocities are smaller than the local sound speed, the structure of the collapsed stellar core is rather simple and can be well described by hydrostatic equilibrium. This allows for an approximate treatment, which is complementary to the approach taken by Burrows and Goshy [18]: Integrating the stellar structure over radius leads to conservation laws for the total mass and energy in the gain layer. The time-dependent radius and velocity of the shock can then be obtained as solutions of an *initial value problem*, which reflects the fact that the destiny of the shock depends on the initial conditions and is controlled by the cumulative effects of neutrino energy deposition and mass accumulation in the gain layer.

Such an analysis also demonstrates the existence of a threshold value for the neutrino luminosity from the neutron star, which is needed to drive shock expansion. This threshold luminosity depends on the rate, $\dot{M}$, of mass infall to the shock, on the neutron star mass and radius, and to some degree also on the shock stagnation radius. Taking into account only the main dependence on $\dot{M}$, it can roughly be written as

$$L_{\nu,\text{crit}}(\dot{M}) \approx L_0 - L_1 \left( \frac{\dot{M}}{M_\odot/\text{s}} \right), \quad (13)$$

with $L_0 \approx 5 \times 10^{52}$ erg s$^{-1}$ and $L_1 \approx 3 \times 10^{52}$ erg s$^{-1}$ for the conditions of Fig. 2.

The neutrino heating in the gain layer is not the only important factor that determines the shock propagation. Energy loss by neutrino emission in the cooling layer has a considerable influence, because it regulates the settling of the matter that is accreted by the nascent neutron star, and therefore the advection of gas through the heating layer. If cooling is inefficient, gas piles up on the neutron star and pushes the shock farther out. If cooling is very efficient, the gas contracts quickly and more gas is dragged downward through the gain radius, extracting mass and energy from the gain layer and thus weakening the support for the shock. This also means that the infall velocity behind the shock increases and the timescale for the gas to stay in the gain layer is reduced. Therefore the efficiency of neutrino energy deposition drops. Such an effect is harmful for shock expansion. It can be diminished by higher $\nu_e$ and $\bar{\nu}_e$ luminosities from the neutrinosphere, which lead to an enhancement of neutrino absorption relative to neutrino emission. On the other hand, muon and tau neutrino and antineutrino production in the accretion layer of the neutron star has a desastrous consequence for the shock, because it is a sink of energy that leaves the star without any significant positive effect above the neutrinosphere, where only $\nu_e$ and $\bar{\nu}_e$ can be absorbed by free nucleons.

These different, competing processes combined explain the slope of the critical line in the left plot of Fig. 2 and in Eq. (13). Shock expansion and acceleration are easier for high mass infall rates, $\dot{M}$, into the shock and for high $\nu_e$ and $\bar{\nu}_e$ luminosities from the nascent neutron star. These luminosities need to be larger when $\dot{M}$ is small. It must be pointed out here, however, that this dependence is a consequence of the fact that the temperature in the cooling layer is considered as a parameter of the discussion. It is assumed to be equal to the neutrinospheric temperature and thus to be mainly determined by the interaction with the neutrino flux from the core of the neutron star, but not by the mass infall and the dynamics in the accretion layer.

Emission of muon and taun neutrinos and antineutrinos from the cooling region is not included in the results displayed in Fig. 2. It would move the critical line to appreciably higher values of the $\nu_e$ plus $\bar{\nu}_e$ luminosity of the collapsed core, which is given along the ordinate.

Neutrino heating is stronger close to the gain radius than right behind the shock. Using an isentropic profile in the gain layer, the evaluation, however, implies very efficient energy transport, e.g., by convective motions in the gain layer. This enhances the postshock pressure and reduces the loss of energy from the gain layer, which is associated with the inward advection of neutrino-heated gas.

Solutions of the toy model for varied parameters show that the energy in the gain layer and therefore the explosion energy of the supernova is limited to some $10^{51}$ erg. The reason for this is the following. Neutrino energy deposition proceeds by $\nu_e$ and $\bar{\nu}_e$ absorption on nucleons. The heated gas expands away from the region of strongest heating as soon as the nucleons have absorbed an energy roughly of the order of the gravitational potential energy, with only a small time lag because of the inertia of the shock, which is confined by the ram pressure of the collapsing stellar material. This does not allow the net energy of the heated gas to become very large. Typically it is of the order of $\sim 5$ MeV per nucleon. With a mass in the gain layer of several $0.01\,M_\odot$ up to $\sim 0.1\,M_\odot$, the total energy does therefore not exceed a few $10^{51}$ erg.

Figure 3 clearly shows this saturation of the explosion energy, which occurs when the gain layer and the shock are expanding. In this sense, neutrino-driven explosions are "self-regulated": Further energy deposition is quenched as the baryons move out of the region of high neutrino fluxes.

The heating rate increases with the neutrino luminosity and the deposited energy is higher for a larger mass in the gain layer (Eq. 5). However, the expansion timescale during which the gas is exposed to high neutrino fluxes, drops when the heating is stronger. Therefore the explosion energy is extremely sensitive to the neutrino luminosity only around the threshold value for getting an explosion.

Neutrino-driven explosions are likely to be "delayed" (up to a few 100 ms after core bounce) rather than "late" (after a few seconds). The density between the gain radius and the shock decreases with time because the proto-neutron star contracts and the gas infall to the shock drops rapidly as time goes on. Therefore the mass $\Delta M$ in the heated region shrinks and the shock must recede to a small radius around or even below 100 km, not favoring a later explosion.

Fulfilling the "explosion criterion" of Fig. 2 once is no guarantee for a successful explosion. The push from the heating region has to be maintained until

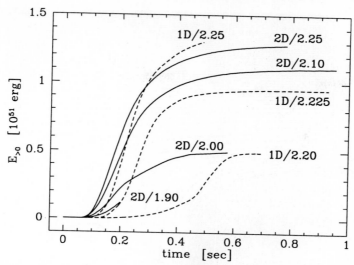

**Fig. 3.** Explosion energies $E_{>0}(t)$ for spherically symmetric models ("1D", dashed lines) and two-dimensional ("2D") supernova models (solid lines) [31,32]. The latter take into account convective overturn between the supernova shock and the neutrino-heating region. The curves display the evolution as a function of time after shock formation for different values of the $\nu_e$ luminosity (labeled in units of $10^{52}$ erg/s). The latter was used as a free parameter at the surface of the nascent neutron star and was roughly equal to the $\bar{\nu}_e$ luminosity. Below the smallest given luminosities, the considered 15 $M_\odot$ star does not explode in 1D and acquires too low an expansion energy in 2D to unbind the whole stellar mantle and envelope.

the material that carries the bulk of the energy is moving ballistically, and a fair fraction of this energy has been converted from internal to kinetic energy. Otherwise, if energy losses by neutrino emission or $PdV$ on the contracting proto-neutron star start to dominate the energy input by neutrino absorption, the pressure-supported expansion can break down again and re-collapse can occur. However, once shock expansion sets in, the conditions for further neutrino heating improve rapidly, and the optical depth of the growing gain layer to neutrinos increases. Provided the neutrino luminosity does not drop, an explosion becomes unavoidable. This requirement favors a high core neutrino luminosity over accretion luminosity to power neutrino-driven explosions.

## 3 Status of Spherical Simulations

Recently, a major shortcoming of previous supernova models has been removed, at least in spherical models. Instead of treating multi-frequency neutrino transport by a flux-limited diffusion approximation [63,64,48,9], the Boltzmann equation can now be solved in connection with hydrodynamical simulations, either by direct discretisation [45] or by a variable Eddington factor technique [53], even

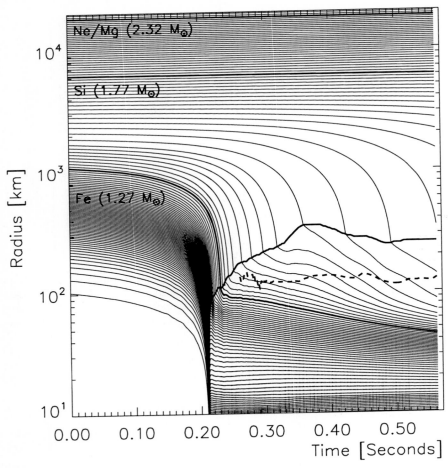

**Fig. 4.** Trajectories of mass shells in the core of a collapsing 15 $M_\odot$ star from a Newtonian simulation with Boltzmann neutrino transport [52,53]. The shells are equidistantly spaced in steps of 0.02 $M_\odot$. The boundaries of the iron core, silicon shell and neon-magnesium shell are indicated by bold lines. The fat, solid curve rising up at 0.21 seconds after the start of the simulation marks the position of the supernova shock, the dashed line denotes the gain radius. In this spherically symmetric simulation, muon and tau neutrinos and antineutrinos were neglected, which favors an explosion. Nevertheless, the shock recedes after having expanded to more than 300 km.

in the general relativistic case [41]. For the first time, the numerical deficiencies of the models are therefore smaller than the uncertainties of the input physics.

The more accurate treatment of the transport, in particular in the semi-transparent neutrino-decoupling region around and outside of the neutrinosphere up to the shock, favors higher energy transfer to the stellar gas in the cooling layer and in the gain layer [42,68]. Nevertheless, the results of these simulations

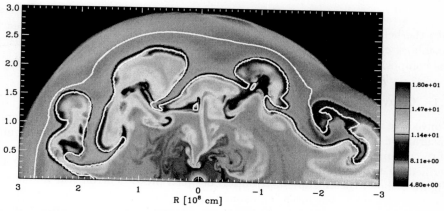

**Fig. 5.** Inhomogeneous distribution of neutrino-heated, hot gas which rises in mushroom-like bubbles, and cooler gas that is accreted through the supernova shock (bumpy discontinuity at about 3000 km) and falls in long, narrows streams towards the newly formed neutron star at the center. The figure shows a snapshot of the entropy at 300 ms after core bounce and shock formation for a two-dimensional simulation of a $15\,M_\odot$ star. The star was exploded by the neutrino-heating mechanism by chosing a suitable value of the neutrino luminosity from the nascent neutron star, which was replaced by an inner boundary condition [39,40]. The white line encompasses the region where radioactive nickel has been formed by nuclear burning in the shock-heated Si layer.

are disappointing. In spherical symmetry, the models do not explode, neither in the Newtonian (Fig. 4), nor in the general relativistic case.

These current models, however, neglect convective effects inside the nascent neutron star as well as in the neutrino-heated region. Since convection has been recognized to be very important, such simulations do not treat the full supernova problem and do not really allow for conclusions about the viability of the neutrino-driven mechanism. Multi-dimensional simulations with Boltzmann neutrino transport are called for.

## 4 Hydrodynamic Instabilities during the Explosion

During the explosion of a supernova, hydrodynamic instabilities and convective processes can occur on different scales in space and time. Convective motions inside the nascent neutron star can speed up the energy transport and raise the neutrino luminosities during a period of seconds (Section 5). In the neutrino-heated region, convective overturn during the first second of the explosion carries hot matter towards the shock front and brings cool gas into the region of strongest neutrino heating near the gain radius. This has important influence on the start of the explosion and the nucleosynthesis of radioactive elements. When the shock propagates through the mantle and envelope of the disrupted star, Rayleigh-Taylor instabilities destroy the onion-shell structure of the progenitor and mix radioactive material with high velocities from near the neutron star into the helium and even hydrogen shells of the star. In this section, the early postshock convection and its interaction with the hydrodynamic instabilities at the composition interfaces of the progenitor star will be discussed.

## 4.1 Convective Overturn in the Neutrino-Heated Region

Convective instabilities in the layers adjacent to the nascent neutron star are a natural consequence of the negative entropy gradient built up by the weakening of the prompt shock prior to its stagnation and by neutrino heating [4]. This was verified by two- and three-dimensional simulations [27,46,28,20,28,57,31,32,44]. Figure 5 shows the entropy distribution between proto-neutron star and supernova shock about 300 ms after core bounce for one such calculation [39,40]. Although there is general agreement about the existence and the growth of hydrodynamic instabilities in the layer between the shock at $R_s$ and the radius of maximum neutrino heating (which is just outside the gain radius, $R_g$), the strength of the convective overturn and its importance for the success of the neutrino-heating mechanism are still a matter of debate.

Two-dimensional simulations with a spectrally averaged, flux-limited diffusion treatment of neutrino transport [28,20], or with the neutrino luminosity being given as a free parameter at the inner boundary, which replaces the neutron star at the center [31,32], found successful explosions in cases where spherically symmetric models fail (Fig. 3). According to these simulations, the convective overturn in the neutrino-heated region has the following effects on the shock propagation. Heated matter from the region close to the gain radius rises outward and at the same time is exchanged with cool gas flowing down from the shock. Since the production reactions of neutrinos ($e^\pm$ capture on nucleons and thermal processes) are very temperature sensitive, the expansion and cooling of rising plasma reduces the energy loss by the reemission of neutrinos. Moreover, the net energy deposition by neutrinos is enhanced as more cool material is exposed to the large neutrino fluxes near the gain radius (the radial dilution of the fluxes goes roughly as $1/r^2$). Since hot matter moves towards the shock, the pressure behind the shock increases, an effect which pushes the shock farther out. This leads to a growth of the gain region and therefore also of the net energy transfer from neutrinos to the stellar gas, favoring an explosion.

The consequences of postshock convection are clearly visible from the results plotted in Fig. 3, where the explosion energy $E_{>0}$ as a function of time is shown for spherically symmetric and two-dimensional calculations of the same post-collapse model, but with different assumed neutrino luminosities from the proto-neutron star [31,32]. $E_{>0}$ is defined to include the sum of internal, kinetic, and gravitational energy for all zones where this sum is positive (the gravitational binding energies of stellar mantle and envelope and additional energy release from nuclear burning are not taken into account). For one-dimensional simulations with $\nu_e$ luminosities (and very similar $\bar{\nu}_e$ luminosities) below $1.9 \cdot 10^{52}$ erg/s explosions could not be obtained when the proto-neutron star was assumed static, and the threshold value of the luminosity was $2.2 \cdot 10^{52}$ erg/s when the

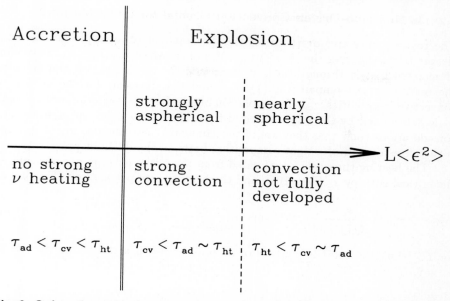

**Fig. 6.** Order scheme for the post-collapse dynamics in dependence of $L_\nu \langle \epsilon_\nu^2 \rangle$, which determines the strength of the neutrino heating outside of the neutrinosphere. The destiny of the star — accretion or explosion — can be understood by the relative size of the timescales of neutrino heating, $\tau_{ht}$, matter advection through the gain region onto the nascent neutron star, $\tau_{adv}$, and growth of convective instabilities, $\tau_{cv}$.

neutron star was contracting. The supporting effects of convective overturn between the gain radius and the shock lead to explosions even below the critical value in spherical symmetry, and to a faster development of the explosion.

Simulations with a better description of the neutrino transport by a multi-energy-group treatment of the neutrino diffusion [44], confirm the existence of such convective processes in the region of neutrino heating, but the associated effects are not strong enough to revive the stalled prompt supernova shock, although the outward motion of the shock is enhanced.

Fully self-consistent, multi-dimensional calculations, however, have not yet been done with a state-of-the-art Boltzmann neutrino transport, which has recently become applicable for spherically symmetric models (see Section 3). The current multi-dimensional simulations therefore demonstrate only the presence and potential importance of convection, but final conclusions on the viability of the neutrino-heating mechanism in the presence of postshock convection are not possible at the moment. A quantitatively meaningful description of the shock revival phase, however, requires an accurate description of the transport as well as a multi-dimensional approach.

## 4.2 Is Neutrino-Driven Convection Crucial for an Explosion?

The role of convective overturn for the development of an explosion becomes clearer by considering the three timescales of neutrino heating, $\tau_{\mathrm{ht}}$, advection of accreted matter through the gain layer into the cooling region and down to the neutron star (compare Fig. 1), $\tau_{\mathrm{ad}}$, and the timescale for the growth of convective instabilities, $\tau_{\mathrm{cv}}$. The evolution of the shock — accretion or explosion — is determined by the relative size of these three timescales. Straightforward considerations show that they are of the same order and the destiny of the star is therefore a result of a tight competition between the different processes.

The heating timescale is estimated from the initial entropy per nucleon, $s_{\mathrm{i}}$, the critical entropy $s_{\mathrm{c}}$ (Eq. (10)), and the heating rate per nucleon (Eq. (4)) as

$$\tau_{\mathrm{ht}} \approx \frac{s_{\mathrm{c}} - s_{\mathrm{i}}}{Q_\nu^+/(k_{\mathrm{B}}T)} \approx 45\,\mathrm{ms} \cdot \frac{s_{\mathrm{c}} - s_{\mathrm{i}}}{5 k_{\mathrm{B}}/N} \frac{R_{\mathrm{g},7}^2 (T/2\mathrm{MeV})\, f}{(L_\nu/4 \cdot 10^{52}\mathrm{erg/s})\langle \epsilon_{\nu,15}^2 \rangle}, \quad (14)$$

for $L_\nu$ being the total luminosity of $\nu_e$ plus $\bar{\nu}_e$. With a postshock velocity of $u_1 = u_0/\beta \approx (\gamma - 1)\sqrt{GM/R_{\mathrm{s}}}/(\gamma + 1)$ the advection timescale is

$$\tau_{\mathrm{ad}} \approx \frac{R_{\mathrm{s}} - R_{\mathrm{g}}}{u_1} \approx 55\,\mathrm{ms} \cdot \left(1 - \frac{R_{\mathrm{g}}}{R_{\mathrm{s}}}\right) \frac{R_{\mathrm{s},200}^{3/2}}{\sqrt{M_1}}, \quad (15)$$

where the gain radius can be determined as

$$R_{\mathrm{g},7} \cong 0.9\, T_{\mathrm{s}}^{3/2} R_{\mathrm{s},200}^{3/2} f^{1/4} \left(\frac{L_\nu}{4 \cdot 10^{52}\mathrm{erg/s}}\right)^{-1/4} \langle \epsilon_{\nu,15}^2 \rangle^{-1/4} \quad (16)$$

from the requirement that the heating rate, Eq. (4), is equal to the cooling rate per nucleon, $Q_\nu^- \approx 2.3\, T^6\,\mathrm{MeV}/(N \cdot \mathrm{s})$, when $R_{\mathrm{s},200}$ is the shock radius in units of 200 km, $Y_n + Y_p \approx 1$ is assumed, and use is made of the power-law behavior of the temperature according to $T(r) \approx T_{\mathrm{s}}(R_{\mathrm{s}}/r)$, with $T_{\mathrm{s}}$ being the postshock temperature in MeV. The growth timescale of convective instabilities in the neutrino-heated region depends on the gradients of entropy and lepton number through the growth rate of Ledoux convection, $\sigma_{\mathrm{L}}$:

$$\tau_{\mathrm{cv}} \approx \frac{\ln(100)}{\sigma_{\mathrm{L}}} \approx 4.6 \left\{ \frac{g}{\rho}\left[\left(\frac{\partial \rho}{\partial s}\right)_{Y_e,P} \frac{ds}{dr} + \left(\frac{\partial \rho}{\partial Y_e}\right)_{s,P} \frac{dY_e}{dr}\right]\right\}^{-1/2}$$

$$\sim 20\,\mathrm{ms} \cdot \left(\frac{R_{\mathrm{s}}}{R_{\mathrm{g}}} - 1\right)^{1/2} \frac{R_{\mathrm{g},7}^{3/2}}{\sqrt{M_1}}. \quad (17)$$

The numerical value was obtained with the gravitational acceleration $g = GM/R_{\mathrm{g}}^2$, $(\partial \rho/\partial s)_P \sim -\rho/s$, and $ds/dr \sim -\frac{1}{2}s/(R_{\mathrm{s}} - R_{\mathrm{g}})$. The term proportional to the gradient of $Y_e$ was assumed to be negligible. $\tau_{\mathrm{cv}}$ of Eq. (17) is sensitive to the detailed conditions between gain radius (close to which $s$ develops a maximum) and the shock. The neutrino heating timescale is shorter for larger values of the neutrino luminosity $L_\nu$ and mean squared neutrino energy

$\langle\epsilon_\nu^2\rangle$. All three timescales, $\tau_{\rm ht}$, $\tau_{\rm ad}$ and $\tau_{\rm cv}$, decrease roughly in the same way with smaller gain radius or shock position.

In order to be a crucial help for the explosion, convective overturn in the neutrino-heated region must develop on a sufficiently short timescale. This happens only in a rather narrow window of $L_\nu\langle\epsilon_\nu^2\rangle$ where $\tau_{\rm cv} < \tau_{\rm ad} \sim \tau_{\rm ht}$ (Fig. 6). For smaller neutrino luminosities the heating is too weak to create a sufficiently large entropy maximum, and rapid convective motions cannot develop before the accreted gas is advected through the gain radius ($\tau_{\rm ad} < \tau_{\rm cv} < \tau_{\rm ht}$). In this case neither with nor without convective processes energetic explosions can occur (Fig. 3). For larger neutrino luminosities the neutrino heating is so strong, and the heating timescale correspondingly short ($\tau_{\rm ht} < \tau_{\rm cv} \sim \tau_{\rm ad}$), that expansion of the postshock layers has set in before the convective activity reaches a significant level. In this case convective overturn is an unavoidable side-effect of the neutrino heating behind the shock, but is not necessary for starting the explosion.

The parametric studies performed by Janka and Müller [31,32] support this discussion, which helps one understanding the seemingly discrepant results obtained by different groups.

## 4.3 Nucleosynthesis and Mixing Instabilities

Besides increasing the efficiency of neutrino energy deposition, convection in the postshock layer has an important influence also on the nucleosynthesis and distribution of radioactive elements. In particular, nickel is not produced by silicon burning in a spherical shell, but is concentrated in dense clumps and pockets between rising bubbles of neutrino-heated matter in the expanding postshock layer (Fig. 5).

The further evolution of the shock until it breaks out of the stellar surface hours later, was recently followed by using adaptive mesh refinement techniques [39,40]. These allow for a dynamic adjustment of the computational grid such that small structures can be treated with high resolution, while the whole computation covers a huge volume.

A few seconds after its formation, the shock has passed the silicon and oxygen layers and propagates through the helium shell of the star. The initial anisotropies have been compressed into a narrow shell, from which new instabilities start to grow. Rayleigh-Taylor mushrooms penetrate into the helium layer and carry O, Si and Ni farther out, while He sinks in. Within minutes, long, dense filaments reach far into the helium shell, associated with them fast-moving knots that contain dominant contributions of different heavy elements from the deeper layers (Fig. 7). Nickel, silicon and oxygen move through helium with velocities up to several 1000 km/s (Fig. 8).

These simulations show that the hydrodynamic instabilities which occur in the first second of the explosion, do not only play a role during the phases of shock rejuvenation and nickel formation. They act also as seed perturbations for the instabilities at the composition interfaces of the progenitor, which finally destroy the onion-shell structure of the pre-collapse star.

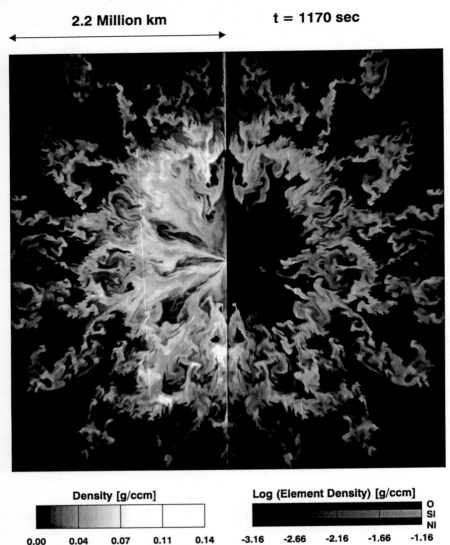

**Fig. 7.** Snapshot from a two-dimensional simulation of the explosion of a blue supergiant star with 15 solar masses at a time 1170 seconds after the stellar core has collapsed to a neutron star [40]. In the left half of the figure the density is shown in a region with a radius of about 2.2 million kilometers, in the right half three color images of the mass densities of radioactive nickel (red and pink), silicon (green, light blue, whitish) and oxygen (deep blue) are superposed. One can see that the ejecta of the explosion are inhomogeneous and anisotropic, and the original onion-shell structure of the exploding star was shredded. Nickel is concentrated in dense, fast-moving clumps along the extended filaments seen in the left plot.

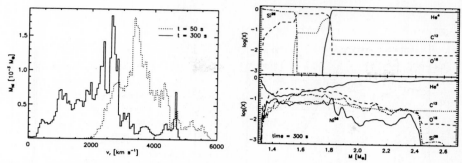

**Fig. 8.** *Left:* Distribution of $^{56}$Ni vs. radial velocity at $t = 50$ s and $t = 300$ s after core bounce. *Right:* Initial composition of the star exterior to the iron core (top) and composition 300 s after core bounce (bottom). C, O, Si, and the newly synthesized Ni have been mixed beyond the inner half of the helium core, and He has been carried inward.

The results are in good agreement with observations of mixing and anisotropies in many Type Ib,c supernovae. In case of Supernova 1987A, a Type II explosion of a massive star which has retained its hydrogen envelope, the observed high nickel velocities in the hydrogen envelope cannot be explained by the models. The nickel clumps are strongly decelerated at the He/H interface, where they enter a dense helium "wall" which builds up after the passage of the shock. The dissipation of the kinetic energy of the clumps does not allow nickel to penetrate into the hydrogen layer with high velocities.

## 5 Neutron Star Formation

Convective energy transport inside the newly formed neutron star can increase the neutrino luminosities considerably [17]. This can be crucial for energizing the stalled supernova shock [63,64].

Convection in the neutron star can be driven by gradients of the entropy and/or proton (electron lepton number) fraction in the nuclear medium [22]. The type of instability which grows most rapidly, e.g., doubly diffusive neutron-finger convection [63,64] or Ledoux convection [17] or quasi-Ledoux convection [35,37], may be a matter of the properties of the nuclear equation of state, which determines the magnitudes and signs of the thermodynamic derivatives [13]. It is also sensitive to the gradients that develop, and thus may depend on the details of the treatment of neutrino transport in the dense interior of the star.

Convection below the neutrinosphere seems to be disfavored during the very early post-bounce evolution by the currently most elaborate supernova models [14,15,43], but can develop deeper inside the nascent neutron star on a longer timescale ($\gtrsim 100$ ms after bounce) and can encompass the whole star within seconds [17,37,35].

Negative lepton number and entropy gradients have been seen in several one-dimensional (spherically symmetric) simulations of the neutrino-cooling phase

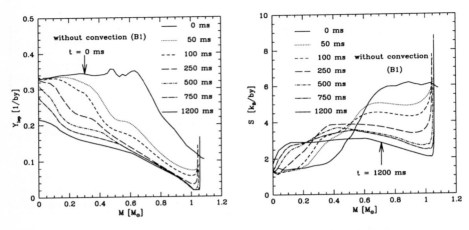

**Fig. 9.** Profiles of the lepton fraction $Y_{\rm lep} = n_{\rm lep}/n_b$ (left) and of the entropy per nucleon, $s$, (right) as functions of enclosed (baryonic) mass for different times in a one-dimensional simulation of the neutrino cooling of a $\sim 1.1\,M_\odot$ proto-neutron star. Negative gradients of lepton number and entropy suggest potentially convectively unstable regions. Time is (roughly) measured from core bounce.

of nascent neutron stars [16,17,36,58] (see also Fig. 9) and have suggested the existence of regions which are potentially unstable against Ledoux convection. Recent calculations [49] with improved neutrino opacities of the nuclear medium, which were described consistently with the employed equation of state, confirm principal aspects of previous simulations, in particular the existence of Ledoux-unstable layers in the neutron star.

## 5.1 Convection inside the Nascent Neutron Star

Two-dimensional, hydrodynamical simulations were performed for the neutrino-cooling phase of a $\sim 1.1\,M_\odot$ proto-neutron star that formed in the core collapse of a $15\,M_\odot$ star [37,35]. The models followed the evolution for a period of more than 1.2 seconds. They demonstrate the development of convection and its importance for the cooling and deleptonization of the neutron star.

The simulations were carried out with the hydrodynamics code *Prometheus*. A general relativistic 1D gravitational potential with Newtonian corrections for asphericities was used, $\Phi \equiv \Phi_{\rm 1D}^{\rm GR} + (\Phi_{\rm 2D}^{\rm N} - \Phi_{\rm 1D}^{\rm N})$, and a flux-limited (equilibrium) neutrino diffusion scheme was applied for each angular bin separately ("$1\tfrac{1}{2}$D").

The simulations showed that convectively unstable surface-near regions (i.e., around the neutrinosphere and below an initial density of about $10^{12}\,{\rm g/cm}^3$) exist only for a short period of a few ten milliseconds after bounce, in agreement with the findings by other groups [14,15,43]. Due to a flat entropy profile and a negative lepton number gradient, convection, however, also starts in a layer deeper inside the star, between an enclosed mass of $0.7\,M_\odot$ and $0.9\,M_\odot$, at

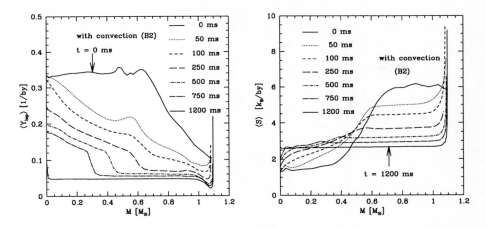

**Fig. 10.** Same as Fig. 9, but for a two-dimensional, hydrodynamical simulation which allowed to follow the development of convection. The plots show angularly averaged quantities in the $\sim 1.1\,M_\odot$ proto-neutron star. In regions with convective activity the gradients of $Y_{\rm lep}$ and $s$ are flattened. The convective layer encompasses an increasingly larger part of the star.

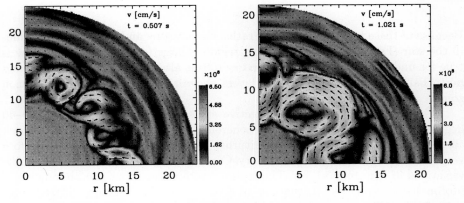

**Fig. 11.** Absolute values of the convective velocity in the proto-neutron star for two instants (about 0.5 s (left) and 1 s (right) after core bounce) as obtained in a two-dimensional, hydrodynamical simulation. The arrows indicate the direction of the velocity field. Note that the neutron star has contracted from a radius of about 60 km initially to little more than 20 km. The growth of the convective region can be seen. Typical velocities of the convective motions are several $10^8$ cm/s.

densities above several $10^{12}$ g/cm$^3$. From there the convective region digs into the star and reaches the center after about one second (Figs. 10, 11, and 13). Convective velocities as high as $5 \cdot 10^8$ cm/s were found (about 10–20% of the local sound speed), corresponding to kinetic energies of up to $1$–$2 \cdot 10^{50}$ erg (Fig. 11).

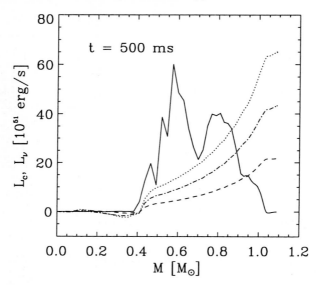

**Fig. 12.** Convective "luminosity" (solid line) and neutrino luminosities (dashed: $L_{\nu_e} + L_{\bar\nu_e}$, dash-dotted: $L_{\nu_\mu}+L_{\bar\nu_\mu}+L_{\nu_\tau}+L_{\bar\nu_\tau}$, dotted: total) as functions of enclosed baryonic mass for the two-dimensional proto-neutron star simulation about 500 ms after core bounce.

Because of these high velocities and rather flat entropy and composition profiles in the star (Fig. 10), the overshooting region is large (see Fig. 13). The same is true for undershooting during the first $\sim$ 100 ms after bounce. Sound waves and perturbations are generated in the layers above and interior to the convection zone.

The coherence lengths of convective structures are of the order of 20–40 degrees (in 2D!) (see Fig. 11) and coherence times are around 10 ms, which corresponds to only one or two overturns. The convective pattern is therefore very time-dependent and nonstationary. Convective motions lead to considerable variations of the composition. The lepton fraction (and thus the abundance of protons) shows relative fluctuations of several 10%. The entropy differences in rising and sinking convective bubbles are much smaller, only a few per cent, while temperature and density fluctuations are typically less than one per cent.

The energy transport in the neutron star is dominated by neutrino diffusion near the center, whereas convective transport plays the major role in a thick intermediate layer where the convective activity is strongest. Radiative transport takes over again when the neutrino mean free path becomes large near the surface of the star (Fig. 12). But even in the convective layer the convective energy flux is only a few times larger than the diffusive flux. This means that neutrino diffusion is not negligibly small in the convective region. This fact has important consequences for the driving mechanism of the convection.

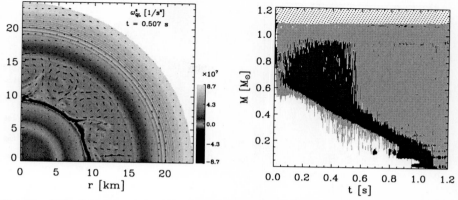

**Fig. 13.** *Left:* Convectively unstable region (corresponding to negative values of the displayed quantity $\omega_{\mathrm{QL}}^2 = -(g/\rho)C_{\mathrm{QL}}$ with $C_{\mathrm{QL}}$ from Eq. (19)) about 500 ms after bounce according to the Quasi-Ledoux criterion which includes non-adiabatic and lepton-transport effects by neutrino diffusion. *Right:* Layer of Quasi-Ledoux convective instability (blue) as function of time for a two-dimensional simulation. The angle-averaged criterion $C_{\mathrm{QL}}^{\mathrm{1D}}(r) \equiv \min_\theta(C_{\mathrm{QL}}(r,\theta)) > 0$ with $C_{\mathrm{QL}}(r,\theta)$ from Eq. (19) is plotted. The dotted area is outside of the computed star, green denotes stable layers where over- and undershooting causes lateral velocities with angularly averaged absolute values of $\langle|v_\theta|\rangle > 10^7\,\mathrm{cm\,s^{-1}}$, and white are convectively "quiet" regions of the star.

## 5.2 Driving Force of Convection

The convective activity in the neutron star cannot be explained by, and considered as ideal Ledoux convection. Applying the Ledoux criterion for local instability,

$$C_{\mathrm{L}}(r,\theta) = \frac{\rho}{g}\sigma_{\mathrm{L}}^2 = \left(\frac{\partial\rho}{\partial s}\right)_{Y_{\mathrm{lep}},P}\frac{\mathrm{d}s}{\mathrm{d}r} + \left(\frac{\partial\rho}{\partial Y_{\mathrm{lep}}}\right)_{s,P}\frac{\mathrm{d}Y_{\mathrm{lep}}}{\mathrm{d}r} > 0\,, \qquad (18)$$

with $\sigma_{\mathrm{L}}$ from Eq. (17) and $Y_e$ replaced by the total lepton fraction $Y_{\mathrm{lep}}$ in the neutrino-opaque interior of the neutron star (for reasons of simplicity, $\nabla s$ was replaced by $\mathrm{d}s/\mathrm{d}r$ and $\nabla Y_{\mathrm{lep}}$ by $\mathrm{d}Y_{\mathrm{lep}}/\mathrm{d}r$), one finds that the convecting region should actually be stable, despite of slightly negative entropy *and* lepton number gradients. In fact, below a critical value of the lepton fraction (e.g., $Y_{\mathrm{lep,c}} = 0.148$ for $\rho = 10^{13}\,\mathrm{g/cm^3}$ and $T = 10.7\,\mathrm{MeV}$) the thermodynamical derivative $(\partial\rho/\partial Y_{\mathrm{lep}})_{s,P}$ changes sign and becomes positive because of nuclear and Coulomb forces in the high-density equation of state [13]. Therefore negative lepton number gradients should stabilize against convection in this regime. However, an idealized assumption of Ledoux convection is not fulfilled in the situations considered here: Because of neutrino diffusion, energy exchange and, in particular, lepton number exchange between convective elements and their surroundings are *not* negligible. Taking the neutrino transport effects on $Y_{\mathrm{lep}}$

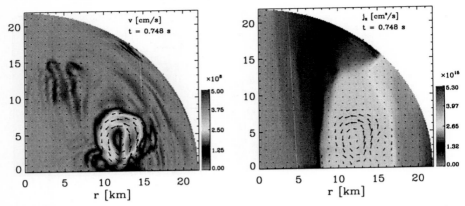

**Fig. 14.** Absolute value of the gas velocity in a convecting, rotating proto-neutron star about 750 ms after bounce (left). Convection is suppressed near the rotation axis (vertical) and develops strongly only near the equatorial plane where a flat distribution of the specific angular momentum $j_z$ (right) has formed.

into account in a modified *"Quasi-Ledoux criterion"* [35],

$$C_{\mathrm{QL}}(r,\theta) \equiv \left(\frac{\partial \rho}{\partial s}\right)_{\langle Y_{\mathrm{lep}}\rangle, \langle P\rangle} \frac{\mathrm{d}\langle s\rangle}{\mathrm{d}r} + \left(\frac{\partial \rho}{\partial Y_{\mathrm{lep}}}\right)_{\langle s\rangle, \langle P\rangle} \left(\frac{\mathrm{d}\langle Y_{\mathrm{lep}}\rangle}{\mathrm{d}r} - \beta_{\mathrm{lep}}\frac{\mathrm{d}Y_{\mathrm{lep}}}{\mathrm{d}r}\right) > 0, \tag{19}$$

one determines instability exactly where the two-dimensional simulation reveals convective activity. In Eq. (19) the quantities $\langle Y_{\mathrm{lep}}\rangle$ and $\langle s \rangle$ mean averages over the polar angles $\theta$, and local gradients have to be distinguished from gradients of angle-averaged quantities which describe the stellar background. The term $\beta_{\mathrm{lep}}(\mathrm{d}Y_{\mathrm{lep}}/\mathrm{d}r)$ with the empirically determined value $\beta_{\mathrm{lep}} \approx 1$ accounts for the change of the lepton concentration along the path of a rising fluid element due to neutrino diffusion. Figure 13 shows that about half a second after core bounce strong, driving forces for convection occur in a narrow ring between 9 and 10 km, where a steep negative gradient of the lepton fraction exists (see Fig. 10). Farther out, convective instability is detected only in finger-like structures of rising, high-$Y_{\mathrm{lep}}$ gas.

### 5.3 Accretion and Rotation

In other two-dimensional models, post-bounce mass accretion and rotation of the forming neutron star were included. Accretion causes stronger convection with larger velocities in a more extended region. This can be explained by the steepening of lepton number and entropy gradients and the increase of the gravitational potential energy when additional matter is added onto the neutron star.

Rotation has very interesting consequences, e.g., leads to a suppression of convective motions near the rotation axis because of a stabilizing stratification of the specific angular momentum (see Fig. 14), an effect which can be understood by applying the (first) Solberg-Høiland criterion for instabilities in rotating, self-gravitating bodies [60]:

$$C_{\rm SH}(r,\theta) \equiv \frac{1}{x^3}\frac{{\rm d}j_z^2}{{\rm d}x} + \frac{a}{\rho}\left[\left(\frac{\partial\rho}{\partial s}\right)_{Y_{\rm lep},P}\nabla s + \left(\frac{\partial\rho}{\partial Y_{\rm lep}}\right)_{s,P}\nabla Y_{\rm lep}\right] < 0 \ . \quad (20)$$

Here, $j_z$ is the specific angular momentum of a fluid element, which is conserved for axially symmetric configurations, $x$ is the distance from the rotation axis, and in case of rotational equilibrium $\boldsymbol{a}$ is the sum of gravitational and centrifugal accelerations, $\boldsymbol{a} = \nabla P/\rho$. Changes of the lepton number in rising or sinking convective elements due to neutrino diffusion were neglected in Eq. (20). Ledoux (or Quasi-Ledoux) convection can only develop where the first term is not too positive. In Fig. 14 fully developed convective motion is therefore constrained to a zone of nearly constant $j_z$ close to the equatorial plane. At higher latitude the convective velocities are much smaller, and narrow, elongated convective cells aligned with cylindrical regions of $j_z = $ const parallel to the rotation axis are visible.

The rotation pattern displayed in Fig. 14 is highly differential with a rotation period of 7.3 ms at $x = 22$ km and of 1.6 ms at $x = 0.6$ km. It has self-consistently developed under the influence of neutrino transport and convection when the neutron star had contracted from an initial radius of about 60 km (with a surface rotation period of 55 ms at the equator and a rotation period of $\sim 5$ ms near the center) to a final radius of approximately 22 km. Due to the differential nature of the rotation, the ratio of rotational kinetic energy to the gravitational potential energy of the star is only 0.78% in the beginning and a few per cent at the end after about 1 s of evolution.

## 5.4 Consequences of Proto-Neutron Star Convection

Convection inside the proto-neutron star can raise the neutrino luminosities within a few hundred ms after core bounce (Fig. 15). In the considered collapsed core of a 15 $M_\odot$ star, $L_{\nu_e}$ and $L_{\bar\nu_e}$ increase by up to 50% and the mean neutrino energies by about 15% at times later than 200–300 ms post bounce. This favors neutrino-driven explosions on timescales of a few hundred milliseconds after shock formation. Also, the deleptonization of the nascent neutron star is strongly accelerated, raising the $\nu_e$ luminosities relative to the $\bar\nu_e$ luminosities during this time. This helps to increase the electron fraction $Y_e$ in the neutrino-heated ejecta and might solve the overproduction problem of $N = 50$ nuclei during the early epochs of the explosion [37]. In case of rotation, the effects of convection on the neutrino emission depend on the direction. Since strong convection occurs only close to the equatorial plane, the neutrino fluxes are convectively enhanced there, while they are essentially unchanged near the poles.

Anisotropic mass motions due to convection in the neutron star lead to gravitational wave emission and anisotropic radiation of neutrinos. The angular variations of the neutrino flux found in the 2D simulations are of the order of

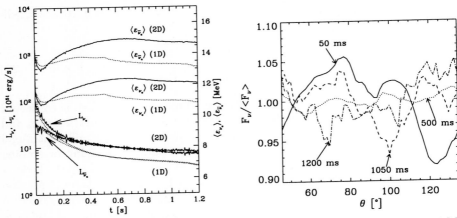

**Fig. 15.** *Left:* Luminosities $L_\nu(t)$ and mean energies $\langle\epsilon_\nu\rangle(t)$ of $\nu_e$ and $\bar\nu_e$ for a 1.1 $M_\odot$ proto-neutron star without ("1D"; dotted) and with convection ("2D"; solid). *Right:* Angular variations of the neutrino flux at different times for the 2D simulation.

5–10% (Fig. 15). With the typical size of the convective cells and the short coherence times of the convective structures, the global anisotropy of the neutrino emission from the cooling proto-neutron star is very small. This implies a kick velocity of the nascent neutron star due to anisotropic neutrino emission of only $\sim 10$ km/s in a 2D simulation (Fig. 16). Because the convective elements are likely to become even smaller in 3D, kick velocities of 300 km/s or even more, as observed for many pulsars, can definitely not be explained by convectively perturbed neutrino emission.

## 5.5 Neutrino Opacities in Nuclear Matter and Neutron Star Convection

Another important issue of interest are the neutrino opacities in the dense and hot nuclear medium of the nascent neutron star. In current supernova models, the description of neutrino-nucleon interactions is incomplete because the standard approximations assume isolated and infinitely massive nucleons [61]. Therefore effects like the fermion phase space blocking of the nucleons, the reduction of the effective nucleon mass by momentum-dependent nuclear interactions in the dense plasma, and nucleon thermal motions and recoil are either neglected completely or approximated in a more or less reliable manner [9,16]. These effects have been recognized to be important [56,33,50,54] for calculations of the neutrino luminosities and spectra, but still await careful inclusion in supernova codes. For this purpose a consistent description of nuclear equation of state and neutrino-matter interactions is desirable.

Many-body (spatial) correlations due to strong interactions [55,29,19,54,67] and multiple-scattering effects by spin-dependent forces between nucleons (temporal spin-density correlations) [51,26] are of particular interest, because they

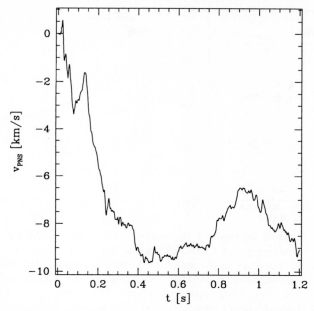

**Fig. 16.** Kick velocity of the neutron star as a function of time, caused by the anisotropic emission of neutrinos due to convection. The two-dimensional simulation was done with a polar grid from 0 to $\pi$.

lead to a reduction of the neutrino opacities in the newly formed neutron star and are associated with additional modes of energy transfer between neutrinos and the nuclear medium.

A reduction of the neutrino opacities implies larger neutrino mean free paths and thus increases the neutrino luminosities. (Fig. 17 and Refs. [38,49,19]). The neutrino diffusion is accelerated most strongly in the very dense core of the nascent neutron star. Convection in the intermediate region between core and outer layers turns out not to be suppressed, but is still the fastest mode of energy transport. Therefore reduced neutrino opacities as well as convective energy transport are important, but the combined effects do not appreciably change the convectively enhanced neutrino emission (Fig. 17, right) [34].

## 6 Summary

Supernova explosions of massive stars are an important phenomenon for applying nuclear and particle physics, in particular neutrino physics. The processes going on in the extremely dense and hot core of the exploding star are accessible to direct measurements only through neutrinos or gravitational waves. Empirical information about the events that cause the explosion and accompany the formation of a neutron star, however, can also be deduced from observable char-

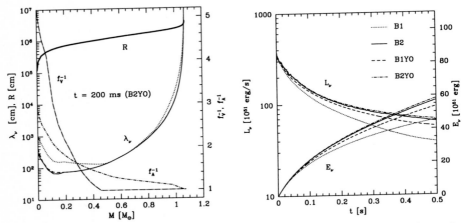

**Fig. 17.** *Left:* Thermal averages of the neutrino mean free paths for $\nu_e$ absorption (dotted line), $\nu_e$ scattering (solid line), and muon and tau neutrino scattering (dashed line), respectively, according to the standard description of the neutrino opacities. $M$ is the baryonic mass enclosed by the radial coordinate $R$ (bold solid line) of the newly formed neutron star about 200 ms after core bounce. Also shown are the factors $f_A^{-1}$ and $f_V^{-1}$ (dash-dotted lines) which give a measure of the increase of the neutrino mean free paths caused by a suppression of the axial-vector and vector current contributions to the neutrino opacities due to in-medium effects [67] (Yamada, personal communication). *Right:* Total neutrino luminosities, $L_\nu$, and integrated energy loss, $E_\nu$, as functions of time for spherically symmetric models without convection (B1 and B1YO) and two-dimensional models with convection (B2 and B2YO). Models B1 and B2 were computed with standard neutrino opacities whereas in B1YO and B2YO in-medium suppression of the neutrino opacities was included [67] (Yamada, personal communication).

acteristics of supernovae, for example their explosion energy or the amount and distribution of radioactive nuclei, and from the properties of neutron stars.

Theoretical models need to establish the link between the core physics and these observables. In the past years it has been recognized that hydrodynamical instabilities and mixing processes on large scales play an important role within the core as well as in the outer layers of the exploding star. Convection can change the cooling of the nascent neutron star, supports the revival of the stalled shock by neutrino heating, and destroys the onion-shell structure of the progenitor star. Multi-dimensional calculations are therefore necessary to understand why and how supernovae explode, and to make predictions for their observable consequences.

Spherically symmetric simulations, Newtonian and general relativistic, with the most advanced treatment of neutrino transport by solving the Boltzmann equation, do not produce explosions. This emphasizes the importance of convection, but may also point to physics still missing in the models. One such weakness of current simulations is an overly simplified description of neutrino interactions with nucleons in the nuclear medium of the neutron star. A kinematically correct treatment of these reactions, taking into account nucleon thermal motions, recoil and fermi blocking, needs only a technical step, but a better understanding of the effects of nucleon correlations and their consistent treatment with the equation of state requires theoretical progress.

The neutrino-heating mechanism, although the favored explanation for the explosion, is still controversial, both because of the status of modeling and because of observations which seem hard to explain. Although significant progress has been made, multi-dimensional simulations with an accurate and reliable handling of neutrino transport and an up-to-date treatment of the input physics are still missing, and definite conclusions can therefore not be drawn at the moment.

**Acknowledgements**

H.-Th. Janka thanks E. Müller and W. Keil for many years of fruitful and enjoyable collaboration. This work was supported by the Sonderforschungsbereich 375 on "Astroparticle Physics" of the Deutsche Forschungsgemeinschaft.

# References

1. E.N. Alexeyev, et al.: Phys. Lett. **B205**, 209 (1988) R.M. Bionta, et al.: Phys. Rev. Lett. **58**, 1494 (1987) K. Hirata, et al.: Phys. Rev. Lett. **58**, 1490 (1987)
2. W. Baade, F. Zwicky: Phys. Rev. **45**, 138 (1934)
3. E. Baron, J. Cooperstein: Astrophys. J. **353**, 597 (1990)
4. H.A. Bethe: Rev. Mod. Phys. **62**, 801 (1990)
5. H.A. Bethe: Astrophys. J. **412**, 192 (1993)
6. H.A. Bethe: Astrophys. J. **449**, 714 (1995)
7. H.A. Bethe: Astrophys. J. **490**, 765 (1997)
8. H.A. Bethe, J.R. Wilson: Astrophys. J. **295**, 14 (1985)
9. S.W. Bruenn: Astrophys. J. Suppl. **58**, 771 (1985)
10. S.W. Bruenn: Phys. Rev. Lett. **59**, 938 (1987)
11. S.W. Bruenn: Astrophys. J. **340**, 955 (1989) Astrophys. J. **341**, 385 (1989)
12. S.W. Bruenn: 'Numerical Simulations of Core Collapse Supernovae'. In: *Nuclear Physics in the Universe*, ed. by M.W. Guidry, M.R. Strayer (IOP, Bristol 1993) p. 31
13. S.W. Bruenn, T. Dineva: Astrophys. J. **458**, L71 (1996)
14. S.W. Bruenn, A. Mezzacappa: Astrophys. J. **433**, L45 (1994)
15. S.W. Bruenn, A. Mezzacappa, T. Dineva: Phys. Rep. **256**, 69 (1995)
16. A. Burrows, J.M. Lattimer: Astrophys. J. **307**, 178 (1986)
17. A. Burrows: Astrophys. J. **318**, L57 (1987) A. Burrows, J.M. Lattimer: Phys. Rep. **163**, 51 (1988)
18. A. Burrows, J. Goshy: Astrophys. J. **416**, L75 (1993)
19. A. Burrows, R.F. Sawyer: Phys. Rev. **C58**, 554 (1998) A. Burrows, R.F. Sawyer: Phys. Rev. **C59**, 510 (1999)
20. A. Burrows, J. Hayes, B.A. Fryxell: Astrophys. J. **450**, 830 (1995)
21. S.A. Colgate, R.H. White: Astrophys. J. **143**, 626 (1966)
22. R.I. Epstein: Mon. Not. R. Astron. Soc. **188**, 305 (1979)
23. C.L. Fryer: Astrophys. J. **522**, 413 (1999)
24. C.L. Fryer, A. Heger: Astrophys. J. **541**, 1033 (2000)

25. C.L. Fryer, V. Kalogera: Astrophys. J., submitted (astro-ph/9911312)
26. S. Hannestad, G. Raffelt: Astrophys. J. **507**, 339 (1998)
27. M. Herant, W. Benz, S.A. Colgate: Astrophys. J. **395**, 642 (1992)
28. M. Herant, W. Benz, W.R. Hix, C.L. Fryer, S.A. Colgate: Astrophys. J. **435**, 339 (1994)
29. C.J. Horowitz, K. Wehrberger: Phys. Lett. **B266**, 236 (1991) C.J. Horowitz, K. Wehrberger: Nucl. Phys. **A531**, 665 (1991)
30. K. Iwamoto, et al.: Nature **395**, 672 (1998) M. Turatto, et al.: Astrophys. J. **534**, L57 (2000) K. Iwamoto, et al.: Astrophys. J. **534**, 660 (2000)
31. H.-Th. Janka, E. Müller: Astrophys. J. **448**, L109 (1995)
32. H.-Th. Janka, E. Müller: Astron. Astrophys. **306**, 167 (1996)
33. H.-Th. Janka, W. Keil, G. Raffelt, D. Seckel: Phys. Rev. Lett. **76**, 2621 (1996)
34. H.-Th. Janka, W. Keil, S. Yamada: in preparation (2001)
35. W. Keil: Konvektive Instabilitäten in entstehenden Neutronensternen. PhD Thesis, Technical University, Munich (1997)
36. W. Keil, H.-Th. Janka: Astron. Astrophys. **296**, 145 (1995)
37. W. Keil, H.-Th. Janka, E. Müller: Astrophys. J. **473**, L111 (1996)
38. W. Keil, H.-Th. Janka, G. Raffelt: Phys. Rev. **D51**, 6635 (1995)
39. K. Kifonidis: Nucleosynthesis and Hydrodynamic Instabilities in Core Collapse Supernovae. PhD Thesis, Technical University, Munich (2001)
40. K. Kifonidis, T. Plewa, H.-Th. Janka, E. Müller: Astrophys. J. **531**, L123 (2000)
41. M. Liebendörfer, A. Mezzacappa, F.-K. Tielemann, O.E.B. Messer, W.R. Hix, S.W. Bruenn: Phys. Rev. D, submitted (astro-ph/0006418)
42. O.E.B. Messer, A. Mezzacappa, S.W. Bruenn, M.W. Guidry: Astrophys. J. **507**, 353 (1998)
43. A. Mezzacappa, A.C. Calder, S.W. Bruenn, J.M. Blondin, M.W. Guidry, M.R. Strayer, A.S. Umar: Astrophys. J. **493**, 848 (1998)
44. A. Mezzacappa, A.C. Calder, S.W. Bruenn, J.M. Blondin, M.W. Guidry, M.R. Strayer, A.S. Umar: Astrophys. J. **495**, 911 (1998)
45. A. Mezzacappa, M. Liebendörfer, O.E.B. Messer, W.R. Hix, F.-K. Tielemann, S.W. Bruenn: Phys. Rev. Lett., submitted (astro-ph/0005366)
46. D.S. Miller, J.R. Wilson, R.W. Mayle: Astrophys. J. **415**, 278 (1993)
47. J.A. Miralles, J.A. Pons, V.A. Urpin: Astrophys. J. **543**, 1001 (2000)
48. E.S. Myra, S.A. Bludman: Astrophys. J. **340**, 384 (1989)
49. J.A. Pons, S. Reddy, M. Prakash, J.M. Lattimer, J.A. Miralles: Astrophys. J. **513**, 780 (1999)
50. M. Prakash, et al.: Phys. Rep. **280**, 1 (1997)
51. G.G. Raffelt, D. Seckel: Phys. Rev. **D52**, 1780 (1995) G.G. Raffelt, D. Seckel, G. Sigl: Phys. Rev. **D54**, 2784 (1996)
52. M. Rampp: Radiation Hydrodynamics with Neutrinos: Stellar Core Collapse and the Explosion Mechanism of Type II Supernovae. PhD Thesis, Technical University, Munich (2000)
53. M. Rampp, H.-Th. Janka: Astrophys. J. **539**, L33 (2000)
54. S. Reddy, M. Prakash M.: Astrophys. J. **478**, 689 (1997) S. Reddy, M. Prakash, J.M. Lattimer: Phys. Rev. **D58**, 013009 (1998)
55. R.F. Sawyer: Phys. Rev. **C40**, 865 (1989)
56. P.J. Schinder: Astrophys. J. Suppl. **74**, 249 (1990)
57. T. Shimizu, S. Yamada, K. Sato: Astrophys. J. **432**, L119 (1994)
58. K. Sumiyoshi, H. Suzuki, H. Toki: Astron. Astrophys. **303**, 475 (1995)
59. F.D. Swesty, J.M. Lattimer, E.S. Myra: Astrophys. J. **425**, 195 (1994)

60. J.-L. Tassoul: *Theory of Rotating Stars*. (Princeton Univ. Press, Princeton 1978)
61. D.L. Tubbs, D.N. Schramm: Astrophys. J. **201**, 467 (1975)
62. J.R. Wilson: 'Supernovae and Post-Collapse Behavior'. In: *Numerical Astrophysics*, ed. by J.M. Centrella, J.M. LeBlanc, R.L. Bowers, J.A. Wheeler (Jones and Bartlett, Boston 1985) p. 422
63. J.R. Wilson, R. Mayle: Phys. Rep. **163**, 63 (1988)
64. J.R. Wilson, R. Mayle: Phys. Rep. **227**, 97 (1993)
65. J.R. Wilson, R. Mayle, S.E. Woosley, T. Weaver: Ann. NY Acad. Sci. **470**, 267 (1986)
66. S.E. Woosley, T.A. Weaver: Astrophys. J. Suppl. **101**, 181 (1995) F.X. Timmes, S.E. Woosley, T.A. Weaver: Astrophys. J. **457**, 834 (1996)
67. S. Yamada: Nucl. Phys. **A662**, 219 (2000) S. Yamada, H. Toki: Phys. Rev. **C61**, 5803 (2000)
68. S. Yamada, H.-Th. Janka, H. Suzuki: Astron. Astrophys. **344**, 533 (1999)

# Evolution of a Neutron Star from Its Birth to Old Age

Madappa Prakash[1], James M. Lattimer[1], Jose A. Pons[1], Andrew W. Steiner[1], and Sanjay Reddy[2]

[1] Department of Physics & Astronomy,
State University of New York at Stony Brook,
Stony Brook, NY-11794-3800, USA
[2] Institute for Nuclear Theory,
University of Washington,
Seattle, WA 98195, USA

**Abstract.** The main stages in the evolution of a neutron star, from its birth as a proto-neutron star, to its old age as a cold, catalyzed configuration, are described. A proto-neutron star is formed in the aftermath of a successful supernova explosion and its evolution is dominated by neutrino diffusion. Its neutrino signal is a valuable diagnostic of its internal structure and composition. During its transformation from a hot, lepton-rich to a cold, catalyzed remnant, the possibility exists that it can collapse into a black hole, which abruptly terminates neutrino emissions. The essential microphysics, reviewed herein, that controls its evolution are the equation of state of dense matter and its associated neutrino opacities. Several simulations of the proto-neutron star evolution, involving different assumptions about the composition of dense matter, are described. After its evolution into a nearly isothermal neutron star a hundred or so years after its birth, it may be observable through its thermal emission in X-rays during its life in the next million years. Its surface temperature will depend upon the rapidity of neutrino emission processes in its core, which depends on the composition of dense matter and whether or not its constituents exhibit superfluidity and superconductivity. Observations of thermal emission offer the best hope of a determination of the radius of a neutron star. The implications for the underlying dense matter equation of state of an accurate radius determination are explored.

## 1 Introduction: The Tale

A proto-neutron star (PNS) is born in the aftermath of a successful supernova explosion as the stellar remnant becomes gravitationally decoupled from the expanding ejecta. Initially, the PNS is optically thick to neutrinos, that is, they are temporarily trapped within the star. The subsequent evolution of the PNS is dominated by $\nu$−diffusion which first results in deleptonization and subsequently in cooling. After a much longer time, photon emissions compete with neutrino emissions in neutron star cooling.

In this paper, we will focus upon the essential microphysical ingredients that govern the macrophysical evolution of neutron stars: the equation of state (EOS) of dense matter and its associated neutrino opacity. Among the characteristics of matter that widely vary among EOS models are their relative compressibilities (important in determining a neutron star's maximum mass), symmetry energies (important in determining the typical stellar radius and in the relative proton fraction) and specific heats (important in determining the local temperature). These characteristics play important roles in determining the matter's composition, in particular the possible presence of additional components (such as hyperons, a pion or kaon condensate, or quark matter), and also significantly affect calculated neutrino opacities and diffusion time scales.

The evolution of a PNS proceeds through several distinct stages [1,2] and with various outcomes, as shown schematically in Fig. 1. Immediately following core bounce and the passage of a shock through the outer PNS's mantle, the star contains an unshocked, low entropy core of mass $M_c \simeq 0.7$ $M_\odot$ in which neutrinos are trapped (the first schematic illustration, labelled (1) in the figure). The core is surrounded by a low density, high entropy ($5 < s < 10$) mantle that is both accreting matter from the outer iron core falling through the shock and also rapidly losing energy due to electron captures and thermal neutrino emission. The mantle extends up to the shock, which is temporarily stationary at a radius of about 200 km prior to an eventual explosion.

After a few seconds (stage 2), accretion becomes less important if the supernova is successful and the shock lifts off the stellar envelope. Extensive neutrino losses and deleptonization will have led to a loss of lepton pressure and the collapse of the mantle. If enough accretion occurs, however, the star's mass could

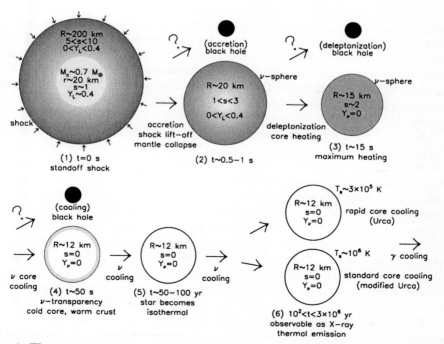

**Fig. 1.** The main stages of evolution of a neutron star. Shading indicates approximate relative temperatures.

increase beyond the maximum mass capable of being supported by the hot, lepton-rich matter. If this occurs, the remnant collapses to form a black hole and its neutrino emission is believed to quickly cease [3].

Neutrino diffusion deleptonizes the core on time scales of 10–15 s (stage 3). The diffusion of high-energy (200–300 MeV) neutrinos from the core to the surface where they escape as low-energy neutrinos (10–20 MeV) generates a large amount of heat within the star (a process akin to joule heating). The core's entropy approximately doubles, producing temperatures in the range of 30–60 MeV, during this time, even as neutrinos continue to be prodigiously emitted from the stars effective surface, known as the $\nu$–sphere.

Strange matter, in the form of hyperons, a Bose condensate, or quark matter, which is suppressed to extremely large densities when neutrinos are trapped in matter, could appear at the end of the deleptonization. The appearance of strange matter leads to a decrease in the theoretical maximum mass that matter is capable of supporting, leading to another possibility for black hole formation [4]. This would occur if the PNS's mass, which must be less than the maximum mass of hot, lepton-rich matter (or else a black hole would already have formed), is greater than the maximum mass of hot, lepton-poor matter. However, if strangeness does not appear, the theoretical maximum mass instead increases during deleptonization and the appearance of a black hole would be unlikely unless accretion in this stage remains significant.

The PNS is now lepton-poor, but it is still hot. While the star has zero net neutrino number, thermally produced neutrino pairs of all flavors are abundant and dominate the emission. Neutrino diffusion continues to cool the star, but the average neutrino energy decreases, and the neutrino mean free path increases. After approximately 50 seconds (stage 4), the mean free path becomes comparable to the stellar radius, and star finally becomes transparent to neutrinos. Since the threshold density for the appearance of strange matter decreases with decreasing temperature, a delayed collapse to a black hole is still possible during this epoch.

Neutrino observations from a galactic supernova will illuminate these stages. The observables will constrain time scales for deleptonization and cooling and the star's binding energy. Dimensionally, diffusion time scales are proportional to $R^2(c\lambda)^{-1}$, where $R$ is the star's radius and $\lambda$ is the effective neutrino mean free path. This generic relation illustrates how both the EOS and the composition, which determine both $R$ and $\lambda$, influence evolutionary time scales. The total binding energy, which is primarily a function of stellar mass and radius (Lattimer & Prakash [5]), should be one of the most accurately measured quantities from neutrino observatories. Currently, Super-Kamiokande and SNO are capable of detecting thousands of neutrinos from a galactic supernova (distance less than 10 kpc). Exciting possibilities lie ahead with many other existing and planned new facilities [6].

Following the onset of neutrino transparency, the core continues to cool by neutrino emission, but the star's crust remains warm and cools less quickly. The crust is an insulating blanket which prevents the star from coming to complete thermal equilibrium and keeps the surface relatively warm ($T \approx 3 \times 10^6$ K) for up to 100 years (stage 5). This timescale is primarily sensitive to the neutron star's radius and the thermal conductivity of the mantle [7], as can be noted from the approximate diffusive relationship $\tau \propto \Delta R^2/\lambda$, where $\Delta R$ is the thickness of the crust. If the rapid decrease in the star's surface temperature predicted to occur when thermal equilibrium is ultimately achieved (see Fig. 16 in Section 6), a valuable constraint on the thickness of the crust, and hence the neutron star radius, could be obtained. The temperature of the surface after the interior of the star becomes isothermal (stage 6) is determined by the rate of neutrino emission in the star's core. The magnitude of the rate is primarily determined by the question of whether or not one or more of the so-called direct Urca processes can occur. The basic Urca process

$$n \to p + e^- + \bar{\nu}_e; \qquad p \to n + e^+ + \nu_e \qquad (1)$$

operates even in degenerate matter because at finite temperature some of the nucleons are in excited states. In addition, direct Urca process involving hyperons, Bose condensates and quarks are also possible. In general, the direct Urca rate is proportional to $T^4$, and is so large that the surface temperatures fall to just a few times $10^5$ K, which becomes very difficult to observe in X-rays except for very nearby stars. A relatively high surface temperature, closer to $10^6$ K, will persist, however, if an Urca process can only occur indirectly with the participation of a spectator nucleon – the modified Urca process, which in the case of nucleons is

$$n + (n,p) \to p + (n,p) + e^- + \bar{\nu}_e; \qquad p + (n,p) \to n + (n,p) + e^+ + \nu_e, \qquad (2)$$

and leads to the so-called standard cooling scenario.

However, there are two circumstances that could prevent the direct Urca process from occurring. First, if the composition of the matter is such that the momentum triangle involving the non-neutrino particles cannot be closed, momentum conservation disallows this process. This occurs, in the case of $n, p, e$, for example, if the $p$ and $e^-$ abundances, which must be equal, are less than $1/8$ the $n$ abundance. This would be the case if the nuclear symmetry energy has a relatively weak density dependence. In addition, direct Urca processes involving hyperons, a Bose condensate, or quarks would not occur, of course, if they are not present. Second, direct Urca processes are suppressed if one of the reactants becomes superfluid. In this case, when the core temperature falls below the superfluid's critical temperature, the rapid cooling is terminated. In the case of a superfluid, the core cooling, and therefore the surface temperature, will be intermediate between those predicted by standard and rapid cooling models. Neutrino emission continues to dominate until neutron stars are approximately 1 million years old, at which point photon cooling from the surface begins to dominate. Unless the interiors cool very rapidly, X-ray emissions from stars remain relatively high until the photon cooling epoch.

Several neutron stars have been suggested to have observable thermal emissions in X-rays. In addition, the nearby neutron star RX J185635-3754, which is the closest known neutron star with a distance of approximately 60 pc, has detectable UV and optical thermal emissions as well. Such objects represent the best chance for measuring a neutron star's radius, especially if the redshift of spectral lines can be determined. Just-launched or proposed X-ray satellites, such as Chandra and XMM offer abundant prospects of observations of photon observations of neutron stars.

The organization of this article is as follows. Section 2 contains a summary of the basic equations of evolution for proto-neutron stars, including a discussion of the equilibrium diffusion approximation. Section 3 details the equation of state of dense matter, taking into account the possibility that neutrinos are trapped in the matter. The possibilities of hyperon-rich matter, kaon condensation and strange quark matter are also discussed here. In the event of a transition to matter containing kaons or quarks, we also consider the possibility that matter could be inhomogeneous with droplets of the strange matter embedded in normal matter. Neutrino-matter interactions are considered in Section 4, which includes discussions of the effects of composition, in-medium dispersion relations, and correlations, in both homogeneous and inhomogeneous phases. In Section 5, we present several simulations of the evolution of proto-neutron stars with different assumptions about the composition of dense matter and highlighting the role of the neutrino opacities. Focus is placed upon predicted neutrino signals and the differences anticipated for varying assumptions about the matter's composition. Section 6 describes the long-term cooling epoch, with a special emphasis on the role of direct Urca processes and superfluidity and a comparison with observations. A discussion of the possibility of detecting superfluidity, including quark color superfluidity, is included. In Section 7, the dependence of the structure of neutron stars on the underlying dense matter equation of state is explored. The relation between the matter's symmetry energy and the radii of neutron stars is highlighted. In addition, the moments of inertia and binding energies of neutron stars are discussed, and observational constraints on the mass and radius of the Vela pulsar from these considerations are elaborated. Section 8 contains our outlook.

## 2   Short-Term Neutrino Cooling: The First Minutes

The cooling of PNSs can be divided into two main regimes: the short-term, lasting perhaps one minute, during which the potential to observe the neutrino signal in terrestrial detectors exists, and the longer term period, lasting perhaps one million years, in which neutrino emissions dominate the cooling but the star is observable only through its thermal, photonic, emissions. This section summarizes the evolution equations relevant for the short-term Kelvin-Helmholtz phase and the estimation of its neutrino signature.

### 2.1   PNS Evolution Equations

The equations that govern the transport of energy and lepton number are obtained from the Boltzmann equation for massless particles[1,8–10]. We will focus on the non-magnetic, spherically symmetric situation. For the PNS problem, fluid velocities are small enough so that hydrostatic equilibrium is nearly fulfilled. Under these conditions, the neutrino transport equations in a stationary metric

$$ds^2 = -e^{2\phi}dt^2 + e^{2\Lambda}dr^2 + r^2 d\theta^2 + r^2 \sin^2\theta\, d\Phi^2 \tag{3}$$

are:

$$\frac{\partial(N_\nu/n_B)}{\partial t} + \frac{\partial(e^\phi 4\pi r^2 F_\nu)}{\partial a} = e^\phi \frac{S_N}{n_B} \tag{4}$$

$$\frac{\partial(J_\nu/n_B)}{\partial t} + P_\nu \frac{\partial(1/n_B)}{\partial t} + e^{-\phi}\frac{\partial(e^{2\phi}4\pi r^2 H_\nu)}{\partial a} = e^\phi \frac{S_E}{n_B}, \tag{5}$$

where $n_B$ is the baryon number density and $a$ is the enclosed baryon number inside a sphere of radius $r$. The quantities $N_\nu$, $F_\nu$, and $S_N$ are the number density, number flux and number source term, respectively, while $J_\nu$, $H_\nu$, $P_\nu$, and $S_E$ are the neutrino energy density, energy flux, pressure, and the energy source term, respectively.

In the absence of accretion, the enclosed baryon number $a$ is a convenient Lagrangian variable. The equations to be solved split naturally into a transport part, which has a strong time dependence, and a structure part, in which evolution is much slower. Explicitly, the structure equations are

$$\frac{\partial r}{\partial a} = \frac{1}{4\pi r^2 n_B e^\Lambda} \quad, \quad \frac{\partial m}{\partial a} = \frac{\rho}{n_B e^\Lambda} \tag{6}$$

$$\frac{\partial \phi}{\partial a} = \frac{e^\Lambda}{4\pi r^4 n_B}(m + 4\pi r^3 P) \quad, \quad \frac{\partial P}{\partial a} = -(\rho + P)\frac{e^\Lambda}{4\pi r^4 n_B}(m + 4\pi r^3 P). \tag{7}$$

The quantities $m$ (enclosed gravitational mass), $\rho$ (mass-energy density), and $P$ (pressure) include contributions from the leptons. To obtain the equations employed in the transport, (4) may be combined with the corresponding equation for the electron fraction

$$\frac{\partial Y_e}{\partial t} = -e^\phi \frac{S_N}{n_B} \tag{8}$$

to obtain

$$\frac{\partial Y_L}{\partial t} + e^{-\phi}\frac{\partial(e^\phi 4\pi r^2 F_\nu)}{\partial a} = 0. \tag{9}$$

Similarly, (5) may be combined with the matter energy equation

$$\frac{dU}{dt} + P\frac{d(1/n_B)}{dt} = -e^\phi \frac{S_E}{n_B}, \tag{10}$$

where $U$ is the specific internal energy and use of the first law of thermodynamics yields

$$e^\phi T \frac{\partial s}{\partial t} + e^\phi \mu_\nu \frac{\partial Y_L}{\partial t} + e^{-\phi}\frac{\partial e^{2\phi}4\pi r^2 H_\nu}{\partial a} = 0, \tag{11}$$

where $s$ is the entropy per baryon.

## 2.2 The Equilibrium Diffusion Approximation

At high density and for temperatures above several MeV, the source terms in the Boltzmann equation are sufficiently strong to ensure that neutrinos are in thermal and chemical equilibrium with the ambient matter. Thus, the neutrino distribution function in these regions is both nearly Fermi-Dirac and isotropic. We can approximate the distribution function as an expansion in terms of Legendre polynomials to $O(\mu)$, which is known as the diffusion approximation. Explicitly,

$$f(\omega,\mu) = f_0(\omega) + \mu f_1(\omega), \quad f_0 = [1 + e^{(\frac{\omega - \mu_\nu}{kT})}]^{-1}, \tag{12}$$

where $f_0$ is the Fermi–Dirac distribution function at equilibrium ($T = T_{mat}$, $\mu_\nu = \mu_\nu^{eq}$), with $\omega$ and $\mu_\nu$ being the neutrino energy and chemical potential, respectively. The main goal is to obtain a relation for $f_1$ in terms of $f_0$. In the diffusion approximation, one obtains [10]

$$f_1 = -D(\omega)\left[e^{-\Lambda}\frac{\partial f_0}{\partial r} - \omega e^{-\Lambda}\frac{\partial \phi}{\partial r}\frac{\partial f_0}{\partial \omega}\right]. \tag{13}$$

The explicit form of the diffusion coefficient $D$ appearing above is given by

$$D(\omega) = \left(j + \frac{1}{\lambda_a} + \kappa_1^s\right)^{-1}. \tag{14}$$

The quantity $j = j_a + j_s$, where $j_a$ is the emissivity and $j_s$ is the scattering contribution to the source term. The absorptivity is denoted by $\lambda_a$ and $\kappa_1^s$ is the scattering contribution to the transport opacity. Substituting

$$\frac{\partial f_0}{\partial r} = -\left(T\frac{\partial \eta_\nu}{\partial r} + \frac{\omega}{T}\frac{\partial T}{\partial r}\right)\frac{\partial f_0}{\partial \omega}, \tag{15}$$

where $\eta_\nu = \mu_\nu/T$ is the neutrino degeneracy parameter, in (13), we obtain

$$f_1 = -D(\omega)e^{-\Lambda}\left[T\frac{\partial \eta}{\partial r} + \frac{\omega}{Te^\phi}\frac{\partial(Te^\phi)}{\partial r}\right]\left(-\frac{\partial f_0}{\partial \omega}\right). \tag{16}$$

Thus, the energy-integrated lepton and energy fluxes are

$$F_\nu = -\frac{e^{-\Lambda}e^{-\phi}T^2}{6\pi^2}\left[D_3\frac{\partial(Te^\phi)}{\partial r} + (Te^\phi)D_2\frac{\partial \eta}{\partial r}\right]$$

$$H_\nu = -\frac{e^{-\Lambda}e^{-\phi}T^3}{6\pi^2}\left[D_4\frac{\partial(Te^\phi)}{\partial r} + (Te^\phi)D_3\frac{\partial \eta}{\partial r}\right]. \tag{17}$$

The coefficients $D_2$, $D_3$, and $D_4$ are related to the energy-dependent diffusion coefficient $D(\omega)$ through

$$D_n = \int_0^\infty dx\, x^n D(\omega) f_0(\omega)(1 - f_0(\omega)), \tag{18}$$

where $x = \omega/T$. These diffusion coefficients depend only on the microphysics of the neutrino-matter interactions (see §4 for details). The fluxes appearing in the above equations are for one particle species. To include all six neutrino types, we redefine the diffusion coefficients in (17):

$$D_2 = D_2^{\nu_e} + D_2^{\bar{\nu}_e}, \quad D_3 = D_3^{\nu_e} - D_3^{\bar{\nu}_e}, \quad D_4 = D_4^{\nu_e} + D_4^{\bar{\nu}_e} + 4D_4^{\nu_\mu}. \tag{19}$$

## 2.3 Neutrino Luminosities

A fair representation of the signal in a terrestrial detector can be found from the time dependence of the total neutrino luminosity and average neutrino energy together with an assumption of a Fermi-Dirac spectrum with zero chemical potential. We will return to discuss the improvements necessary to obtain more accurate information about the spectra.

The total neutrino luminosity is the time rate of change of the star's gravitational mass, and is therefore primarily a global property of the evolution. This luminosity, due to energy conservation, must also equal

$$L_\nu = e^{2\phi} 4\pi r^2 H_\nu \tag{20}$$

at the edge of the star. This relation serves as a test of energy conservation, at least for all times greater than about 5 ms, when the star comes into radiative equilibrium. For times greater than about 5 ms, initial transients become quite small and the predicted luminosities should be relatively accurate compared to full transport simulation. Estimate of the average energy of neutrinos is made from the temperature $T_\nu$ of the matter at the neutrinosphere $R_\nu$, defined to be the location in the star where the flux factor $\xi_H = 0.25$. However, since the spectrum may not be Fermi-Dirac at the neutrinosphere, a diffusion scheme cannot give a very precise value for the average energy. We use the average energy $<E_\nu> \approx 3T_\nu$, where $T_\nu$ is a mass average in the outermost zone. Because it is a globally determined quantity, the luminosity $L_\nu$ is necessarily more accurately determined than either $R_\nu$ or $T_\nu$.

# 3 The Equation of State of Neutrino Trapped Matter

The rationale for considering different possibilities for the composition of dense matter is largely due to the fact that QCD at finite baryon density remains unsolved. Effective QCD based models have raised intriguing possibilities concerning the composition of dense matter including the presence of hyperons, pion or kaon condensates, and quark matter (see [4] for extensive references). It is also important to have predictions for the plain-vanilla case of nucleons alone. The contrast can be dramatic, since additional components offer the possibility of BH formation during the evolution of a PNS. In what follows, the symbols $np$ refer to matter with nucleons alone, $npH$ to matter including hyperons, $npK$ to matter with nucleons and kaons, and $npQ$ to matter with nucleons and quarks. In all cases, leptons in beta equilibrium are included.

## 3.1 Matter with Nucleons and Hyperons

The masses and radii of neutron stars depend upon the matters' compressibility, the composition of matter at high density, and the nuclear symmetry energy (e.g., [4]). In the PNS problem, the finite temperature aspects of the EOS also play an important role. During the early evolution the entropy in the central regions is moderately high, $s \sim 1-2$ (in units of Boltzmann's constant), which correspond to temperatures in the range $T = 20 - 50$ MeV. These features may be explored by employing a finite temperature field-theoretical model in which the interactions between baryons are mediated by the exchange of $\sigma, \omega$, and $\rho$ mesons.[1] The hadronic Lagrangian density is given by a generalization [12] of relativistic mean field theory [13]

$$L_H = \sum_i \overline{B_i}(-i\gamma^\mu \partial_\mu - g_{\omega i}\gamma^\mu \omega_\mu - g_{\rho i}\gamma^\mu \mathbf{b}_\mu \cdot \mathbf{t} - M_i + g_{\sigma i}\sigma)B_i$$
$$- \frac{1}{4}W_{\mu\nu}W^{\mu\nu} + \frac{1}{2}m_\omega^2 \omega_\mu \omega^\mu - \frac{1}{4}\mathbf{B}_{\mu\nu}\mathbf{B}^{\mu\nu} + \frac{1}{2}m_\rho^2 b_\mu b^\mu$$
$$+ \frac{1}{2}\partial_\mu \sigma \partial^\mu \sigma - \frac{1}{2}m_\sigma^2 \sigma^2 - U(\sigma) \qquad (21)$$

Here, $B$ are the Dirac spinors for baryons and $\mathbf{t}$ is the isospin operator. The sums include baryons $i = n, p, \Lambda, \Sigma$, and $\Xi$. The field strength tensors for the $\omega$ and $\rho$ mesons are $W_{\mu\nu} = \partial_\mu \omega_\nu - \partial_\nu \omega_\mu$ and $\mathbf{B}_{\mu\nu} = \partial_\mu \mathbf{b}_\nu - \partial_\nu \mathbf{b}_\mu$, respectively. The potential $U(\sigma)$ represents the self-interactions of the scalar field and is taken to be of the form [14]

$$U(\sigma) = \frac{1}{3}bM_n(g_{\sigma N}\sigma)^3 + \frac{1}{4}c(g_{\sigma N}\sigma)^4 \,. \qquad (22)$$

The partition function $Z_H$ for the hadronic degrees of freedom is evaluated in the mean field approximation. The total partition function $Z_{total} = Z_H Z_L$, where $Z_L$ is the standard noninteracting partition function of the leptons. Using $Z_{total}$, the thermodynamic quantities can be obtained in the standard way. The additional conditions needed to obtain a solution are provided by the charge neutrality requirement, and, when neutrinos are trapped, the set of equilibrium chemical potential relations required by the general condition

$$\mu_i = b_i\mu_n - q_i(\mu_l - \mu_{\nu_\ell}) \,. \qquad (23)$$

where $b_i$ is the baryon number of particle $i$ and $q_i$ is its charge. The introduction of additional variables, the neutrino chemical potentials, requires additional constraints, which we supply by fixing the lepton fractions, $Y_{L\ell}$, appropriate for

---

[1] Note that the couplings in these models may be chosen to reproduce the results of numerically more intensive microscopic potential models, such as that of Akmal and Pandharipande [11], so that the gross features of the zero temperature thermodynamics can be reproduced. Additional advantages to this approach are that the effects of finite temperature and arbitrary proton fraction may be incorporated more easily.

conditions prevailing in the evolution of the PNS. In addition to models containing only nucleonic degrees of freedom (GM1np & GM3np) we investigate models that allow for the presence of hyperons (GM1npH & GM3npH). For the determination of the various coupling constants appearing in $Z_H$ see [4].

The lepton chemical potentials influence the deleptonization epoch. For np models a lower nuclear symmetry energy favors a larger $\nu_e$ fraction and has little effect on the $e^-$ fraction at $Y_L = 0.4$. Models with hyperons lead to significantly larger $\mu_{\nu_e}$ and lower $\mu_e$, both of which influence the diffusion of electron neutrinos. The electron chemical potentials in neutrino free matter are reduced to a greater extent by changes in composition and symmetry energy as there are no neutrinos to compensate for changes in $\hat{\mu} = \mu_n - \mu_p$.

## 3.2 Matter with a Kaon Condensate

The contents of this section are extracted from Pons et al. [15]. For the kaon sector, we use a Lagrangian that contains the usual kinetic energy and mass terms along with the meson interactions [18]. Kaons are coupled to the meson fields through minimal coupling; specifically,

$$L_K = \mathcal{D}_\mu^* K^+ \mathcal{D}^\mu K^- - m_K^{*2} K^+ K^-, \qquad (24)$$

where the vector fields are coupled via the standard form

$$\mathcal{D}_\mu = \partial_\mu + ig_{\omega K}\omega_\mu + ig_{\rho K}\gamma^\mu \mathbf{b}_\mu \cdot \mathbf{t} \qquad (25)$$

and $m_K^* = m_K - \frac{1}{2}g_{\sigma K}\sigma$ is the effective kaon mass.

In the mean field approach, the thermodynamic potential per unit volume in the kaon sector is [15]

$$\begin{aligned}\frac{\Omega_K}{V} &= \tfrac{1}{2}(f\theta)^2(m_K^{*2} - (\mu + X_0)^2) \\ &+ T \int_0^\infty \frac{d^3p}{(2\pi)^3}\left[\ln(1 - e^{-\beta(\omega^- - \mu)}) + \ln(1 - e^{-\beta(\omega^+ + \mu)})\right],\end{aligned} \qquad (26)$$

where $X_0 = g_{\omega K}\omega_0 + g_{\rho K}b_0$, the Bose occupation probability $f_B(x) = (e^{\beta x} - 1)^{-1}$, $\omega^\pm = \sqrt{p^2 + m_K^{*2}} \pm X_0$, $f = 93$ MeV is the pion decay constant and the condensate amplitude, $\theta$, can be found by extremization of the partition function. This yields the solution $\theta = 0$ (no condensate) or, if a condensate exists, the equation

$$m_K^* = \mu_K + X_0 \qquad (27)$$

where $\mu_K$ is the kaon chemical potential. In beta-stable stellar matter the conditions of charge neutrality

$$\sum_B q_B n_B - n_e - n_K = 0 \qquad (28)$$

and chemical equilibrium

$$\mu_i = b_i\mu_n - q_i(\mu_l - \mu_{\nu_\ell}) \tag{29}$$

$$\mu_K = \mu_n - \mu_p \tag{30}$$

are also fulfilled.

The kaon condensate is assumed to appear by forming a mixed phase with the baryons satisfying Gibbs' rules for phase equilibrium [19]. Matter in this mixed phase is in mechanical, thermal and chemical equilibrium, so that

$$p^I = p^{II}, \quad T^I = T^{II}, \quad \mu_i^I = \mu_i^{II}, \tag{31}$$

where the superscripts I and II denote the nucleon and kaon condensate phases, respectively. The conditions of global charge neutrality and baryon number conservation are imposed through the relations

$$\begin{aligned}\chi q^I + (1-\chi)q^{II} &= 0 \\ \chi n_B^I + (1-\chi)n_B^{II} &= n_B,\end{aligned} \tag{32}$$

where $\chi$ denotes the volume fraction of nucleonic phase, $q$ the charge density, and $n_B$ the baryon density. We ignore the fact that the phase with the smallest volume fraction forms finite-size droplets [16,17]; in general, this would tend to decrease the extent of the mixed phase region. Further general consequences of imposing Gibbs' rules in a multicomponent system are that the pressure varies continuously with density in the mixed phase and that the charge densities must have opposite signs in the two phases to satisfy global charge neutrality. We note, however, that not all choices of nucleon-nucleon and kaon-nucleon interactions permit the Gibbs' rules to be satisfied (for an example of such an exception, see [15]). The models chosen in this work *do* allow the Gibbs' rules to be fulfilled at zero and finite temperatures and in the presence of trapped neutrinos.

The nucleon-meson couplings are determined by adjusting them to reproduce the properties of equilibrium nucleonic matter at $T = 0$. We use the numerical values used by [14], i.e., equilibrium density $n_0 = 0.153$ fm$^{-3}$, equilibrium energy per particle of symmetric nuclear $E/A = -16.3$ MeV, effective mass $M^* = 0.78M$, compression modulus $K_0 = 240$ MeV, and symmetry energy $a_{sym} = 32.5$ MeV. These values yield the coupling constants $g_\sigma/m_\sigma = 3.1507$ fm, $g_\omega/m_\omega = 2.1954$ fm, $g_\rho/m_\rho = 2.1888$, $b = 0.008659$, and $c = -0.002421$.

The kaon-meson couplings $g_{\sigma K}$ and $g_{\omega K}$ are related to the magnitude of the kaon optical potential $U_K$ at the saturation density $n_0$ of isospin symmetric nuclear matter:

$$U_K(n_0) = -g_{\sigma K}\sigma(n_0) - g_{\omega K}\omega_0(n_0). \tag{33}$$

Fits to kaonic atom data have yielded values in the range $-(50-200)$ MeV [20–24]. We use $g_{\omega K} = g_{\omega N}/3$ and $g_{\rho K} = g_{\rho N}/2$ on the basis of simple quark and isospin counting. Given the uncertainty in the magnitude of $|U_K|$, consequences for several values of $|U_K|$ were explored in [15]. Moderate values of $|U_K|$

generally produce a second order phase transition and, therefore, lead to moderate effects on the gross properties of stellar structure. Values in excess of 100 MeV were found necessary for a first order phase transition to occur; in this case kaon condensation occurs at a relatively low density with an extended mixed phase region, which leads to more pronounced effects on the structure due to a significant softening of the EOS.

The phase boundaries of the different phases are displayed in Fig. 2 in a $Y_L$–$n_B$ plane for an optical potential $U_K$ of –100 MeV (left) and –120 MeV (right), respectively. The nucleonic phase, the pure kaon matter phase, and the mixed phase are labelled I, II, and III, respectively. Solid lines mark the phase transition at zero temperature and dashed lines mark the phase transition at an entropy per baryon of $s = 1$. Note that finite entropy effects are small and do not affect significantly the phase transition density. The dash-dotted line shows the electron fraction $Y_e$ as a function of density in cold, catalyzed matter (for which $Y_L = Y_e$), which is the final evolutionary state. The region to the left of this line corresponds to negative neutrino chemical potentials and cannot be reached during normal evolutions. The solid and dashed lines, which separate the pure phases from the mixed phase, vary roughly linearly with the lepton fraction. Also notice the large, and nearly constant, densities of the boundary between the mixed phase III and the pure kaon phase II. These densities, for the cases shown, lie above the central densities of the maximum mass stars, so that region II does not generally exist in proto-neutron stars (see [15]). The effect of increasing the lepton number is to reduce the size of the mixed phase (which in fact shrinks to become a second order phase transition for $Y_L > 0.4$

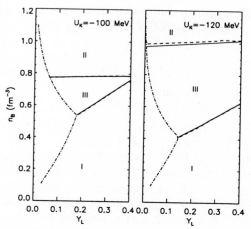

**Fig. 2.** Phase boundaries between pure nucleonic matter (I), pure kaon condensed matter (II) and a mixed phase (III) in the $Y_L$–$n_B$ plane for $U_K = -100$ MeV (left panel) and $U_K = -120$ MeV (right panel). The solid line corresponds to $s = 0$ and the dashed line to $s = 1$. The dashed-dotted line shows the baryon density as a function of the lepton fraction for $s = 0$, neutrino-free ($Y_L = Y_e$) matter

and $U_K = -100$ MeV) and to shift the critical density to higher densities. A similar effect is produced by decreasing the magnitude of the optical potential.

## 3.3 Matter with Quarks

The discussion in this section follows the work of Steiner, Prakash & Lattimer [25]. The thermodynamic potential of the quark phase is $\Omega = \Omega_{\text{FG}} + \Omega_{\text{Int}}$, where

$$\frac{\Omega_{\text{FG}}}{V} = 2N_c T \sum_{i=u,d,s} \int \frac{d^3p}{(2\pi)^3} \left[\ln\left(1 - f_i\right) + \ln\left(1 - \bar{f}_i\right)\right] \quad (34)$$

denotes the Fermi gas contribution arising from quarks. We consider three flavors, $i = u, d, s$ and three colors, $N_c = 3$ of quarks. The distribution functions of fermions and anti-fermions are $f_i = [1 + \exp(\beta(E_i - \mu_i))]^{-1}$ and $\bar{f}_i = [1 + \exp(\beta(E_i + \mu_i))]^{-1}$, where $E_i$ and $\mu_i$ are the single particle energy and chemical potential, respectively, of quark species $i$. To explore the sensitivity of the quark model, we contrast the results of the MIT bag and the Nambu Jona-Lasinio (henceforth NJL) models for $\Omega_{\text{Int}}$.

In the MIT bag model, the Fermi gas contribution is calculated using current, as opposed to dynamical, quark masses. We will restrict ourselves to the simplest bag model and keep only the constant cavity pressure term. The results are qualitatively similar to what is obtained by including perturbative corrections, if the bag constant $B$ is slightly altered [26].

Several features of the Lagrangian of Quantum Chromo-Dynamics (QCD), including the spontaneous breakdown of chiral symmetry, are exhibited by the Nambu Jona-Lasinio (NJL) model, which shares many symmetries with QCD. In its commonly used form, the NJL Lagrangian reads

$$\mathcal{L} = \bar{q}(i\slashed{\partial} - \hat{m}_0)q + G\sum_{k=0}^{8}\left[(\bar{q}\lambda_k q)^2 + (\bar{q}i\gamma_5\lambda_k q)^2\right] \\ - K\left[\det{}_f(\bar{q}(1+\gamma_5)q) + \det{}_f(\bar{q}(1-\gamma_5)q)\right]. \quad (35)$$

The determinant operates over flavor space, $\hat{m}_0$ is the $3 \times 3$ diagonal current quark mass matrix, $\lambda_k$ represents the 8 generators of SU(3), and $\lambda_0$ is proportional to the identity matrix. The four-fermion interactions stem from the original formulation of this model [27], while the flavor mixing, determinental interaction is added to break $U_A(1)$ symmetry [28]. Since the coupling constants $G$ and $K$ are dimensionful, the quantum theory is non-renormalizable. Therefore, an ultraviolet cutoff $\Lambda$ is imposed, and results are considered meaningful only if the quark Fermi momenta are well below this cutoff. The coupling constants $G$ and $K$, the strange quark mass $m_{s,0}$, and the three-momentum ultraviolet cutoff parameter $\Lambda$, are fixed by fitting the experimental values of $f_\pi$, $m_\pi$, $m_K$ and $m_{\eta'}$. We use the values of [29], namely $\Lambda = 602.3$ MeV, $G\Lambda^2 = 1.835$, $K\Lambda^5 = 12.36$, and $m_{0,s} = 140.7$ MeV, obtained using $m_{0,u} = m_{0,d} = 5.5$ MeV. The subscript "0" denotes current quark masses. Results of the gross properties of PNSs obtained by the alternative parameter sets of [30] and [31] are similar to the results quoted below.

In the mean field approximation at finite temperature and at finite baryon density, the thermodynamic potential due to interactions among quarks is given by [31]:

$$\frac{\Omega_{\text{Int}}}{V} = -2N_c \sum_{i=u,d,s} \int \frac{d^3p}{(2\pi)^3} \left( \sqrt{m_i^2 + p^2} - \sqrt{m_{0,i}^2 + p^2} \right)$$
$$+ 2G\langle \bar{q}_i q_i \rangle^2 - 4K \langle \bar{q}_u q_u \rangle \langle \bar{q}_d q_d \rangle \langle \bar{q}_s q_s \rangle . \tag{36}$$

In both (34) and (36) for the NJL model, the quark masses are dynamically generated as solutions of the gap equation obtained by requiring that the potential be stationary with respect to variations in the quark condensate $\langle \bar{q}_i q_i \rangle$:

$$m_i = m_{0,i} - 4G\langle \bar{q}_i q_i \rangle + 2K \langle \bar{q}_j q_j \rangle \langle \bar{q}_k q_k \rangle , \tag{37}$$

$(q_i, q_j, q_k)$ representing any permutation of $(u, d, s)$. The quark condensate $\langle \bar{q}_i q_i \rangle$ and the quark number density $n_i = \langle q_i^\dagger q_i \rangle$ are given by:

$$\langle \bar{q}_i q_i \rangle = -2N_c \int \frac{d^3p}{(2\pi)^3} \frac{m_i}{E_i} \left[ 1 - f_i - \bar{f}_i \right]$$
$$n_i = \langle q_i^\dagger q_i \rangle = 2N_c \int \frac{d^3p}{(2\pi)^3} \left[ f_i - \bar{f}_i \right] . \tag{38}$$

A comparison between the MIT bag and NJL models is facilitated by defining an effective bag pressure in the NJL model to be [32] $B_{eff} = \Omega_{\text{int}}/V - B_0$ with $B_0 V = \Omega_{\text{int}}|_{n_u=n_d=n_s=0}$ a constant value which makes the vacuum energy density zero. In this way, the thermodynamic potential can be expressed as $\Omega = B_{eff} V + \Omega_{\text{FG}}$ which is to be compared to the MIT bag result $\Omega = BV + \Omega_{\text{FG}}$. Note, however, that $\Omega_{FG}$ in the NJL model is calculated using the dynamical quark masses from (37).

The temperature as a function of baryon density for fixed entropy and net lepton concentration is presented in Fig. 3, which compares the cases $(s = 1, Y_{L_e} = 0.4)$ and $(s = 2, Y_{\nu_e} = 0)$ both including and ignoring quarks. The temperature for a multicomponent system in a pure phase can be analyzed with the relation for degenerate Fermi particles

$$T = \frac{s}{\pi^2} \left( \frac{\sum_i p_{F_i} \sqrt{p_{F_i}^2 + (m_i^*)^2}}{\sum_i p_{F_i}^3} \right)^{-1} , \tag{39}$$

where $m_i^*$ and $p_{F_i}$ are the effective mass and the Fermi momentum of component $i$, respectively. This formula is quite accurate since the hadronic and quark Fermi energies are large compared to the temperature. The introduction of hyperons or quarks lowers the Fermi energies of the nucleons and simultaneously increases the specific heat of the matter, simply because there are more components. In the

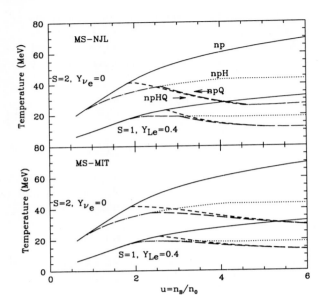

**Fig. 3.** Temperature versus density in units of $n_0$ for two PNS evolutionary snapshots. The upper (lower) panel displays results for the NJL (MIT bag) Lagrangian. The parameters $\zeta = \xi = 0$ in the Müller-Serot (MS) hadronic Lagrangian [33] are chosen. Results are compared for matter containing only nucleons (np), nucleons plus hyperons (npH), nucleons plus quarks (npQ) and nucleons, hyperons and quarks (npHQ). Bold curves indicate the mixed phase region

case of quarks, a further increase, which is just as significant, occurs due to the fact that quarks are rather more relativistic than hardens. The combined effects for quarks are so large that, in the case $M_0^* = 0.6M$ shown in Fig. 3, an actual reduction of temperature with increasing density occurs along an adiabatic. The effect is not necessarily as dramatic for other choices of $M_0^*$, but nevertheless indicates that the temperature will be smaller in a PNS containing quarks than in stars without quarks. The large reduction in temperature might also influence neutrino opacities, which are generally proportional to $T^2$. However, the presence of droplet-like structures in the mixed phase, not considered here, will modify the specific heat. In addition, these structures may dominate the opacity in the mixed phase [34]. However, a PNS simulation is necessary to consistently evaluate the thermal evolution, since the smaller pressure of quark-containing matter would tend to increase the star's density and would oppose this effect.

This last point is highlighted in Fig. 4 which shows phase diagrams for the mixed phase in the baryon density-neutrino fraction plane. The upper and lower boundaries of the mixed phase region are displayed as bold lines, while the central densities of the maximum mass configurations are shown as light lines. In no case, for either quark model and whether or not hyperons are included, are pure quark stars possible. The high-density phase boundaries are always well

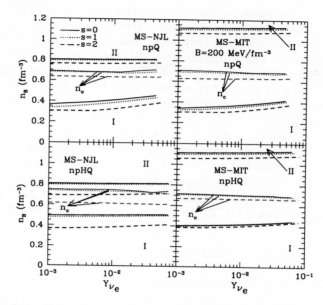

**Fig. 4.** The phase diagram of the quark-hadronic transition in the baryon number density - neutrino concentration plane for three representative snapshots during the evolution of a proto-neutron star. The left (right) panels are for the NJL (MIT bag) quark EOS, and hyperons are (are not) included in the bottom (top) panels. The parameters $\zeta = \xi = 0$ in the Muller-Serot (MS) hadronic Lagrangian are chosen. The lower- and upper-density boundaries of the mixed phase are indicated by bold curves. The central densities of maximum mass configurations are shown by thin curves

above the central densities. While in the optimum case, in which the parameters of both the hadronic and quark EOSs are fine-tuned, it is possible for a pure quark core to form if $B < 150$ MeV fm$^{-3}$, the maximum mass decreases below 1.44 $M_\odot$ if $B < 145$ MeV fm$^{-3}$. This narrow window, which further decreases or disappears completely if the hadronic EOS is altered, suggests that pure quark configurations may be unlikely.

## 3.4 Inhomogeneous Phases

It is widely believed that some type of phase transition will occur in nuclear matter at high densities. For example, a transition to unconfined quark matter should exist at sufficiently high density, and at lower densities, a first-order transition to a Bose condensate phase might exist. Such phase transitions are expected to soften the equation of state, leading to changes in the mass-radius relation and lowering the maximum mass. Phase transitions can also influence transport and weak interaction rates in matter.

Glendenning has shown that, due to the existence of two conserved charges (baryon number and charge) instead of just one, first order phase transitions can lead to a large mixed phase region in the neutron star interior [35]. The mixed phase consists of high baryon density, negatively charged, matter coexisting with lower density, positively charged, baryonic matter. The situation is entirely analogous to the well-known situation involving the mixed phase consisting of nuclei and a surrounding nucleonic vapor that occurs below nuclear saturation density [36]. The occurrence of a mixed phase, as opposed to a Maxwell construction, results in a wider transition in which bulk thermodynamic properties such as pressure vary less rapidly but are softer over a wider density range. In addition, the propagation of neutrinos whose wavelength is greater than the typical droplet size and less than the inter-droplet spacing will be greatly affected by the heterogeneity of the mixed phase, as a consequence of the coherent scattering of neutrinos from the matter in the droplet. The thermodynamics and the effect on neutron star structure of two situations have been studied in some detail: first order kaon condensation [18,37,38] and the quark-hadronic transition [39,40].

## 4 Neutrino-Matter Interaction Rates

One of the important microphysical inputs in PNS simulations is the neutrino opacity at supra-nuclear density [1,41–45]. Although it was realized over a decade ago that the effects due to degeneracy and strong interactions significantly alter the neutrino mean free paths, it is only recently that detailed calculations have become available [34,46–52]. The scattering and absorption reactions that contribute to the neutrino opacity are

$$\nu_e + B \to e^- + B', \quad \bar{\nu}_e + B \to e^+ + B', \tag{40}$$

$$\nu_X + B \to \nu_X + B', \quad \nu_X + e^- \to \nu_X + e^-, \tag{41}$$

where the scattering reactions are common to all neutrino species and the dominant source of opacity for the electron neutrinos is due to the charged reaction. The weak interaction rates in hot and dense matter are modified due to many in-medium effects. The most important of these are:

(1) *Composition*: The neutrino mean free paths depend sensitively on the composition which is sensitive to the nature of strong interactions. First, the different degeneracies of the different Fermions determines the single-pair response due to Pauli blocking. Second, neutrinos couple differently to different baryonic species; consequently, the net rates will depend on the individual concentrations.

(2) *In-medium dispersion relations*: At high density, the single-particle spectra are significantly modified from their noninteracting forms due to effects of strong interactions. Interacting matter features smaller effective baryon masses and energy shifts relative to non-interacting matter.

(3) *Correlations*: Repulsive particle-hole interactions and Coulomb interactions generally result in a screened dielectric response and also lead to collective excitations in matter. These effects may be calculated using the Random Phase Approximation (RPA), in which ring diagrams are summed to all orders. Model

calculations [46,48,50,53–58] indicate that at high density the neutrino cross sections are suppressed relative to the case in which these effects are ignored. In addition, these correlations enhance the average energy transfer in neutrino-nucleon collisions. Improvements in determining the many-body dynamic form factor and assessing the role of particle-particle interactions in dense matter at finite temperature are necessary before the full effects of many-body correlations may be ascertained.

The relative importance of the various effects described above on neutrino transport is only beginning to be studied systematically. As a first step, we will focus on effects due to modifications (1) through (3) above. To see how this is accomplished, we start with a general expression for the differential cross section [59,60]

$$\frac{1}{V}\frac{d^3\sigma}{d^2\Omega_3 dE_3} = -\frac{G_F^2}{128\pi^2}\frac{E_3}{E_1}\left[1 - \exp\left(\frac{-q_0 - (\mu_2 - \mu_4)}{T}\right)\right]^{-1}$$
$$\times (1 - f_3(E_3)) \text{ Im } (L^{\alpha\beta}\Pi^R_{\alpha\beta}), \qquad (42)$$

where the incoming neutrino energy is $E_1$ and the outgoing electron energy is $E_3$. The factor $[1 - \exp((-q_0 - \mu_2 + \mu_4)/T)]^{-1}$ maintains detailed balance, for particles labeled '2' and '4' which are in thermal equilibrium at temperature $T$ and in chemical equilibrium with chemical potentials $\mu_2$ and $\mu_4$, respectively. The final state blocking of the outgoing lepton is accounted for by the Pauli blocking factor $(1 - f_3(E_3))$. The lepton tensor $L_{\alpha\beta}$ is given by

$$L^{\alpha\beta} = 8[2k^\alpha k^\beta + (k \cdot q)g^{\alpha\beta} - (k^\alpha q^\beta + q^\alpha k^\beta) \mp i\epsilon^{\alpha\beta\mu\nu}k_\mu q_\nu] \qquad (43)$$

The target particle retarded polarization tensor is

$$\text{Im}\Pi^R_{\alpha\beta} = \tanh\left(\frac{q_0 + (\mu_2 - \mu_4)}{2T}\right)\text{Im } \Pi_{\alpha\beta}, \qquad (44)$$

where $\Pi_{\alpha\beta}$ is the time ordered or causal polarization and is given by

$$\Pi_{\alpha\beta} = -i\int \frac{d^4p}{(2\pi)^4}\text{Tr }[T(G_2(p)J_\alpha G_4(p+q)J_\beta)]. \qquad (45)$$

Above, $k_\mu$ is the incoming neutrino four-momentum and $q_\mu$ is the four-momentum transfer. In writing the lepton tensor, we have neglected the electron mass term, since typical electron energies are of the order of a few hundred MeV. The Green's functions $G_i(p)$ (the index $i$ labels particle species) describe the propagation of baryons at finite density and temperature. The current operator $J_\mu$ is $\gamma_\mu$ for the vector current and $\gamma_\mu\gamma_5$ for the axial current. Effects of strong and electromagnetic correlations may be calculated by utilizing the RPA polarization tensor

$$\Pi^{RPA} = \Pi + \Pi^{RPA}D\Pi, \qquad (46)$$

where $D$ denotes the interaction matrix, in (42) (see [50] for more details).

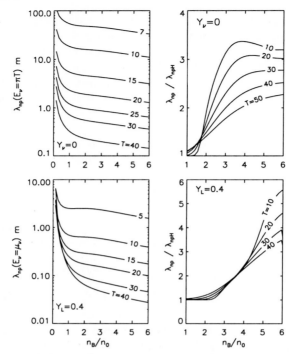

**Fig. 5.** Neutrino mean free paths in matter with nucleons only (left panels). Right panels show ratios of mean free paths in matter without and with hyperons. Abscissa is baryon density $n_B$ ($n_0$ is the nuclear equilibrium density). Top panels show scattering mean free paths common to all neutrino species. The bottom panels show results for electron neutrino mean free paths where absorption reactions are included. The neutrino content is labelled in the different panels

### 4.1 Neutrino Mean Free Paths

The differential cross section (42) is needed in multi-energy group neutrino transport codes. However, more approximate neutrino transport algorithms (as in Section 2) require the total cross section as a function of the neutrino energy for the calculation of diffusion coefficients. The cross section per unit volume of matter (or equivalently the inverse mean free path) is obtained by integrating $E_3$ and $\Omega_3$ in (42).

Under degenerate conditions even modest changes to the composition significantly alter the neutrino scattering and absorption mean free paths. In Fig. 5, the neutrino scattering and absorption mean free paths are shown for models GM3np and GM3npH relevant to the deleptonization and cooling epochs. The top panels show the scattering mean free paths common to all neutrino species in neutrino free matter. The scattering mean free paths for thermal neutrinos ($E_\nu = \pi T$) is shown in the left panel for various temperatures. To study the influence of hyperons, the ratio of the $\lambda_{np}/\lambda_{npH}$ is shown in the right panels. The presence of hyperons significantly increase the scattering cross sections, by

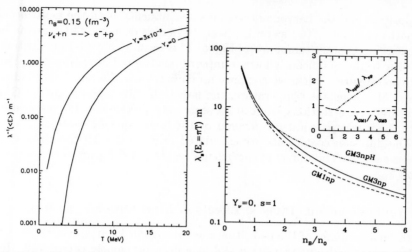

**Fig. 6.** Left: Charged current inverse neutrino mean free paths versus temperature. Right: Comparison of scattering mean free paths in neutrino poor matter at fixed entropy for different EOSs in matter containing nucleons and also hyperons

a factor $\sim (2-3)$. Similar results for the absorption cross sections are shown in the lower panels for $Y_L = 0.4$. Again we notice a significant enhancement (right panel) when hyperons appear, the factor here could be as large as 5.

During the deleptonization stage, lepton number transport is sensitive to charged current reactions which dominate scattering reactions. At zero temperature, charged current reactions $\nu + n \leftrightarrow e + p$ depend sensitively on the proton fraction $Y_p$ [61]. Kinematic restrictions require $Y_p$ to be larger than $11 - 14\%$ (direct Urca threshold). At early times, a finite neutrino chemical potential favors a large $Y_p$ throughout the star, which enables these reactions to proceed without any hindrance. Toward the late stages, however, $Y_p$ decreases with decreasing $\mu_\nu$ and charged current reactions may be naively expected to become inoperative. The threshold density for the charged current reaction when $\mu_\nu = 0$ and $T = 0$ depends sensitively on the density dependence of the nuclear symmetry energy. In field-theoretical models, in which the symmetry energy is largely given by contributions due to $\rho$-meson exchange, the critical density is typically $n_B = 2 \sim 3 n_0$. However, finite temperatures favor larger $Y_p$'s and increase the average neutrino energy enabling the charged current reactions to proceed even below these densities. Fig. 6 shows that this is the case even at relatively low temperatures $(T \sim 3-5)$ MeV for a baryon density $n_B = 0.15$ fm$^{-3}$. The sharp rise with temperature, which occurs even for $Y_\nu = 0$, clearly indicates that this reaction dominates the $\nu_e$ opacity even during the late deleptonization era. Thus, charged current reactions cannot be simply turned off when the neutrino chemical potential becomes small enough as was done in prior PNS simulations [1].

The EOS and neutrino mean free paths are intimately related, which is best illustrated by comparing the results shown in Fig. 5 with those shown in Fig. 6.

Composition and the baryon effective masses influence both the neutrino mean free paths and the matter's specific heat. Hyperons decrease the neutrino mean free paths at constant temperature Fig. 5. This trend is reversed at constant entropy due to the significantly lower temperatures favored in npH matter. Similar effects are apparent when we compare np models with different baryon effective masses. At a constant temperature, the larger effective mass in model GM3np favors larger cross sections, while at constant entropy this trend is again reversed due to the lower temperatures favored by the larger specific heat.

The diffusion coefficients are calculated using (18) with the cross sections discussed above. The diffusion coefficients $D_2, D_3$, and $D_4$ are functions of $n_B$, $T$, and $Y_{\nu_e}$.

## 4.2 Inhomogeneous Phases: Effects of First Order Transitions

The thermodynamics of the two situations, first order kaon condensation [18,37] and the quark-hadronic transition [40], has been previously considered. Reddy, Bertsch and Prakash [34] have studied the effects of inhomogeneous phases on $\nu$-matter interactions. Based on simple estimates of the surface tension between nuclear matter and the exotic phase, typical droplet sizes range from $5-15$ fm [38], and inter-droplet spacings range up to several times larger. The propagation of neutrinos whose wavelength is greater than the typical droplet size and less than the inter-droplet spacing, i.e., 2 MeV $\lesssim E_\nu \lesssim$ 40 MeV, will be greatly affected by the heterogeneity of the mixed phase, as a consequence of the coherent scattering of neutrinos from the matter in the droplet.

The Lagrangian that describes the neutral current coupling of neutrinos to the droplet is

$$\mathcal{L}_W = \frac{G_F}{2\sqrt{2}} \, \bar{\nu}\gamma_\mu(1-\gamma_5)\nu \, J_D^\mu \,, \tag{47}$$

where $J_D^\mu$ is the neutral current carried by the droplet and $G_F = 1.166 \times 10^{-5}$ Gee$^{-2}$ is the Fermi weak coupling constant. For non-relativistic droplets, $J_D^\mu = \rho_W(x) \, \delta^{\mu 0}$ has only a time like component. Here, $\rho_W(x)$ is the excess weak charge density in the droplet. The total weak charge enclosed in a droplet of radius $r_d$ is $N_W = \int_0^{r_d} d^3x \, \rho_W(x)$ and the form factor is $F(q) = (1/N_W) \int_0^{r_d} d^3x \, \rho_W(x) \, \sin qx/qx$. The differential cross section for neutrinos scattering from an isolated droplet is then

$$\frac{d\sigma}{d\cos\theta} = \frac{E_\nu^2}{16\pi} G_F^2 N_W^2 (1+\cos\theta) F^2(q) \,. \tag{48}$$

In the above equation, $E_\nu$ is the neutrino energy and $\theta$ is the scattering angle. Since the droplets are massive, we consider only elastic scattering for which the magnitude of the momentum transfer is $q = \sqrt{2}E_\nu(1-\cos\theta)$.

We must embed the droplets into the medium to evaluate the neutrino transport parameters. The droplet radius $r_d$ and the inter-droplet spacing are determined by the interplay of surface and Coulomb energies. In the Winger-Seitz approximation, the cell radius is $R_W = (3/4\pi N_D)^{1/3}$ where the droplet density is $N_D$. Except for one aspect, we will neglect coherent scattering from more than one droplet. If the droplets form a lattice, Brags scattering will dominate and our description would not be valid. But for low density and a liquid phase, interference from multiple droplets affects scattering only at long wavelengths. If the ambient temperature is not small compared to the melting temperature, the droplet phase will be a liquid and interference effects arising from scattering off different droplets are small for neutrino energies $E_\nu \gtrsim (1/R_W)$. However, multiple droplet scattering cannot be neglected for $E_\nu \lesssim 1/R_W$. The effects of other droplets is to cancel scattering in the forward direction, because the interference is destructive except at exactly zero degrees, where it produces a change in the index of refraction of the medium. These effects are usually incorporated by multiplying the differential cross section (48) by the static form factor of the medium. The static form factor, defined in terms of the radial distribution function of the droplets, $g(r)$, is

$$S(q) = 1 + N_D \int d^3r \exp i\mathbf{q}.\mathbf{r}\, (g(r) - 1). \tag{49}$$

The droplet correlations, which determine $g(r)$, arise due to the Coulomb force and is measured in terms of the dimensionless Coulomb number

$$\Gamma = Z^2 e^2/(8\pi R_W k T).$$

Due to the long-range character of the Coulomb force, the role of screening and the finite droplet size, $g(r)$ cannot be computed analytically. We use a simple ansatz for the radial distribution function $g(r < R_W) = 0$ and $g(r > R_W) = 1$. For this choice, the structure factor is independent of $\Gamma$. Monte Carlo calculations [62] of the liquid structure function of a simple one component plasma indicate that our choice of $S(q)$ is conservative for typical neutrino energies of interest.

The simple ansatz for $g(r)$ is equivalent to subtracting, from the weak charge density $\rho_W$, a uniform density which has the same total weak charge $N_W$ as the matter in the Winger-Seine cell. Thus, effects due to $S(q)$ may be incorporated by replacing the form factor $F(q)$ by

$$F(q) \to \tilde{F}(q) = F(q) - 3\,\frac{\sin qR_W - (qR_W)\cos qR_W}{(qR_W)^3}. \tag{50}$$

The neutrino–droplet differential cross section per unit volume then follows:

$$\frac{1}{V}\frac{d\sigma}{d\cos\theta} = N_D\,\frac{E_\nu^2}{16\pi} G_F^2 N_W^2 (1 + \cos\theta)\tilde{F}^2(q). \tag{51}$$

Note that even for small droplet density $N_D$, the factor $N_W^2$ acts to enhance the droplet scattering. To quantify the importance of droplets as a source of opacity, we compare with the standard scenario in which matter is uniform and composed of neutrons. The dominant source of opacity is then due to scattering from thermal fluctuations and

$$\frac{1}{V}\frac{d\sigma}{d\cos\theta} = \frac{G_F^2}{8\pi}\left(c_V^2(1+\cos\theta) + (3-\cos\theta)c_A^2\right) E_\nu^2 \times \frac{3}{2}\,n_n\left[\frac{k_B T}{E_{fn}}\right], \tag{52}$$

where $c_V$ and $c_A$ are respectively the vector and axial coupling constants of the neutron, $n_n$ is the neutron number density, $E_{fn}$ is the neutron Fermi energy and $T$ is the matter temperature [63].

The transport cross sections that are employed in studying the diffusive transport of neutrinos in the core of a neutron star are differential cross sections weighted by the angular factor $(1 - \cos\theta)$. The transport mean free path $\lambda(E_\nu)$ for a given neutrino energy $E_\nu$ is given by

$$\frac{1}{\lambda(E_\nu)} = \frac{\sigma_T(E_\nu)}{V} = \int d\cos\theta\, (1 - \cos\theta) \left[\frac{1}{V}\frac{d\sigma}{d\cos\theta}\right]. \quad (53)$$

Models of first order phase transitions in dense matter provide the weak charge and form factors of the droplets and permit the evaluation of $\nu$–droplet scattering contributions to the opacity of the mixed phase. For the models considered, namely the first order kaon condensate and the quark-hadronic phase transition, the neutrino mean free paths in the mixed phase are shown in the left and right panels of Fig. 7, respectively. The results are shown for the indicated values of the baryon density $n_B$ and temperature $T$ where the model predicts a mixed phase exists. The kaon droplets are characterized by radii $r_d \sim 7$ fm and inter-droplet spacings $R_W \sim 20$ fm, and enclose a net weak vector charge $N_W \sim 700$. The quark droplets are characterized by $r_d \sim 5$ fm and $R_W \sim 11$ fm, and an enclosed weak charge $N_W \sim 850$. For comparison, the neutrino mean free path in uniform neutron matter at the same $n_b$ and $T$ are also shown. It is apparent that there is a large coherent scattering-induced reduction in the mean free path for the typical energy $E_\nu \sim \pi T$. At much lower energies, the inter-droplet correlations tend to screen the weak charge of the droplet, and at higher energies the coherence is attenuated by the droplet form factor.

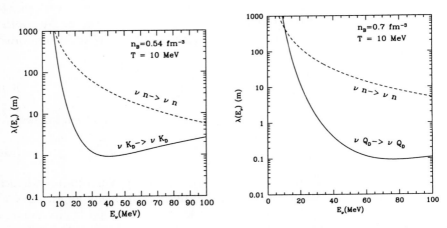

**Fig. 7.** Neutrino mean free paths as a function of neutrino energy. Solid lines are for matter in a mixed phase containing kaons (left panel) and quarks (right panel), and dashed curves are for uniform matter.

The large reduction in neutrino mean free path found here implies that the mixed phase will cool significantly slower than homogeneous matter. Consequently, the observable neutrino luminosity at late times might be affected as it is driven by the transport of energy from the deep interior. The reduced mean free path in the interior will tend to prolong the late time neutrino emission phase.

## 4.3 Effects of Quark Superconductivity and Superfluidity

Recent theoretical works [64,65] suggest that quarks form Cooper pairs in medium, a natural consequence of attractive interactions destabilizing the Fermi surface. Although the idea of quark pairing in dense matter is not new [64,66], it has recently seen renewed interest in the context of the phase diagram of QCD [65]. Model calculations, mostly based on four-quark effective interactions, predict the restoration of spontaneously broken chiral symmetry through the onset of color superconductivity at low temperatures. They predict an energy gap of $\Delta \sim 100$ MeV for a typical quark chemical potential of $\mu_q \sim 400$ MeV. As in BCS theory, the gap will weaken for $T > 0$, and at some critical temperature $T_c$ there is a (second-order) transition to a "standard" quark-gluon plasma. During cooling from an initial temperature in excess of $T_c$, the formation of a gap in the fermionic excitation spectrum in quark matter will influence various transport properties of the system. Carter and Reddy have studied its influence on the transport of neutrinos [67].

The differential neutrino scattering cross section per unit volume in an infinite and homogeneous system of relativistic fermions as calculated in linear response theory is given by (42). The medium is characterized by the quark polarization tensor $\Pi_{\alpha\beta}$. In the case of free quarks, each flavor contributes a term of the form

$$\Pi_{\alpha\beta}(q) = -i\text{Tr}_c \int \frac{d^4p}{(2\pi)^4} \text{Tr}\left[S_0(p)\Gamma_\alpha S_0(p+q)\Gamma_\beta\right], \tag{54}$$

where $S_0(p)$ is the free quark propagator at finite chemical potential and temperature. The outer trace is over color and simplifies to a $N_c = 3$ degeneracy. The inner trace is over spin, and the $\Gamma_\alpha$ are the neutrino-quark vertex functions which determine the spin channel. Specifically, the vector polarization is computed by choosing $(\Gamma_\alpha, \Gamma_\beta) = (\gamma_\alpha, \gamma_\beta)$. The axial and mixed vector-axial polarizations are similarly obtained from $(\Gamma_\alpha, \Gamma_\beta) = (\gamma_\alpha\gamma_5, \gamma_\beta\gamma_5)$ and $(\Gamma_\alpha, \Gamma_\beta) = (\gamma_\alpha, \gamma_\beta\gamma_5)$, respectively.

The free quark propagators in (54) are naturally modified in a superconducting medium. As first pointed out by Bardeen, Cooper, and Schrieffer several decades ago, the quasi-particle dispersion relation is modified due to the presence of a gap in the excitation spectrum. In calculating these effects, we will consider the simplified case of QCD with two quark flavors which obey $\text{SU}(2)_L \times \text{SU}(2)_R$ flavor symmetry, given that the light $u$ and $d$ quarks dominate low-energy phenomena. Furthermore we will assume that, through some unspecified effective interactions, quarks pair in a manner analogous to the BCS mechanism [68]. The relevant consequences of this are the restoration of chiral symmetry (hence all quarks are approximately massless) and the existence of an energy gap at zero temperature, $\Delta_0$, with approximate temperature dependence,

$$\Delta(T) = \Delta_0 \sqrt{1 - \left(\frac{T}{T_c}\right)^2}. \tag{55}$$

The critical temperature $T_c \simeq 0.57\Delta_0$ is likewise taken from BCS theory; this relation has been shown to hold for perturbative QCD [69] and is thus a reasonable assumption for non-perturbative physics.

Breaking the $SU_c(3)$ color group leads to complications not found in electrodynamics. In QCD the superconducting gap is equivalent to a diquark condensate, which can at most involve two of the three fundamental quark colors. The condensate must therefore be colored. Since the scalar diquark (in the $\bar{3}$ color representation) appears to always be the most attractive channel, we consider the anomalous (or Gorkov) propagator [70]

$$F(p)_{abfg} = \langle q_{fa}^T(p) C \gamma_5 q_{gb}(-p) \rangle$$
$$= -i\epsilon_{ab3}\epsilon_{fg}\Delta \left( \frac{\Lambda^+(p)}{p_o^2 - \xi_p^2} + \frac{\Lambda^-(p)}{p_o^2 - \bar{\xi}_p^2} \right) \gamma_5 C. \tag{56}$$

Here, $a, b$ are color indices, $f, g$ are flavor indices, $\epsilon_{abc}$ is the usual anti-symmetric tensor and we have conventionally chosen 3 to be the condensate color. This propagator is also antisymmetric in flavor and spin, with $C = -i\gamma_0\gamma_2$ being the charge conjugation operator.

The color bias of the condensate forces a splitting of the normal quark propagator into colors transverse and parallel to the diquark. Quarks of color 3, parallel to the condensate in color space, will be unaffected and propagate freely, with

$$S_0(p)_{af}^{bg} = i\delta_a^b \delta_f^g \left( \frac{\Lambda^+(p)}{p_o^2 - E_p^2} + \frac{\Lambda^-(p)}{p_o^2 - \bar{E}_p^2} \right) (p_\mu \gamma^\mu - \mu\gamma_0). \tag{57}$$

This is written in terms of the particle and anti-particle projection operators $\Lambda^+(p)$ and $\Lambda^-(p)$ respectively, where $\Lambda^\pm(p) = (1 \pm \gamma_0 \boldsymbol{\gamma} \cdot \hat{p})/2$. The excitation energies are simply $E_p = |\boldsymbol{p}| - \mu$ for quarks and $\bar{E}_p = |\boldsymbol{p}| + \mu$ for anti-quarks.

On the other hand, transverse quark colors 1 and 2 participate in the diquark and thus their quasi-particle propagators are given as

$$S(p)_{af}^{bg} = i\delta_a^b \delta_f^g \left( \frac{\Lambda^+(p)}{p_o^2 - \xi_p^2} + \frac{\Lambda^-(p)}{p_o^2 - \bar{\xi}_p^2} \right) (p_\mu \gamma^\mu - \mu\gamma_0). \tag{58}$$

The quasi-particle energy is $\xi_p = \sqrt{(|\boldsymbol{p}| - \mu)^2 + \Delta^2}$, and for the anti-particle $\bar{\xi}_p = \sqrt{(|\boldsymbol{p}| + \mu)^2 + \Delta^2}$.

The appearance of an anomalous propagator in the superconducting phase indicates that the polarization tensor gets contributions from both the normal

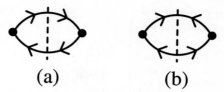

**Fig. 8.** Standard loop (a) and anomalous loop (b) diagrams contributing to the quark polarization operator

quasi-particle propagators (58) and anomalous propagator (56). Thus, to order $G_F^2$, (54) is replaced with the two contributions corresponding to the diagrams shown in Fig. 8, and written

$$\Pi_{\alpha\beta}(q) = -i \int \frac{d^4p}{(2\pi)^4} \{\text{Tr}\,[S_0(p)\Gamma_\alpha S_0(p+q)\Gamma_\beta] \\ + 2\text{Tr}\,[S(p)\Gamma_\alpha S(p+q)\Gamma_\beta] + 2\text{Tr}\,[F(p)\Gamma_\alpha \bar{F}(p+q)\Gamma_\beta]\}\,. \quad (59)$$

The remaining trace is over spin, as the color trace has been performed. Fig. 8(a) corresponds to the first two terms, which have been decomposed into one term with ungapped propagators (57) and the other with gapped quasi-particle propagators (58). Fig. 8(b) represents the third, anomalous term.

For neutrino scattering we must consider vector, axial, and mixed vector-axial channels, all summed over flavors. The full polarization, to be used in evaluating (42), may be written

$$\Pi_{\alpha\beta} = \sum_f \left[(C_V^f)^2 \Pi_{\alpha\beta}^V + (C_A^f)^2 \Pi_{\alpha\beta}^A - 2 C_V^f C_A^f \Pi_{\alpha\beta}^{VA}\right]\,. \quad (60)$$

The coupling constants for up quarks are $C_V^u = \frac{1}{2} - \frac{4}{3}\sin^2\theta_W$ and $C_A^u = \frac{1}{2}$, and for down quarks, $C_V^d = -\frac{1}{2} + \frac{2}{3}\sin^2\theta_W$ and $C_A^d = -\frac{1}{2}$, where $\sin^2\theta_W \simeq 0.23$ is the Weinberg angle.

The differential cross section, (42) and the total cross section are obtained by integrating over all neutrino energy transfers and angles. Results for the neutrino mean free path, $\lambda = V/\sigma$, are shown in Fig. 9 as a function of incoming neutrino energy $E_\nu$ (for ambient conditions of $\mu_q = 400$ MeV and $T = 30$ MeV). They show the same energy dependence found previously for free relativistic and degenerate fermionic matter [49]; $\lambda \propto 1/E_\nu^2$ for $E_\nu \gg T$ and $\lambda \propto 1/E_\nu$ for $E_\nu \ll T$. The results indicate that this energy dependence is not modified by the presence of a gap when $\Delta \sim T$. Thus, the primary effect of the superconducting phase is a much larger mean free path. This is consistent with the suppression found in the vector-longitudinal response function, which dominates the polarization sum (60), at $q_0 < q$.

We now consider the cooling of a macroscopic sphere of quark matter, a toy approximation for the core of a neutron star with a mixed quark phase, as it becomes superconducting. As in the preceding calculation, we consider the relatively simple case of two massless flavors with identical chemical potentials

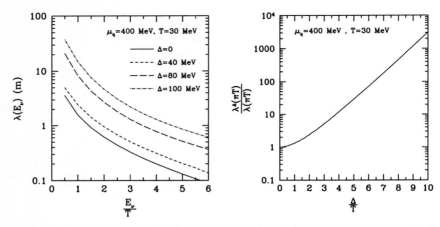

**Fig. 9.** Left panel: Neutrino mean free path as a function of neutrino energy $E_\nu$. Right panel: Neutrino mean free paths for $E_\nu = \pi T$ as a function of $\Delta/T$. These results are virtually independent of temperature for $T \lesssim 50$ MeV.

and disregard the quarks parallel in color to the condensate; *i.e.* we consider a background comprised exclusively of quasi-quarks.

The cooling of a spherical system of quark matter from $T \sim T_c \sim 50$ MeV is driven by neutrino diffusion, for the neutrino mean free path is much smaller than the dimensions of system of astrophysical size, and yet several orders of magnitude larger than the mean free path of the quarks. The diffusion equation for energy transport by neutrinos in a spherical geometry is

$$C_V \frac{dT}{dt} = -\frac{1}{r^2} \frac{\partial L_\nu}{\partial r}, \tag{61}$$

where $C_V$ is the specific heat per unit volume of quark matter, $T$ is the temperature, and $r$ is the radius. The neutrino energy luminosity for each neutrino type, $L_\nu$, depends on the neutrino mean free path and the spatial gradients in temperature and is approximated by an integral over neutrino energy $E_\nu$

$$L_\nu \cong 6 \int dE_\nu \frac{c}{6\pi^2} E_\nu^3 r^2 \lambda(E_\nu) \left|\frac{\partial f(E_\nu)}{\partial r}\right|. \tag{62}$$

We assume that neutrino interactions are dominated by the neutral current scattering which is common to all neutrino types, accounting for the factor 6 in (62).

The solution to the diffusion equation will depend on the initial temperature gradients. However, being primarily interested in a qualitative description of cooling through a second-order phase transition to superconducting matter, the temporal behavior can be characterized by a time scale $\tau_c$ which is proportional to the inverse cooling rate. This characteristic time is

$$\tau_c(T) = C_V(T) \frac{R^2}{c\langle\lambda(T)\rangle}, \tag{63}$$

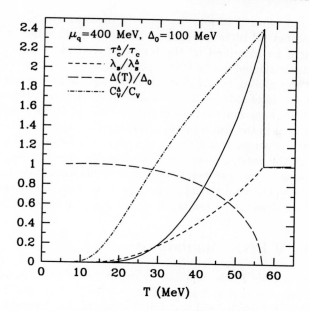

**Fig. 10.** The extent to which different physical quantities are affected by a superconducting transition. Ratios of the cooling time scale (solid curve), the inverse mean free path (short-dashed curve) and the matter specific heat (dot-dashed curve) in the superconducting phase to that in the normal phase is shown as a function of the matter temperature. The ratio of the gap to its zero temperature value $\Delta_0$ is also shown (long-dashed curve).

and is a strong function of the temperature since it depends on both the specific heat $C_V$ and the energy-averaged neutrino mean free path, $\langle \lambda(T) \rangle$. The latter quantity is here approximated by $\lambda(E_\nu = \pi T)$ since the neutrinos are in thermal equilibrium. Using BCS theory, as described in the previous section, $\langle \lambda(T) \rangle$ depends on the gap $\Delta$ as shown in the right panel of Fig. 9. The results indicate that for small $\Delta/T$, $\lambda$ is not strongly modified, but as $\Delta/T$ increases so too does $\lambda$, non-linearly at first and then exponentially for $\Delta/T \gtrsim 5$. Also, in the BCS theory, the temperature dependencies of $C_V$ (dashed curve) and $\Delta$ (dot-dashed curve) are shown in Fig. 10. Finally, the ratio $\tau_c^\Delta(T)/\tau_c(T)$, a measure of the extent to which the cooling rate is changed by a gap, is shown by the solid line in Fig. 10. We note that the diffusion approximation is only valid when $\lambda \ll R$ and will thus fail for very low temperatures, when $\lambda \lesssim R$.

These results are readily interpreted. The cooling rate around $T_c$ is influenced mainly by the peak in the specific heat associated with the second order phase transition, since the neutrino mean free path is not strongly affected when $T \geq \Delta$. Subsequently, as the matter cools, both $C_V$ and $\lambda^{-1}$ decrease in a non-linear fashion for $T \sim \Delta$. Upon further cooling, when $T \ll \Delta$, both $C_V$ and $\lambda^{-1}$ decrease exponentially. Both effects accelerate the cooling process.

We conclude that if it were possible to measure the neutrino luminosity from the hypothetical object described here, a second order superconducting transition might be identified by the temporal characteristics of the late time supernova neutrino signal from a PNS. Specifically, there might be a brief interval during which the cooling would slow when the core temperature falls below $T_c$, signified by a period of reduced neutrino detection. However, this effect might be obscured by $\nu$-opaque matter outside the star's core. If the neutrino opacity of these outer regions of the star is large, it is likely that any sharp temporal feature associated with neutrino transport in the core will be diluted as the neutrino diffuse through the outer regions. Nevertheless, the main finding, which is that phase transitions in the core can have a discernible impact on the transport of neutrinos and suggests that the late time supernova neutrino signal is a promising probe of the high density and low temperature region of the QCD phase diagram.

## 5 Results of PNS Simulations

Neutrino signals from PNSs depends on many stellar properties, including the mass; initial entropy, lepton fraction and density profiles; and neutrino opacities. Pons et al. [10] carried out a detailed study of the dependence of neutrino emission on PNS characteristics. They verified the generic results of Burrows & Lattimer [1] that both neutrino luminosities and average energies increased with increasing mass (see Fig. 11). In addition, they found that variations in initial entropy and lepton fraction profiles in the outer regions of the PNS caused only transient (lasting a few tenths of a second) variations in neutrino luminosities and energies. Variations in the central lepton fraction and entropy were found to produce modest changes in neutrino luminosities that persisted to late times. The central values of lepton fraction and entropy are established during core collapse, and will depend upon the initial properties of the star as well as the EOS and neutrino transport during the collapse.

### 5.1 Baseline Results

Properties of the dense matter EOS that affect PNS evolution include the compressibility, symmetry energy, specific heat, and composition. Pons et al. [10] employed a field theoretical EOS [4], with which the results due to some differences in stellar size (due to variations in nuclear interactions) and composition were studied. Some results are summarized in Fig. 11. Overall, both the average energies and luminosities of stars containing hyperons were larger compared to those without. In addition, for stars without hyperons, those stars with smaller radii had higher average emitted neutrino energy, although the predicted luminosities for early times ($t < 10$ s) were insensitive to radii. This result only holds if the opacities are calculated consistently with each EOS [49,50]; otherwise rather larger variations in evolutions would have been found [45,51]. The same held true for models which allowed for the presence of hyperons, except

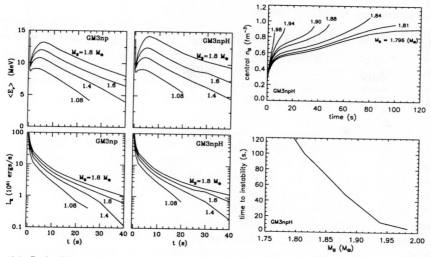

**Fig. 11.** Left: The evolution of the neutrino average energy and total neutrino luminosity is compared for several assumed PNS baryon masses and EOSs. The EOSs in the left panels contain only baryons and leptons while those in the right panels also contain hyperons. Top right panel: Evolution of the central baryon number density for different baryonic mass stars containing hyperons (model GM3npH) which are metastable. Bottom right panel: Time required by stars shown in the top panel to reach the unstable configuration

when the initial proto-neutron star mass was significantly larger than the maximum mass for cold, catalyzed matter. Another new result was that the average emitted neutrino energy of all flavors increased during the first 2-5 seconds of evolution, and then decreased nearly linearly with time. For times larger than about 10 seconds, and prior to the occurrence of neutrino transparency, the neutrino luminosities decayed exponentially with a time constant that was sensitive to the high-density properties of matter. Significant variations in neutrino emission occurred beyond 10 seconds: it was found that neutrino luminosities were larger during this time for stars with smaller radii and with the inclusion of hyperons in the matter. Finally, significant regions of the stars appeared to become convectively unstable during the evolution, as several works have found [71-75].

The right panel of Fig. 11 shows the time development of the central baryon density (top panel) and also the time to the collapse instability as a function of baryon mass (bottom panel). The larger the mass, the shorter the time to instability, since the PNS does not have to evolve in lepton number as much. Above 2.005 $M_\odot$, the metatstability disappears because the GM3npH initial model with the lepton and entropy profiles we chose is already unstable. Below about 1.73 $M_\odot$, there is no metastability, since this is the maximum mass of the cold, catalyzed npH star for GM3. The signature of neutrino emission from a metastable PNS should be identifiable and it is discussed in Section 5.4.

## 5.2 Influence of Many-Body Correlations

The main effect of the larger mean free paths produced by RPA corrections [50] is that the inner core deleptonizes more quickly. In turn, the maxima in central temperature and entropy are reached on shorter timescales. In addition, the faster increase in thermal pressure in the core slows the compression associated with the deleptonization stage, although after 10 s the net compressions of all models converge.

The relatively large, early, changes in the central thermodynamic variables do not, however, translate into similarly large effects on observables such as the total neutrino luminosity and the average radiated neutrino energy, relative to the baseline simulation. The luminosities for the different models are shown as a function of time in Fig. 12. The left panel shows the early time development in detail. The exploratory models agree with the results reported in [51,52]. However, the magnitude of the effects when full RPA corrections are applied is somewhat reduced compared to the exploratory models. It is especially important that at and below nuclear density, the corrections due to correlations are relatively small. Since information from the inner core is transmitted only by the neutrinos, the time scale to propagate any high density effect to the neutrinosphere is the neutrino diffusion time scale. Since the neutrinosphere is at a density approximately 1/100 of nuclear density, and large correlation corrections occur only above 1/3 nuclear density where nuclei disappear, we find that correlation corrections calculated here have an effect at the neutrinosphere only after

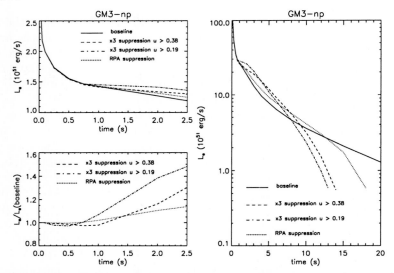

**Fig. 12.** Left: The upper panel shows the total emitted neutrino luminosity for the PNS evolutions described in Reddy et al. [50]. The lower panel shows the ratio of the luminosities obtained in the three models which contain corrections to the baseline (Hartree approximation) model. Right: Emitted neutrino luminosity for long-term PNS evolutions

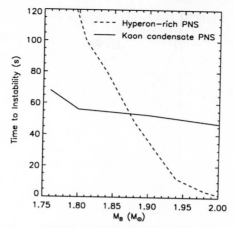

**Fig. 13.** Lifetimes of metastable stars as a function of the stellar baryon mass. Solid lines show results for PNSs containing kaon-condensates and dashed lines show the results of Pons, et al. [10] for PNSs containing hyperons

1.5 s. Moreover, the RPA suppression we have calculated is considerably smaller than those reported in [51,52], reaching a maximum of about 30% after 5 s, compared to a luminosity increase of 50% after only 2 s. However, the corrections are still very important during the longer-term cooling stage (see Fig. 12), and result in a more rapid onset of neutrino transparency compared to the Hartree results.

### 5.3 Signals in Detectors

In Fig. 13 the lifetimes versus $M_B$ for stars containing hyperons ($npH$) and $npK$ stars are compared [76]. In both cases, the larger the mass, the shorter the lifetime. For kaon-rich PNSs, however, the collapse is delayed until the final stage of the Kelvin-Helmholtz epoch, while this is not necessarily the case for hyperon-rich stars.

In Fig. 14 the evolution of the total neutrino energy luminosity is shown for different models. Notice that the drop in the luminosity for the stable star (solid line), associated with the end of the Kelvin-Helmholtz epoch, occurs at approximately the same time as for the metastable stars with somewhat higher masses. In all cases, the total luminosity at the end of the simulations is below $10^{51}$ erg/s. The two upper shaded bands correspond to SN 1987A detection limits with KII and IMB, and the lower bands correspond to detection limits in SNO and SuperK for a future galactic supernova at a distance of 8.5 kpc. The times when these limits intersect the model luminosities indicate the approximate times at which the count rate drops below the background rate $(dN/dt)_{BG} = 0.2$ Hz.

The poor statistics in the case of SN 1987A precluded a precise estimate of the PNS mass. Nevertheless, had a collapse to a black hole occurred in this

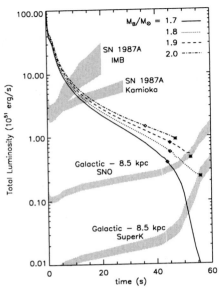

**Fig. 14.** The evolution of the total neutrino luminosity for stars of various baryon masses. Shaded bands illustrate the limiting luminosities corresponding to a count rate of 0.2 Hz in all detectors, assuming a supernova distance of 50 kpc for IMB and Kamioka, and 8.5 kpc for SNO and SuperK. The width of the shaded regions represents uncertainties in the average neutrino energy from the use of a diffusion scheme for neutrino transport

case, it must have happened after the detection of neutrinos ended. Thus the SN 1987A signal is compatible with a late kaonization-induced collapse, as well as a collapse due to hyperonization or to the formation of a quark core. More information would be extracted from the detection of a galactic SN with the new generation of neutrino detectors.

In SNO, about 400 counts are expected for electron antineutrinos from a supernova located at 8.5 kpc. The statistics would therefore be improved significantly compared to the observations of SN 1987A. A sufficiently massive PNS with a kaon condensate becomes metastable, and the neutrino signal terminates, before the signal decreases below the assumed background. In SuperK, however, up to 6000 events are expected for the same conditions (because of the larger fiducial mass) and the effects of metastability due to condensate formation in lower mass stars would be observable.

### 5.4 What Can We Learn From Neutrino Detections?

The calculations of Pons, et al. [76] show that the variations in the neutrino light curves caused by the appearance of a kaon condensate in a stable star are small, and are apparently insensitive to large variations in the opacities assumed for them. Relative to a star containing only nucleons, the expected signal differs by an amount that is easily masked by an assumed PNS mass difference of 0.01 − 0.02 $M_\odot$. This is in spite of the fact that, in some cases, a first order phase transition appears at the star's center. The manifestations of this phase transition are minimized because of the long neutrino diffusion times in the star's core and the Gibbs' character of the transition. Both act in tandem to prevent either a "core-quake" or a secondary neutrino burst from occurring during the Kelvin-Helmholtz epoch.

Observable signals of kaon condensation occur only in the case of metastable stars that collapse to a black hole. In this case, the neutrino signal for a star closer than about 10 kpc is expected to suddenly stop at a level well above that of the background in a sufficiently massive detector with a low energy threshold such as SuperK. This is in contrast to the signal for a normal star of similar mass for which the signal continues to fall until it is obscured by the background. The lifetime of kaon-condensed metastable stars has a relatively small range, of order 50–70 s for the models studied here, which is in sharp contrast to the case of hyperon-rich metastable stars for which a significantly larger variation in the lifetime (a few to over 100 s) was found. This feature of kaon condensation suggests that stars that destabilize rapidly cannot do so because of kaons.

Pons, et al. [76] determined the minimum lifetime for metastable stars with kaons to be about 40 s by examining the most favorable case for kaon condensation, which is obtained by maximizing the magnitude of the optical potential. The maximum optical potential is limited by the binary pulsar mass constraint, which limits the star's maximum gravitational mass to a minimum value of 1.44 $M_\odot$. Therefore, should the neutrino signal from a future supernova abruptly terminate sooner than 40 s after the birth of the PNS, it would be more consistent with a hyperon- or quark-induced instability than one due to kaon condensation.

It is important to note that the collapse to a black hole in the case of kaon condensation is delayed until the final stages of the Kelvin-Helmholtz epoch, due to the large neutrino diffusion time in the inner core. Consequently, to distinguish between stable and metastable kaon-rich stars through observations of a cessation of a neutrino signal from a galactic supernovas is only possible using sufficiently massive neutrino detectors with low energy thresholds and low backgrounds, such as the current SNO and SuperK, and future planned detectors including the UNO.

## 5.5 Expectations From Quark Matter

Strangeness appearing in the form of a mixed phase of strange quark matter also leads to metastability. Although quark matter is also suppressed by trapped neutrinos [40,25], the transition to quark matter can occur at lower densities than the most optimistic kaon case, and the dependence of the threshold density on $Y_L$ is less steep than that for kaons. Thus, it is an expectation that metastability due to the appearance of quarks, as for the case of hyperons, might be able to occur relatively quickly. Steiner, et al. [25] have demonstrated that the temperature along adiabats in the quark-hadronic mixed phase is much smaller than what is found for the kaon condensate-hadronic mixed phase. Calculations of PNS evolution with a mixed phase of quark matter, including the possible effects of quark matter superfluidity [67] are currently in progress and will be reported separately.

# 6 Long-Term Cooling: The Next Million Years

Following the transparency of the neutron star to neutrinos, the only observational link with these objects is through photon emissions, either as a pulsar or through thermal emissions or both. Thermal emissions of course are controlled by the temperature evolution of the star, and this depends sensitively upon its internal composition. The tabulation of temperatures and ages for a set of neutron stars would go a long way to deciding among several possibilities.

## 6.1 Thermal Evolution

The cooling of a young (age $< 10^5$ yr) neutron star is mainly governed by $\nu$–emission processes and the specific heat [77]. Due to the extremely high thermal conductivity of electrons, a neutron star becomes nearly isothermal within a time $t_w \approx 1 - 100$ years after its birth, depending upon the thickness of the crust [7]. After this time its thermal evolution is controlled by energy balance:

$$\frac{dE_{th}}{dt} = C_V \frac{dT}{dt} = -L_\gamma - L_\nu + H, \tag{64}$$

where $E_{th}$ is the total thermal energy and $C_V$ is the specific heat. $L_\gamma$ and $L_\nu$ are the total luminosities of photons from the hot surface and $\nu$s from the interior, respectively. Possible internal heating sources, due, for example, to the decay of the magnetic field or friction from differential rotation, are included in $H$. Our cooling simulations were performed by solving the heat transport and hydrostatic equations including general relativistic effects (see [77]). The surface's effective temperature $T_e$ is much lower than the internal temperature $T$ because of a strong temperature gradient in the envelope. Above the envelope lies the atmosphere where the emerging flux is shaped into the observed spectrum from which $T_e$ can be deduced. As a rule of thumb $T_e/10^6$ K $\approx \sqrt{T/10^8}$ K, but modifications due to magnetic fields and chemical composition may occur.

## 6.2 Rapid vs. Slow Cooling

The simplest possible $\nu$ emitting processes are the direct Urca processes $f_1 + \ell \to f_2 + \nu_\ell, f_2 \to f_1 + \ell + \overline{\nu_\ell}$, where $f_1$ and $f_2$ are either baryons or quarks and $\ell$ is either an electron or a muon. These processes can occur whenever momentum conservation is satisfied among $f_1, f_2$ and $\ell$ (within minutes of birth, the $\nu$ chemical potential vanishes). If the unsuppressed direct Urca process for *any* component occurs, a neutron star will rapidly cool because of enhanced emission: the star's interior temperature $T$ will drop below $10^9$ K in minutes and reach $10^7$ K in about a hundred years. $T_e$ will hence drop to less than

300,000 K after the crustal diffusion time $t_w$ [7,78,79]. This is the so-called rapid cooling paradigm. If no direct Urca processes are allowed, or they are all suppressed, cooling instead proceeds through the significantly less rapid modified Urca process in which an additional fermion enables momentum conservation. This situation could occur if no hyperons are present, or the nuclear symmetry energy has a weak density dependence [61,80]. The $\nu$ emission rates for the nucleon, hyperon, and quark Urca and modified Urca processes can be found in [81].

## 6.3 Superfluid and Superconducting Gaps

Pairing is unavoidable in a degenerate Fermi liquid if there is an attractive interaction in *any* channel. The resulting superfluidity, and in the case of charged particles, superconductivity, in neutron star interiors has a major effect on the star's thermal evolution through suppressions of neutrino ($\nu$) emission processes and specific heats [77,78]. Neutron ($n$), proton ($p$) and $\Lambda$-hyperon superfluidity in the $^1S_0$ channel and $n$ superfluidity in the $^3P_2$ channel have been shown to occur with gaps of a few MeV or less [82,83]. However, the density ranges in which gaps occur remain uncertain. At large baryon densities for which perturbative QCD applies, pairing gaps for like quarks have been estimated to be a few MeV [66]. However, the pairing gaps of unlike quarks ($ud$, $us$, and $ds$) have been suggested to be several tens to hundreds of MeV through non-perturbative studies [65] kindling interest in quark superfluidity and superconductivity [84,85] and their effects on neutron stars [86,87].

The effect of the pairing gaps on the emissivities and specific heats for massive baryons are investigated in [88] and are here generalized to the case of quarks. The principal effects are severe suppressions of both the emissivity and specific heat when $T \ll \Delta$, where $\Delta$ is the pairing gap. In a system in which several superfluid species exist the most relevant gap for these suppressions is the smallest one. The specific heat suppression is never complete, however, because leptons remain unpaired. Below the critical temperature $T_c$, pairs may recombine, resulting in the emission of $\nu\bar{\nu}$ pairs with a rate that exceeds the modified Urca rate below $10^{10}$ K [89]; these processes are included in our calculations.

The baryon and quark pairing gaps we adopt are shown in Fig. 15. Note that gaps are functions of Fermi momenta ($p_F(i)$, $i$ denoting the species) which translates into a density dependence. For $p_F(n,p) \lesssim 200 - 300$ MeV/c, nucleons pair in the $^1S_0$ state, but these momenta correspond to densities too low for enhanced $\nu$ emission involving nucleons to occur. At higher $p_F$'s, baryons pair in higher partial waves. The $n$ $^3P_2$ gap has been calculated for the Argonne $V_{18}$, CD-Bonn and Nijmegen I & II interactions [82]. This gap is crucial since it extends to large $p_F(n)$ and can reasonably be expected to occur at the centers of neutron stars. For $p_F(n) > 350$ MeV/c, gaps are largely uncertain because of both experimental and theoretical uncertainties [82]. The curves [a], [b] and [c] in Fig. 15 reflect the range of uncertainty. The $p$ $^3P_2$ gap is too small to be of interest. Gaps for the $^1S_0$ pairing of $\Lambda$, taken from [83] and shown as dotted curves, are highly relevant since $\Lambda$s participate in direct Urca emission as soon

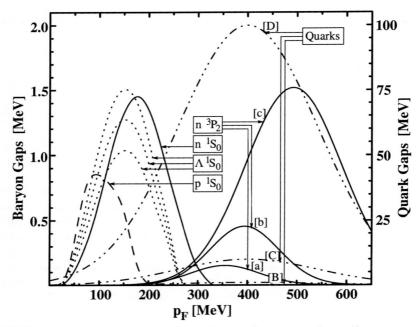

**Fig. 15.** Pairing gaps adopted for neutron $^1S_0$ and $^3P_2$, proton $^1S_0$, $\Lambda$ $^1S_0$, and quarks. The n $^3P_2$ gaps are anisotropic; plotted values are angle-averaged. The $\Lambda$ gaps correspond, in order of increasing $\Delta$, to background densities $n_B = 0.48, 0.64$ and $0.8$ fm$^{-3}$, respectively. The s-wave quark gaps are schematic; see text for details

as they appear [80]. Experimental information beyond the $^1S_0$ channel for $\Lambda$ is not available. $\Delta$s for $\Sigma$−hyperons remain largely unexplored. The quark ($q$) gaps are taken to be Gaussians centered at $p_F(q) = 400$ MeV/$c$ with widths of 200 MeV/$c$ and heights of 100 MeV [model D], 10 MeV [C], 1 MeV [B] and 0.1 MeV [A], respectively. The reason for considering quark gaps much smaller than suggested in [65,66] is associated with the multicomponent nature of charge-neutral, beta-equilibrated, neutron star matter as will become clear shortly.

## 6.4 Effects of Composition

We consider four generic compositions: charge-neutral, beta equilibrated matter containing nucleons only ($np$), nucleons with quark matter ($npQ$), nucleons and hyperons ($npH$), and nucleons, hyperons and quarks ($npHQ$). In the cases involving quarks, a mixed phase of baryons and quarks is constructed by satisfying Gibbs' phase rules for mechanical, chemical and thermal equilibrium [35]. The phase of pure quark matter exists only for very large baryon densities, and rarely occurs in our neutron star models. Baryonic matter is calculated using a field-theoretic model at the mean field level [90]; quark matter is calculated using either a bag-like model or the Nambu-Jona-Lasinio quark model [40,25]. The equation of state (EOS) is little affected by the pairing phenomenon, since the energy density gained is negligible compared to the ground state energy densities without pairing.

Additional particles, such as quarks or hyperons, have the effect of softening the EOS and increasing the central densities of stars relative to the $np$ case. For the $npQ$ model studied, a mixed phase appears at the density $n_B = 0.48$ fm$^{-3}$. Although the volume fraction of quarks is initially zero, the quarks themselves have a significant $p_F(q)$ when the phase appears. The $p_F$s of the three quark flavors become the same at extremely high density, but for the densities of interest they are different due to the presence of negatively charged leptons. In particular, $p_F(s)$ is much smaller than $p_F(u)$ and $p_F(d)$ due to the larger $s$-quark mass. Use of the Nambu–Jona-Lasinio model, in which quarks acquire density-dependent masses resembling those of constituent quarks, exaggerates the reduction of $p_F(s)$. This has dramatic consequences since the pairing phenomenon operates at its maximum strength when the Fermi momenta are exactly equal; even small asymmetries cause pairing gaps to be severely reduced [85,91]. In addition, one may also expect p-wave superfluidity, to date unexplored, which may yield gaps smaller than that for the s-wave. We therefore investigate pairing gaps that are much smaller than those reported for the case of s-wave superfluidity and equal quark $p_F$'s.

The introduction of hyperons does not change these generic trends. In the case $npH$, the appearance of hyperons changes the lepton and nucleon $p_F$'s similarly to the appearance of quarks although with less magnitude. While the appearance of quarks is delayed by the existence of hyperons, at high densities the $p_F$'s of nucleons and quarks remain similar to those of the $npQ$ case. The existence of either hyperons or quarks, however, does allow the possibility of additional direct Urca processes involving themselves as well as those involving nucleons by decreasing $p_F(n) - p_F(p)$. For the $npQ$ and $npHQ$ models studied, the maximum masses are $\cong 1.5 M_\odot$, the central baryon densities are $\cong 1.35$ fm$^{-3}$, and the volume fractions of quarks at the center are $\cong 0.4$.

## 6.5 Examples of Results

Cooling simulations of stars without hyperons and with hyperons are compared, in Figs. 16 and 17, respectively, to available observations of thermal emissions from pulsars. Sources for the observational data can be found in [92]. However, at the present time, the inferred temperatures must be considered as upper limits because the total flux is contaminated, and in some cases dominated, by the pulsar's magnetospheric emission and/or the emission of a surrounding synchrotron nebula. Furthermore, the neutron star surface may be reheated by magnetospheric high energy photons and particles; late-time accretion for non-pulsing neutron stars is also possible. Other uncertainties arise in the temperature estimates due to the unknown chemical composition and magnetic field strength in the surface layers, and in the age, which is based upon the observed spin-down time. In these figures, the bolder the data symbol the better the data.

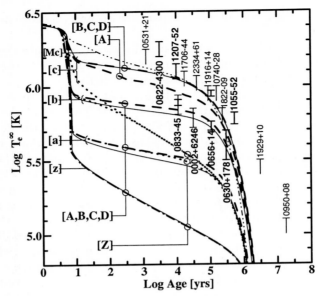

**Fig. 16.** Cooling of 1.4M$_\odot$ stars with $np$ matter (continuous curves) and $npQ$ matter (dashed and dotted curves). The curves labelled as [a], [b], and [c] correspond to $n$ $^3P_2$ gaps as in Fig. 15; [z] corresponds to zero $n$ gap. Models labelled [A], [B], [C] and [D] correspond to quark gaps as in Fig. 15; [Z] corresponds to zero quark gap

## $np$ and $npQ$ Matter

The $np$ case is considered in Fig. 16, in which solid lines indicate the temperature evolution of a 1.4 M$_\odot$ star for quarkless matter: case [z] is for no nucleon pairing at all, and cases [a], [b] and [c] correspond to increasing values for the neutron $^3P_2$ gap, according to Fig. 15. The field-theoretical model employed for the nucleon interactions allows the direct nucleon Urca process, which dominates the cooling. The unimpeded direct Urca process carries the temperature to values well below the inferred data. Pairing suppresses the cooling for $T < T_c$, where $T$ is the interior temperature, so $T_e$ increases with increasing $\Delta$. If the direct Urca process is not allowed, the range of predicted temperatures is relatively narrow due to the low emissivity of the modified Urca process. We show an example of such cooling (curve [Mc]) using the $n$ $^3P_2$ gap [c] for a 1.4M$_\odot$ with an EOS [93] for which the direct Urca cooling is not allowed.

The other curves in the figure illustrate the effects of quarks upon the cooling. The dotted curves [Z] are for vanishingly small quark gaps; the dashed curves ([A], [B], [C] and [D]) are for quark gaps as proposed in Fig. 15. For nonexistent ([z]) or small ([a]) nucleon gaps, the quark Urca process is irrelevant and the dependence on the existence or the size of the quark gaps is very small. However, for large nucleon gaps ([b] and [c]), the quark direct Urca process quickly dominates the cooling as the nucleon direct Urca process is quenched. It is clear that for quark gaps of order 1 MeV or greater ([B], [C] or [D]) the effect of quarks is again

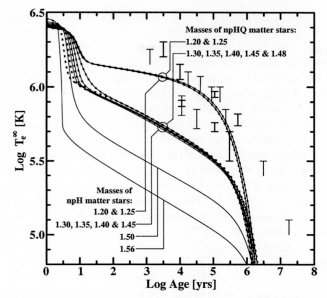

**Fig. 17.** Cooling of stars with $npH$ (continuous lines) and $npHQ$ matter (dotted lines) for various stellar masses (in $M_\odot$). $n\ ^3P_2$ gaps are from case [c] while quark gaps, when present, are from model [C] of Fig. 15

very small. There is at most a slight increase in the stars temperatures at ages between $10^1$ to $10^{5-6}$ years due to the reduction of $p_F(n)$ and the consequent slightly larger gap (Fig. 15). Even if the quark gap is quite small ([A]), quarks have an effect only if the nucleon gap is very large ([b] or [c]), i.e., significantly larger than the quark gap: the nucleon direct Urca process is suppressed at high temperatures and the quark direct Urca process has a chance to contribute to the cooling. We find that the effects of changing the stellar mass $M$ are similar to those produced by varying the baryon gap, so that only combinations of $M$ and $\Delta$ might be constrained by observation.

### $npH$ and $npHQ$ Matter

The thermal evolution of stars containing hyperons has been discussed in [94,95], but we obtain qualitatively different results here. Hyperons open new direct Urca channels: $\Lambda \to p + e + \overline{\nu}_e$ and $\Lambda + e \to \Sigma^- + \nu_e$ if $\Sigma^-$'s are present, with their inverse processes. Previous results showed that the cooling is naturally controlled by the smaller of the $\Lambda$ or $n$ gap. However, this is significantly modified when the $\Lambda$ gap is more accurately treated. At the $\Lambda$ appearance threshold, the gap must vanish since $p_F(\Lambda)$ is vanishingly small. We find that a very thin layer, only a few meters thick, of unpaired or weakly paired $\Lambda$'s is sufficient to control the cooling. This effect was overlooked in previous works perhaps because they lacked adequate zonal resolution.

In Fig. 17 we compare the evolution of stars of different masses made of either $npH$ or $npHQ$ matter. We find that all stars, except the most massive $npH$ ones, follow two distinctive trajectories depending on whether their central density is below or above the threshold for $\Lambda$ appearance (= 0.54 fm$^{-3}$ in our model EOS, the threshold star mass being 1.28M$_\odot$). In the case of $npH$ matter, stars with $M > 1.50$M$_\odot$ are dense enough so that the $\Lambda$ $^1S_0$ gap vanishes and hence undergo fast cooling, while stars made of $npHQ$ matter do not attain such high densities. The temperatures of $npH$ stars with masses between 1.3 and 1.5 M$_\odot$ are below the ones obtained in the models of Fig. 16 with the same $n$ $^3P_2$ gap [b], which confirms that the cooling is dominated by the very thin layer of unpaired $\Lambda$'s (the slopes of these cooling curves are typical of direct Urca processes). Only if the $n$ $^3P_2$ gap $\lesssim$ 0.3 MeV do the cooling curves fall below what is shown in Fig. 17. Notice, moreover, that in the mass range 1.3 – 1.48 M$_\odot$ the cooling curves are practically indistinguishable from those with unpaired quark matter shown in Fig. 16. In these models with $npH$ or $npHQ$ matter, there is almost no freedom to "fine-tune" the size of the gaps to attain a given $T_e$: stars with $\Lambda$'s will all follow the same cooling trajectory, determined by the existence of a layer of unpaired or weakly paired $\Lambda$'s, as long as the $n$ $^3P_2$ gap is not smaller. It is, in some sense, the same result as in the case of $np$ and $npQ$ matter: the smallest gap controls the cooling and now the control depends on how fast the $\Lambda$ $^1S_0$ gap increases with increasing $p_F(\Lambda)$.

## 6.6 Implications

Our results indicate that observations could constrain combinations of the smaller of the neutron and $\Lambda$−hyperon pairing gaps and the star's mass. Deducing the sizes of quark gaps from observations of neutron star cooling will be extremely difficult. Large quark gaps render quark matter practically invisible, while vanishing quark gaps lead to cooling behaviors which are nearly indistinguishable from those of stars containing nucleons or hyperons. Moreover, it also appears that cooling observations by themselves will not provide definitive evidence for the existence of quark matter itself.

# 7 The Structure of Catalyzed Stars

In this section, we explore from a theoretical perspective, how the structure of neutron star depends upon the assumed EOS. This study is crucial if new observations of masses and radii are to lead to effective constraints of the EOS of dense matter. Two general classes of stars can be identified: *normal* stars in which the density vanishes at the stellar surface, and *self-bound* stars in which the density at the surface is finite. Normal stars originate from nuclear force models which can be conveniently grouped into three broad categories: non-relativistic potential models, relativistic field theoretical models, and relativistic Dirac-Brueckner-Hartree-Fock models. In each of these approaches, the presence of additional softening components such as hyperons, Bose condensates or quark matter, can be incorporated. Details of these approaches have been further considered in Lattimer et al. [96] and Prakash et al. [4]. A representative sample, and some general attributes, including references and typical compositions, of equations of state employed here are summarized in Table 1.

For normal matter, the EOS is that of interacting nucleons above a transition density of 1/3 to 1/2 $n_s$. Below this density, the ground state of matter consists of heavy nuclei in equilibrium with a neutron-rich, low-density gas of nucleons. However, for most of the purposes of this paper, the pressure in the region $n < 0.1$ fm$^{-3}$ is not relevant as it does not significantly affect the mass-radius relation or other global aspects of the star's structure. Nevertheless, the value of the transition density, and the pressure there, are important ingredients for the determination of the size of the superfluid crust of a neutron star that is believed to be involved in the phenomenon of pulsar glitches (Link, Epstein, & Lattimer [97]).

Four equations of state are taken from Akmal & Pandharipande [11]. These are: AP1 (the AV18 potential), AP2 (the AV18 potential plus $\delta v_b$ relativistic boost corrections), AP3 (the AV18 potential plus a three-body UIX potential ), and AP4 (the AV18 potential plus the UIX potential plus the $\delta v_b$ boost). Three equations of state from Muller & Serbs [33], labelled MS1–3, correspond to different choices of the parameters $\xi$ and $\zeta$ which determine the strength of the nonlinear vector and isovector interactions at high densities. The numerical values used are $\xi = \zeta = 0; \xi = 1.5, \zeta = 0.06;$ and $\xi = 1.5, \zeta = 0.02$, respectively. Six EOSs come from the phenomenological non-relativistic potential model of Prakash, Ainsworth & Lattimer [98], labelled PAL1–6, which have different choices of the symmetry energy parameter at the saturation density, its density dependence, and the bulk nuclear matter incompressibility parameter $K_s$. The incompressibilities of PAL1–5 were chosen to be $K_s = 180$ or 240 MeV, but PAL6 has $K_s = 120$ MeV. Three interactions denoted GM1–3 come from the field-theoretical model of Glendenning & Moszkowski [14]. Two interactions come from the field-theoretical model of Glendenning & Schaffner-Bielich [18]: GL78 with $U_K(\rho_0) = -140$ MeV and TM1 with $U_K = -185$ MeV. The labels denoting the other EOSs in Table 1 are identical to those in the original references.

The rationale for exploring a wide variety of EOSs, even some that are relatively outdated or in which systematic improvements are performed, is two-fold. First, it provides contrasts among widely different theoretical paradigms. Second, it illuminates general relationships that exist between the pressure-density relation and the macroscopic properties of the star such as the radius. For example, AP4 represents the most complete study to date of Akmal & Pandharipande [11], in which many-body and special relativistic corrections are progressively incorporated into prior models, AP1–3. AP1–3 are included here because they represent different pressure-energy density-baryon density relations, and serve to reinforce correlations between neutron star structure and microscopic physics observed using alternative theoretical paradigms. Similarly, several different parameter sets for other EOSs are chosen.

**Table 1.** Approach refers to the underlying theoretical technique. Composition refers to strongly interacting components (n=neutron, p=proton, H=hyperon, K=kaon, Q=quark); all models include leptonic contributions.

| Symbol | Reference | Approach | Composition |
|---|---|---|---|
| FP | Friedman & Pandharipande [99] | Variational | np |
| PS | Pandharipande & Smith [100] | Potential | $n\pi^0$ |
| WFF(1-3) | Wiringa, Fiks & Fabrocini [101] | Variational | np |
| AP(1-4) | Akmal & Pandharipande [11] | Variational | np |
| MS(1-3) | Muller & Serbs [33] | Field Theoretical | np |
| MPA(1-2) | Müther, Prakash & Ainsworth [102] | Dirac-Brueckner HF | np |
| ENG | Engvik et al. [103] | Dirac-Brueckner HF | np |
| PAL(1-6) | Prakash, Ainsworth & Lattimer [98] | Schematic Potential | np |
| GM(1-3) | Glendenning & Moszkowski [14] | Field Theoretical | npH |
| GS(1-2) | Glendenning & Schaffner-Bielich [37] | Field Theoretical | npK |
| PCL(1-2) | Prakash, Cooke & Lattimer [40] | Field Theoretical | npHQ |
| SQM(1-3) | Prakash, Cooke & Lattimer [40] | Quark Matter | $Q(u,d,s)$ |

In all cases, except for PS (Pandharipande & Smith [100]), the pressure is evaluated assuming zero temperature and beta equilibrium without trapped neutrinos. PS only contains neutrons among the baryons, there being no charged components. We chose to include this EOS, despite the fact that it has been superseded by more sophisticated calculations by Pandharipande and coworkers, because it represents an extreme case producing large radii neutron stars.

The pressure-density relations for some of the selected EOSs are shown in Fig. 18 which displays three significant features to note for normal EOSs. First, there is a fairly wide range of predicted pressures for beta-stable matter in the density domain $n_s/2 < n < 2n_s$. For the EOSs displayed, the range of pressures covers about a factor of five, but this survey is by no means exhaustive. That such a wide range in pressures is found is somewhat surprising, given that each of the EOSs provides acceptable fits to experimentally-determined nuclear matter properties. Clearly, the extrapolation of the pressure from symmetric matter to nearly pure neutron matter is poorly constrained. Second, the *slopes* of the pressure curves are rather similar. A polytropic index of $n \simeq 1$, where $P = Kn^{1+1/n}$, is implied. Third, in the density domain below $2n_s$, the pressure-density relations seem to fall into two groups. The higher pressure group is primarily composed of relativistic field-theoretical models, while the lower pressure group is primarily composed of non-relativistic potential models. It is significant that relativistic field-theoretical models generally have symmetry energies that increase proportionately to the density while potential models have much less steeply rising symmetry energies.

A few of the plotted normal EOSs have considerable softening at high densities, especially PAL6, GS1, GS2, GM3, PS and PCL2. PAL6 has an abnormally small value of incompressibility ($K_s = 120$ MeV). GS1 and GS2 have phase tran-

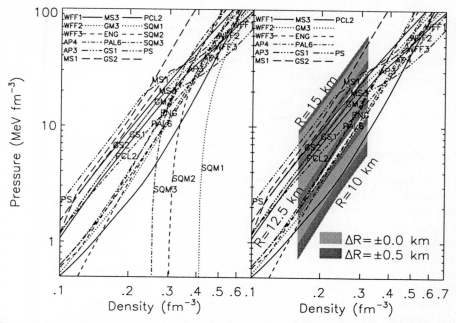

**Fig. 18.** Left: The pressure-density relation for a selected set of EOSs contained in Table 1. The pressure is in units of MeV fm$^{-3}$ and the density is in units of baryons per cubic fermi. The nuclear saturation density is approximately 0.16 fm$^{-3}$. Right: The pressures inferred from the empirical correlation (65), for three hypothetical radius values (10, 12.5 and 15 km) overlaid on the pressure-density relations shown on the left. The light shaded region takes into account only the uncertainty associated with $C(n, M)$; the dark shaded region also includes a hypothetical uncertainty of 0.5 km in the radius measurement. The neutron star mass was assumed to be 1.4 M$_\odot$.

sitions to matter containing a kaon condensate, GM3 has a large population of hyperons appearing at high density, PS has a phase transition to a neutral pion condensate and a neutron solid, and PCL2 has a phase transition to a mixed phase containing strange quark matter. These examples are representative of the kinds of softening that could occur at high densities.

The best-known example of self-bound stars results from Witten's [104] conjecture (also see Fahri & Jaffe [105], Haensel, Zdunik & Schaeffer [106], Alcock & Olinto [107], and Prakash et al. [108]) that strange quark matter is the ultimate ground state of matter. In this paper, the self-bound EOSs are represented by strange-quark matter models SQM1–3, using perturbative QCD and an MIT-type bag model, with parameter values given in Table 2. The existence of an energy ceiling equal to the baryon mass, 939 MeV, for zero pressure matter requires that the bag constant $B \leq 94.92$ MeV fm$^{-3}$. This limiting value is chosen, together with zero strange quark mass and no interactions ($\alpha_c = 0$), for the model SQM1. The other two models chosen, SQM2 and SQM3, have bag constants adjusted so that their energy ceilings are also 939 MeV.

**Table 2.** Parameters for self-bound strange quark stars. Numerical values employed in the MIT bag model as described in Fahri & Jaffe [105].

| Model | $B$ (MeV fm$^{-3}$) | $m_s$ (MeV) | $\alpha_c$ |
|---|---|---|---|
| SQM1 | 94.92 | 0 | 0 |
| SQM2 | 64.21 | 150 | 0.3 |
| SQM3 | 57.39 | 50 | 0.6 |

## 7.1 Neutron Star Radii

Fig. 19 displays the resulting mass-radius relations for catalyzed matter. Rhoades & Ruffini [109] demonstrated that the assumption of causality beyond a fiducial density $\rho_f$ sets an upper limit to the maximum mass of a neutron star: $4.2\sqrt{\rho_s/\rho_f}$ M$_\odot$. Lattimer et al. [96] have shown that the causality constraint also sets a lower limit to the radius: $R \gtrsim 1.52 R_s$, where $R_s = 2GM/c^2$, which is shown in Fig. 19. For a 1.4 M$_\odot$ star, this is about 4.5 km. The most reliable estimates of neutron star radii in the near future will likely stem from observations of thermal emission from their surfaces. Such estimates yield the so-called "radiation radius" $R_\infty = R/\sqrt{1-R_s}$, a quantity resulting from redshifting the star's luminosity and temperature. A given value of $R_\infty$ implies that $R < R_\infty$ and $M < 0.13(R_\infty/\text{km})$ M$_\odot$. Contours of $R_\infty$ are also displayed. With the exception of model GS1, the EOSs used to generate Fig. 19 result in maximum masses greater than 1.442 M$_\odot$, the limit obtained from PSR 1913+16. From a theoretical perspective, it appears that values of $R_\infty$ in the range of 12–20 km are possible for normal neutron stars whose masses are greater than 1 M$_\odot$.

One observes that *normal* neutron stars have minimum masses of about 0.1 M$_\odot$ that are primarily determined by the EOS below $n_s$. At the minimum mass, the radii are generally in excess of 100 km. Self-bound stars have no minimum mass and the maximum mass self-bound stars have nearly the largest radii possible for a given EOS. If the strange quark mass $m_s = 0$ and interactions are neglected ($\alpha_c = 0$), the maximum mass is related to the bag constant $B$ in the MIT-type bag model by $M_{max} = 2.033\,(56\text{ MeV fm}^{-3}/B)^{1/2}$ M$_\odot$. Prakash et al. [108] and Lattimer et al. [96] showed that the addition of a finite strange quark mass and/or interactions produces larger maximum masses. The constraint that $M_{max} > 1.44$ M$_\odot$ is thus automatically satisfied by the condition that the energy ceiling is 939 MeV, and non-zero values of $m_s$ and $\alpha_c$ yield larger radii for every mass. The locus of maximum masses is given simply by $R \cong 1.85 R_s$ (Lattimer et al. [96]) as shown in the right-hand panel of Fig. 19. Strange quark stars with electrostatically supported normal-matter crusts (Glendenning & Weber [110]) have larger radii than those with bare surfaces. Coupled with the additional constraint $M > 1$M$_\odot$ from protoneutron star models, MIT-model strange quark stars cannot have $R < 8.5$ km or $R_\infty < 10.5$ km. These values are comparable to the smallest possible radii for a Bose (pion or kaon) condensate EOS.

One striking feature of Fig. 19 is that in the mass range from 1–1.5 M$_\odot$ or more the radius has relatively little dependence upon the stellar mass. The major

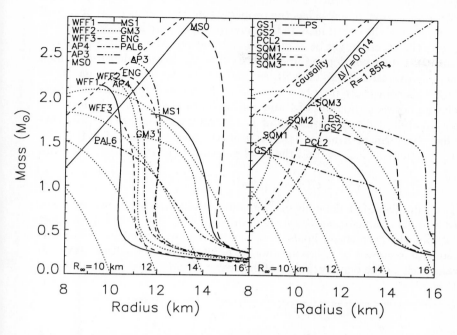

**Fig. 19.** Mass-radius curves for several EOSs listed in Table 1. The left panel is for stars containing nucleons and, in some cases, hyperons. The right panel is for stars containing more exotic components, such as mixed phases with kaon condensates or strange quark matter, or pure strange quark matter stars. In both panels, the lower limit causality places on $R$ is shown as a dashed line, a constraint derived from glitches in the Vela pulsar is shown as the solid line labelled $\Delta I/I = 0.014$, and contours of constant $R_\infty = R/\sqrt{1 - 2GM/Rc^2}$ are shown as dotted curves. In the right panel, the theoretical trajectory of maximum masses and radii for pure strange quark matter stars is marked by the dot-dash curve labelled $R = 1.85 R_s$

exceptions illustrated are the model GS1, in which a mixed phase containing a kaon condensate appears at a relatively low density, and the model PAL6, which has an extremely small nuclear incompressibility (120 MeV). Both of these have considerable softening and a large increase in central density for $M > 1 \, M_\odot$.

While it is generally assumed that a stiff EOS implies both a large maximum mass and a large radius, many counter examples exist. For example, GM3, MS1 and PS have relatively small maximum masses but have large radii compared to most other EOSs with larger maximum masses. Also, not all EOSs with extreme softening have small radii for $M > 1 \, M_\odot$ (e.g., GS2, PS). Nonetheless, for stars with masses greater than 1 $M_\odot$, only models with a large degree of softening (including strange quark matter configurations) can have $R_\infty < 12$ km.

To understand the relative insensitivity of the radius to the mass for normal neutron stars, it is relevant that a Newtonian polytrope with $n = 1$ has the property that the stellar radius is independent of both the mass and central density. An $n = 1$ polytrope also has the property that the radius is proportional to the square root of the constant $K$ in the polytropic pressure law $P = K\rho^{1+1/n}$. This suggests that there might be a quantitative relation between the radius and the pressure that does not depend upon the EOS at the highest densities, which determines the overall softness or stiffness (and hence, the maximum mass).

In fact, this conjecture may be verified. Fig. 20 shows a remarkable empirical correlation between the radii of 1 and 1.4 $M_\odot$ normal stars and the matter's pressure evaluated at fiducial densities of 1, 1.5 and 2 $n_s$. Numerically, the correlation has the form of a power law between the radius $R_M$, defined as the radius at a particular mass $M$, and the total pressure $P(n)$ evaluated at a given density:

$$R_M \simeq C(n, M) \, [P(n)]^{0.23-0.26} \,. \tag{65}$$

$C(n, M)$ is a number that depends on the density $n$ at which the pressure was evaluated and the stellar mass $M$. An exponent of $1/4$ was chosen for display in Fig. 20, but the correlation holds for a small range of exponents about this value. Using an exponent of $1/4$, and ignoring points associated with EOSs with phase transitions in the density ranges of interest, we find values for $C(n, M)$, in

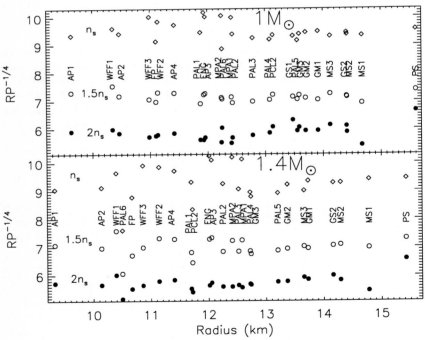

**Fig. 20.** Empirical relation between pressure, in units of MeV fm$^{-3}$, and $R$, in km, for EOSs listed in Table 1. The upper panel shows results for 1 $M_\odot$ (gravitational mass) stars; the lower panel is for 1.4 $M_\odot$ stars. The different symbols show values of $RP^{-1/4}$ evaluated at three fiducial densities

**Table 3.** The quantity $C(n, M)$ of (65) The quantity $C(n, M)$, in units of km fm$^{3/4}$ MeV$^{-1/4}$, which relates the pressure (evaluated at density $n$) to the radius of neutron stars of mass $M$. The errors are standard deviations

| $n$ | 1 M$_\odot$ | 1.4 M$_\odot$ |
|---|---|---|
| $n_s$ | $9.53 \pm 0.32$ | $9.30 \pm 0.60$ |
| $1.5 n_s$ | $7.14 \pm 0.15$ | $7.00 \pm 0.31$ |
| $2 n_s$ | $5.82 \pm 0.21$ | $5.72 \pm 0.25$ |

units of km fm$^{3/4}$ MeV$^{-1/4}$, which are listed in Table 3. The error bars are taken from the standard deviations. The correlation is seen to be somewhat tighter for the baryon density $n = 1.5 n_s$ and $2 n_s$ cases. Despite the relative insensitivity of radius to mass for a particular EOS in this mass range, the nominal radius $R_M$ has a variation $\sim 5$ km. The largest deviations from this correlation occur for EOSs with extreme softening or for configurations very near their maximum mass. This correlation is valid only for cold, catalyzed neutron stars, i.e., not for protoneutron stars which have finite entropies and might contain trapped neutrinos.

If a measurement of $P$ near $n_s$ can be deduced in this way, an important clue about the symmetry properties of matter will be revealed. The energy per particle and pressure of cold, beta stable nucleonic matter is

$$E(n, x) \simeq E(n, 1/2) + S_v(n)(1 - 2x)^2,$$
$$P(n, x) \simeq n^2[E'(n, 1/2) + S'_v(n)(1 - 2x)^2], \qquad (66)$$

where $E(n, 1/2)$ is the energy per particle of symmetric matter and $S_v(n)$ is the bulk symmetry energy (which is density dependent). Primes denote derivatives with respect to density. If only one term in this expansion is important, as noted by Prakash, Ainsworth & Lattimer [98], then

$$S_v(n) \simeq \frac{1}{2} \frac{\partial^2 E(n, x)}{\partial x^2} \simeq E(n, 0) - E(n, 1/2). \qquad (67)$$

At $n_s$, the symmetry energy can be estimated from nuclear mass systematics and has the value $S_v \equiv S_v(n_s) \approx 27-36$ MeV. Attempts to further restrict this range from consideration of fission barriers and the energies of giant resonances provide constraints between $S_v$ and $S_v(n)$ primarily by providing correlations between $S_v$ and $S_s$, the surface symmetry parameter. Lattimer & Prakash [111] detail how $S_s$ is basically a volume integral of the quantity $1 - S_v/S_v(n)$ through the nucleus. However, both the magnitude of $S_v$ and its density dependence $S_v(n)$ remain uncertain. Part of the bulk symmetry energy is due to the kinetic energy for noninteracting matter, which for degenerate nucleonic matter is proportional to $n^{2/3}$, but the remainder of the symmetry energy, due to interactions, is also expected to contribute significantly to the overall density dependence.

Leptonic contributions must to be added to (66) to obtain the total energy and pressure; the electron energy per baryon is $(3/4)\hbar c x (3\pi^2 n x)^{1/3}$. Matter in neutron stars is in beta equilibrium, i.e., $\mu_e = \mu_n - \mu_p = -\partial E/\partial x$, which permits the evaluation of the equilibrium proton fraction. The pressure at the saturation density becomes

$$P_s = n_s(1 - 2x_s)[n_s S'_v(1 - 2x_s) + S_v x_s], \tag{68}$$

where $S'_v \equiv S'_v(n_s)$ and the equilibrium proton fraction at $n_s$ is

$$x_s \simeq (3\pi^2 n_s)^{-1}(4S_v/\hbar c)^3 \simeq 0.04, \tag{69}$$

for $S_v = 30$ MeV. Due to the small value of $x_s$, we find that $P_s \simeq n_s^2 S'_v$.

Were we to evaluate the pressure at a larger density, contributions featuring other nuclear parameters, including the nuclear incompressibility $K_s = 9(dP/dn)|_{n_s}$ and the skewness $K'_s = -27n_s^3(d^3 E/dn^3)|_{n_s}$, also contribute. However, the $K_s$ and $K'_s$ terms largely cancel, up to $2n_s$, so the symmetry term dominates.

At present, experimental guidance concerning the density dependence of the symmetry energy is limited and mostly based upon the division of the nuclear symmetry energy between volume and surface contributions. Upcoming experiments involving heavy-ion collisions which might sample densities up to $\sim (3-4)n_s$, will be limited to analyzing properties of the nearly symmetric nuclear matter EOS through a study of matter, momentum, and energy flow of nucleons. However, the parity-violating experiment [112] to accurately determine the thickness of the neutron skin in $^{208}$Pb at Jefferson Lab will provide important constraints. The neutron skin thickness is directly proportional to $S_s/S_v$. In addition, studies of heavy nuclei far off the neutron drip lines using radioactive ion beams might also provide useful constraints.

## 7.2 Moments of Inertia

Besides the stellar radius, other global attributes of neutron stars are potentially observable, including the moment of inertia and the binding energy. These quantities depend primarily upon the ratio $M/R$ as opposed to details of the EOS (Lattimer & Prakash [5]).

The moment of inertia, for a star uniformly rotating with a very small or zero angular velocity $\Omega$, is [113]

$$I = (8\pi/3) \int_0^R r^4(\rho + P/c^2) e^{(\lambda - \nu)/2} dr. \tag{70}$$

Useful approximations which are valid for three analytic, exact, solutions to GR, the incompressible fluid (Inc), the Tolman VII (Tolman [114]; VII) solution, and Buchdahl's [115] solution (Buch), are

$$I_{Inc}/MR^2 \simeq 2(1 - 0.87\beta - 0.3\beta^2)^{-1}/5, \tag{71}$$
$$I_{Buch}/MR^2 \simeq (2/3 - 4/\pi^2)(1 - 1.81\beta + 0.47\beta^2)^{-1}, \tag{72}$$
$$I_{TVII}/MR^2 \simeq 2(1 - 1.1\beta - 0.6\beta^2)^{-1}/7. \tag{73}$$

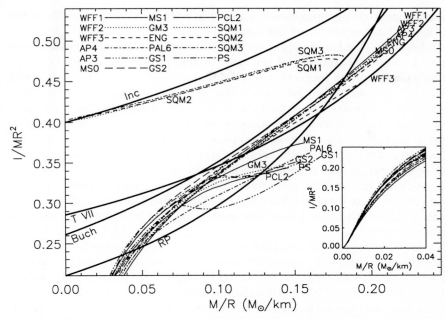

**Fig. 21.** The moment of inertia $I$, in units of $MR^2$, for several EOSs listed in Table 1. The curves labelled "Inc", "T VII", "Buch" and "RP" are for an incompressible fluid, the Tolman [114] VII solution, the Buchdahl [115] solution, and an approximation of Ravenhall & Pethick [116], respectively. The inset shows details of $I/MR^2$ for $M/R \to 0$

Fig. 21 indicates that the T VII approximation is a rather good approximation to most EOSs without extreme softening at high densities, for $M/R \geq 0.1$ $M_\odot$/km. The EOSs with softening fall below this trajectory.

Another interesting result from Fig. 21 concerns the moments of inertia of strange quark matter stars. Such stars are relatively closely approximated by incompressible fluids, this behavior becoming exact in the limit of $\beta \to 0$. This could have been anticipated from the $M \propto R^3$ behavior of the $M-R$ trajectories for small $\beta$ strange quark matter stars as observed in Fig. 19.

## 7.3 Crustal Fraction of the Moment of Inertia

A new observational constraint involving $I$ concerns pulsar glitches. Occasionally, the spin rate of a pulsar will suddenly increase (by about a part in $10^6$) without warning after years of almost perfectly predictable behavior. However, Link, Epstein & Lattimer [97] argue that these glitches are not completely random: the Vela pulsar experiences a sudden spinup about every three years, before returning to its normal rate of slowing. Also, the size of a glitch seems correlated with the interval since the previous glitch, indicating that they represent self-regulating instabilities for which the star prepares over a waiting time. The angular momentum requirements of glitches in Vela imply that $\geq 1.4\%$ of the star's moment of inertia drives these events.

Glitches are thought to represent angular momentum transfer between the crust and another component of the star. In this picture, as a neutron star's crust spins down under magnetic torque, differential rotation develops between the stellar crust and this component. The more rapidly rotating component then acts as an angular momentum reservoir which occasionally exerts a spin-up torque on the crust as a consequence of an instability. A popular notion at present is that the freely spinning component is a superfluid flowing through a rigid matrix in the thin crust, the region in which dripped neutrons coexist with nuclei, of the star. As the solid portion is slowed by electromagnetic forces, the liquid continues to rotate at a constant speed, just as superfluid He continues to spin long after its container has stopped. This superfluid is usually assumed to locate in the star's crust, which thus must contain at least 1.4% of the star's moment of inertia.

The high-density boundary of the crust is naturally set by the phase boundary between nuclei and uniform matter, where the pressure is $P_t$ and the density $n_t$. The low-density boundary is the neutron drip density, or for all practical purposes, simply the star's surface since the amount of mass between the neutron drip point and the surface is negligible. One can utilize . (70) to determine the moment of inertia of the crust alone with the assumptions that $P/c^2 \ll \rho$, $m(r) \simeq M$, and $\omega j \simeq \omega_R$ and $P \propto \rho^{4/3}$ in the crust (Lattimer & Prakash [5]):

$$\frac{\Delta I}{I} \simeq \frac{28\pi P_t R^3}{3Mc^2} \frac{(1 - 1.67\beta - 0.6\beta^2)}{\beta} \left[1 + \frac{2P_t(1 + 5\beta - 14\beta^2)}{n_t m_b c^2 \beta^2}\right]^{-1}. \quad (74)$$

In general, the EOS parameter $P_t$, in the units of MeV fm$^{-3}$, varies over the range $0.25 < P_t < 0.65$ for realistic EOSs. The determination of this parameter requires a calculation of the structure of matter containing nuclei just below nuclear matter density that is consistent with the assumed nuclear matter EOS. Unfortunately, few such calculations have been performed. Like the fiducial pressure at and above nuclear density which appears in Eq. (65), $P_t$ should depend sensitively upon the behavior of the symmetry energy near nuclear density.

Link, Epstein & Lattimer [97] established a lower limit to the radius of the Vela pulsar by using . (74) with $P_t$ at its maximum value and the glitch constraint $\Delta I/I \geq 0.014$:

$$R > 3.9 + 3.5 M/M_\odot - 0.08(M/M_\odot)^2 \text{ km}. \quad (75)$$

As shown in Fig. 22, this constraint is somewhat more stringent than one based upon causality. Better estimates of the maximum value of $P_t$ should make this constraint more stringent.

## 7.4 Binding Energies

The binding energy formally represents the energy gained by assembling $N$ baryons. If the baryon mass is $m_b$, the binding energy is simply $BE = Nm_b - M$

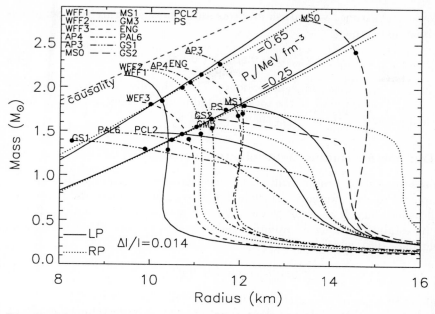

**Fig. 22.** Mass-radius curves for selected EOSs from Table 1, comparing theoretical contours of $\Delta I/I = 0.014$ from approximations developed in this paper, labelled "LP", and from Ravenhall & Pethick [116], labelled "RP", to numerical results (solid dots). Two values of $P_t$, the transition pressure demarking the crust's inner boundary, which bracket estimates in the literature, are employed. The region to the left of the $P_t = 0.65$ MeV fm$^{-3}$ curve is forbidden if Vela glitches are due to angular momentum transfers between the crust and core, as discussed in Link, Epstein & Lattimer [97]. For comparison, the region excluded by causality alone lies to the left of the dashed curve labelled "causality" as determined by Lattimer et al. [96] and Glendenning [117]

in mass units. However, the quantity $m_b$ has various interpretations in the literature. Some authors take it to be 939 MeV/$c^2$, the same as the neutron or proton mass. Others take it to be about 930 MeV/$c^2$, corresponding to the mass of $C^{12}/12$ or $Fe^{56}/56$. The latter choice would be more appropriate if $BE$ was to represent the energy released in by the collapse of a white-dwarf-like iron core in a supernova explosion. The difference in these definitions, 10 MeV per baryon, corresponds to a shift of $10/939 \simeq 0.01$ in the value of $BE/M$. This energy, $BE$, can be deduced from neutrinos detected from a supernova event; indeed, it might be the most precisely determined aspect of the neutrino signal.

Lattimer & Yahil [118] suggested that the binding energy could be approximated as

$$BE \approx 1.5 \cdot 10^{51}(M/M_\odot)^2 \text{ ergs} = 0.084(M/M_\odot)^2 \text{ M}_\odot. \quad (76)$$

Prakash et al. [4] also concluded that such a formula was a reasonable approximation, based upon a comparison of selected non-relativistic potential and field-theoretical models, good to about ±20 %.

However, Lattimer & Prakash [5] proposed a more accurate representation of the binding energy:

$$BE/M \simeq 0.6\beta/(1-0.5\beta),  \quad (77)$$

which incorporates some radius dependence. Thus, the observation of supernova neutrinos, and the estimate of the total radiated neutrino energy, will yield more accurate information about $M/R$ than about $M$ alone.

### 7.5 Outlook for Radius Determinations

Any measurement of a radius will have some intrinsic uncertainty. In addition, the empirical relation we have determined between the pressure and radius has a small uncertainty. It is useful to display how accurately the equation of state might be established from an eventual radius measurement. This can be done by inverting . (65), which yields

$$P(n) \simeq [R_M/C(n,M)]^4. \quad (78)$$

The inferred ranges of pressures, as a function of density and for three possible values of $R_{1.4}$, are shown in the right panel of Fig. 18. It is assumed that the mass is 1.4 $M_\odot$, but the results are relatively insensitive to the actual mass. Note from Table 3 that the differences between $C$ for 1 and 1.4 $M_\odot$ are typically less than the errors in $C$ itself. The light shaded areas show the pressures including only errors associated with $C$. The dark shaded areas show the pressures when a hypothetical observational error of 0.5 km is also taken into account. These results suggest that a useful restriction to the equation of state is possible if the radius of a neutron star can be measured to an accuracy better than about 1 km.

The reason useful constraints might be obtained from just a single measurement of a neutron star radius, rather than requiring a series of simultaneous mass-radii measurements as Lindblom [119] proposed, stems from the fact that we have been able to establish the empirical correlation, . (65). In turn, it appears that this correlation exists because most equations of state have slopes $d\ln P/d\ln n \simeq 2$ near $n_s$.

## 8 Tasks and Prospects

There are several topics that will merit attention from theoretical and experimental perspectives. Among those dealing with $\nu$-matter interactions are:

### Dynamic Structure Functions from Microscopic Calculations

Neutrino production and propagation in the hot and dense matter of interest in astrophysics depends crucially on the excitation spectrum of the nuclear medium to spin and spin-isospin probes. This is because, for momentum transfers small compared to the nucleon mass, the dominant coupling of neutrinos to nucleons is through the weak, axial-vector current. For non-relativistic nucleons, neutral current neutrino scattering probes the strength of the nuclear matrix elements of the spin operator, while the charged current absorption reactions probe the strength of the matrix elements of the spin-isospin operator. These matrix elements, evaluated in the nuclear many body basis $\Psi$, satisfy well known sum rules. In particular, the response function associated with the operator $\mathcal{O}(\boldsymbol{q})$

$$S_\mathcal{O}(q_0, \boldsymbol{q}) = \int_{-\infty}^{\infty} dt \, \exp(-iq_o t) \, \langle \Psi | \mathcal{O}^\dagger(\boldsymbol{q}) \, \mathcal{O}(\boldsymbol{q}) | \Psi \rangle \qquad (79)$$

obeys the energy weighted sum rule [59]

$$\int_{-\infty}^{\infty} dq_0 \, q_0 \, S_\mathcal{O}(q_o, \boldsymbol{q}) = \frac{1}{2} \langle |\Psi \left[ [\mathcal{O}^\dagger(\boldsymbol{q}), \mathcal{H}], \mathcal{O}(\boldsymbol{q}) \right] |\Psi \rangle \qquad (80)$$

where $\mathcal{H}$ is the interaction Hamiltonian. The structure of the double commutator in the above equation clearly underscores the need to better understand the role of nuclear interactions, such as the tensor and spin-orbit interactions, that do not conserve nucleon spin. These interactions are therefore of particular relevance to the study of the spin and spin-isospin response of dense nuclear systems.

Friman & Maxwell [120] first emphasized the importance of tensor correlations in the process

$$\nu_e + n + n \to e^- + p + n, \qquad (81)$$

and noted that their neglect underestimates the rate of $\nu_e$ absorption by as much as an order of magnitude. In their study, they used a hard core description of the short range correlations and a one pion exchange model for the medium and long range ones. Sawyer & Soni [121] and Haensel & Jerzak [122], who used additional correlations based on a Reid soft core potential, confirmed that large reductions were possible in degenerate matter for non-degenerate neutrinos.

Since these earlier works, many-body calculations have vastly improved (e.g. [11]) and have been well-tested against data on light nuclei and nuclear matter. Much better tensor correlations are now available, so that we may better pin down the rate of absorption due to the above process. Detailed calculations to include arbitrary matter and neutrino degeneracies encountered in in many astrophysical applications are necessary.

**Axial Charge Renormalization**

In dense matter, the axial charge of the baryons is renormalized [123–125], which alters the neutrino-baryon couplings from their vacuum values. Since the axial contribution to the scattering and absorption reactions is typically three times larger than the vector contributions, small changes in the axial vector coupling constants significantly affect the cross sections. The calculation of this renormalization requires a theoretical approach which treats the pion and chiral symmetry breaking explicitly. So far, this has been done in isospin symmetric nuclear matter [126], but not for neutron matter or for beta-equilibrated neutron star matter. Substantial reductions may be expected in the $\nu$-matter cross sections from this in-medium effect.

### Multi-Pair Excitations

Neutrinos can also excite many-particle states in an interacting system, inverse bremsstrahlung being an example of a two-particle excitation [127]. These excitations provide an efficient means of transferring energy between the neutrinos and baryons which are potentially significant in low-density matter. However, multigroup neutrino transport will be needed to include this effect. In addition, such calculations require source terms for neutrino processes such as bremsstrahlung and neutrino pair production. The latter process has been accurately treated in [128].

### $\pi^-$ and $K^-$ Dispersion Relations Through $\nu$-Nucleus Reactions

The experimental program that would do the most to illuminate theoretical issues permeating neutrino interactions in dense matter would be studies of neutrino reactions on heavy nuclei, the only direct way of probing the matrix elements of the axial current in nuclear matter. Pioneering suggestions in this regard have been put forth by Sawyer & Soni [129,130], Ericson [131], and Sawyer [132]. The basic idea is to detect positively charged leptons ($\mu^+$ or $e^+$) produced in inclusive experiments

$$\bar{\nu} + X \to \mu^+ \text{ ( or } e^+) + \pi^- \text{ ( or } K^-) + X \qquad (82)$$

which is kinematically made possible when the in-medium $\pi^-$ or $K^-$ dispersion relation finds support in space-like regions. The sharp peaks at forward angles in the differential cross section versus lepton momentum survive the 100-200 MeV width in the incoming Gee or so neutrinos from accelerator experiments. Calculations of the background from quasi-elastic reactions indicate that the signal would be easily detectable.

### $\nu$-Matter Interactions at Sub-Nuclear Density

Analogous to the effects of inhomogeneities for $\nu$-matter interactions discussed in the case of a first-order kaon condensate or quark matter transition is the case of coherent scattering of neutrinos from closely-spaced nuclei at sub-nuclear densities. The sizes and separations of nuclei are similar to those of the droplets discussed for the kaon and quark situations, so the range of neutrino energies most affected will be similar. These will be important in reshaping the $\nu$ spectrum from PNSs, and are of potential importance in the supernova mechanism itself due to the large energy dependence of $\nu$-matter cross sections behind the shock.

In addition to these, several topics of interest from an astrophysical perspective include:

### Improvements in PNS Simulations

These include a) an adequate treatment of convection coupled with neutrino transport appear to be necessary based upon large regions that are potentially convectively unstable; b) the consideration of other softening components in dense matter that might produce effects dissimilar to those found when considering hyperons and kaons, i.e., quarks; c) improved transport calculations with many energy groups, especially in the transparent regime; d) a self-consistent treatment of accretion, which is known to significantly contribute to the early $\nu$ emission. The latter two items necessitate the coupling of a multi-group transport scheme with a hydrodynamical code of the type generally used for supernova simulations.

## Determination of the Neutron Skin of Neutron-rich Nuclei

The Jefferson Lab experiment [112] (PREX) is anticipated to yield accurate measurements of the neutron-skin thickness of $^{208}$Pb. This quantity, from a theoretical viewpoint, is the volume integral of the inverse of the symmetry energy throughout a nucleus, and represents how the nuclear symmetry energy is split between volume and surface contributions. Not coincidentally, the density dependence of the symmetry energy is also implicated in the predicted neutron star radius [5].

## Determination of the Radius of a Neutron Star

The best prospect for measuring a neutron star's radius may be the nearby object RX J185635-3754. Parallax information [133] indicates its distance to be about 60 pc. In addition, it may be possible to identify spectral lines with the Chandra and XMM X-ray facilities that would not only yield the gravitational redshift, but would identify the atmospheric composition. Not only would this additional information reduce the uncertainty in the deduced value of $R_\infty$, but, *both* the mass and radius for this object might thereby be estimated. It is also possible that an estimate of the surface gravity of the star can be found from further comparisons of observations with atmospheric modelling, and this would provide a further check on the mass and radius.

## Acknowledgements

Research support from NSF grant INT-9802680 (for MP and JML) and DOE grants FG02-88ER-40388 (for MP and AS) and FG02-87ER40317 (for JML and JAP) and FG06-90ER40561 (for SR) is gratefully acknowledged. It is a pleasure to acknowledge collaborations with George Bertsch, Greg Carter, Paul Ellis, Juan Miralles, and Dany Page who have contributed significantly to the material presented in this article.

# References

1. A. Burrows, J.M. Lattimer: Astrophys. J. **307**, 178 (1986)
2. A. Burrows: Ann. Rev. Nucl. Sci. **40**, 181 (1990)
3. A. Burrows: Astrophys. J. **334**, 891 (1988)

4. M. Prakash, I. Bombaci, Manju Prakash, P.J. Ellis, J.M. Lattimer, R. Knorren: Phys. Rep. **280**, 1 (1997) P.J. Ellis, J.M. Lattimer, M. Prakash: Comm. Nucl. Part. Phys. **22**, 63 (1996)
5. J.M. Lattimer, M. Prakash: Astrophys. J. *in press* (2000)
6. *Next Generation Nucleon Decay and Neutrino Dectector*, ed. by N. Diwan, C.K. Jung (AIP, New York 2000)
7. J.M. Lattimer, K.A. van Riper, M. Prakash, Manju Prakash: Astrophys. J. **425**, 802 (1994)
8. R.W. Lindquist: Ann. Phys. **37**, 478 (1966)
9. K.S. Thorne: Mon. Not. R. Astron. Soc. **194**, 439 (1981)
10. J.A. Pons, S. Reddy, M. Prakash, J.M. Lattimer, J.A. Miralles: Astrophys. J. **513**, 780 (1999)
11. A. Akmal, V.R. Pandharipande: Phys. Rev. C **56**, 2261 (1997)
12. N. K. Glendenning, Astrophys. J. **293** 470 (1985)
13. B.D. Serot, J.D. Walecka: In *Advances in Nuclear Physics* Vol. 16, ed. by J.W. Negele, E. Vogt, (Plenum, New York 1986) p. 1
14. N.K. Glendenning, S. Moszkowski: Phys. Rev. Lett. **67**, 2414 (1991)
15. J.A. Pons, S. Reddy, P.J. Ellis, M. Prakash, J.M. Lattimer: Phys. Rev. C **62** 035803 (2000)
16. N. K. Glendenning, Nuclear Physics B (Proc. Suppl.) **24B** (1991) 110; Phys. Rev. D **46** 1274 (1992)
17. N. K. Glendenning and S. Pei, Phys. Rev. C **52** 2250 (1995)
18. N.K. Glendenning, J. Schaffner-Bielich: Phys. Rev. C **60** 025803 (1999)
19. J.W. Gibbs: Tran. Conn. Acad. **III**, 108 (1876)
20. E. Friedman, A. Gal, C.J. Batty: Nucl. Phys. A **579**, 578 (1994)
21. E. Friedman, A. Gal, J. Mares, A. Cieply: Phys. Rev. C **60**, 024314 (1999)
22. T. Waas, W. Weise: Nucl. Phys. A **625**, 287 (1997)
23. A. Ramos, E. Oset: Nucl. Phys. A **671**, 481 (2000)
24. A. Baca, C. Garcia-Recio, J. Nieves: Nucl. Phys. A **673**, 335 (2000)
25. A.W. Steiner, M. Prakash, J.M. Lattimer: Phys. Lett. B **486** 239 (2000)
26. Manju Prakash, E. Baron, M. Prakash: Phys. Lett. B, **243**, 175 (1990)
27. Y. Nambu, G. Jona-Lasinio: Phys. Rev. **122**, 345 (1961)
28. G. 't Hooft: Phys. Rep. **142**, 357 (1986)
29. P. Rehberg, S.P. Klevansky, J. Hüfner: Phys. Rev. C **53**, 410 (1996)
30. T. Kunihiro: Phys. Lett. B **219**, 363 (1989)
31. T. Hatsuda, T. Kunihiro: Phys. Rep. **247**, 221 (1994)
32. M. Buballa, M. Oertel: Phys. Lett. B **457**, 261 (1999)
33. H. Muller, B.D. Serot: Nucl. Phys. A **606** (1996) 508
34. S. Reddy, G.F. Bertsch, M. Prakash: Phys. Lett. B **475**, 1 (2000)
35. N.K. Glendenning: Phys. Rev. D **46**, 1274 (1992)
36. J.M. Lattimer, D.G. Ravenhall: Astrophys. J. **223**, 314 (1978)
37. N.K. Glendenning, J. Schaffner-Bielich: Phys. Rev. Lett. **81**, 4564 (1998)
38. M. Christiansen, N.K. Glendenning, J. Schaffner-Bielich: Phys. Rev. C **62**, 025804 (2000) T. Norsen, S. Reddy: nucl-th/0010075 (2000)
39. N.K. Glendenning, S. Pei: Phys. Rev. C **52**, 2250 (1995)
40. M. Prakash, J.R. Cooke, J.M. Lattimer: Phys. Rev. D **52**, 661 (1995)
41. S.W. Bruenn: Astrophys. J. Supp. **58**, 771 (1985)
42. A. Mezzacappa, S.W. Bruenn: Astrophys. J. **405**, 637 (1993)
43. J.R. Wilson, R.W. Mayle: In *The Nuclear Equation of State* Part A, ed. W. Greiner, H. Stöcker (Plenum, New York 1989) p. 731

44. H. Suzuki, K. Sato: In *The Structure and Evolution of Neutron Stars*, ed. D. Pines, R. Tamagaki, S. Tsuruta (Addison-Wesley, New York 1992) p. 276
45. W. Keil, H.T. Janka: Astron. & Astrophys. **296**, 145 (1995)
46. S. Reddy, J. Pons, M. Prakash, M., J.M. Lattimer: In *Stellar Evolution, Stellar Explosions and Galactic Chemical Evolution*, ed. T. Mezzacappa (IOP Publishing, Bristol 1997) p. 585
47. S. Reddy, M. Prakash: Astrophys. J. **423**, 689 (1997)
48. M. Prakash, S. Reddy: In *Nuclear Astrophysics*, ed. M. Buballa, W. Nörenberg, J. Wambach, A. Wirzba (GSI: Darmstadt 1997), p. 187
49. S. Reddy, M. Prakash, J.M. Lattimer: Phys. Rev. D **58**, 013009 (1998)
50. S. Reddy, M. Prakash, J.M. Lattimer, J.A. Pons: Phys. Rev. C **59**, 2888 (1999)
51. A. Burrows, R.F. Sawyer: Phys. Rev. C **58**, 554 (1998)
52. A. Burrows, R.F. Sawyer: Phys. Rev. C **59**, 510 (1999)
53. R.F. Sawyer: Phys. Rev. D **11**, 2740 (1975)
54. R.F. Sawyer: Phys. Rev. C **40**, 865 (1989)
55. N. Iwamoto, C.J. Pethick: Phys. Rev. D **25**, 313 (1982)
56. C.J. Horowitz, K. Wehrberger: Nucl. Phys. A **531**, 665 (1991) Phys. Rev. Lett. **66**, 272 (1991) Phys. Lett. B **226**, 236 (1992)
57. G. Raffelt, D. Seckel: Phys. Rev. D **52**, 1780 (1995)
58. G. Sigl: Phys. Rev. Lett **76**, 2625 (1996)
59. A.L. Fetter, J.D. Walecka: In *Quantum Theory of Many Particle Systems* (McGraw-Hill, New York 1971)
60. S. Doniach, E.H. Sondheimer: In *Green's Functions for Solid State Physicists* (The Benjamin/Cummings Publishing Company, Inc., Reading 1974)
61. J.M. Lattimer, C.J. Pethick, M. Prakash, P. Haensel: Phys. Rev. Lett. **66**, 2701 (1991)
62. C.J. Horowitz: Phys. Rev. D **55**, 4577 (1997)
63. N. Iwamoto, C.J. Pethick: Phys. Rev. D **25**, 313 (1982)
64. B.C. Barrois: Nucl. Phys. B **129**, 390 (1977) S.C. Frautschi: In *Workshop on Hadronic Matter at Extreme Energy Density* (Erice, Italy 1978)
65. M. Alford, K. Rajagopal, F. Wilczek: Phys. Lett. B **422**, 247 (1998) Nucl. Phys. B **357**, 443 (1999) *ibid.* **558**, 219 (1999) R. Rapp, T. Schäffer, E.V. Shuryak, M. Velkovsky: Phys. Rev. Lett. **81**, 53 (1998) Ann. Phys. **280**, 35 (2000)
66. D. Bailin, A. Love: Phys. Rept. **107**, 325 (1984)
67. G. W. Carter, S. Reddy: Phys. Rev. D **62** 103002 (2000)
68. J. Bardeen, L.N. Cooper, J.R. Schrieffer: Phys. Rev. **108**, 1175 (1957)
69. R.D. Pisarski, D.H. Rischke: Phys. Rev. D **61**, 051501 (2000)
70. R.D. Pisarski, D.H. Rischke: Phys. Rev. D **60**, 094013 (1999)
71. A. Burrows, B. Fryxell: Astrophys. J. Lett. **413**, L33 (1993)
72. M. Herant, W. Benz, J. Hicks, C. Fryer, S.A. Colgate: Astrophys. J. **425**, 339 (1994)
73. W. Keil, T.H. Janka, E. Müller: Astrophys. J. Lett. **473**, L111 (1995)
74. A. Mezzacappa, A.C. Calder, S.W. Bruenn, J.M. Blondin, M.W. Guidry, M.R. Strayer, A.S. Umar: Astrophys. J. **495**, 911 (1998)
75. J.A. Miralles, J.A. Pons, V.A. Urpin: Astrophys. J. **543**, 1001 (2000)
76. J.A. Pons, J.A. Miralles, M. Prakash, J.M. Lattimer: Astrophys. J. (2000) submitted; astro-ph/0008389
77. S. Tsuruta: Phys. Rep. **292**, 1 (1998) D.Page: in *The Many Faces of Neutron Stars*, ed. by R. Bucheri, J. van Paradijs, M.A. Alpar (Kluwer Academic Publishers, Dordrecht, 1998) p. 539

78. D. Page, J.H. Applegate: Astrophys. J. **394**, L17 (1992)
79. The relation between $T_e$ and $T$ is increased by about 50% if H or He dominates the atmosphere's composition. Our qualitative results concerning superfluidity, however, will be unaffected by the atmospheric composition.
80. M. Prakash, Manju Prakash, J.M. Lattimer, C.J. Pethick: Astrophys. J. **390**, L77 (1992)
81. C.J. Pethick: Rev. Mod. Phys. **64**, 1133 (1992) M. Prakash: Phys. Rep. **242**, 297 (1994) B. Friman, O.V. Maxwell: Astrophys. J. **232**, 541 (1979) N. Iwamoto: Phys. Rev. Lett. **44**, 1637 (1980)
82. M. Baldo, O. Elgaroy, L. Engvik, M. Hjorth-Jensen, H.-J. Schulze: Phys. Rev. C **58**, 1921 (1998)
83. S. Balberg, N. Barnea: Phys. Rev. C **57**, 409 (1997)
84. D.T. Son: Phys. Rev. D **59**, 094019 (1999) R. Pisarski, D. Rischke: Phys. Rev. D **61**, 051501 (2000) *ibid.* **61**, 074017 (2000) J. Berges, K. Rajagopal: Nucl. Phys. B **538**, 215 (1999) M. Alford, J. Berges, K. Rajagopal: Phys. Rev. D **D60**, 074014 (1999) T. Schäffer: Nucl. Phys. A **642**, 45 (1998) T. Schäffer, F. Wilczek: Phys. Rev. Lett. **82**, 3956 (1999) Phys. Rev. D **60**, 074014 (1999) G.W. Carter, D. Diakonov: Phys. Rev. D **60**, 16004 (1999)
85. P.F. Bedaque: hep-ph/9910247
86. D. Blaschke, T. Klähn, D.N. Voskresensky: Astrophys. J. **533**, 406 (2000)
87. D. Page, M. Prakash, J.M. Lattimer, A.W. Steiner: Phys. Rev. Lett. **85**, 2048 (2000)
88. K.P. Levenfish, D.G. Yakovlev: Astronomy Rep. **38**, 247 (1994) Astronomy Lett. **20**, 43 (1994) Astronomy Lett. **22**, 49 (1996)
89. E. Flowers, M. Ruderman, P. Sutherland: Astrophys. J. **205**, 541 (1976)
90. J. Zimanyi, S.A. Moszkowski: Phys. Rev. C **42**, 416 (1990)
91. T. Alm, G. Röpke, A. Sedrakian, F. Weber: Nucl. Phys. A **604**, 491 (1996)
92. D. Page: Astrophys. J. Lett. **479**, L43 (1997)
93. A. Akmal, V.R. Pandharipande, D.G. Ravenhall: Phys. Rev. C **58**, 1804 (1998)
94. D. Page: in *Neutron Stars and Pulsars: Thirty Years after the Discovery*, ed. by N. Shibazaki, et al. (Universal Academy Press, Tokyo 1998) p. 183
95. C. Schaab, S. Balberg, J. Schaffner-Bielich: Astrophys. J. Lett. **504**, L99 (1998)
96. J.M. Lattimer, M. Prakash, D. Masak, A. Yahil: Astrophys. J. **355** (1990) 241
97. B. Link, R.I. Epstein, J.M. Lattimer: Phys. Rev. Lett. **83** 3362 (1999)
98. M. Prakash, T.L. Ainsworth, J.M. Lattimer: Phys. Rev. Lett. **61** 2518 (1988)
99. B. Friedman, V.R. Pandharipande: Nucl. Phys. A **361** 502 (1981)
100. V.R. Pandharipande, R.A. Smith: Nucl. Phys. A **237** 507 (1975)
101. R.B. Wiringa, V. Fiks, A. Fabrocine: Phys. Rev. C **38** 1010 (1988)
102. H. Müther, M. Prakash, T.L. Ainsworth: Phys. Lett. **199** 469 (1987)
103. L. Engvik, M. Hjorth-Jensen, E. Osnes, G. Bao, E. Østgaard: Phys. Rev. Lett. **73** 2650 (1994)
104. E. Witten: Phys. Rev. **D30** 272 (1984)
105. E. Fahri, R. Jaffe: Phys. Rev. **D30** 2379 (1984)
106. P. Haensel, J.L. Zdunik, R. Schaefer: Astron. & Astrophys. **217** 137 (1986)
107. C. Alcock, A. Olinto: Ann. Rev. Nucl. Sci. **38** 161 (1988)
108. Manju Prakash, E. Baron, M. Prakash: Phys. Lett. B **243** 175 (1990)
109. C.E. Rhoades, R. Ruffini: Phys. Rev. Lett. **32** 324 (1974)
110. N.K. Glendenning, F. Weber: Astrophys. J. **400** 672 (1992)
111. J.M. Lattimer, M. Prakash: *to be published*
112. C.J. Horowitz, S.J. Pollock, P.A. Souder, R. Michaels: nucl-th/9912038

113. J. B. Hartle, Astrophys. J. **150** 1005 (1967)
114. R.C. Tolman: Phys. Rev. **55** 364 (1939)
115. H.A. Buchdahl: Astrophys. J. **147** 310 (1967)
116. D.G. Ravenhall, C.J. Pethick: Astrophys. J. **424** 846 (1994)
117. N.K. Glendenning: Phys. Rev. D **46** 4161 (1992)
118. J.M. Lattimer, A. Yahil: Astrophys. J. **340** 426 (1989)
119. L. Lindblom: Astrophys. J. **398** 569 (1992)
120. B. Friman, O. Maxwell: Astrophys. J. 232, 541 (1979)
121. R.F. Sawyer, A. Soni: Astrophys. J. **230**, 859 (1979)
122. P. Haensel, A.J. Jerzak: Astron. & Astrophys. **179**, 127 (1987)
123. D.H. Wilkinson: Phys. Rev. C **7**, 930 (1973)
124. M. Rho: Nucl. Phys. A **231**, 493 (1974)
125. G.E. Brown, M. Rho: Phys. Rev. Lett. **66**, 2720 (1991)
126. G. Carter, P.J. Ellis, S. Rudaz: Nucl. Phys. A **603**, 367 (1996)
127. S. Hannestad, G. Raffelt: Astrophys. J. **507**, 339 (1998)
128. J.A. Pons, J.A. Miralles, J.-M. Ibañez: Astron. & Astrophys. Supp. **129**, 343 (1998)
129. R.F. Sawyer, A. Soni: Phys. Rev. Lett. **38**, 1383 (1977)
130. R.F. Sawyer, A. Soni: Phys. Rev. C **18**, 898 (1978)
131. M. Ericson: Nucl. Phys. A **518**, 116 (1990)
132. R.F. Sawyer: Phys. Rev. Lett. **73**, 3363 (1994)
133. F.M. Walter: Astrophys. J. *in press* (2001)

# Neutron Star Kicks and Asymmetric Supernovae

Dong Lai

Center for Radiophysics and Space Research, Department of Astronomy
Cornell University, Ithaca, NY 14853, USA
Email: dong@astro.cornell.edu

**Abstract.** Observational advances over the last decade have left little doubt that neutron stars received a large kick velocity (of order a few hundred to a thousand km s$^{-1}$) at birth. The physical origin of the kicks and the related supernova asymmetry is one of the central unsolved mysteries of supernova research. We review the physics of different kick mechanisms, including hydrodynamically driven, neutrino – magnetic field driven, and electromagnetically driven kicks. The viabilities of the different kick mechanisms are directly related to the other key parameters characterizing nascent neutron stars, such as the initial magnetic field and the initial spin. Recent observational constraints on kick mechanisms are also discussed.

## 1 Evidence for Neutron Star Kicks and Supernova Asymmetry

It has long been recognized that neutron stars (NSs) have space velocities much greater (by about an order of magnitude) than their progenitors'. (e.g., Gunn & Ostriker 1970). A natural explanation for such high velocities is that supernova explosions are asymmetric, and provide kicks to the nascent NSs. In the last few years, evidence for NS kicks and supernova asymmetry has become much stronger. The observational facts and considerations that support (or even require) NS kicks fall into three categories:

**(1) Large NS Velocities ($\gg$ the progenitors' velocities $\sim 30$ km s$^{-1}$):**
• Recent studies of pulsar proper motion give $200 - 500$ km s$^{-1}$ as the mean 3D velocity of NSs at birth (e.g., Lyne and Lorimer 1994; Lorimer et al. 1997; Hansen & Phinney 1997; Cordes & Chernoff 1998), with possibly a significant population having velocities greater than 1000 km s$^{-1}$. While velocity of $\sim 100$ km s$^{-1}$ may in principle come from binary breakup in a supernova (without kick), higher velocities would require exceedingly tight presupernova binary. Statistical analysis seems to favor a bimodal pulsar velocity distribution, with peaks around 100 km s$^{-1}$ and 500 km s$^{-1}$ (Arzoumanian et al. 2001; see also Hansen & Phinney 1997; Cordes & Chernoff 1998).
• Direct evidence for pulsar velocities $\gtrsim 1000$ km s$^{-1}$ has come from observations of the bow shock produced by the Guitar Nebula pulsar (B2224+65) in the interstellar medium (Cordes, Romani & Lundgren 1993).
• The studies of neutron star – supernova remnant associations have, in many cases, indicated large NS velocities (e.g., Frail et al. 1994), although identifying the association can be tricky sometimes (e.g. Kaspi 1999; Gaensler 2000). Of special interest is the recent studies of magnetar–SNR associations: the SGR 0526-66

- N49 association, implying $V_\perp \sim 2900\,(3\,\mathrm{kyr}/t)\,\mathrm{km\,s^{-1}}$, and the possible association of SGR 1900+14 with G42.8+0.6, implying $V_\perp \sim 1800\,(10\,\mathrm{kyr}/t)\,\mathrm{km\,s^{-1}}$. (However, the proper motion of SGR 1806-20 may be as small as 100 km s$^{-1}$, and AXP 1E2259+586, AXJ 1845-0258, and AXP 1E1841-045 lie close to the centers of their respective remnants, CTB 109, G29.6+0.1, and Kes 73) (see Gaensler 2000).

**(2) Characteristics of NS Binaries (Individual Systems and Populations):** While large space velocities can in principle be accounted for by binary break-up (as originally suggested by Gott et al. 1970; see Iben & Tutukov 1996), many observed characteristics of NS binaries demonstrate that binary break-up can not be solely responsible for pulsar velocities, and that kicks are required (see also Tauris & van den Heuvel 2000). Examples include:

• The detection of geodetic precession in binary pulsar PSR 1913+16 implies that the pulsar's spin is misaligned with the orbital angular momentum; this can result from the aligned pulsar-He star progenitor only if the explosion of the He star gave a kick to the NS that misalign the orbit (Cordes et al. 1990; Kramer 1998; Wex et al. 1999).

• The spin-orbit misalignment in PSR J0045-7319/B-star binary, as manifested by the orbital plane precession (Kaspi et al. 1996; Lai et al. 1995) and fast orbital decay (which indicates retrograde rotation of the B star with respect to the orbit; Lai 1996a) require that the NS received a kick at birth (see Lai 1996b).

• The observed system radial velocity (430 km s$^{-1}$) of X-ray binary Circinus X-1 requires $V_{\mathrm{kick}} \gtrsim 500\,\mathrm{km\,s^{-1}}$ (Tauris et al. 1999).

• High eccentricities of Be/X-ray binaries cannot be explained without kicks (Verbunt & van den Heuvel 1995).

• Evolutionary studies of NS binary population (in particular the double NS systems) imply the existence of pulsar kicks (e.g., Deway & Cordes 1987; Fryer & Kalogera 1997; Fryer et al. 1998).

**(3) Observations of SNe and SNRs:** There are many direct observations of nearby supernovae (e.g., spectropolarimetry: Wang et al. 2000, Leonard et al. 2000; X-ray and gamma-ray observations and emission line profiles of SN1987A: McCray 1993, Utrobin et al. 1995) and supernova remnants (e.g., Morse, Winkler & Kirshner 1995; Aschenbach et al. 1995) which support the notion that supernova explosions are not spherically symmetric.

Finally it is of interest to note that recent study of the past association of the runaway star $\zeta$ Oph with PSR J1932+1059 (Hoogerwerf et al. 2000) or with RX 185635-3754 (Walter 2000) also implies a kick to the NS.

## 2 The Problem of Core-Collapse Supernovae and Neutron Star Kicks

The current paradigm for core-collapse supernovae leading to NS formation is that these supernovae are neutrino-driven (see Burrows 2000, Janka 2000 for recent review): As the central core of a massive star collapses to nuclear density, it rebounds and sends off a shock wave, leaving behind a proto-neutron star. The shock stalls at several 100's km because of neutrino loss and nuclear dissociation in the shock. A fraction of the neutrinos emitted from the proto-neutron star get absorbed by nucleons behind the shock, thus reviving the shock, leading to an explosion on the timescale several 100's ms — This is the so-called "Delayed Mechanism". However, 1D simulations with detailed neutrino transport seem to indicate that neutrino heating of the stalled shock, by itself, does not lead to an explosion or produce the observed supernova energetics (see Rampp & Janka 2000). It has been argued that neutrino-driven convection in the proto-neutron star (which tends to increase the neutrino flux) and that in the shocked mantle (which tends to increase the neutrino heating efficiency) are central to the explosion mechanism, although there is no consensus on the robustness of these convections (e.g., Herant et al. 1994; Burrows et al. 1995; Janka & Müller 1996; Mezzacappa et al. 1998). What is even more uncertain is the role of rotation and magnetic field on the explosion (see Mönchmeyer et al. 1991; Rampp, Müller & Ruffert 1998; Khokhlov et al. 1999; Fryer & Heger 2000 for simulations of collapse/explosion with rotation, and Thompson & Duncan 1993 and Thompson 2000a for discussion of possible dynamo processes and magnetic effects).

It is clear that despite decades of theoretical investigations, our understanding of the physical mechanisms of core-collapse supernovae remains significantly incomplete. The prevalence of neutron star kicks poses a significant mystery, and indicates that large-scale, global deviation from spherical symmetry is an important ingredient in our understanding of core-collapse supernovae (see Burrows 2000).

In the following sections, we review different classes of physical mechanisms for generating NS kicks (§§3-5), and then discuss possible observational constraints and astrophysical implications (§6).

## 3   Hydrodynamically Driven Kicks

The collapsed stellar core and its surrounding mantle are susceptible to a variety of hydrodynamical (convective) instabilities (e.g., Herant et al. 1994; Burrows et al. 1995; Janka & Müller 1996; Keil et al. 1996; Mezzacappa et al. 1998). It is natural to expect that the asymmetries in the density, temperature and velocity distributions associated with the instabilities can lead to asymmetric matter ejection and/or asymmetric neutrino emission. Numerical simulations, however, indicate that the local, post-collapse instabilities are not adequate to account for kick velocities $\gtrsim 100$ km s$^{-1}$ (Janka & Müller 1994; Burrows & Hayes 1996; Janka 1998; Keil 1998) — These simulations were done in 2D, and it is expected that the flow will be smoother on large scale in 3D simulations, and the resulting kick velocity will be even smaller.

There is now a consensus that global asymmetric perturbations in presupernova cores are required to produce the observed kicks hydrodynamically (Goldreich, Lai & Sahrling 1996; Burrows & Hayes 1996). Numerical simulations by Burrows & Hayes (1996) demonstrate that if the precollapse core is mildly asymmetric, the newly formed NS can receive a kick velocity comparable to the observed values. (In one simulation, the density of the collapsing core exterior to $0.9M_\odot$ and within $20^0$ of the pole is artificially reduced by 20%, and the resulting kick is about 500 km s$^{-1}$.) Asymmetric motion of the exploding material (since the shock tends to propagate more "easily" through the low-density region) dominates the kick, although there is also contribution (about 10 – 20%) from asymmetric neutrino emission. The magnitude of kick velocity is proportional to the degree of initial asymmetry in the imploding core. Thus the important question is: What is the origin of the initial asymmetry?

## 3.1 Presupernova Perturbations

Goldreich et al. (1996) suggested that overstable g-mode oscillations in the presupernova core may provide a natural seed for the initial asymmetry. These overstable g-modes arise as follows. A few hours prior to core collapse, a massive star ($M \gtrsim 8M_\odot$) has gone through a successive stages of nuclear burning, and attained a configuration with a degenerate iron core overlaid by an "onion skin" mantle of lighter elements. The rapidly growing iron core is encased in and fed by shells of burning silicon and oxygen, and the entire assemblage is surrounded by a thick convection zone. The nearly isothermal core is stably stratified and supports internal gravity waves. These waves cannot propagate in the unstably stratified convection zone, hence they are trapped and give rise to core g-modes in which the core oscillates with respect to the outer parts of the star. The overstability of the g-mode is due to the "$\varepsilon$-mechanism" with driving provided by temperature sensitive nuclear burning in Si and O shells surrounding the core before it implodes. It is simplest to see this by considering a $l = 1$ mode: If we perturb the core to the right, the right-hand-side of the shell will be compressed, resulting in an increase in temperature; Since the shell nuclear burning rate depends sensitively on temperature (power-law index $\sim$ 47 for Si burning and $\sim$ 33 for O burning), the nuclear burning is greatly enhanced; this generates a large local pressure, pushing the core back to the left. The result is an oscillating g-mode with increasing amplitude.

The main damping mechanism comes from the leakage of mode energy. The local (WKB) dispersion relation for nonradial waves is

$$k_r^2 = (\omega^2 c_s^2)^{-1}(\omega^2 - L_l^2)(\omega^2 - N^2), \tag{1}$$

where $k_r$ is the radial wavenumber, $L_l = \sqrt{l(l+1)}c_s/r$ ($c_s$ is the sound speed) and $N$ are the acoustic cut-off (Lamb) frequency and the Brunt-Väisälä frequency, respectively. Since acoustic waves whose frequencies lie above the acoustic cutoff can propagate through convective regions, each core g-mode will couple to an outgoing acoustic wave, which drains energy from the core g-modes (see Fig. 1). This leakage of mode energy can be handled with an outgoing propagation boundary condition in the mode calculation. Also, neutrino cooling tends to damp the mode. Since the nuclear energy generation rate depends more sensitively on temperature than pair neutrino emission (power law index $\sim$ 9), cooling

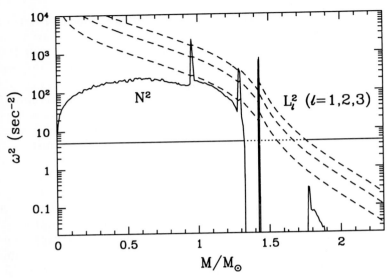

**Fig. 1.** Propagation diagram computed for a $15 M_\odot$ presupernova model of Weaver and Woosley (1993). The solid curve shows $N^2$, where $N$ is the Brunt-Väisälä frequency; the dashed curves show $L_l^2$, where $L_l$ is the acoustic cutoff frequency, with $l = 1, 2, 3$. The spikes in $N^2$ result from discontinuities in entropy and composition. The iron core boundary is located at $1.3 M_\odot$, the mass-cut at $1.42 M_\odot$. Convective regions correspond to $N = 0$. Gravity modes (with mode frequency $\omega$) propagate in regions where $\omega < N$ and $\omega < L_l$, while pressure modes propagate in regions where $\omega > N$ and $\omega > L_l$. Note that a g-mode trapped in the core can lose energy by penetrating the evanescent zones and turning into an outgoing acoustic wave (see the horizontal line). Also note that g-modes with higher $n$ (the radial order) and $l$ (the angular degree) are better trapped in the core than those with lower $n$ and $l$.

is never comparable to nuclear heating locally. Instead, thermal balance is mediated by the convective transport of energy from the shells, where the rate of nuclear energy generation exceeds that of neutrino energy emission, to the cooler surroundings where the bulk of the neutrino emission takes place. Calculations (based on the $15 M_\odot$ and $25 M_\odot$ presupernova models of Weaver & Woosley 1993) indicate that a large number of g-modes are overstable, although for low-order modes (small $l$ and $n$) the results depend sensitively on the detailed structure and burning rates of the presupernova models. The typical mode periods are $\gtrsim 1$ s, the growth time $\sim 10 - 50$ s, and the lifetime of the Si shell burning is $\sim$ hours (Lai & Goldreich 2000b, in preparation).

Our tentative conclusion is that overstable g-modes can potentially grow to large amplitudes prior to core implosion, although a complete understanding of the global pre-collapse asymmetries is probably out of reach at present, given the various uncertainties in the presupernova models. For example, the O-Si burning shell is highly convective, with convective speed reaching 1/4 of the sound speed, and hydrodynamical simulation may be needed to properly modeled such convection zones (see Bazan & Arnett 1998, Asida & Arnett 2000). Alternatively, it has been suggested that the convection itself may provide the seed of asymmetry in the presupernova core (Bazan & Arnett 1998), although it is not clear whether the perturbations have sufficiently large scales to be relevant to supernova kicks.

## 3.2 Amplification of Perturbation During Core Collapse

Core collapse proceeds in a self-similar fashion, with the inner core shrinking subsonically and the outer core falling supersonically at about half free-fall speed (Yahil 1983). The inner core is stable to non-radial perturbations because of the significant role played by pressure in its subsonic collapse. Pressure is less important in the outer region, making it more susceptible to large scale instability. A recent stability analysis of Yahil's self-similar collapse solution (which is based on Newtonian theory and a polytropic equation of state $P \propto \rho^\Gamma$, with $\Gamma \sim 1.3$) does not reveal any unstable global mode before the proto-neutron star forms (Hanawa & Matsumoto 2000; Lai 2000). However, during the subsequent accretion of the outer core (involving 15% of the core mass) and envelope onto the proto-neutron star, nonspherical Lagrangian perturbations can grow according to $\Delta\rho/\rho \propto r^{-1/2}$ (independent of $l$) or even $\Delta\rho/\rho \propto r^{-1}$ (for $l = 1$ when the central collapsed object is displaced from the origin of the converging flow) (Lai & Goldreich 2000) The asymmetric density perturbations seeded in the presupernova star, especially those in the outer region of the iron core, are therefore amplified (by a factor of 5-10) during collapse. The enhanced asymmetric density perturbation may lead to asymmetric shock propagation and breakout, which then give rise to asymmetry in the explosion and a kick velocity to the NS (see Burrows & Hayes 1996).

## 4 Neutrino – Magnetic Field Driven Kicks

The second class of kick mechanisms rely on asymmetric neutrino emission induced by strong magnetic fields. Since 99% of the NS binding energy (a few times $10^{53}$ erg) is released in neutrinos, tapping the neutrino energy would appear to be an efficient means to kick the newly-formed NS. The fractional asymmetry $\alpha$ in the radiated neutrino energy required to generate a kick velocity $V_{\rm kick}$ is $\alpha = MV_{\rm kick}c/E_{\rm tot}$ (= 0.028 for $V_{\rm kick} = 1000$ km s$^{-1}$, NS mass $M = 1.4\,M_\odot$ and total neutrino energy radiated $E_{\rm tot} = 3 \times 10^{53}$ erg).

### 4.1 Effect of Parity Violation

Because weak interaction is parity violating, the neutrino opacities and emissivities in a magnetized nuclear medium depend asymmetrically on the directions of neutrino momenta with respect to the magnetic field, and this can give rise to asymmetric neutrino emission from the proto-neutron star. Chugai (1984) (who gave an incorrect expression for the electron polarization in the relativistic, degenerate regime) and Vilenkin (1995) considered neutrino-electron scattering, but this is less important than neutrino-nucleon scattering in determining neutrino transport in proto-neutron stars. Dorofeev et al. (1985) considered neutrino emission by Urca processes, but failed to recognize that in the bulk interior of the star the asymmetry in neutrino emission is cancelled by that associated with neutrino absorption (Lai & Qian 1998a).

Horowitz & Li (1998) suggested that large asymmetries in the neutrino flux could result from the cumulative effect of multiple scatterings of neutrinos by slightly polarized nucleons (see also Lai & Qian 1998a; Janka 1998). However, it can be shown that, although the scattering cross-section is asymmetric with respect to the magnetic field for individual neutrinos, detailed balance requires that there be no cumulative effect associated with multiple scatterings in the bulk interior of the star where thermal equilibrium is maintained to a good approximation (Arras & Lai 1999a; see also Kusenko et al. 1998). For a given neutrino species, there is a drift flux of neutrinos along the magnetic field in addition to the usual diffusive flux. This drift flux depends on the deviation of the neutrino distribution function from thermal equilibrium. Thus asymmetric neutrino flux can be generated in the outer region of the proto-neutron star (i.e., above the neutrino-matter decoupling layer, but below the neutrinosphere) where the neutrino distribution deviates significantly from thermal equilibrium. While the drift flux associated with $\nu_\mu$'s and $\nu_\tau$'s is exactly canceled by that associated with $\bar\nu_\mu$'s and $\bar\nu_\tau$'s, there is a net drift flux due to $\nu_e$'s and $\bar\nu_e$'s. Arras & Lai (1999b) found that the asymmetry parameter for the $\nu_e$-$\bar\nu_e$ flux is dominated for low energy neutrinos ($\lesssim 15$ MeV) by the effect of ground (Landau) state electrons in the absorption opacity, $\epsilon_{\rm abs} \simeq 0.6 B_{15}(E_\nu/1 \text{ MeV})^{-2}$, where $B_{15} = B/(10^{15} \text{ G})$, and for high energy neutrinos by nucleon polarization ($\epsilon \sim \mu_m B/T$). Averaging over all neutrino species, the total asymmetry in neutrino flux is of order $\alpha \sim 0.2 \epsilon_{\rm abs}$, and the resulting kick velocity $V_{\rm kick} \sim 50 \, B_{15}$ km s$^{-1}$. There is probably a factor of 5 uncertainty in this estimate. To firm up this estimate requires solving the neutrino transport equations in the presence of parity violation for realistic proto-neutron stars.

## 4.2 Effect of Asymmetric Field Topology

A different kick mechanism relies on the asymmetric magnetic field distribution in proto-neutron stars (see Bisnovatyi-Kogan 1993; however, he considered neutron decay, which is not directly relevant for neutrino emission from proto-neutron stars). Since the cross section for $\nu_e$ ($\bar\nu_e$) absorption on neutrons (protons) depends on the local magnetic field strength due to the quantization of energy levels for the $e^-$ ($e^+$) produced in the final state, the local neutrino fluxes emerged from different regions of the stellar surface are different. Calculations indicate that to generate a kick velocity of $\sim 300$ km s$^{-1}$ using this mechanism alone would require that the difference in the field strengths at the two opposite poles of the star be at least $10^{16}$ G (Lai & Qian 1998b). Note that unlike the kick due to parity violation (see §4.1), this mechanism does not require the magnetic field to be ordered, i.e., only the magnitude of the field matters.

## 4.3 Dynamical Effect of Magnetic Fields

A superstrong magnetic field may also play a dynamical role in the proto-neutron star. For example, it has been suggested that a locally strong magnetic field can induce "dark spots" (where the neutrino flux is lower than average) on the stellar surface by suppressing neutrino-driven convection (Duncan & Thompson 1992). While it is difficult to quantify the kick velocity resulting from an asymmetric distribution of dark spots, order-of-magnitude estimate indicates that a local magnetic field of at least $10^{15}$ G is needed for this effect to be of importance. Much work remains to be done to quantify the magnetic effects (especially when coupled with rotation) on the dynamics of the proto-neutron star and the supernova explosion (see, e.g., LeBlanc & Wilson 1970; Thompson & Duncan 1993).

## 4.4 Exotic Neutrino Physics

There have also been several ideas of pulsar kicks which rely on nonstandard neutrino physics. It was suggested (Kusenko & Segre 1996) that asymmetric $\nu_\tau$ emission could result from the Mikheyev-Smirnov-Wolfenstein flavor transformation between $\nu_\tau$ and $\nu_e$ inside a magnetized proto-neutron star because a magnetic field changes the resonance condition for the flavor transformation. This mechanism requires neutrino mass of order 100 eV. A similar idea (Akhmedov et al. 1997) relies on both the neutrino mass and the neutrino magnetic moment to facilitate the flavor transformation (resonant neutrino spin-flavor precession; see also Grasso et al. 1998). More detailed analysis of neutrino transport (Janka & Raffelt 1998), however, indicates that even with favorable neutrino parameters (such as mass and magnetic moment) for neutrino oscillation, the induced pulsar kick is much smaller than previously estimated (i.e., $B \gg 10^{15}$ G is required to obtain a 100 km s$^{-1}$ kick).

It is clear that all the kick mechanisms discussed in this section (§4) are of relevance only for $B \gtrsim 10^{15}$ G. While recent observations have lent strong support that some neutron stars ("magnetars") are born with such a superstrong magnetic field (e.g., Thompson & Duncan 1993; Vasisht & Gotthelf 1997; Kouveliotou et al. 1998,1999; Thompson 2000b), it is not clear (perhaps unlikely) that ordinary radio pulsars (for which large velocities have been measured) had initial magnetic fields of such magnitude (see als §6).

## 5 Electromagnetically Driven Kicks

Harrison & Tademaru (1975) show that electromagnetic (EM) radiation from an off-centered rotating magnetic dipole imparts a kick to the pulsar along its spin axis. The kick is attained on the initial spindown timescale of the pulsar (i.e., this really is a gradual acceleration), and comes at the expense of the spin kinetic energy. We (Lai, Chernoff & Cordes 2001) have reexamined this effect and found that the force on the pulsar due to asymmetric EM radiation is larger than the original Harrison & Tademaru expression by a factor of four. If the dipole is displaced by a distance $s$ from the rotation axis, and has components $\mu_\rho, \mu_\phi, \mu_z$ (in cylindrical coordinates), the force is given by (to leading order in $\Omega s/c$)

$$F = \frac{8}{15}\left(\frac{\Omega s}{c}\right)\frac{\Omega^4 \mu_z \mu_\phi}{c^4}. \tag{2}$$

(The sign is such that negative $F$ implies $\mathbf{V}_{\rm kick}$ parallel to the spin $\mathbf{\Omega}$.) The dominant terms for the spindown luminosity give

$$L = \frac{2\Omega^4}{3c^3}\left(\mu_\rho^2 + \mu_\phi^2 + \frac{2\Omega^2 s^2 \mu_z^2}{5c^2}\right). \tag{3}$$

For a "typical" situation, $\mu_\rho \sim \mu_\phi \sim \mu_z$, the asymmetry parameter $\epsilon \equiv F/(L/c)$ is of order $0.4(\Omega s/c)$. For a given $\Omega$, the maximum $\epsilon_{\rm max} = \sqrt{0.4} = 0.63$ is achieved for $\mu_\rho/\mu_z = 0$ and $\mu_\phi/\mu_z = \sqrt{0.4}\,(\Omega s/c)$. From $M\dot{V} = \epsilon(L/c) = -\epsilon(I\Omega\dot{\Omega})/c$, we obtain the kick velocity

$$V_{\rm kick} \simeq 260\, R_{10}^2 \left(\frac{\bar{\epsilon}}{0.1}\right)\left(\frac{\nu_i}{1\,\rm kHz}\right)^2 \left[1 - \left(\frac{\nu}{\nu_i}\right)^2\right]\,\rm km\, s^{-1}, \tag{4}$$

where $R = 10 R_{10}$ km is the neutron star radius, $\nu_i$ is the initial spin frequency, $\nu$ is the current spin frequency of the pulsar, and $\bar{\epsilon} = (\Omega_i^2 - \Omega^2)^{-1}\int\epsilon\,d\Omega^2$. For the "optimal" condition, with $\mu_\rho = 0$, $\mu_\phi/\mu_z = \sqrt{0.4}\,(\Omega_i s/c)$, and $\epsilon = \sqrt{0.4}\,[2\Omega_i\Omega/(\Omega^2 + \Omega_i^2)]$, we find

$$V_{\rm kick}^{(\rm max)} \simeq 1400\, R_{10}^2 \left(\frac{\nu_i}{1\,\rm kHz}\right)^2 \,\rm km\, s^{-1}. \tag{5}$$

Thus, if the NS was born rotating at $\nu_i \gtrsim 1$ kHz, it is possible, in principle, to generate spin-aligned kick of a few hundreds km s$^{-1}$ or even 1000 km s$^{-1}$.

Equations (4) and (5) assume that the rotational energy of the pulsar entirely goes to electromagnetic radiation. Recent work has shown that a rapidly rotating ($\nu \gtrsim 100$ Hz) NS can potentially lose significant angular momentum to gravitational waves generated by unstable r-mode oscillations (e.g., Andersson 1998; Lindblom, Owen & Morsink 1998; Owen et al. 1998; Andersson, Kokkotas & Schutz 1999; Ho & Lai 2000). If gravitational radiation carries away the rotational energy of the NS faster than the EM radiation does, then the electromagnetic rocket effect will be much diminished (Gravitational radiation can also carry away linear momentum, but the effect for a NS is negligible). In the linear regime, the r-mode amplitude $\alpha \sim \xi/R$ (where $\xi$ is the surface Lagrangian displacement; see the references cited above for more precise definition of $\alpha$) grows due to gravitational radiation reaction on a timescale $t_{\rm grow} \simeq 19\,(\nu/1\,{\rm kHz})^{-6}$ s. Starting from an initial amplitude $\alpha_i \ll 1$, the mode grows to a saturation level $\alpha_{\rm sat}$ in time $t_{\rm grow}\ln(\alpha_{\rm sat}/\alpha_i)$ during which very little rotational energy is lost. After saturation, the NS spins down due to gravitational radiation on a timescale

$$\tau_{\rm GR} = \left|\frac{\nu}{\dot\nu}\right|_{\rm GR} \simeq 100\,\alpha_{\rm sat}^{-2}\left(\frac{\nu}{1\,\rm kHz}\right)^{-6}\,\rm s, \tag{6}$$

(Owen et al. 1998). By contrast, the spindown time due to EM radiation alone is

$$\tau_{\rm EM} = \left|\frac{\nu}{\dot{\nu}}\right|_{\rm EM} \simeq 10^7 \, B_{13}^{-2} \left(\frac{\nu}{1\,{\rm kHz}}\right)^{-2} \, {\rm s}, \tag{7}$$

where $B_{13}$ is the surface dipole magnetic field in units of $10^{13}$ G. Including gravitational radiation, the kick velocity becomes

$$V_{\rm kick} \simeq 260 \, R_{10}^2 \left(\frac{\bar{\epsilon}}{0.1}\right) \left(\frac{\nu_i}{1\,{\rm kHz}}\right)^2 \frac{1}{\beta} \ln\left[\frac{1+\beta}{1+\beta\,(\nu/\nu_i)^2}\right] \, {\rm km \, s}^{-1}, \tag{8}$$

where in the second equality we have replaced $\epsilon$ by constant mean value $\bar{\epsilon}$, and $\beta$ is defined by

$$\beta \equiv \left(\frac{\tau_{\rm EM}}{\tau_{\rm GR}}\right)_i \simeq \left(\frac{\alpha_{\rm sat}}{10^{-2.5}}\right)^2 \left(\frac{\nu_i}{1\,{\rm kHz}}\right)^4 B_{13}^{-2}. \tag{9}$$

For $\beta \ll 1$, equation (8) becomes eq. (4); for $\beta \gg 1$, the kick is reduced by a factor $1/\beta$.

Clearly, for the EM rocket to be viable as a kick mechanism at all requires $\beta \lesssim 1$. The value of $\alpha_{\rm sat}$ is unknown. Analogy with secularly unstable bar-mode in a Maclaurin spheroid implies that $\alpha_{\rm sat} \sim 1$ is possible (e.g., Lai & Shapiro 1995). It has been suggested that turbulent dissipation in the boundary layer near the crust (if it exists early in the NS's history) may limit $\alpha_{\rm sat}$ to a small value of order $10^{-2}$-$10^{-3}$ (Wu, Matzner & Arras 2000). The theoretical situation is not clear at this point (see Lindblom et al. 2000 for recent simulations of nonlinear r-modes).

## 6 Astrophysical Constraints on Kick Mechanisms

In §§3-5 we have focused on the *physics* of different kick mechanisms. All these mechanisms still have intrinsic physics uncertainties and require more theoretical work. For example:

(1) For the hydrodynamical driven kicks, one needs to better understand the structure of pre-SN core in order to determine whether overstable g-modes can grow to large amplitudes; more simulation would be useful to pin down the precise relationship between the magnitudes of the initial asymmetry and the kick velocity;

(2) For the neutrino–magnetic field driven kicks, more elaborate neutrino transport calculation is necessary to determine (to within a factor of 2) the value of $B$ needed to generate (say) $V_{\rm kick} = 300 \, {\rm km \, s}^{-1}$;

(3) For the electromagnetically driven kicks, the effect of gravitational radiation (especially the r-mode amplitude) needs to be better understood.

We now discuss some of the astrophysical/observational constraints on the kick mechanisms.

## 6.1 Initial Magnetic Field of NS

The neutrino-magnetic field driven kicks (§4) require initial $B \gtrsim 10^{15}$ G to be of interest. While magnetars may have such superstrong magnetic fields at birth (e.g., Thompson & Duncan 1993; Kouveliotou et al. 1998,1999; Thompson 2000b), the situation is not clear for ordinary radio pulsars, whose currently measured magnetic fields are of order $10^{12}$ G. It is difficult for an initial large-scale $10^{15}$ G to decay (via Ohmic diffusion or ambipolar diffusion) to the canonical $10^{12}$ G on the relevant timescale of $10^3 - 10^7$ years. However, one cannot rule out the possibility that in the proto-neutron star phase, a convection-initiated dynamo generates a "transient" superstrong magnetic field, lasting a few seconds, and then the field gets destroyed by "anti-dynamo" as the convection ceases. Obviously if we can be sure that this is not possible, then we can discard the mechanisms discussed in §4.

## 6.2 Initial Spin of NS

To produce sufficient velocity, the electromagnetic rocket effect (§5) requires the NS initial spin period $P_i$ to be less than $1 - 2$ ms. It is widely thought that radio pulsars are not born with such a rapid spin, but rather with a more modest $P_i = 0.02 - 0.5$ s (e.g., Lorimer et al. 1993). The strongest argument for this comes for the energetics of pulsar nebulae (particularly Crab). But this is not without uncertainties. For example, a recent analysis of the energetics of the Crab Nebula suggests an initial spin period $\sim 3 - 5$ ms followed by fast spindown on a time scale of 30 yr (Atoyan 1999). As for the Vela pulsar, the energetics of the remnant do not yield an unambiguous constraint on the initial spin. Also, the recent discovery of the 16 ms X-ray pulsar (PSR J0537-6910) associated with the Crab-like supernova remnant N157B (Marshall et al. 1998) in the Large Magellanic Cloud implies that at least some NSs are born with spin periods in the millisecond range. So at this point it may be prudent to consider $P_i \sim 1$ ms as a possibility (see also §6.5).

## 6.3 Natal vs. Post-Natal Kicks

There is a qualitative difference between natal kicks (including the hydrodynamical driven and neutrino–magnetic field driven kicks) and post-natal kicks. Because it is a slow process, the Harrison-Tademaru effect may have difficulty in explaining some of the characteristics of NS binaries (even if the physics issues discussed in §5 work out to give a large $V_{\rm kick}$), such as the spin-orbit misalignment in PSR J0045-7319 (Kaspi et al. 1996) and PSR 1913+16 needed to produce the observed precessions. For example, in the case of PSR J0045-7319 – B star binary: if we assume that the orbital angular momentum of the presupernova binary is aligned with the spin of the B star, then the current spin-orbit misalignment can only be explained by a fast kick with $\tau_{\rm kick}$ less than the post-explosion orbital period $P_{\rm orb}$. Similarly, a slow kick (with $\tau_{\rm kick} \gtrsim P_{\rm orb}$) may be inconsistent with the NS binary populations (e.g., Dewey & Cordes 1987; Fryer

& Kalogera 1997; Fryer et al. 1998). However, note that $\tau_{\rm kick} \sim \tau_{\rm EM}$ for the Harrison-Tademaru effect depends on value of $B$ [see eq. (7)], thus can be made much smaller than $P_{\rm orb}$ (which typically ranges from hours to several months or a few years at most for relevant binaries) if $B$ is large.

## 6.4 Correlations Between Velocity and Other Properties of NS?

Despite some earlier claims to the contrary, statistical studies of pulsar population have revealed no correlation between $V_{\rm kick}$ and magnetic field strength, or correlation between the kick direction and the spin axis (e.g., Cordes & Chernoff 1998; Deshpande et al. 1999). Given the large systematic uncertainties, the statistical results, by themselves, cannot reliably constrain any kick mechanism. For example, the magnetic field strengths required for the neutrino-driven mechanisms are $\gtrsim 10^{15}$ G, much larger than the currently inferred dipolar surface fields of typical radio pulsars; there are large uncertainties in using the polarization angle to determine the pulsar spin axis; differential galactic rotation is important for distant NSs and cannot be accounted for unless the distance is known accurately and the NS has not moved far from its birth location; several different mechanisms (including binary breakup) may contribute to the observed NS velocities (see Lai, Chernoff & Cordes 2001).

Recent observations of the Vela pulsar and the surrounding compact X-ray nebula with the Chandra X-ray Observatory reveal a two sided asymmetric jet at a position angle coinciding with the position angle of the pulsar's proper motion (Pavlov et al. 2000; see http://chandra.harvard.edu/photo/cycle1/vela/ for image) The symmetric morphology of the nebula with respect to the jet direction strongly suggests that the jet is along the pulsar's spin axis. Analysis of the polarization angle of Vela's radio emission corroborates this interpretation. Similar evidence for spin-velocity alignment also exists for the Crab pulsar. Thus, while statistical analysis of pulsar population neither support nor rule out any spin-kick correlation, at least for the Vela and Crab pulsars, the proper motion and the spin axis appear to be aligned. Interestingly, both Crab and Vela pulsars have relatively small transverse velocities (of order $100\,{\rm km\,s^{-1}}$).

## 6.5 The Effect of Rotation and Spin-Kick Alignment?

The apparent alignment between the spin axis and proper motion for the Crab and Vela pulsar raises an interesting question: Under what conditions is the spin-kick alignment expected for different kick mechanisms? Let us look at the three classes of mechanisms discussed in §§3-5 (see Lai, Chernoff & Cordes 2001).

(1) Electromagnetically Driven Kicks: The spin-kick slignment is naturally produced. (Again, note that $P_i \sim 1-2$ ms is required to generate sufficiently large $V_{\rm kick}$).

(2) Neutrino–Magnetic Field Driven Kicks: The kick is imparted to the NS near the neutrinosphere (at 10's of km) on the neutrino diffusion time, $\tau_{\rm kick} \sim$ 10 seconds. As long as the initial spin period $P_i$ is much less than a few seconds, spin-kick alignment is naturally expected.

(3) Hydrodynamically Driven Kicks: The low-order g-modes trapped in the presupernova core ($M \simeq 1.4 M_\odot$, $R \simeq 1500$ km) have periods of 1-2 seconds, much shorter than the rotation period of the core (unless the core possesses a dynamically important angular momentum after collapse), thus the g-modes are not affected by rotation. Also, since the rotational speed of the core is typically less than the speed of convective eddies ($\simeq$1000-2000 km s$^{-1}$, about 20% of the sound speed) in the burning shell surrounding the iron core, rotation should not significantly affect the shell convection either. Thus the development of large-scale presupernova (dipolar) asymmetry is not influenced by the core rotation. But even though the primary thrust to the NS (upon core collapse) does not depend on spin, the net kick will be affected by rotational averaging if the asymmetry pattern (near the shock breakout) rotates with the matter at a period shorter than the kick timescale $\tau_{\rm kick}$. Here the situation is more complicated because the primary kick to the NS is imparted at a large radius, $r_{\rm shock} \gtrsim 100$ km (since this the location of the stalled shock). To obtain effective spin averaging, we require the rotation period at $r_{\rm shock}$ to be shorter than $\tau_{\rm kick} \sim 100$ ms (this $\tau_{\rm kick}$ is the shock travel time at speed of $10^4$ km s$^{-1}$ across $\sim 1000$ km, the radius of the mass cut enclosing $1.4 M_\odot$). Assuming angular momentum conservation, this translates into the requirement that the final NS spin period $P_i \lesssim 1$ ms. We thus conclude that if rotation is dynamically unimportant for the core collapse and explosion (corresponding to $P_s \gg 1$ ms), then rotational averaging is inefficient and the hydrodynamical mechanism does not produce spin-kick alignment.

The discussion above is based on the standard picture of core-collapse supernovae, which is valid as long as rotation does *not* play a dynamically important role (other than rotational averaging) in the supernova. If, on the other hand, rotation is dynamically important, the basic collapse and explosion may be qualitatively different (e.g., core bounce may occur at subnuclear density, the explosion is weaker and takes the form of two-sided jets; see Mönchmeyer et al. 1991; Rampp, Müller & Ruffert 1998; Khokhlov et al. 1999). The possibility of a kick in such systems has not been studied, but it is conceivable that an asymmetric dipolar perturbation may be coupled to rotation, thus producing spin-kick alignment.

It has been suggested that the presupernova core has negligible angular momentum and the pulsar spin may be generated by off-centered kicks when the NS forms (Spruit & Phinney 1998). It is certainly true that even with zero precollapse angular momentum, some rotation can be produced in the proto-neutron star (Burrows et al. 1995 reported a rotation period of order a second generated by stochastic torques in their 2D simulations of supernova explosions), although $P_i \lesssim 30$ ms seems difficult to get. In this picture, the spin will generally be perpendicular to the velocity; aligned spin-kick may be possible if the kick is the result of many small thrusts which are appropriately oriented (Spruit & Phinney 1998) — this might apply if small-scale convection were responsible for the kick. But as discussed in §3, numerical simulations indicate that such convection alone does not produce kicks of sufficient amplitude. Therefore, spin-kick alignment requires that the proto-neutron star have a "primordial" rotation (i.e., with angular momentum coming from the presupernova core).

Clearly, *if* spin-kick alignment is a generic feature for all NSs, it can provide strong constraints on the kick mechanisms, supernova explosion mechanisms, as well as initial conditions of NSs.

**Acknowledgement**

I thank my collaborators Phil Arras, David Chernoff, Jim Cordes, Peter Goldreich, and Yong-Zhong Qian for their important contributions and insight. This work is supported by NASA grants NAG 5-8356 and NAG 5-8484 and by a fellowship from the Alfred P. Sloan foundation. I thank the organizers for the invitation, and the European Center for Theoretical Physics for travel support to participate the "Neutron Star Interiors" Workshop (Trento, Italy, June-July 2000). This paper is also based on a lecture given at the ITP conference on "Spin and Magnetism in Neutron Stars" (Santa Barbara, October 2000). I acknowledge ITP at UCSB (under NSF Grant PHY99-07949) for support while this paper was being completed.

# References

1. Akhmedov, E. K., Lanza, A., & Sciama, D. W. 1997, Phys. Rev. D, 56, 6117
2. Andersson, N. 1998, ApJ, 502, 708
3. Andersson, N., Kokkotas, K.D., & Schutz, B.F. 1999, ApJ, 510, 846
4. Arras, P., & Lai, D. 1999a, ApJ, 519, 745
5. Arras, P., & Lai, D. 1999b, Phys. Rev. D60, 043001
6. Arzoumanian, Z., Chernoff, D.F., & Cordes, J.M. 2001, in preparation.
7. Aschenbach, B., Egger, R., & Trumper, J. 1995, Nature, 373, 587.
8. Asida, S.M., & Arnett, D. 2000, ApJ, in press (astro-ph/0006451).
9. Atoyan, A. M. 1999, AAp, 346, L49
10. Bazan, G., & Arnett, D. 1998, ApJ, 496, 316
11. Bisnovatyi-Kogan, G. S. 1993, Astron. Astrophys. Trans., 3, 287
12. Burrows, A. 2000, Nature, 403, 727
13. Burrows, A., Hayes, J., & Fryxell, B.A. 1995, ApJ, 450, 830
14. Burrows, A., & Hayes, J. 1996, Phys. Rev. Lett., 76, 352
15. Chugai, N.N. 1984, Sov. Astron. Lett., 10, 87
16. Cordes, J.M., Romani, R.W., & Lundgren, S.C. 1993, Nature, 362, 133
17. Cordes, J.M., Wasserman, I., & Blaskiewicz, M. 1990, ApJ, 349, 546
18. Cordes, J.M., & Chernoff, D.F. 1998, ApJ, 505, 315
19. Deshpande, A.A., Ramachandran, R., & Radhakrishnan, V. 1999, A&A, 351, 195
20. Deway, R. J., & Cordes, J. M. 1987, ApJ, 321, 780
21. Dorofeev, O.F., et al. 1985, Sov. Astron. Lett., 11, 123
22. Duncan, R.C., & Thompson, C. 1992, ApJ, 392, L9.
23. Frail, D.A., Goss, W.M., & Whiteoak, J.B.Z. 1994, ApJ, 437, 781
24. Fryer, C., Burrows, A., & Benz, W. 1998, ApJ, 498, 333
25. Fryer, C. L., & Heger, A. 2000, 541, 1033
26. Fryer, C., & Kalogera, V. 1997, ApJ, 489, 244 (erratum: ApJ, 499, 520)

27. Gaensler, B.M. 2000, in "Pulsar Astronomy — 2000 and Beyond" (ASP Conf. Proceedings) (astro-ph/9911190).
28. Goldreich, P., Lai, D., & Sahrling, M. 1996, in "Unsolved Problems in Astrophysics", ed. J.N. Bahcall and J.P. Ostriker (Princeton Univ. Press)
29. Gott, J.R., Gunn, J.E., & Ostriker, J.P. 1970, ApJ, 160, L91.
30. Grasso, D., Nunokawa, H., & Valle, J.W.F. 1998, Phys. Rev. Lett., 81, 2412
31. Gunn, J.E., & Ostriker, J.P. 1970, ApJ, 160, 979
32. Hanawa, T., & Matsumoto, T. 2000, PASJ, in press.
33. Hansen, B.M.S., & Phinney, E.S. 1997, MNRAS, 291, 569
34. Herant, M., et al. 1994, ApJ, 435, 339
35. Harrison, E.R., & Tademaru, E. 1975, ApJ, 201, 447
36. Ho, W.C.G., & Lai, D. 2000, ApJ, 543, 386
37. Hoogerwerf, R., de Bruijne, J.H.J., & de Zeeuw, P.T. 2000, ApJ, in press (astro-ph/0007436)
38. Horowitz, C.J., & Li, G. 1998, Phys. Rev. Lett., 80, 3694
39. Iben, I., & Tutukov, A. V. 1996, ApJ, 456, 738
40. Janka, H.-T. 1998, in "Neutrino Astrophysics", ed. M. Altmann et al. (Garching: Tech. Univ. München) (astro-ph/9801320)
41. Janka, H.-T. 2000, submitted to A&A (astro-ph/0008432)
42. Janka, H.-Th., & Müller, E. 1994, A&A, 290, 496
43. Janka, H.-T., & Müller, E. 1996, A&A, 306, 167
44. Janka, H.-T., & Raffelt, G.G. 1998, Phys. Rev. D59, 023005
45. Johnston, H. M., Fender, R. P., & Wu, K. 1999, MNRAS, 308, 415.
46. Kaspi, V.M., et. al. 1996, Nature, 381, 583
47. Kaspi, V.M. 1999, in Pulsar Astronomy (ASP Conference Series), ed. M. Kramer, N. Wex and R. Wielebinski (astro-ph/9912284)
48. Keil, W. 1998, in Proc. of the 4th Ringberg Workshop on Neutrino Astrophysics (Munich, MPI).
49. Keil, W., Janka, H.-Th., Müller, E. 1996, ApJ, 473, L111
50. Khokhlov, A.M., et al. 1999, ApJ, 524, L107
51. Kouveliotou, C., et al. 1998, Nature, 393, 235
52. Kouveliotou, C., et al. 1999, ApJ, 510, L115
53. Kramer, M. 1998, ApJ, 509, 856
54. Kusenko, A., & Segré, G. 1996, Phys. Rev. Lett., 77, 4872
55. Kusenko, A., Segré, G., & Vilenkin, A. astro-ph/9806205
56. Lai, D. 1996a, ApJ, 466, L35
57. Lai, D. 1996b, in Proceedings of the 18th Texas Meeting on Relativistic Astrophysics, ed. A.V. Olinto et al. (World Scientific), p. 634 (astro-ph/9704134)
58. Lai, D. 2000, ApJ, 540, 946
59. Lai, D., Bildsten, L., & Kaspi, V.M. 1995, ApJ, 452, 819
60. Lai, D., Chernoff, D.F., & Cordes, J.M. 2001, ApJ, in press (astro-ph/0007272)
61. Lai, D., & Goldreich, P. 2000a, ApJ, 535, 402
62. Lai, D., & Goldreich, P. 2000b, in preparation
63. Lai, D., & Qian, Y.-Z. 1998a, ApJ, 495, L103 (erratum: 501, L155)
64. Lai, D., & Qian, Y.-Z. 1998b, ApJ, 505, 844
65. LeBlanc, J.M., & Wilson, J.R. 1970, ApJ, 161, 541
66. Leonard, D.C., Filippenko, A.V., & Ardila, D.R. 2000, ApJ, submitted (astro-ph/0009285)
67. Lindblom, L., Owen, B.J., & Morsink, S.M. 1998, Phys. Rev. Lett., 80, 4843
68. Lindblom, L., Tohline, J.E., & Vallisneri, M. 2000, astro-ph/0010653

69. Lorimer, D.R., Bailes, M., Dewey, R.J., & Harrison, P.A. 1993, MNRAS, 263, 403
70. Lorimer, D.R., Bailes, M., & Harrison, P.A. 1997, MNRAS, 289, 592
71. Lyne, A.G., & Lorimer, D.R. 1994, Nature, 369, 127
72. Marshall, F. E., et al. 1998, ApJ, 499, L179
73. McCray, R. 1993, ARA&A, 31, 175
74. Mezzacappa, A., et al. 1998, ApJ, 495, 911.
75. Mönchmeyer, R., Schäfer, G., Müller, E., Kates, R.E. 1991, A&A, 246, 417
76. Morse, J.A., Winkler, P.F., & Kirshner, R.P. 1995, AJ, 109, 2104
77. Owen, B.J., Lindblom, L., Cutler, C., Schutz, B.F., Vecchio, A., & Andersson, N. 1998, Phys. Rev. D, 58, 084020
78. Pavlov, G. G., et al. 2000, BAAS, 32, 733
79. Rampp, M., & Janka, H.-Th 2000, ApJ, 539, L33
80. Rampp, M., Müller, E., & Ruffert, M. 1998, A&A, 332, 969
81. Spruit, H., & Phinney, E.S. 1998, Nature, 393, 139
82. Tauris, T., et al. 1999, MNRAS, 310, 1165
83. Tauris, T., & van den Heuvel, E.P.J. 2000, astro-ph/0001015
84. Thompson, C. 2000a, ApJ, 534, 915
85. Thompson, C. 2000b, in Proceedings NATO Advanced Study Institute "The Neutron Star-Black Hole Connection", ed. V. Connaughton et al. (astro-ph/0010016)
86. Thompson, C., & Duncan, R.C. 1993, ApJ, 408, 194.
87. Thompson, C., & Duncan, R. C. 1996, ApJ, 473, 322
88. Utrobin, V.P., Chugai, N.N., & Andronova, A.A. 1995, A & A, 295, 129
89. van den Heuvel, E.P.J., & van Paradijs, J. 1997, ApJ, 483, 399.
90. Vasisht, G., & Gotthelf, E.V. 1997, ApJ, L129
91. Verbunt, F., & van den Heuvel, E.P.J. 1995, in X-ray Binaries, ed. W.H.G. Lewin et al (Cambridge Univ. Press), p.457.
92. Vilenkin, A. 1995, ApJ, 451, 700
93. Walter, F.M. 2000, ApJ, in press (astro-ph/0009031)
94. Wang, L., Howell, D.A., Höflich, P., & Wheeler, J.C. 2000, ApJ, submitted (astro-ph/9912033)
95. Weaver, T.A., & Woosley, S.E. 1993, Phys. Rep., 227, 65
96. Wex, N., Kalogera, V., & Kramer, M. 2000, ApJ, 528, 401
97. Wu, Yanqin, Matzner, C. D., & Arras, P. 2000, ApJ, submitted (astro-ph/0006123)

# Spin and Magnetism in Old Neutron Stars

Monica Colpi[1], Andrea Possenti[2], Serge Popov[3], and Fabio Pizzolato[4]

[1] Università di Milano Bicocca, Dip. di Fisica, P.zza della Scienza 3, Milano
[2] Osservatorio Astronomico di Bologna, Via Ranzani 1, 40127 Bologna
[3] Sternberg Astronomical Institute, Universitetskii Pr. 13, 119899, Moscow
[4] Istituto di Fisica Cosmica G. Occhialini, Via Bassini 15, 20133 Milano

**Abstract.** The thermal, spin and magnetic evolution of neutron stars in the old low mass binaries is first explored. Recycled to very short periods via accretion torques, the neutron stars lose their magnetism progressively. If accretion proceeds undisturbed for 100 Myrs these stars can rotate close to break up with periods far below the minimum observed of 1.558 ms. We investigate their histories using population synthesis models to show that a tail should exist in the period distribution below 1.558 ms. The search of these ultrafastly spinning neutron stars as pulsars can help discriminating among the various equations of state for nuclear matter, and can shed light into the physics of binary evolution. The evolution of isolated neutron stars in the Galaxy is explored beyond the pulsar phase. Moving through the tenuous interstellar medium, these old solitary neutron stars lose their rotational energy. Whether also their magnetism fades is still a mystery. A population synthesis model has revealed that only a tiny fraction of them is able to accrete from the interstellar medium, shining in the X-rays. There is the hope that these solitary stars will eventually appear as faint sources in the *Chandra* sky survey. This might give insight on the long term evolution of the magnetic field in isolated objects.

## 1 Introduction

The amount of rotation and the strength of the magnetic field determine many of the neutron star's observational properties. Over the neutron star lifetime, the spin and the field change and the study of their evolution provides important clues into the physics of the stellar interior.

Of the billion neutron stars in the Galaxy, we shall be mainly concerned with the evolution of two distinct populations: The millisecond pulsars and the isolated neutron stars both aging, the first in low-mass binaries, the second as field stars moving in the interstellar medium of our Milky Way. We will show that the interaction with their surroundings may profoundly alter their spin and magnetic field. The extent of these changes and the modes vary in the two scenarios. It is in exploring this diversity that we wish to infer the nature of the equation of state and to provide a unified view of field decay. In particular, the existence of "unconventional" sources, such as *sub-millisecond pulsars*, stars rotating close to their break up limit, and as *solitary neutron stars accreting the interstellar medium,* is a crucial test for our studies. Their discovery is an observational and theoretical challenge.

In Section 2 we trace the evolution of a neutron star in the Ejector, Propeller and Accretion phases, described briefly using simple background arguments.

Section 3 surveys the physical models for the evolution of the magnetic field both in isolated and accreting systems. In Section 4 we explore possible individual pathways that may lead to the formation, in low mass binaries, of neutron stars spinning very close to their mass shedding threshold, a limit sensitive to the equation of state for nuclear matter. Pathways of isolated neutron stars follow. The first four sections set the general background used to construct, in Sections 5 and 6, statistical models aimed at determining the presence and abundance of sub-millisecond pulsars in the Galaxy, and of solitary neutron stars shining in the X-rays. Specifically, in Section 5 we explore, within the recycling scenario, the star's spin and magnetic evolution using physical models and the role played by disc instabilities in affecting the latest phases of binary evolution toward the Ejector state of sub-millisecond pulsars. In Section 6 we carry out the first stellar census. We then establish how elusive neutron stars can be as accreting sources from the interstellar medium due to their large velocities and to magnetic field decay.

## 2 The Ejector, Propeller and Accretion Phases

Over the stellar lifetime, magnetic and hydrodynamic torques acting on the neutron star (NS hereafter) induce secular changes in its spin rate. Four physical parameters determine the extent of the torques: the magnetic field strength $B$, the rotational period $P$, the density of the surrounding (interstellar) medium $n$, and the rate of mass inflow $\dot{M}$ toward the NS. According to the magnitude of these quantities, a NS experiences three different evolutionary paths: Ejector ($\mathcal{E}$), Propeller ($\mathcal{P}$), and Accretion ($\mathcal{A}$) [1]. In phase $\mathcal{E}$ the NS braking results from the loss of magneto-dipole radiation as in an ordinary pulsar. The implied spin-down rate is

$$\frac{dP}{dt} = \frac{2\pi^2}{c^3 I} \frac{\mu^2}{P} \;, \tag{1}$$

where $I$ is the NS moment of inertia and $\mu = B_s R_s^3/2$ the NS magnetic moment, function of the stellar radius $R_s$ and polar magnetic field $B_s$. The torque decreases with increasing $P$, and at the current period $P$ (much longer than the initial period) the time spent in phase $\mathcal{E}$ is

$$\tau_\mathcal{E} = \frac{c^3 I P^2}{4\pi^2 \mu^2}. \tag{2}$$

Resulting from the emission of electromagnetic waves and charged particles, the rotational energy loss of equation (1) can proceed well beyond the active radiopulsar phase and the magneto-dipolar outflow creates a hollow cavern nesting the NS. Phase $\mathcal{E}$ remains active as long as the characteristic radius $r_{st}$ of this cavern (the *stopping radius* [1] [2]) is larger than both the *light cylinder* radius

$$r_{lc} = \frac{cP}{2\pi} \tag{3}$$

and the *gravitational radius*

$$r_G = \frac{2GM}{c_s^2 + v_{\rm rel}^2} \qquad (4)$$

where $c_s$ is the sound speed of the interstellar medium (ISM) and $v_{\rm rel}$ the NS velocity relative to the medium. Phase $\mathcal{E}$ ends when matter leaks through $r_{\rm st}$, and this occurs when the period exceeds

$$P_{\mathcal{E}\to\mathcal{P}} = \frac{2\pi^{3/4}}{c\sqrt{2GM}} \mu^{1/2} v^{1/2} \varrho^{-1/4} \qquad (5)$$

where $v = (c_s^2 + v_{\rm rel}^2)^{1/2}$ and $\varrho$ is the mass density of the ISM [3]. At $P_{\mathcal{E}\to\mathcal{P}}$ the outflowing pulsar momentum flux is unable to balance the ram pressure of the gravitationally captured matter at $r_G$ [3]. Matter initiates its infall and approaches the *magnetospheric radius*

$$r_A = \left(\frac{\mu^4}{2GM\dot{M}^2}\right)^{1/7}. \qquad (6)$$

Here the magnetic pressure of the dipolar stellar field ($\propto r^{-6}$) balances the ram pressure of the infalling material ($\propto r^{-5/2}$), accreting at a rate $\dot{M}$. At $r_A$ the steeply rising magnetic field would thread the flow, enforcing it to corotation. However, not always can matter trespass this edge: penetration is prevented whenever the rotational speed $\Omega$ of the uniformly rotating magnetosphere exceeds the local Keplerian velocity $\omega_K$ $(= [GM/r_A^3]^{1/2})$ at $r_A$. This condition translates into a comparison between the *corotation radius*

$$r_{\rm cor} = \left(\frac{GM}{\Omega^2}\right)^{1/3} \qquad (7)$$

and $r_A$. When $r_{\rm cor}$ lies inside $r_A$ the magnetosphere centrifugally lifts the plasma above its escape velocity, inhibiting its further infall toward the NS surface: this is the propeller phase $\mathcal{P}$. The magnetically driven torques lead to a secular spin-down of the NS occurring at a rate

$$\frac{d}{dt}(I\Omega) = \xi \dot{M} r_{\rm cor}^2 [\omega_K(r_A) - \Omega], \qquad (8)$$

where $\dot{M}$ is the accretion rate at $r_A$, and $\xi$ a numerical factor dependent on the accretion pattern, ranging from $\xi \simeq 1$ for disc accretion to $\xi \simeq 10^{-2}$ for spherical accretion. The effectiveness of the propeller in spinning down the NS and in ejecting matter far out from the magnetosphere is still largely model dependent [4]. Recently, evidence that this mechanism is at work in binaries of high [5] and low mass [6] has come from X-ray observations.

Phase $\mathcal{P}$ terminates only when the corotation radius $r_{\rm cor}$ increases above $r_A$; thereon the NS is able to accept matter directly onto its surface: this is the accretion phase $\mathcal{A}$. The transition $\mathcal{P} \to \mathcal{A}$ occurs at a critical NS spin rate $P_{\mathcal{P}\to\mathcal{A}}$ obtained by equating $r_A$ and $r_{\rm cor}$:

$$P_{\mathcal{P}\to\mathcal{A}} = \frac{2\pi}{\sqrt{GM}} \left(\frac{\mu^4}{2GM\dot{M}^2}\right)^{3/14} \propto \frac{B^{6/7}}{\dot{M}^{3/7}}. \quad (9)$$

In isolated NS, the accretion flow is almost spherically symmetric and the NS period $P$, after the onset of $\mathcal{A}$, may vary erratically due to the positive and negative random torques resulting from the turbulent ISM [7,8] [1]

In a disc-like geometry (in binaries) the spin evolution during accretion is guided by the advective, magnetic and viscous torques present in the disc. Depending on their relative magnitude, these torques can either spin the NS up or down. The rate of change of the NS angular momentum is obtained integrating the different contributions over the whole disc, yielding

$$\frac{d}{dt}(I\Omega) \simeq \dot{M} r_A^2 \omega_K(r_A) + \int_{r_A}^{\infty} B_z^2 \frac{R^3 h}{\eta}[\omega_K(R) - \Omega] dR , \quad (10)$$

where $\eta$ is the plasma electrical diffusivity, $h$ the disc scale thickness and $R$ the radial cylindrical coordinate (in eq. 10 we neglect the local viscous contribution as it vanishes at $r_A$). The first term in the rhs of equation (10) describes the advection of angular momentum by accretion. Magnetic torques are instead non-local, resulting from an integral that extends over the whole disc: different regions can give either positive (spin-up) or negative contributions (spin-down). When $\Omega \gg \omega_K(r_A)$, the integrand in (10) is negative yielding a spin-down magnetic torque. On the other hand, if the star rotates very slowly ($\Omega \ll \omega_K(r_A)$) the overall magnetic torque is positive, leading to a secular spin-up: when the total torque is positive the spin-up process is termed *recycling*. At a critical (intermediate) value of $\Omega$, termed *equilibrium angular velocity* $\omega_{\rm eq}$, the magnetic torque is negative, and its value exactly offsets the (positive) advective torque. In this regime the NS accretes matter from the disc without exchanging angular momentum at all. The precise value of $\omega_{\rm eq}$ is rather uncertain, but it ought to lie between $0.71 - 0.95$ times the critical frequency $\omega_{\rm cr} = 2\pi/P_{\mathcal{P}\to\mathcal{A}}$, in the mid of the propeller and accretor regimes (see [9] and references therein). Since $\omega_{\rm cr}$ and $\omega_{\rm eq}$ have near values, they are often confused in the literature. There is a narrow range for $\Omega$ between $\omega_{\rm eq}$ and $\omega_{\rm cr} = 2\pi/P_{\mathcal{P}\to\mathcal{A}}$ within which the NS accretes mass but spins down.

Once the star has reached its equilibrium period $P_{\rm eq} = 2\pi/\omega_{\rm eq}$, further changes in the spin period $P$ are possible only if the magnetic field and/or $\dot{M}$ vary. Generally, field decay causes the star to slide along the so called *spin-up line* ($P = P_{\rm eq}(\mu, \dot{M})$) at the corresponding accretion rate $\dot{M}$. The spin period decreases on the time scale $\tau_\mu$, as $P_{\rm eq}$ is proportional to $\mu^{6/7}$. A large increase in the mass transfer rate can induce further spin up since the magnetosphere at

---
[1] Old accreting isolated NSs are thus expected to show strong fluctuations of $\dot{P}$ over a time scale comparable to the crossing time $\tau_{\rm fluct} \sim min(r_{\rm ISM}, r_G)/v_{\rm rel}$, where $r_{\rm ISM}$ is the spatial scale of the ISM inhomogeneities.

$r_A$ is squeezed in a region of higher Keplerian velocity. A decrease in $\dot{M}$ has the opposite effect and if there is a decline in the mass transfer rate (due to disc instabilities and/or perturbations in the atmosphere of the donor star), the NS can transit from $\mathcal{A} \to \mathcal{P} \to \mathcal{E}$. The critical period for the last transition does not coincides with equation (5), but occurs when the Alfven radius $r_A$ (function of the mass transfer rate $\dot{M}$) exceeds the light cylinder radius $r_{lc}$ (eqs. 3 and 6), i.e., when

$$P_{\mathcal{P}\to\mathcal{E}} = \frac{2\pi}{c}\left(\frac{\mu^4}{2GM\dot{M}^2}\right)^{1/7}, \qquad (11)$$

(note that this transition is not symmetric relative to $\mathcal{E} \to \mathcal{P}$). Over the stellar lifetime, NSs can trace loops moving through the various phases ($\mathcal{P} \leftrightarrow \mathcal{A}$ or/and $\mathcal{P} \leftrightarrow \mathcal{E}$). It may happen that $r_A$ increases above $r_G$ when $\dot{M}$ is very low (as in the ISM): in this case the star is in the so called Georotator state ($\mathcal{G}$).

The major drivers of the NS evolution are magnetic field decay (though amplification is an interesting possibility) and variations in the mass transfer rate: their study is thus the subject of §3 and §4..

## 3 Magnetic Field Evolution

In the $\mu$ versus $P$ diagram of pulsars (PSRs) we clearly find NSs endowed by intense magnetic fields (the canonical pulsars) and the "millisecond" pulsars (MSPs), NSs with a much lower magnetic moment. The first are isolated objects and largely outnumber the millisecond pulsars often living in binaries with degenerate companion stars. The origin of this observational dichotomy seems understood but the physical mechanisms driving the field evolution remain still uncertain. We review here current ideas for the evolution of the $B$ field. They are applied later when investigating (i) the existence of sub-millisecond pulsars, and (ii) the nature of the six isolated NSs discovered by *Rosat*.

### 3.1 Historical and Observational Outline

Soon after the discovery of PSRs, Ostriker & Gunn [10] proposed that their surface magnetic field should not remain constant but decay. Under this assumption they explained the absence of PSRs with periods much longer than one second (the decay of $\mu$ would bring the objects below the death line, turning off the active radio emission [11]) and the claimed dimming of the radio luminosity $L_R$ proportional to $\mu^2$ [12]. Their pioneering statistical study suggested that the NSs at birth have very high magnetic moments $\mu \simeq 10^{29} \div 10^{31}$ G cm$^3$, whose values decay exponentially due to ohmic dissipation in the NS interior, on a typical time-scale [13]

$$\tau_\mu = 4\sigma R^2/\pi c^2, \qquad (12)$$

where $\sigma$ is the conductivity, taken as uniform. Ostriker & Gunn evaluated $\tau_\mu$ using the electrical conductivity for the crystallized crust ($\simeq 0.6 \times 10^{23}$ sec$^{-1}$ [14]), thereby assuming implicitly that the magnetic field resides in the outer layers ($\rho < 10^{14}$ g cm$^{-3}$) of the star, and obtained $\tau_\mu \sim 4 \times 10^6$ yr. But in the same year, Baym, Pethick & Pines [15] argued that the magnetic field pervades the entire star and so one had to adopt the much higher $\sigma$ of the core, thus predicting a very long $\tau_\mu$ exceeding the Hubble time, a result confirmed [16] in 1971 when different conductivities in the NS interior were taken into account. When a larger number of pulsars became available for statistics, many authors compared the observations with population synthesis Monte Carlo models. At first [17] data seemed to support evidence for field decay, but subsequent investigations [18–21] have indicated that the observed PSR population is compatible with no decay or with decay on time-scales $\tau_\mu > 10^8$ yr longer than the pulsar phase, beyond which the NSs become practically invisible. There is a way to probe indirectly the possible decay of the field by searching for those old NSs moving with low speed ($< 40$ km s$^{-1}$) that can gravitationally capture the interstellar medium and shine as dim accreting sources. As discussed in §6, their presence in the Galaxy depends on the evolution of the field.

As far as the old MSPs are concerned, these sources possess very low values of $\mu \sim 7 \times 10^{25} - 10^{27}$ G cm$^3$ [22] at the moment they appear as pulsars. The observations are again consistent with no field decay over their radio active phase, a result inferred from the old age of the dwarf companion stars, that implies a $\tau_\mu$ at least comparable to their true galactic age [23–26]. This does not preclude from the possibility that field decay has occurred during their previous evolution: this is discussed and described in §3.4.

In the next sections we survey the models for the decay of the magnetic field both for isolated NSs and for the NSs in binaries.

## 3.2 Field Evolution in Isolated Objects: Spontaneous Decay?

Models for magnetic field decay in isolated NSs flourished over the last decade. A first improvement was to abandon the hypothesis of a uniform conductivity in the stellar crust and core; $\sigma$ increases when moving from the outer liquid layers to the deeper crystallized crust [27]. Sang & Chanmugam [28] noticed that the decay of a field residing only in the crust is not strictly exponential because of its inward diffusion toward regions of higher $\sigma$. Moreover, the decay rate depends on the (highly uncertain) depth penetrated by the initial field: the decay is more rapid if, at birth, the field resides in the outer lower density regions of the crust both because $\sigma$ is lower and because $\tau_\mu$ depends on the length scale of field gradients (less than a tenth of the stellar radius $R$ if $\mu$ is confined in the crust).

The inclusion, in the calculation, of the cooling history of the NS was a second major improvement [29–31]. It confirmed the non exponential behavior of the decay, showing in addition that a slower cooling would accelerate the decay (as a warmer star has a lower conductivity). A first short phase ( $\lesssim 10^6$ yr) of comparatively rapid decay was found, during which $\mu$ is reduced from values $\sim 6 \div 30 \times 10^{30}$ G cm$^3$, typical of young pulsars in supernova remnants, to more canonical values $\sim 1 \div 3 \times 10^{30}$ G cm$^3$ typical of normal PSRs. (During this phase the electron-phonon interaction dominates $\sigma$). A second phase of no decay was found lasting 10÷300 Myr, followed later by a power-law decay (dominated by the interaction of the electron upon the impurities) and eventually an exponential decay.

If the field penetrates the whole star, the core ohmic decay time-scale would be too long to allow decay [32]. Thus, an alternative scenario was proposed based on the hypothesis that NSs have superfluid and superconducting cores. The angular momentum of the core of a NS is believed to be carried by [33] $N_{\text{vortex}} \simeq 2 \times 10^{16}/P_{\text{sec}}$ neutron vortex lines, parallel to the spin axis; the magnetic flux should be confined in $N_{\text{fluxoid}}$ quantized proton flux tubes, where $N_{\text{fluxoid}} \simeq 10^{32} \, B_{C,13}$, with $B_{C,13}$ being the average field strength in the core, expressed in units of $10^{13}$ Gauss. In 1990, Srinivasan et al. [34] suggested that the inter-pinning [35] between these quantized entities causes the fluxoids to be carried and deposited in the crust as the NS spins down. The magnetic flux penetrated into the crust will then decay due to ohmic dissipation and the surface magnetic field of the NS keeps decreasing until it relaxes at the value of the residual field in the core. In this framework the *spin history* of the star drives and controls the changes of the magnetic field. Assuming the same radial velocity for both the fluxoids and the vortices this model implies a time-evolution

$$\mu(t) = \mu_0 \left( 1 + t/\tau_{\text{sdd}} \right)^{-1/4} \tag{13}$$

occurring on a spin down time

$$\tau_{\text{sdd}} \sim 8 \times 10^6 \, P_0^2 \, I_{45} \mu_{0,30}^{-2} \quad \text{yr,} \tag{14}$$

where index 0 denotes values at the onset of phase $\mathcal{E}$, and $I_{45}$ the NS moment of inertia in units of $10^{45}$ g cm$^2$. The equation (13) models the simplest version of the *spin down driven* (sdd) decay of $\mu$ in isolated NSs. It can be refined, accounting for the non instantaneous relaxation of the surface magnetic field to the value in the core [36].

Disappointedly, until now it is not possible to perform a reliable statistical test of all the models. They predict clearly distinct evolutionary pathways only during the early stage of the pulsar life, namely during the first $10^6$ yr after birth. Therefore a large sample of young NSs is requested for a comparison. Up to now, the pulsar catalog lists only an handful of such objects, as the observed PSR population is dominated by elder sources, with characteristic time of few $10^7$ yr. In such situation, there is no statistical clue for rejecting the hypothesis of a non decaying field. Some of the surveys in progress, as the one running at Parkes [37], dedicated to the search of young PSRs, could ultimately solve this problem.

## 3.3 Magnetic Field Evolution in Binaries: Accretion Driven Scenario

None of the models for spontaneous decay, proposed so far for the isolated objects, can explain magnetic moments of $\sim 10^{26}$ G cm$^3$, which are typical of the millisecond PSRs. In order to attain these values of $\mu$ over a Hubble time, the crustal models would require currents located only in a narrow layer below a density $\rho < 6 \times 10^{10}$ gcm$^{-3}$, a very unlikely possibility. On the other hand, the "sdd" models would predict only a modest decrease of $\mu$, at most of two orders of magnitude with respect to the initial values. However, one common feature groups the low-$\mu$ objects, i.e., that *all the PSRs with low magnetic moment spent a part of their life in a multiple stellar system*. Guided by this fact, many authors have tried to relate the low values of $\mu$ with the interaction between the NS and its companion star(s). Two different scenarios have been proposed so far: (i) The *accretion driven scenario* developed under the hypothesis of a *crustal magnetic field*; (ii) The *spin driven scenario* lying on the hypothesis of a *core magnetic field*.

In this section we consider scenario (i) taking the hypothesis that the currents generating the field are located deep in the crust [2]. A first observational suggestion about the effects of the accretion on a (crustal) magnetic field was given in 1986 by Taam & van den Heuvel [40]: they noticed an approximate dependence of the surface magnetic field $B_s$ on the amount of mass accreted onto the neutron star. Shibazaki et al. [41] later presented an empirical formula for the decay in presence of accretion:

$$\mu = \frac{\mu_0}{1 + \Delta M_{\rm acc}/m_*} \quad (15)$$

where $\mu_0$ represents the magnetic moment at the onset of accretion, $\Delta M_{\rm acc}$ is the amount of accreted mass and $m_*$ a parameter to be fitted with the observations. These authors claimed that $m_* \sim 10^{-4} M_\odot$ could reproduce the Taam & van den Heuvel's correlation. Zhang, Wu & Yang [42,43] gave physical foundation to (15) assuming that the compressed accreted matter has ferromagnetic permeability. However, using a larger database Wijers [44] showed that these models are not fully consistent with the available data both on X-ray binaries and recycled pulsars. Romani [45] first introduced the accretion rate $\dot{M}$ as a second parameter in driving the magnetic moment decay in addition to $\Delta M_{\rm acc}$. He pointed out that the accretion produces two major effects: (i) heating of the crust (depending on $\dot{M}$), which determines a reduction of the conductivity (and in turn a hastening of the ohmic decay) and (ii) advection of the field screened by the diamagnetic accreted material. As a result, the final value of $\mu$ depends both on $\Delta M_{\rm acc}$ and on $\dot{M}$. Moreover, the advection of the field lines stops when $\mu \lesssim 10^{27}$ G cm$^3$, resulting in an asymptotic value for the magnetic moment (in agreement with the observations).

The existence of a bottom field (with a predicted scaling $\mu \propto \dot{M}^{1/2}$) is also possible if the currents sustaining the surface magnetic field of the NS are neutralized by the currents developed in the diamagnetic blobs of accreted

---

[2] The onset of a thermo-magnetic instability, which transforms heat into magnetic energy at the moment of NS formation, is an effective mechanism to produce strong fields in the crust of a NS [38,39]. Although this instability is not yet completely explored for poloidal fields, it is a plausible mechanism for the origin of a crust field which does not depend on special assumptions.

matter [46]. In this case the decay of $\mu$ typically ends when the accretion disc skims the NS surface [47].

The first fully consistent theoretical calculations of the accretion-induced-decay of the crustal magnetic field of a NS were performed by Urpin, Geppert & Konenkov [48–51]. The basic physical mechanism is the diffusion (ohmic decay of currents) and advection of the magnetic field sustained by currents circulating in the non superconducting crust. The magnetic field evolution is calculated according to the induction equation:

$$\frac{\partial \boldsymbol{B}}{\partial t} = -\frac{c^2}{4\pi} \nabla \wedge \left(\frac{1}{\sigma} \nabla \wedge \boldsymbol{B}\right) + \nabla \wedge (\boldsymbol{v} \wedge \boldsymbol{B}) \qquad (16)$$

where $\boldsymbol{B}$ is the magnetic induction and $\boldsymbol{v}$ is the velocity of the moving fluid ($= 0$ in a non accreting NS, $|\boldsymbol{v}| = \dot{M}/4\pi r^2 \rho$ for a radial approximation of the accreted fluid flow). In their calculation, the superconducting NS core is assumed to expel the magnetic field (in the following we will refer to this boundary condition for the field at the crust-core interface as BCI); the induction equation is solved for a dipolar field and $\sigma$ is given as in [52].

Accretion affects the ohmic decay in two ways: (i) *heating the crust*, so reducing $\sigma$; (ii) *transporting matter and currents* toward the core-crust boundary. The relative importance of these two effects depends on other physical parameters, such as the initial depth penetrated by the currents, the impurity content ($Q$), the accretion rate, the duration of the accretion phase and the equation of state of the nuclear matter. Coupling the magnetic history of the star to its spin evolution, Urpin, Geppert & Konenkov [53] found that *neutron stars born with standard magnetic moments and spin periods can evolve to low-field rapidly rotating objects*, as illustrated in Figure 1.

The behavior of currents and the response of nuclear matter to an interior field, at the interface between the crust and the core of a NS (typical density $\simeq 2 \times 10^{14}$ g cm$^3$), is still an open issue for the theorists. In view of these uncertainties, Konar & Bhattacharya [54] chose a different boundary condition at the crust-core interface for solving the equation (16). They noticed that the deposition of accreted matter on top of the crust can imply the assimilation of original current-carrying crustal layers into the superconducting core of the NS. In particular they assumed [54] the *newly formed superconducting material to retain the magnetic flux* in the form of Abrikosov fluxoids [15] rather than to expel it through the Meissner effect (in the following this boundary condition will be labeled as BCII). Within this model, accretion produces a third effect on the evolution of $\mu$: (iii) the *assimilation of material into the core*, where the conductivity is huge, freezes the decay.

As illustrated in Figure 2, also *this hypothesis leads to a reduction of the surface magnetic field of more than 4 orders of magnitude* in $10^8$ yr, explaining the low $\mu$ and its possible freezing in the millisecond pulsar population. Likewise for BC I, it emerges that the decay of $\mu$ depends not only on the total accreted mass, but on the accretion rate itself. However, in this case the higher the accretion rate is, the stronger is the pull of crustal material into the core resulting in

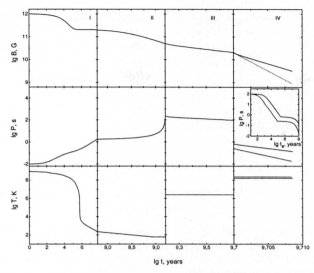

**Fig. 1.** Magnetic, rotational and thermal evolution vs time of a 1.4 $M_\odot$ NS with $\mu = 10^{30}$ G cm$^3$ at birth, as calculated in [53], under BCI. The adopted equation of state is PS (a stiff one). The initial depth of the currents is $10^{13}$ g cm$^{-3}$ and the impurity parameter $Q = 0.01$. Phase I and II refers to $\mathcal{E}$ and $\mathcal{P}$ respectively. The value of $\dot{M}$ from the stellar wind (referred to as phase III of $\mathcal{A}$) is $2/3 \times 10^{-8} \dot{M}_E$, and the lifetime of the companion star on the main sequence is $5 \times 10^9$ yr. During the disc accretion phase (as referred to IV), $\dot{M}$ is $2/3 \times 10^{-3} \dot{M}_E$ (*solid line*) or $2/3 \times 10^{-2} \dot{M}_E$ (*dotted line*). Phase IV lasts $10^8$ yr. The inserted panel enlarges the spin evolution during the very short phase at the boundary between phase III and IV, that is when disc accretion sets in.

an earlier freezing of the decay. In a subsequent paper [55], it was claimed that such a positive correlation between $\dot{M}$ and the final value of $\mu$ is supported by some observational evidence [56] [57].

Other refinements of the accretion-induced decay scenario include the effects of a non spherical symmetric accretion [58], the evolution of multipoles at the NS surface [58], the post-accretion increase of $\mu$ due to the re-diffusion of the buried field towards the surface [59], and the relativistic corrections to the decaying history [60]. All together, they produce only slight changes to the aforementioned results.

### 3.4 Magnetic Field Evolution in Binaries: Spin Driven Scenario

Both the simple idea of the flux conservation during the gravitational collapse and the action of a dynamo in the convective proto-neutron star [61] lead to the existence of a magnetic field penetrating the whole NS. Due to the huge conductivity of the core matter, no ohmic decay occurs during a Hubble time and so other effects have been invoked to account for the low field NSs.

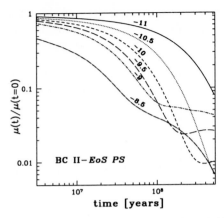

**Fig. 2.** Evolution, under BCII, of the surface magnetic field of a 1.4 $M_\odot$ NS undergoing accretion at six different levels. From top to bottom $\dot{M}_{acc} = 2/3$ of $10^{-5}; 10^{-4}; 10^{-3}; 10^{-2} \dot{M}_E$, and $1.3 \times 10^{-2}; 2/3 \times 10^{-1}$ $\dot{M}_E$. The flattening of the curves for the higher $\dot{M}$ is clearly visible [54].

Besides magneto-dipole braking, a NS experiences a longer and more significant phase of spin-down, phase $\mathcal{P}$ (described in §2) which enhances the effect of the spin-down driven scenario sketched in §3.2. This is particularly important when the NS lives in a binary system. Miri & Bhattacharya [36] and Miri [62] explored the case of low mass systems showing that at the end of $\mathcal{P}$ and $\mathcal{A}$ phases the magnetic moment $\mu$ has decayed, relative to its initial value $\mu_0$, by a factor scaling with $P_0/P_{\max}$ where $P_0$ is the initial rotational period and $P_{\max}$ the longest period attained during the phase where $\mathcal{P}$ is active ($\mathcal{P}$ results from the interaction of the companion wind with the NS). Assuming that the star spins-down to $P \lesssim 10^3$ sec, a residual magnetic moment is

$$\mu_{\text{final}} \simeq \frac{0.1 \div 1 \text{ sec}}{1000 \text{ sec}} \mu_0 \simeq 10^{-4} \div 10^{-3} \, \mu_0 \simeq 10^8 \div 10^9 \text{ G cm}^3 \qquad (17)$$

compatible with the values observed for MSP population.

Recently, Konar & Bhattacharya [63] re-examined this scenario in the case of accreting NSs, incorporating the microphysics of the crust and the material movements due to the accretion. They concluded that the model can reproduce the values of $\mu$ observed in the low-mass systems only for large values of the unknown impurity parameter $Q \gtrsim 0.05$ (in contrast with the *accretion driven decay* models demanding for a $Q \lesssim 0.01$). If the wind accretion phase is short or absent, $Q$ should even exceed unity. A more serious objection to this model was risen by Konenkov & Geppert [64] who pointed out that the motion of the proton flux tubes in the core leads to a distortion of the field structure near the crust-core interface and this in turn creates a back-reaction of the crust on the fluxoids expulsion: it results that (i) the sdd can be adequate for describing the $\mu$-evolution only for weak initial magnetic moments ($\mu_0 \lesssim 10^{29}$ G cm$^3$) and (ii) the predicted correlation $\mu_{\text{final}} \propto P_{\max}^{-1}$ is no more justified.

Finally, we have to mention another class of models in which the magnetic evolution is strictly related with the spin history of the star: in 1991 Ruderman [65,66] proposed that the coupling among $P$ and $\mu$ could take place via crustal plate tectonics. The rotational torques acting on the neutron star cause the crustal plates to migrate, thus dragging the magnetic poles anchored in them. As a consequence the effective magnetic dipole moment can strongly vary, either in intensity (decreasing or growing) and in direction, suggesting a tendency to produce an overabundance of both orthogonal (spin axis $\perp$ magnetic axis) and nearly aligned rotators (spin axis $\parallel$ magnetic axis). Current observation of disc population MSPs seem compatible with this feature [67]. Moreover the crust-cracking events could account for the glitches seen in young PSRs [68]. A more extended description of the variety of the spin driven decay models can be found for example in [69].

## 4 Evolutionary Pathways in Various Environment

As illustrated in the previous sections, the magnetic field and spin evolution in phases $\mathcal{E}$, $\mathcal{P}$ or $\mathcal{A}$ is a sensitive function of the environment. For the isolated NSs, the intrinsic velocity distribution plays, in addition, an important role. Thus, we here describe the NS evolution in the two main environments: in low mass binaries and in the ISM medium.

### 4.1 Binaries

The NSs in binaries experience a complex evolution which is tightly coupled to the orbital and internal evolution of the companion star [70]. The powerful pulsar wind initially sweeps the stellar wind away and the NS spends its lifetime in the $\mathcal{E}$ phase. With the weakening of the electro-magnetic pressure with increasing period, an extended $\mathcal{P}$ phase establishes (lasting $10^8 - 10^9$ yr) during which the NS is spun down further to periods of $\sim 10^2 - 10^4$ seconds in its interaction with the stellar wind of the companion star. When the period $P_{\mathcal{P} \to \mathcal{A}}$ is attained, accretion sets in down to the NS surface and we could possibly trace this phase identifying the X-ray emission from the wind fed NS. Here the NS can accrete with no exchange of angular momentum with the incoming matter unless the magnetic field decays due to either accretion induced or spin induced decay. The NS can slide along the corresponding equilibrium spin up line. As soon as the donor star evolve into a giant state, it can fill its Roche lobe (and matter overflows from the inner Lagrangian point forming an accretion disc. When a disc establishes (on the viscous time scale) large accretion rates are available and from this moment on recycling starts. In the Roche lobe overflow phase (RLO) the NS can be re-accelerated to millisecond periods while the field decays down to values of $10^8 - 10^9$ G cm$^3$. The mass transfer and orbital evolution are complex to model and it is now believed that a significant fraction of the mass lost by the donor does not accrete onto the NS, but is ejected from the system [71].

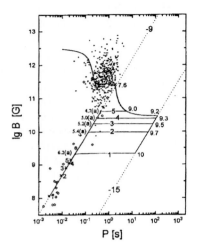

**Fig. 3.** The evolutionary tracks of NSs in the plane $B - P$ for different durations of the evolution (labeled by the numbers on the right of the curves, in $\log_{10}$ yr) during phases I($\mathcal{E}$) + II($\mathcal{P}$) + III ($\mathcal{A}$) + IV($\mathcal{A}$). The initial surface magnetic field is $3 \times 10^{12}$ G, $Q = 0.03$, and the other parameters as in Figure 1 [51].

Once reached the spin-up line at the current accretion rate, the NS loads matter acquiring angular momentum at the rate of field decay. Recycling ceases when the donor star has evolved beyond the red giant phase and a dwarf remnant is left over stellar evolution, or when the binary as become detached, so that the donor underfills its Rocche lobe [70]. If accretion ends abruptly, the NS may avoid phase $\mathcal{P}$, transiting directly to $\mathcal{E}$ and possibly re-appearing as an active radio pulsar. In low mass binaries where the donor star is a low main sequence star, evolution proceeds in a rather ordered way: from $\mathcal{E}$ to $\mathcal{P}$ (from the stellar wind), to $\mathcal{A}$ (from wind fed accretion), to $\mathcal{A}$ again (from a Keplerian disc; RLO), as shown in Figure 3. Eventually the NS tranits to $\mathcal{E}$ or $\mathcal{P}/\mathcal{E}$ when accretion halts. Observational and theoretical considerations hint for a large decay of the magnetic moment $\mu$ in phase $\mathcal{A}$, particularly during RLO, as outlined also in Figures 1 and 3. The recycling scenario will be applied in §5 to address the problem of the existence of sub-millisecond pulsars.

### 4.2 Isolated Neutron Stars

At birth, isolated NSs experience phase $\mathcal{E}$. Further evolution depends on the NS velocity relative to the ISM, the value of the ISM density and the magnetic field intensity. The large mean kick velocities acquired at birth ($\langle V \rangle \sim 300$ km s$^{-1}$, [20,21,72,73]) can reduce the extent of magnetic torques in phase $\mathcal{P}$ and can impede accretion fully (as $\dot{M} \propto v_{\rm rel}^{-3}$ and here $v_{\rm rel}$ coincides with $\langle V \rangle$). It is thus possible that phase $\mathcal{E}$ never ends. This occurs when the NS speed $V$ is as high as $V > 100 \, \mu_{30} \, n^{1/2}$ km s$^{-1}$ : This speed is derived estimating the duration of phase $\mathcal{E}$ at $P_{\mathcal{E}\to\mathcal{P}}$ (eqs. 2 and 5)

$$\tau_{\mathcal{E}\to\mathcal{P}} = \tau_{\mathcal{E}}(P_{\mathcal{E}\to\mathcal{P}}) \sim 10^9 n^{-1/2} V_{10} \mu_{30}^{-1} \text{ yr} \qquad (18)$$

imposing $\tau_{\mathcal{E}\to\mathcal{P}} \sim 10^{10}$ yr (in eq. 18, $V_{10}$ and $\mu_{30}$ are the velocity and the magnetic moment in units of 10 km s$^{-1}$ and $10^{30}$ G cm$^3$, respectively; hereon we will denote the normalizations as subscripts, for simplicity). Clearly, only the slowest NSs can loop into the accretion phase under typical ISM conditions ( where $n \lesssim 1$ cm$^{-3}$). (Molecular clouds seem a privileged site for phase $\mathcal{A}$ [74], but the probability of crossing these high density regions is relatively low and shortlived.)

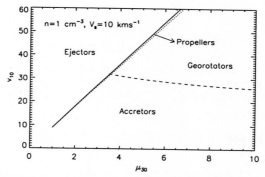

**Fig. 4.** Phases $\mathcal{E}, \mathcal{P}, \mathcal{A}$ and $\mathcal{G}$, in the $V_{10} - \mu_{30}$ plane for isolated NSs moving in a ISM of density $n = 1$ cm$^{-3}$ and a sound speed of 10 km s$^{-1}$, from [101].

Typically, in the ISM

$$P_{\mathcal{E}\to\mathcal{P}} \sim 10\, n^{-1/4} V_{10}^{1/2} \mu_{30}^{-1}\, \text{s} \quad \text{and} \quad P_{\mathcal{P}\to\mathcal{A}} \sim 2.5 \times 10^3 n^{-3/4} V_{10}^{1/7} \mu_{30}^{6/7}\, \text{s}, \qquad (19)$$

if the magnetic field is constant in time. As indicated in equations (18) and (19), the NS evolution is guided by the two key parameters, $V$ and $\mu$ (at fixed ISM's density). This is quantitatively illustrated in Figure 4 where the NS's phases are identified in the $V$ vs $B$ plane. A strong constant magnetic field ($\mu \sim 10^{30}$ G cm$^3$) implies large magneto-dipole losses that decelerate rapidly the NS to favor its entrance in phase $\mathcal{A}$ (even when $V \sim 300$ km s$^{-1}$), despite the long period of the $\mathcal{P} \to \mathcal{A}$ transition (note eq. 19).

What are the consequences of field decay on the evolution? Whether field decay enhances or reduces the probability of a transition beyond $\mathcal{E}$ is a subtle question that has recently been partly addressed [75–78]. Two competing effects come into play when $B$ decays: (i) the spin-down rate slows down because of the weakening of $B$, causing the NS to persist longer in state $\mathcal{E}$; (ii) the periods $P_{\mathcal{E}\to\mathcal{P}}$ and $P_{\mathcal{P}\to\mathcal{A}}$ instead decrease with $B$, and this acts in the opposite sense.

To explore this delicate interplay, Popov & Prokhorov [78] studied a toy model where a field is exponentially decaying on a scale $\tau_\mu$ from its initial value $\mu_0$ down to a bottom value $\mu_{bo}$. Under this hypothesis, the Ejector time scale

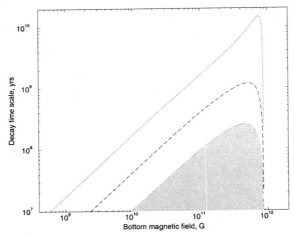

**Fig. 5.** Loci, in the $\tau_\mu - B_{\mathrm{bo}}$ plane, where phase $\mathcal{E}$ lasts longer than $10^{10}$ yr. The curves refer (from top to bottom) to initial fields of $5 \times 10^{11}$ (dotted line), $10^{12}$ (dashed line), $2 \times 10^{12}$ G (hatched region), for $v_{\mathrm{rel}} + c_s = 40$ km s$^{-1}$, and ISM density $n = 1$ cm$^{-3}$.

(with a decaying $\mu$) can be estimated analytically to give

$$\tau_{\mathcal{E},\mu} = \begin{cases} -\tau_\mu \ln\left[\dfrac{\tau_\mathcal{E}}{\tau_\mu}\left(1 + \dfrac{\tau_\mu^2}{\tau_\mathcal{E}^2}\right)^{1/2} - 1\right] & \tau_{\mathcal{E},\mu} < \tau_{\mathrm{bo}} \\ \tau_{\mathrm{bo}} + \dfrac{\mu_0}{\mu_{\mathrm{bo}}}\tau_\mathcal{E} - \dfrac{1}{2}\tau_\mu \dfrac{\mu_0}{\mu_{\mathrm{bo}}}\left(1 - e^{2\tau_{\mathrm{bo}}/\tau_\mu}\right) & \tau_{\mathcal{E},\mu} > \tau_{\mathrm{bo}} \end{cases} \quad (20)$$

where $\tau_{\mathrm{bo}} = \tau_d \ln(\mu_0/\mu_{\mathrm{bo}})$ is the time when the bottom field is attained and $\tau_\mathcal{E}$ is from equation (2) for a constant field. Figure 5 shows the loci where the time spent in phase $\mathcal{E}$ equals the age of the Galaxy, i.e., 10 Gyr, as a function of the bottom field $B_{\mathrm{bo}}$ and of the decay time-scale $\tau_\mu$. The "forbidden" region lies just below each curve. As it appears from the Figure, the interval over which the NS never leaves stage $\mathcal{E}$ is non negligible: It widens when $\mu_0$ is decreased, due to the weakness of the magneto-dipole torque. Large as well as very low bottom fields do not constrain $\tau_\mu$ and permit entrance to phase $\mathcal{A}$. When $\mu_{\mathrm{bo}}$ is very low, there is a turn-off of all magnetospheric effects on the inflowing matter and matter accretes promptly.

Similarly, Colpi et al. [75] considered a model in which the spin evolution causes the core field to migrate to the crust where dissipation processes drive ohmic field decay (Sect. 3). In this circumstance, the entrance to phase $\mathcal{A}$ is less likely if $\tau_\mu \sim 10^8$ yr, but possible otherwise. In summary, field decay can hinder the stars in phase $\mathcal{E}$ if its decay is somewhat "fine-tuned", while a fast decay would drive them into $\mathcal{A}$ promptly. The strongest theoretical argument against phase $\mathcal{A}$ remains however the high kick velocity that the NSs acquire at birth (see the review of Lai in this book). There remain nevertheless open the possibility that weak field accreting NSs exist in the Sun's vicinity.

**Table 1.** Disc accretion onto a 1.4 $M_\odot$ neutron star

| EoS | Fate | $T/W_{fin}$ | $\Delta M_B$ | $M_{G,fin}$ | $P_{fin}$ | $M_{G,max}^{static}$ | $M_{G,max}^{rotating}$ | $P_{min}^{abs}$ |
|---|---|---|---|---|---|---|---|---|
| F | Collapse | 0.034 | 0.172 | 1.52 | 0.72 | 1.46 | 1.67 | 0.47 |
| A | MassShed | 0.120 | 0.428 | 1.77 | 0.60 | 1.66 | 1.95 | 0.47 |
| E | MassShed | 0.115 | 0.414 | 1.76 | 0.66 | 1.75 | 2.05 | 0.48 |
| AU | MassShed | 0.126 | 0.446 | 1.79 | 0.70 | 2.13 | 2.55 | 0.47 |
| D | MassShed | 0.111 | 0.405 | 1.76 | 0.73 | 1.65 | 1.95 | 0.57 |
| FPS | MassShed | 0.114 | 0.416 | 1.76 | 0.75 | 1.80 | 2.12 | 0.53 |
| UT | MassShed | 0.119 | 0.429 | 1.78 | 0.75 | 1.84 | 2.19 | 0.54 |
| UU | MassShed | 0.121 | 0.436 | 1.78 | 0.78 | 2.20 | 2.61 | 0.50 |
| C | MassShed | 0.103 | 0.389 | 1.74 | 0.89 | 1.86 | 2.17 | 0.59 |
| N* | MassShed | 0.130 | 0.484 | 1.84 | 1.08 | 2.64 | 3.22 | 0.68 |
| L | MassShed | 0.116 | 0.443 | 1.80 | 1.25 | 2.70 | 3.27 | 0.76 |
| M | MassShed | 0.091 | 0.367 | 1.74 | 1.49 | 1.80 | 2.10 | 0.81 |

The second and the third columns contain the fate of the NS at the end point of recycling and the resulting ratio of kinetic energy over gravitational energy. The fourth and fifth columns report the total accreted baryonic mass and the final gravitational mass (in units of solar masses). The sixth column lists the final attained rotational periods. The seventh column collects the values of the maximum mass for a non-rotating spherical configuration, the eighth for a maximally rotating star and its corresponding minimum period for stability (ninth column). The table is from Cook, Shapiro & Teukolsky [79].

This section is devoted to the statistical study of the old NSs in binaries in the aim at exploring the recycling process and the possibility that ultra fastly spinning NSs can form in these systems. A similar approach is presented in §6 for the isolated NSs in the Galaxy.

PSR1937+21, at present, is the NS having the shortest rotational period $P_{min} = 1.558$ ms ever detected. Despite its apparent smallness, $P_{min}$ is not a critical period for NS rotation: as shown by Cook, Shapiro & Teukolsky [79], the period $P_{min}$ is longer than the limiting period, $P_{sh}$, below which the star becomes unstable to mass shedding at its equator, irrespective to the adopted equation of state (EoS) for the nuclear matter. Table 1 just shows the values of $P_{sh}$ for a set of EoS and the corresponding values of accreted baryonic mass. It shows how sensitive is $P_{sh}$ to the EoS, indicating that its determination is important for our understanding of nuclear processes in dense matter.

The other important processes which intervene in accelerating a NS up to $P_{sh}$ are the evolution of the mass transfer rate from the low mass companion star and the evolution of the NS magnetic moment $\mu$ [80]. Within the recycling scenario [81][82] all schemes suggested for the origin and the evolution of $\mu$ allow the accreting NSs to reduce their $\mu$ and $P$ to the values that are characteristic of MSPs. However, due to the large volume of the parameters in the mass transfer scenario and to the strong coupling between it and the field evolution of the NS in a low mass binary (LMB), only through a statistical approach is it possible to establish how efficient the recycling process is in spinning a NS to $P < P_{\min}$. This was recognized first by Possenti et al. [83,84], who carried on statistical analyses of NSs in the millisecond and sub-millisecond range, using a Monte Carlo population synthesis code with $\sim$ 3000 particles. It accounts for the evolution during the early phases when the NS in the LMB behaves as if isolated ($\mathcal{E}$ and $\mathcal{P}$) and later when fed by wind accretion ($\mathcal{A}$). The mass transfer during RLO (again $\mathcal{A}$) is modelled considering a range of accretion rates which is close to the one observed. The population synthesis code follows also the last radio-pulsar phase, whose duration is chosen from a flat probability distribution in the logarithm of time (see Table 2 for a summary of the parameters used in the population synthesis code). The model incorporates the detailed physics of the evolution of a crustal magnetic field (as discussed in §3.3, using BCI and BCII to mimic expulsion or assimilation of the field in the NS core), and includes the relativistic corrections [80] necessary to describe the spin-up process.

**Table 2.** Population syntheses parameters

| Physical quantity | Distribution | Values | Units |
|---|---|---|---|
| NS period at $t_0^{RLO}$ (*) | Flat | 1 $\rightarrow$ 100 | sec |
| NS $\mu$ at $t_0^{RLO}$ (*) | Gaussian | Log$<\mu_0>$=28.50 ; $\sigma$=0.32 | G cm$^3$ |
| $\dot{m}$ in RLO phase (#) | Gaussian | Log$<\dot{m}>=1.00$ ; $\sigma$=0.50 | $\dot{M}_E$ |
| Minimum accreted mass | One-value | 0.01 | $M_\odot$ |
| RLO accretion phase time (†) | Flat in Log | $10^6 \rightarrow \tau_{RLO}^{max}$(‡) | year |
| MSP phase time | Flat in Log | $10^8 \rightarrow 3 \times 10^9$ | year |

(*) $t_0^{RLO}$ = initial time of the Roche Lobe Overflow phase
(#) baryonic accretion rate during the Roche Lobe Overflow phase
(†) a Maximum accreted Mass of $0.5 M_\odot$ is permitted during the RLO phase
(‡) max duration of the RLO phase; explored values: $5\times10^7$ yr - $10^8$ yr - $5\times10^8$ yr

Evolution is followed also beyond the RLO $\mathcal{A}$ phase when accretion terminates. The increasing evidence that NSs in LMBs may suffer phases of transient accretion (perhaps due to thermal-viscous instabilities in an irradiation dominated disc [85,86]) is suggestive that mass transfer onto a NS may not stop suddenly: the star probably undergoes a progressive reduction of the mean accretion rate, modulated by phases of higher and lower accretion. This in turn can start a cycle of $\mathcal{P}$ phases which in principle could vanish the effect of the previous spin-up. With the aim of exploring the effect of a decaying $\dot{M}$ on the population of fastly spinning objects, two possibilities have been investigated: a persistent accretion for a time $\tau_{RLO}$ or a persistent accretion for a shorter time followed by a transient phase mimicking the quenching of the mass transfer. The quenching of accretion has been modelled as a power law decay for $\dot{M}$ with index $\Gamma$ varying from $1 \rightarrow 10$ (the last value is representative of an almost sudden switch off).

Figure 6 collects the fractional distribution of the recycled model NSs using the set of parameters reported in Table 2 with $\tau_{RLO}^{max} = 5 \times 10^8$ yr. Guided by the values of $P_{min}$ and $\mu_{min}$ (the shortest rotational period observed in PSR 1937+21 and the weakest magnetic moment observed in PSRJ2317+1439), the particles are divided in four groups. Those filling the first quadrant ($P \geq P_{min}$ and $\mu \geq \mu_{min}$) behave as the known MSPs. Also the objects belonging to the second quadrant ($P < P_{min}$ and $\mu \geq \mu_{min}$) should shine as PSRs [87]. The effective observability of the objects in the third quadrant ($P < P_{min}$ and $\mu < \mu_{min}$) as radio sources represents instead a challenge for the modern pulsar surveys. Most of them will be above the "death-line" [88], and might have a bolometric luminosity comparable to that of the known MSPs. Thereafter we shortly refer to *sub*-MSPs as to all objects having $P < P_{min}$ and $\mu$ above the "death-line". Objects in the fourth quadrant ($P \geq P_{min}$ and $\mu < \mu_{min}$) are probably radio quiet NSs, because they tend to be closer to the theoretical "death-line", and they are in a period range which was already searched with good sensitivity by the radio surveys.

The four crosses are for two representative EoSs (a very stiff EoS with a break-up period 1.4 ms and a mildly soft one, with $P_{sh} \simeq 0.7$ ms) and for two different boundary conditions (BCI & BCII) for the magnetic field at the crust-core interface. It appears that objects with periods $P < P_{min}$ are present in

|  | BC I | | BC II | | |
|---|---|---|---|---|---|
| **stiff** | 2% | 60% | 1% | 99% | $10^{26.8}$ Gcm³ |
|  | 6% | 32% | 0% | 0% |  |
| **soft** | 4% | 42% | 11% | 57% | $10^{25.8}$ Gcm³ |
|  | 34% | 20% | 32% | 0% |  |
|  | 1.558 ms | | 1.558 ms | | |
|  | | standard evolution | | | |

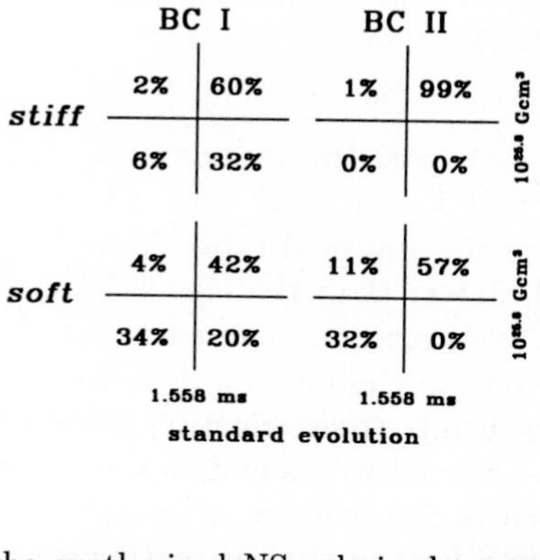

**Fig. 6.** Distributions of the synthesized NSs, derived normalizing the sample to the total number of model stars with $P < 10$ ms and arbitrary value of $\mu$. The $\mu - P$ plane is divided in four regions. As a guideline the upper left number in each cross gives the percentage of objects having $P < P_{min}$ and $\mu > \mu_{min}$ (the typical variance is about 1%).

a statistically significant number. In effect, *a tail of potential sub-MSPs always emerges* for any reliable choice of the parameters listed in Table 2.

In the synthetic sample for the *mild-soft* EoS, the "barrier" at $P_{\rm sh} \simeq 0.7$ ms is clearly visible in Figure 7 (solid lines): the *mild-soft EoS gives rise to period-distributions that increase rather steeply toward values smaller than 2 ms*, irrespective to the adopted BCs. Instead, the boundary condition affects the distribution on $\mu$: BCII produces a smaller number of objects with low field, as the field initially decays but, when the currents are advected toward the crust-core boundary, their decay is halted and the field reaches a bottom value.

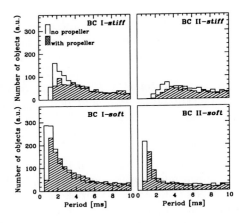

**Fig. 7.** Calculated distribution of millisecond NSs as a function of the spin period $P$ ($\mu$ is let vary over the whole range). *Solid line* denotes the distribution in absence of propeller, whilst *dashed area* denotes the case of a strong propeller effect. The absolute number of objects is in arbitrary units.

Even the *very stiff* EoS permits periods $P < P_{\rm min}$, but the "barrier" of mass shedding (at $P_{\rm sh} \simeq 1.4$ ms) is so close to $P_{\rm min} = 1.558$ ms that only few NSs reach these extreme rotational rates. Moreover the period distribution for the stiff EoS is much flatter than that for the soft-EoS, displaying a broad maximum at $P \sim 3$ ms. It was recently claimed that X-ray sources in LMBs show rotational periods clustering in the interval $2 \to 4$ ms [89,90]. This effect could be explained introducing a fine tuned relation between $\mu$ and $\dot{M}$ ($\mu \propto \dot{M}^{1/2}$ [89]). Alternatively, gravitational waves emission has been invoked [91,92]. Here we notice that such a clustering can be a natural statistical outcome of the recycling process if the EoS for the nuclear matter is stiff enough.

Many physical ingredients necessary to fully describe the propeller induced spin-down of a NS at the end of the RLO phase are poorly known or difficult to assess (e.g. the exact law for the decrease of the mass transfer rate or the efficiency in the extraction of the angular momentum from the NS to the propelled matter). Parametrizing the most uncertain quantities, the largest effect occurs for a propeller phase lasting $\sim 50\%$ of the RLO phase with a power-law index $\Gamma \sim 8$. Figure 7 reports the period distributions (dashed areas) when such a strong propeller phase is included. We note that *a strong propeller can threaten the formation of NSs with $P < P_{min}$ and $\mu > \mu_{min}$* in the case of the very stiff EoS, whilst *for the mild-soft EoS the distributions preserve a maximum just about $P_{min}$*.

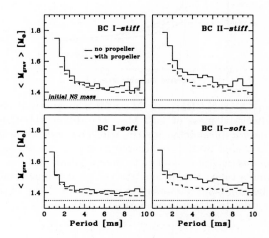

**Fig. 8.** Average gravitational masses of the re-accelerated NSs as a function of their final spin period $P$. $\mu$ varies over the whole range and the synthesized NSs are binned in 0.5 ms wide intervals. The initial mass of the static NSs is set equal to $1.35 M_\odot$ in all the cases. *Thick solid line* denotes the mass-distribution in absence of propeller, whereas *thin dashed line* with a strong propeller included.

This statistical analysis provides also information on the NS mass distribution as a function of $P$ at the end of evolution. The initial NS gravitational mass, in the evolutionary code, was set at $M_G = 1.35$ $M_\odot$ (according to the narrow Gaussian distribution with $\sigma = 0.04$ $M_\odot$ resulting from measuring the mass of the NSs in five relativistic NS-NS binary systems [93]). Figure 8 clearly shows that the observed millisecond population should have undergone a mass load of $\lesssim 0.1 M_\odot$ during recycling. This is consistent with the few estimates of the masses of millisecond pulsars in low mass binaries [94] and raises the problem of explaining how the NS can get rid of the bulk of the mass (0.5 − 1.5 $M_\odot$) released by the companion during the RLO phase [71].

Figure 8 suggests that the mass function steepens toward high values only when $P$ falls below $\sim P_{min}$, approaching $M \sim 1.7 \div 1.8$ $M_\odot$. That is a straight consequence of a results already pointed out by Burderi et al. [80]: a large mass deposition (at least $\gtrsim 0.25 M_\odot$) is required to spin a NS to ultra short periods, as illustrated in Figure 9. The action of the propeller during the evolution has the

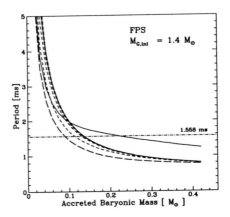

**Fig. 9.** Period versus Accreted Baryonic Mass for a magnetized NS using FPS-EoS and an initial gravitational mass of 1.40 $M_\odot$. The different pathways (a sample is represented by the *dashed lines*) define a strip, which narrows towards shorter periods. The strip is upper bounded by the evolutionary path for an unmagnetized NS (*bold solid line*). The *bold long dashed line* refers to the evolution along the spin-up line and is calculated assuming a tuned torque function (which maximizes the efficiency of the spin-up process). The *thin solid line* refers to a very slow decay of $\mu$, implying larger mass depositions in order to attain very short spin periods.

effect of reducing the mass infall: the mass distribution is only slightly affected for the mild-soft EoS, while for the stiff EoS the difference is more pronounced.

In summary, the wandering among the various phases can account for the interesting phenomenology of the MSPs observed and opens the possibility that even more extreme objects like the *sub*-MSPs exist in the Galaxy.

## 5 The NS Census in the Milky Way

The isolated NSs follow a completely different evolution pattern. Generally, over the Hubble time, they cover large portion of the Milky Way and thus explore regions where the ISM is inhomogeneous looping among the various phases erratically. With a population synthesis model, Popov *et al.*[101] traced their evolution in the ISM, and in the Milky Way potential. The model NSs, born in the galactic plane at a rate proportional to the square of the ISM density, have initially short spin periods, magnetic fields clustered around $10^{12}$ G, and spatial (kick) velocities that can be drawn for a Maxwellian distribution (with mean velocity modulus $\langle V \rangle$ treated as a parameter).

The collective properties are illustrated in Figure 10 for two initial values of $\mu_0 = 0.5 - 1 \times 10^{30}$ G cm$^3$: For a non evolving field most of isolated NSs spend their lives as Ejectors and there is no possibility to observe them. Propellers are shortlived [7], and Georotators are rare. A tiny fraction (a few percents) on the

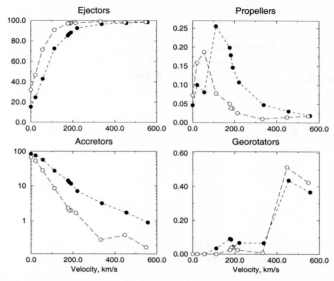

**Fig. 10.** Fractions of NSs (in percents) in phases $\mathcal{E}, \mathcal{P} \mathcal{A}$ and $\mathcal{G}$ versus the mean kick velocity $\langle V \rangle$, for a constant magnetic moment $\mu = 0.5 \times 10^{30}\,\mathrm{G\,cm^3}$ (open circles) and $\mu = 10^{30}\,\mathrm{G\,cm^3}$ (filled circles); typical statistical uncertainty for $\mathcal{E}$ and $\mathcal{A} \sim$ 1-2%.

NSs are in the Accretor stage if $\langle V \rangle$ is above 200 km s$^{-1}$. Only in the unrealistic case of a low mean velocity, the bulk of the population would be in stage $\mathcal{A}$.

As illustrated in Figure 11, in phase $\mathcal{A}$, the mean value of $\dot{M}$ is a few $10^9$ g s$^{-1}$ and, among the Accretors, the typical velocity clusters around 50 km s$^{-1}$, and the luminosity $L \sim (GM/R)\dot{M}$ around $10^{29}$ erg s$^{-1}$. Accreting isolated NSs are "visible", but they would be extremely dim objects emitting predominantly in the soft X-rays, with a polar cap black body effective temperature of 0.6 eV. What can we learn when comparing the theoretical census with the observations? We can discover if field decays in isolated objects, if the population is devoid of slow objects, and if the stars effectively spin down due to the unavoidable interaction with the tenuous ISM. The search of accreting isolated NSs is thus compelling [102–104,77].

## 5.1 Accreting Isolated Neutron Stars in the *Rosat* Sky?

Despite intensive observational campaigns, no irrefutable identification of an isolated accreting NS has been presented so far. Six soft sources have been found in *Rosat* field [77], identified as isolated NSs from the optical and X-ray data. Observations, however, do not permit to unveil the origin of their emission, yet. These sources could be powered either by accretion or by the release of internal energy in relatively young ($\approx 10^6$ yr) cooling NSs [3]. Although relatively bright

---

[3] see [77] for an updated review, the description of the sources and a complete reference list.

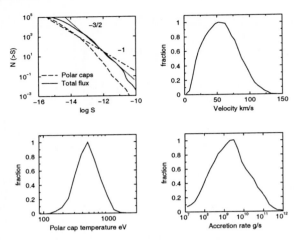

**Fig. 11.** Upper left panel: the log N-log S distribution for Accretors within 5 kpc from the Sun. Dashed (solid) curve refers to emission from polar cap (entire NS surface) in the range 0.5-2.4 keV: straight lines with slopes -1 and -3/2 are included for comparison. Fro top right to bottom right: the velocity $V$, effective polar cap temperature and accretion rate $\dot{M}$ distributions for Accretors [106]

(up to $\approx 1$ cts s$^{-1}$), the proximity of the sources (inferred from the column density in the X-ray spectra) makes their luminosity ($L \approx 10^{31}$ erg s$^{-1}$) near to that expected from either a close-by cooling NS or from an accreting NS, among the most luminous. Their X-ray spectrum is soft and thermal, again as predicted for both Accretors and Coolers.

Can a comparison with theoretical expectations help in discriminating among the two hypothesis? First, the paucity of very soft X-ray sources in the *Rosat* fields (in comparison with earlier expectation [103,104]) is indicating that Accretors are rare objects. If these six sources are indeed accreting, this implies that the average velocity of the isolated NSs population at birth has to exceed $\sim 200$ km s$^{-1}$ (a figure, which is consistent with that derived from radio pulsars statistics [105]). In addition, since observable accretion–powered isolated NSs are (intrinsically) slow objects, these results exclude the possibility that the present velocity distribution of NSs is richer in low–velocity objects with respect to a Maxwellian. Thus, despite the fact that in our Galaxy there are many old NSs (about $10^9$), the young ($\sim 10^6$ yr) cooling NSs seems to outnumber, at the *Rosat* counts, those in phase $\mathcal{A}$. This is what emerges also from the calculation of the log N-log S distribution both of cooling and accreting stars (the last carried on with "census"; [106]). At the bright counts, the local population of cooling NSs would dominate over the log N-log S of the dimmer and warmer (less absorbed) Accretors. There is the hope that *Chandra* and *Newton* will detect them, at the flux limit of $10^{-16}$ erg cm$^{-2}$ s$^{-1}$ [106]. In support of the cooling hypothesis there is also the recent measurement[107,108] of the velocity (200 km s$^{-1}$) of RXJ1856 (a member of this class) indicating that this source is, most likely, not powered by accretion having a high velocity. Interestingly, the cooling hypothesis (when explored in the log N-log S plane) implies that the NS birth rate in the Solar vicinity, over the last $10^6$ yr, is higher than that inferred from radiopulsar observations [106,109]. This might be a crucial point as it may indicate that most of the young NSs may not share the observational properties of the canonical PSRs, hinting for the presence of a background of "anomalous" young cooling NSs.

About field decay in the accretion hypothesis? Decay of the field to extreme low values would imply a large number of Accretors [101], not observed, thus excluding this possibility. This indicates that, in crustal models [28,30], electric currents need to be located deep into the crust and that the impurity parameter needs to be not exceedingly large [63], in spin-down induced models. What is still uncertain is whether the paucity is due mainly to a velocity effect than to a "fine-tuned" decay, a problem not solved yet, statistically.

While our theoretical expectation hints in favor of the cooling hypothesis there remain yet a puzzle: the long period of one of these sources, RXJ0720 [110]. If powered by accretion, RXJ0720 would have a "weak" field NS ($\mu \gtrsim 10^{26}$ G cm$^3$) and this would be a rather direct prove of some field decay, in isolated NSs [111,112]. Can a young cooling object have such a slow rotation? Do we have to change or view that NSs come to birth with ultra short periods? A new challenging hypothesis has been put forward [110,113,114] that RXJ0720 is just the descendant of a highly magnetized NS (a *Magnetar*) that have suffered a severe spin down accompanied by a nonlinear decay of the field whose energy is powering the X-ray luminosity [115]. This issue remains one of the new problems of the NS physics, making the debate on the nature of these sources an even more exciting problem.

# 6 Conclusions

In this review we have traced the evolution of old NSs transiting through the Ejector, Propeller and Accretor phases. The NSs of our review are far from being "canonical" MSPs in light binaries, or "canonical" PSRs in the field. The ultra fastly spinning NSs, that we have recycled in binaries, are rather extreme relativistic heavy NSs: If discovered, they will unable us to probe the stellar interior in an unprecedented way, and to constrain the physics driving magnetic field decay, in interacting systems. As regard to the field NSs, the discovery of the six *Rosat* sources has just opened the possibility of unveiling "unconventional" NSs, evolving in isolation. Whether they are accreting or cooling objects is still a mystery and even more mysterious and fascinating is their possible link with "Magnetars". The study of these "unusual" NSs can open new frontiers in this already active field of research.

## Acknowledgments

The authors would like to thank the European Center for Theoretical Physics and the organizers of the conference "Physics of Neutron Star Interiors" for kind ospitality during the workshop.

## References

1. Lipunov, V.M., *Astrophysics of Neutron Stars* (Springer-Verlag)
2. Shvartsman, V.F. 1971, *Soviet Astr. AJ.* 15, 337
3. Treves, A., Colpi, M. & Lipunov, V.M. 1993 *A&A*, 319
4. Lovelace, R.V.E., Romanova, M.M. & Bisnovatyi-Kogan, G. S. 1999, *ApJ*, 514, L368
5. Stella, L., White N.E. & Rosner R. 1986, *ApJ*, 308, 669
6. Campana S. et al. 1998 *ApJ*, 499, L65
7. Lipunov, V.M. & Popov, S.B. 1995, *AZh*, 71, 711
8. Konenkov, D. Yu. & Popov, S.B. 1997, *AZh*, 23 ,569
9. Li, X.-D. 1999, astro-ph/9903190
10. Ostriker, J.P. & Gunn, J.E. 1969, *ApJ*, 157, 1395
11. Ruderman, M. & Sutherland, P. 1975, *ApJ*, 196, 51
12. Lyne, A.G., Manchester, R.N. & Taylor, J.H. 1985, *MNRAS*,
13. Lamb, H. 1883, *Phil,Trans.Roy.Soc.London*, 174, 519
14. Canuto, V. 1969, *ApJ*, 159, 651
15. Baym, G., Pethick, C. & Pines, D. 1969, *Nature*, 224, 674
16. Chanmugam, G. & Gabriel, M. 1971, *A&A*, 16, 149
17. Narayan, R. & Ostriker, J.P. 1990, *ApJ*, 352, 222
18. Bhattacharya, D., Wijers, R.A.M.J., Hartman, J.W. & Verbunt, *A&A*, 254, 198
19. Wakatsuki, S., Hikita, A., Sato, N. & Itoh, N. 1992, *ApJ*, 392, 628
20. Lorimer, D.R. 1994, *Ph.D.Thesis*, University of Manchester, U.K.
21. Hartman, J.W., Verbunt, F., Bhattacharya, D., Wijers, R.A.M.J. 1997, *A&A*, 322, 477
22. Toscano, M. et al. 1999, *MNRAS*, 307, 924
23. Bhattacharya, D. & Srinivasan, G. 1986, *Curr.Sci.*, 55, 327
24. Kulkarni, S.R. 1986, *ApJ*, 306, L85
25. van den Heuvel, E.P.J., van Paradijs, J.A. & Taam, R.E. 1986, *Nature*, 322, 153
26. Verbunt, F., Wijers, R.A. & Burm, H. 1990, *A&A*, 234, 195
27. Yakovlev, D., & Urpin, V. 1980, *SvA*, 24, 303
28. Sang, Y. & Chanmugam, G. 1987, *ApJ*, 323, L61
29. Sang, Y. & Chanmugam, G. & Tsuruta, S. 1990, in *Neutron stars and their birth events*, ad. W.Kundt (Dordrecht: Kluwer), 127
30. Urpin, V.A. & Muslimov, A.G. 1992, *MNRAS*, 256, 261
31. Urpin, V.A., & Konenkov, D. 1998, *MNRAS*, 292, 167
32. Pethick, C. & Sahrling, M. 1995, *ApJ*, 453, L29
33. Ruderman, M. 1972, *ARA&A*, 10, 427
34. Srinivasan, G., Bhattacharya, D., Muslimov, A.G. & Tsygan, A.I. 1990, *Curr.Sci.*, 59, 31
35. Muslimov, A.G. & Tsygan, A.I. 1985, *Ap&SS*, 115, 43
36. Miri, M.J. & Bhattacharya, D. 1994, *MNRAS*, 269, 455
37. Camilo, F. et al. 2000, in *Pulsar Astronomy - 2000 and beyond*, Proc. of IAU Colloquium 177, ASP Conf.Series, in press

38. Urpin, V.A., Levshakov, S.A., & Yakovlev, D.G. 1986, *MNRAS* 219, 703
39. Wiebicke, H.J. & Geppert, U. 1996, *A&A*, 309, 203
40. Taam, R.E. & van den Heuvel, E.P.J. 1986, *ApJ*, 305, 235
41. Shibazaki, N., Murakami, T., Shaham, J. & Nomoto, K. 1989, *Nature*, 342, 656
42. Zhang, C.M., Wu, X.J. & Yang, G.C. 1994, *A&A*, 283, 889
43. Zhang, C.M. 1998, *A&A*, 330, 195
44. Wijers, R.A.M.J. 1997, *MNRAS*, 287, 607
45. Romani, R.W. 1990, *Nature*, 347, 741
46. Arons, J. & Lea, S.M. 1980, *ApJ*, 235, 1016
47. Burderi, L., King, A. & Wynn, G.A. 1996, *MNRAS*, 283, L63
48. Geppert, U., & Urpin, V. 1994, *MNRAS*, 271, 490
49. Urpin, V.A., & Geppert, U. 1996, *MNRAS*, 275, 1117
50. Urpin, V.A., & Geppert, U. 1996, *MNRAS*, 278, 471
51. Urpin, V.A., & Geppert, U. & Konenkov 1998, *A&A*, 331, 244
52. Possenti, A. 2000, *Ph.D. Thesis*, University of Bologna
53. Urpin, V.A., & Geppert, U. & Konenkov 1998, *MNRAS*,
54. Konar, S., & Bhattacharya, D., 1997, *MNRAS*, 284, 311
55. Konar, S., & Bhattacharya, D. 1999, *MNRAS*, 303, 588
56. Psaltis, D. & Lamb, F.K. 1997, Proc. Symp. *Neutron Stars and Pulsars*, Tokyo, Ed. N. Shibazaki, N. Kawai, S. Shibata, T. Kifune, Universal Academy Press, Inc.
57. White, N.E., & Zhang, W. 1997, *ApJL*, 490, L87
58. Sahrling, M. 1998, astro-ph/9804047
59. Young, E.J. & Chanmugam, G., *ApJ*, 442, L53
60. Sengupta, S. 1998, *ApJ*, 501, 792
61. Thompson, Ch. & Duncan, R. 1993, *ApJ*, 408, 194
62. Miri, M.J. 1996, *MNRAS*, 283, 1214
63. Konar, S., & Bhattacharya, D. 1999, *MNRAS*, 308, 795
64. Konenkov, D. & Geppert, U., astro-ph/9910492
65. Ruderman, M. 1991, *ApJ*, 382, 576
66. Ruderman, M. 1991, *ApJ*, 382, 587
67. Chen, K., Ruderman, M. & Zhu, T. 1998, *ApJ*, 493, 397
68. Ruderman, M., Zhu, T. & Chen, K. 1998, *ApJ*, 492, 267
69. Ruderman, M. 1995, *JA&A*, 16, 207
70. Bhattacharya, D., & van den Heuvel, E.P.J. 1991, Phys. Rep., 203, 1
71. Tauris, T.M., & Savonije, G.J, 1999, A&A, 350, 928
72. Hansen, B.M.S., & Phinney, E.S. 1997, MNRAS, 291, 569
73. Cordes, J.M., & Chernoff, D.F. 1997, ApJ, 482, 971
74. Colpi, M., Campana, S., & Treves, A. 1993, A&A, 278, 161
75. Colpi, M., Turolla, R., Zane, S., & Treves, A. 1998, ApJ, 501, 252
76. Livio, M., Xu, C., $ Frank, J. 1998, ApJ, 492, 298
77. Treves, A., Turolla, R., Zane, S., & Colpi, M. 2000, PASP, 112, 297
78. Popov, S.B., & Prokhorov, M.E. 2000, A&A, 357, 164
79. Cook, G.B., Shapiro S.L., & Teukolsky, S.A. 1994, ApJL, 423, L117
80. Burderi, L., Possenti, A., Colpi, M., Di Salvo, T., & D'Amico, N. 1999,
81. Alpar, M.A., Cheng, A.F., Ruderman, M.A., & Shaham, J. 1982, Nature, 300, 728
82. Bhattacharya, D. 1995, in X-ray Binaries, ed. W.H.G. Lewin, J. van Paradijs & E.P.J. van den Heuvel (Cambridge Univ. Press), 5
83. Possenti, A., Colpi, M., D'Amico, N., & Burderi, L. 1998, ApJL, 497, L97
84. Possenti, A., Colpi, M., Geppert, U., Burderi, L., & D'Amico, N. 1999, ApJS, 125, 463

85. van Paradijs, J. 1996, A&A, 464, L139
86. King, A.R., Frank, J., Kolb, U., & Ritter, H. 1997, ApJ, 484, 844
87. Burderi, L., & D'Amico, N. 1997, ApJ, 490, 343
88. Chen, K. & Ruderman, M. 1993, ApJ, 402, 264
89. White, N.E., & Zhang, W. 1997, ApJL, 490, L87
90. van der Klijs, M. 1998, in *The Many Faces of Neutron Stars*, eds. Alpar, Buccheri, van Paradijs, Dordrecht: Kluwer, 337
91. Andersson, N., Kokkotas, K.D., & Stergioulas, N. 1999, ApJ, 516, 307
92. Bildsten, L. 1998, ApJ, 501, L89
93. Thorsett, S.E., Chakrabarty, D. 1999, ApJ, 512, 288
94. Nice, D.J., Splaver, M.E., & Stairs, I.H. 2001, ApJ, in press (astro-ph/0010489)
95. Konar, S., & Bhattacharya, D. 1999, MNRAS, 303, 588
96. Miri, M.J., & Bhattacharya, D. 1994, MNRAS, 269, 455
97. Ruderman, M.A., Zhu, T.,& Chen, K 1998, ApJ, 492, 267
98. Sang, Y., & Chunmugam, G. 1987, ApJ, 323, L61
99. Srinivasan, G., *et al.* 1990, Curr.Sci. 59, 31
100. Urpin, V.A., Geppert U., & Konenkov, D. 1998, MNRAS, 295, 907
101. Popov, S.B., Colpi, M., Treves, A., Turolla, R., Lipunov, V.M., & Prokhorov, M.E. 2000, ApJ, 530, 896
102. Ostriker, J.P., Rees, M.J., & Silk, J. 1970, Astrophys. Lett., 6, 179
103. Treves, A.,& Colpi, M. 1991, A&A 241, 107
104. Blaes, O.,& Madau, P. 1993, ApJ, 403, 690
105. Lyne, A.G., Lorimer, D.R. 1994, Nature, 369, 127
106. Popov, S.B., Colpi, M., Prokhorov, M.E., Treves, A., & Turolla, R. 2000, to appear in ApJL
107. Walter, F., Matthews, L.D. 1997, Nature, 389, 358
108. Walter, F.M. 2000, astro-ph/0009031
109. Neühauser, R., & Trümper, J.E. 1999, A&A, 343, 151
110. Haberl, F. et al. 1997, A&A, 326, 662
111. Konenkov, D. Yu., & Popov S.B. 1997, Astronomy Letters, 23, 569
112. Wang, J. 1997, ApJ, 486, L119
113. Heyl, J.S., & Kulkarni, S.R. 1998, ApJ, 506, L61
114. Colpi, M., Geppert, U., & Page, D. 2000, ApJ, 529, L29
115. Thompson, C., & Duncan, R.C. 1996, ApJ, 473, 322

# Neutrino Cooling of Neutron Stars: Medium Effects

Dmitri N. Voskresensky

Moscow Institute for Physics and Engineering,
Russia, 115409 Moscow, Kashirskoe shosse 31
Gesellschaft für Schwerionenforschung GSI,
P.O.Box 110552, D-64220 Darmstadt, Germany

**Abstract.** This review demonstrates that the neutrino emission from the dense hadronic component in neutron stars is subject to strong modifications due to collective effects in nuclear matter. With the most important in-medium processes incorporated in the cooling code an overall agreement with available soft $X$ ray data can be easily achieved. With these findings so called *"standard"* and *"non-standard"* cooling scenarios are replaced by one general *"nuclear medium cooling scenario"* which relates slow and rapid neutron star coolings to the star masses (interior densities). In-medium effects play an important role also in the early hot stage of the neutron star evolution by decreasing the neutrino opacity for less massive and increasing it for more massive neutron stars. A formalism for the calculation of the neutrino radiation from nuclear matter is presented that treats on equal footing one-nucleon and multiple-nucleon processes as well as reactions with resonance bosons and condensates. The cooling history of neutron stars with quark cores is also discussed.

## 1 Introduction

The EINSTEIN, EXOSAT and ROSAT observatories have measured surface temperatures of certain neutron stars (NS) and put upper limits on the surface temperatures of some other NS (cf. [1–3] and further references therein). The data for some supernova remnants indicate rather slow cooling, while the data for several pulsars point to an essentially more rapid cooling.

Physics of NS cooling is based on a number of ingredients, among which the neutrino emissivity of the high density hadronic matter in the star core plays a crucial role. Except for the first minutes/hours when the temperature of the star is above 1 MeV, the star is transparent to neutrinos, i.e. once produced inside the star, neutrinos and antineutrinos leave without further interactions (collisions) with the ambient matter, thus carrying away the star's energy; in other words, the condition $\lambda_\nu, \lambda_{\bar\nu} \gg R$, where $\lambda_\nu, \lambda_{\bar\nu}$ are the neutrino and antineutrino mean free paths and $R$ is star radius is satisfied below the temperatures of the order of MeV. In the so called *"standard scenario"* of the NS cooling (scenario for slow cooling) the most important channel of neutrino radiation down to temperatures $T \sim 10^9$K is the modified Urca (MU) process $nn \to npe\bar\nu$. First estimates of the emissivity (energy radiation per unit time) for this reaction were carried out in [4,5]. The [6,7] recalculated the emissivity of this process in a model, where the nucleon-nucleon (NN) interaction was treated with in terms of slightly modified free one-pion exchange (FOPE).

This important result for the emissivity of the modified Urca (MU) reaction, $\varepsilon_\nu$[FOPE], turned out to be by an order of magnitude larger than the previous estimates. It was used in several computer simulations, which led to emergence of the *"standard scenario"* of neutron star cooling, see e.g. [8–10].

Besides the MU process, in the framework of the *"standard scenario"*, numerical codes included also the processes of nucleon (neutron and proton) bremsstrahlung (NB) $nn \to nn\nu\bar\nu$ and $np \to np\nu\bar\nu$, which, however, contribute less to the total luminosity of the star compared to the MU process, see [11,6]. Medium effects enter the above two-nucleon (MU and NB) rates mainly through the effective mass of the nucleons which has a smooth density dependence. Therefore in the FOPE model the density dependence of the reaction rates is rather weak and the neutrino losses of a NS depend only very weakly on its mass. As a result the *"standard scenario"*, (based on the reaction rates derived in [6]) predicts NS temperatures which agree well with the measured temperatures of several slowly cooling NS, but fails to explain the temperatures of more rapidly cooling stars. In addition the *"standard scenario"* includes processes of neutrino emission in the NS crust which become important at low temperatures.

The *non-standard scenario* of NS cooling is based on different types of direct Urca-like processes involving exotic agents (so called exotica), such as the pion Urca (PU) [12] and kaon Urca (KU) [13,14] $\beta$–decay processes and direct Urca (DU) on nucleons and hyperons [15] which are allowed only in sufficiently dense interiors of rather massive NS. The main difference in the cooling efficiency driven by the DU-like processes on the one hand and the MU and NB processes on the other hand lies in the quite different phase spaces associated with these reactions. In the case of MU and NB reactions the available phase space is the one corresponding to a two-fermion interaction (in the initial and final states of the reactions there are two baryons) while in the pion (kaon) $\beta$–decay and DU on nucleons and hyperons the phase space corresponds to one-fermion decays (as there is a single baryon in the initial and final state of the reaction).

The critical density of pion condensation in NS matter is $\varrho_{c\pi} \simeq (1 \div 3)\varrho_0$ depending on the type of condensation (neutral or charged) and the model, see [16–19]. The critical density of kaon condensation is $\varrho_{cK} \simeq (2 \div 6)\varrho_0$ depending on the type ($K^-$ or $\bar K^0$, $S$ or $P$ wave) and the model, see [20,21]. Critical density for the DU process is $\varrho_{cU} \simeq (2 \div 6)\varrho_0$ depending on the model for the equation of state (EoS), see [15,18]. Recent calculations [19] estimated the critical density of neutral pion condensation to be $2.5\varrho_0$, while for the charged pion condensation - $1.7\varrho_0$. On the other hand based on a variational calculation [18] argued that the critical densities are even for smaller ($\simeq 1.3\varrho_0$ for $\pi^0$ condensation). At the same time the EoS of [18] allows for DU process only at $\varrho > 5\varrho_0$.

There is no bridge between *"standard"* and *"non-standard"* scenarios due to complete ignorance of in-medium modifications of $NN$ interaction such as strong polarization of the soft modes (like virtually dressed pion and kaon modes mediating a part of in-medium baryon–baryon interaction). *A pion condensate may* appear only due to an enhancement of the medium polarization with increasing baryon density and it seems thereby quite inconsistent to ignore the effects of softening of the pion modes for $\varrho < \varrho_{c\pi}$, and suddenly switch on a fully developed condensate for $\varrho > \varrho_{c\pi}$.

Now let us on the basis of the results of [22–25,16,26] briefly discuss a general *"nuclear medium cooling scenario"* which treats the obvious shortcoming of the two above mentioned scenarios. First of all one observes [22] that in the nuclear matter many new reaction channels open up as compared to the vacuum. Standard Feynman technique is inadequate for calculating the in-medium reaction rates if particle widths are important, since there are no free particle asymptotic states in the matter. The summation of all perturbative Feynman diagrams where free Green functions are replaced by the in-medium ones leads to a double counting due to multiple repetitions of some processes (for an extensive discussion of this defect see [27]). This failure calls for a formalism dealing with closed diagrams (integrated over all possible in-medium particle states) constructed from full non-equilibrium Green functions. Such a formalism was elaborated in [23,24] first within quasiparticle approximation (QPA) for nucleons and was named in [24] *"the optic theorem formalism (OTF) in non-equilibrium diagram technique"*. It was demonstrated that standard calculations of the rates via squared reaction matrix elements and calculation using OTF coincide within QPA picture for the fermions. In [28] the formalism was generalized to include the effects of arbitrary particle widths. The latter formalism treats on an equal footing one-nucleon and multiple-nucleon processes as the resonance reaction contributions of the opsonic origin as well as processes with participation of zero sounds and reactions on boson condensates. Each diagram in the series with full Green functions is free from the infrared divergences. Both, the correct quasiparticle (QP) and quasiclassical limits are recovered.

Except for very early stages of NS evolution (minutes - hours) typical averaged lepton energy ($\gtrsim T$) is larger then the nucleon particle width $\Gamma_N \sim T^2/\varepsilon_{FN}$ and the nucleons can be treated within the QPA. This observation simplifies much the calculations since one can use an intuitive way of separation of the processes according to the available phase space. The one-nucleon processes have the largest emissivity (if they are not forbidden by the energy-momentum conservation laws), then two-nucleon processes provide the dominant contribution to the emissivity, etc.

In the temperature interval $T_c < T < T_{opac}$ ($T_c$ is a typical temperature for the nucleon pairing and $T_{opac}$ is typical temperature at which neutrino/antineutrino mean free path $\lambda_\nu/\lambda_{\bar\nu}$ is approximately equal to the star radius $R$) the neutrino emission is dominated by the medium modified Urca (MMU) and medium nucleon bremsstrahlung (MNB) processes if one-nucleon reactions like DU, PU and KU are forbidden, as it is the case for $\varrho < \varrho_{cU}, \varrho_{c\pi}, \varrho_{cK}$. Corresponding diagrams for MMU process are schematically shown in Fig. 1. The references [22–25,16] considered $NN$ interaction within Fermi liquid Landau–Migdal approach. They incorporated the softening of the medium one-pion exchange (MOPE) mode, other medium polarization effects, like nucleon-nucleon

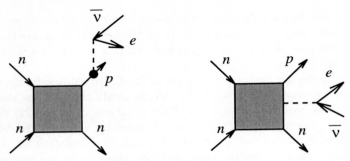

**Fig. 1.** Antineutrino emission from a nucleon leg (left graph) and from intermediate scattering states (right) in MMU process. Full dot includes weak coupling vertex renormalization.

correlations in the vertices, renormalization of the local part of $NN$ interaction by the loops, as well as the possibility of the neutrino emission from the intermediate reaction states and resonance DU-like reactions involving zero sound quanta and boson condensates.

The references [23,25,16] have demonstrated that for $\varrho \gtrsim \varrho_0$ second diagram of Fig. 1 gives the main contribution to the emissivity of MMU process rather than the first one, whose contribution has been earlier evaluated in the framework of FOPE model in [6]. This fact essentially modifies the absolute value the rate of the reaction $nn \to npe\bar{\nu}$ as well as its density, which becomes very steep. Thereby, for stars with masses larger than the solar mass the resulting emissivities were proved to be substantially larger than those values calculated in FOPE model. With an increase of the star mass (central density) pion mode continues to soften and MMU and MNB rates increase further. At $\varrho > \varrho_{c\pi}$ pion condensation begins to contribute. Actually, the condensate droplets may exist already at a smaller density if a mixed phase appears, as suggested in [29]. At $T > T_{melt}$, where $T_{melt}$ is the melting temperature, roughly $\sim$ several MeV, the mixed phase is in a liquid state and PU processes involving independent condensate droplets would be possible. At $T < T_{melt}$ condensate droplets arrange themselves in a crystalline lattice which substantially suppresses underlying neutrino processes.

The reference [12] considered the reaction $n \to p\pi_c^- e\bar{\nu}$, whereas [22,23] included other possible pion $\pi^\pm$, $\pi^0$ condensate processes on charged and neutral currents (e.g., like $n\pi_c^0 \to pe\bar{\nu}$, $n\pi_c^+ \to p\nu\bar{\nu}$ and $n\pi_c^0 \to n\nu\bar{\nu}$) as well as resonance reactions on zero sound modes which are possible also when $\varrho < \varrho_{c\pi}$. Due to the $NN$ correlations all pion condensate rates are significantly suppressed (by factors $\sim 10 - 100$ compared to the first estimate of [12], see also [30,22,23,31]. At $\varrho \sim \varrho_{c\pi}$ both the MMU and PU processes are of the same order of magnitude [23] demonstrating a smooth transition at higher densities (star masses) which is absent in the "standard" and "non-standard" scenarios.

For $T < T_c$ the reactions of neutrino pair radiation from superfluid nucleon pair breaking and formation (NPBF) shown in Fig. 2 become the dominant neutrino loss processes. The neutron pair breaking and formation (nPBF) process

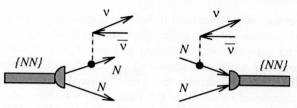

**Fig. 2.** Neutrino–antineutrino emission from Cooper pair-breaking (left graph) and pair-formation processes (right graph).

for the case of the $^1S_0$ pairing was first calculated in [32] using the standard Bogolyubov technique. Later this process was independently calculated in [24,25] to demonstrate the efficiency of OTF within the closed non-equilibrium diagram technique developed there. Moreover [24] calculated emissivity of the corresponding process involving protons (pPFB) taking into account strong coupling $p\nu\bar{\nu}$ vertex renormalization (see first diagram (18) below). This vertex renormalization increases the pPFB emissivity (for $\varrho \sim (1.5 \div 3)\rho_0$) by one-two orders of magnitude compared to what would be found with the vacuum vertex. As a result both nPFB and pPFB emissivities turn out to be of the same order of magnitude. The emissivities of NPBF processes have the same suppression factor $\sim \exp(-2\Delta/T)$ as MU, NB, MMU and MNB at $T < T_c$ but compared to the latter the NPBF processes have a large one nucleon phase space volume. [24] and then [16] also sketched how to incorporate $^3P_2$ pairing (the developed formalism easily allows to do it) and indicated that the exponential suppression of the specific heat and the emissivity should be replaced by only a power law suppression when the gap vanishes at the Fermi sphere poles. This is precisely the case for the $^3P_2(|m_J|=2)$ pairing, where $m_J$ is the projection of the total pair momentum onto quantization axis. This idea was then worked out in [33–35] without reference.

Let me make a remark about the history of the development of the understanding of these processes, since in several works (see, e.g., [36,35]) a surprise was expressed as to why these processes were not noticed/included in the cooling simulations over many years. Despite the fact that the first calculation by [32], found a correct analytic expression for the emissivity in the case of $^1S_0$ pairing of neutrons, it underestimated the numerical value of the emissivity by an order of magnitude. The authors made no statements about the possible dominance of this process over the MU and hence its importance for cooling simulations. The asymptotic behavior of their expression for emissivity

$$\varepsilon[\text{nPBF}] \sim 10^{20} T_9^7 \exp(-2\Delta/T)$$

when $T \ll \Delta$, $T_9 = T/10^9$K, [as follows from expression (1b) of [32] and from their rough asymptotic estimate of the integral (see below (13b))], does not reveal a large nucleon phase space factor (which is $\sim 10^{28}$) nor appropriate temperature behavior. Most probably the underestimation of the numerical rate of the reaction in [32] (we again point out that their analytic expression (1a) is correct) which obscured the dominance of this process over MU, became the main reason why this important result remained unnoticed during many years. The reference [24] overlooked the sign of the polarization diagram involving anomalous Green's functions for the case of $S$-wave pairing; as a result a contribution $\propto g_A^2$ ($g_A$ is axial-vector coupling constant) appeared, which should be absent for $S$-wave pairing. It gives, nevertheless, the main contribution in the case of $3_2^P$ pairing. A reasonable numerical estimate of the emissivity was presented valid for both $S$ and $P$ pairings (except case when $|m_J|= 2$) including $NN$ correlation effects. Uncertainty of this estimate is within a factor $(0.5 \div 2)$ which is allowed by variation of not very well known correlation factors. Later this estimate was then incorporated in a cooling simulation code in [26]. The correct asymptotic behavior of the emissivity is

$$\varepsilon[\text{nPFB}, \text{pPFB}] \sim 10^{28}(\Delta/\text{MeV})^7 (T/\Delta)^{1/2} \exp(-2\Delta/T)$$

for $T \ll \Delta$, which shows a large one-nucleon phase space factor and very moderate $T$-dependence of the pre-factor. In this manner the value of the reaction's rate was related in [24] to the value of the pairing gap. The possibility of the dominant role of this process (even compared to enhanced MMU and PU rather than only to MU) was unambiguously stressed. Unfortunately [24] contained a number of obvious misprints which were partially corrected in subsequent papers; although the authors take the full responsibility for these misprints, an attentive reader would have easily detected them. The importance of the NPFB processes was once more stressed in the review [16]. The first quotation of the [32] was given in [26]. The latter reference incorporated the most important in-medium effects in a cooling code, among them the nPBF and pPBF as equally important processes. The latter (pPBF) process was then rediscovered in [34], where the authors claim that its contribution to the neutrino emissivity of the star is negligible, however this is an artefact of their ignorance of correlations in weak interaction vertices. The work of [36] supported the conclusion of [26] that the NPBF processes are the dominant cooling process when $T < T_c$.

The medium modifications of all the above mentioned rates result in a pronounced density dependence (for NPBF processes mainly via dependence of the pairing gaps on the density and dependence on the $NN$ correlation factors), which links the cooling behavior of a neutron star decisively to its mass [22,23,16,26]. As a result, the above mentioned medium modifications lead to a more rapid cooling than found in the *"standard scenario"*. Hence they provide a possible explanation for the observed deviations of some of the pulsar temperatures from the predictions of the *"standard"* cooling scenario. In particular, they provide a smooth transition from the *"standard"* to the *"non-standard"* cooling with increasing central densities of the star, i.e., star masses. Thus by means of taking into account of most important in-medium effects in the reaction rates one is indeed able to achieve an appropriate agreement with both the high as well as the low observed pulsar temperatures that leads to the new *"nuclear medium cooling scenario"*. Using a collection of modern EoS for nuclear matter, which covered both relativistic as well as non-relativistic models, [26] also demonstrated the relative robustness of these in-medium cooling mechanisms against variations in the EoS of dense NS matter (for EoS that allow for a wide dense hadronic region in NS).

At the initial stage ($T > T_{opac}$) a newly formed hot NS is opaque for neutrinos/antineutrinos. Within FOPE model the value $T_{opac}$ was estimated in [6]. Elastic scatterings were included in [37,38] and pion condensation effect on the opacity was discussed in [39]. Medium effects dramatically affect the neutrino/antineutrino mean free paths, since $\lambda_{\nu(\bar{\nu})} \propto 1/ \mid M \mid^2$, where $M$ is the reaction matrix element. Thereby, one-nucleon elastic scattering processes, like $N\nu \to N\nu$, are suppressed for energy and momentum transfers $\omega < qv_{FN}$ by $NN$ correlations [24,16,40,41]. Neutrino/antineutrino absorption in two-nucleon MMU and MNB processes is substantially increased with increasing density (since $\mid M \mid^2$ for MMU and MNB processes increases with the density) [23,24,16]. Thus more massive NS are opaque for neutrinos up to lower temperatures that also results in a delay of neutrino pulse. Within the QPA for the nucleons the value $T_{opac}$ was estimated with taking into account of medium effects in [23,16]. The [42,43,16] considered possible consequences of such a delay for supernova explosions. On the other hand, at $T > (1 \div 2)$MeV one should take care of the neutrino/antineutrino radiation in multiple $NN$ scatterings (Landau–Pomeranchuk–Migdal (LPM) effect) when averaged neutrino–antineutrino energy, $\omega_{\nu\bar{\nu}} \sim$ several $T$, becomes to be smaller than the nucleon width $\Gamma_N$ [28]. Numerical evaluations of $\Gamma_N$ in application to MNB processes were done in [44] and [45] within the Brückner scheme and the Bethe–Salpeter equation, respectively. The LPM effect suppresses the rates of the neutrino elastic scattering processes on nucleons and it also suppresses MNB rates. For NS of rather low mass ($\lesssim M_\odot$) the suppression of the rates of neutral current processes due to the multiple collision coherence effect prevails over the enhancement due to the pion softening, and for sufficiently massive NS ($\gtrsim 1.4 M_\odot$) the enhancement prevails the suppression. MMU emissivity remains almost unaffected by the LPM effect since averaged $\bar{\nu}e$ energy $\sim p_{Fe}$ is rather large ($\gg \Gamma_N$).

Besides pion and kaon condensates, the possibility of the presence of the quark matter at sufficiently high baryon densities is under current discussion. These phases are expected to exist in the most massive NS. In addition, some discussion is devoted to the possibility of the existence of less massive, stable or metastable, dense self-bound stars glued by condensates or by strange quarks, cf. [46,47,16,48,49]. Uncertainty in these predictions are due to the poorly known EoS of dense matter at high baryon densities. Several possible types of star models which differ by the values of the MIT bag constant $B$ were discussed in the literature: ordinary NS without any quark core, hybrid neutron stars (HNS) with quark matter present only in their deep interiors (for intermediate values of $B$), NS with large quark cores (QCNS) surrounded by a narrow hadronic layer and a typical crust for a NS, and, finally, quark stars (QS) with a tiny crust of normal matter and/or without any crust (both for low values of $B$). If these objects are produced in supernova explosions within ordinary mechanism of blowing off the mantle, they must be rather massive HNS. If an extra support for the blowing off the matter arises, there may appear less massive objects like QCNS, QS and even objects of arbitrary size.

Special interest in cooling of HNS, QCNS, and QS is motivated by recent works [50,51] which demonstrated the possibility of diquark condensates characterized by large pairing gaps ($\Delta_q \lesssim 100$ MeV) in dense quark matter. The two-flavor (2SC) and the three-flavor (3SC) color superconducting phases allow for unpaired quarks of one color whereas in the color-flavor locking (CFL) phase all the quarks are paired. The presence of large quark gap may significantly affect the cooling history of HS, QCNS and QS. Thus it is interesting to confront this possibility to the observational data. Cooling of QCNS, and QS was discussed in [52,53] while the case of HNS was considered in [54,53]. In contrast to [52,54] more recent work [53] incorporated the heat transport which turns out to be important over a longer timescales compared to the case where the color superconducting quark matter is absent.

This review is organized as follows. Sect. 2 discusses basic ideas of the Fermi liquid approach for description of nuclear matter. The $NN$ interaction amplitude is constructed with an explicit treatment of the long-ranged soft pion mode and vertex renormalizations due to $NN$ correlations. The meaning of the pion softening effect is clarified and a comparison of MOPE and FOPE models is given. Also a renormalization of the weak interaction in NS matter is performed. Sect. 3 discusses the cooling of NS at $T < T_{opac}$. A comparison of the emissivities of MMU and MU processes shows a significant enhancement of the in-medium reaction rates. Then we discuss DU-like processes and demonstrate the medium effect due to vertex renormalizations. The role played by the in-medium neutrino radiation mechanisms in the cooling evolution of NS is then demonstrated within a realistic cooling simulation. Next we consider influence of the in-medium effects on the neutrino mean free path at initial stage of NS cooling. The essential role of the multiple $NN$ collisions is discussed. Sect. 4 presents OTF in non-equilibrium closed diagram technique in the framework of QPA for the nucleons as well as beyond the QPA incorporating genuine particle width effects. In the last Sect. 5, following the [53], we review recent results on the cooling of HNS.

## 2 Nuclear Fermi liquid description

### 2.1 $NN$ interaction. Separation of hard and soft modes

At temperatures of our interest ($T \ll \varepsilon_{Fn}$) neutrons are only slightly excited above their Fermi sea and all the processes occur in a narrow region in the vicinity of $\varepsilon_{Fn}$. In such a situation the Fermi liquid approach seems to be the most efficient one. Within this approach the diagrams which are important on the large length-scales are treated explicitly whereas those diagrams which are important on short-scales are represented by local quantities given by phenomenological, so called, Landau-Migdal (LM) parameters. Thus using the argumentation of the Fermi liquid theory [55–57,16] the retarded $NN$ interaction amplitude is presented as follows (see also [58])

$$\blacksquare = \varnothing + \varnothing\!-\!\!\blacksquare + \varnothing\!-\!\!\blacksquare \qquad (1)$$

where

$$\varnothing = \otimes + \succ\!\!\sim\!\!\prec \qquad (2)$$

The solid line represents a nucleon, whereas double-line, a $\Delta$ isobar. The double-wavy line corresponds to an exchange of a free pion with inclusion of the contributions of the residual $S$ wave $\pi NN$ interaction and $\pi\pi$ scattering, i.e. the residual irreducible interaction to the nucleon particle-holes and delta-nucleon holes. The full particle-hole, delta-nucleon hole and pion irreducible block (first block in (2)) is by its construction essentially more local than contributions given by explicitly presented graphs. Therefore, it is parameterized with the help of the LM parameters. In the standard Landau Fermi liquid theory fermions are supposed to be at their Fermi surface and the Landau parameters are further expanded in Legendry polynomials with respect to the angle between fermionic momenta. Fortunately, only zero and first harmonics enter in the physical quantities. The momentum dependence of the residual part of nuclear forces is expected to be not as pronounced and one can avoid performing this expansion. Then these parameters, i.e. $f_{nn}$, $f_{np}$ and $g_{nn}$, $g_{np}$ in scalar and spin channels respectively, are considered as weakly momentum dependent quantities. In principle, they should be calculated as functions of the density, neutron and proton concentrations, energy and momentum but as a first approximation one can use those extracted from the analysis of the experimental data on atomic nuclei.

The part of the interaction involving $\Delta$ isobar is constructed in a similar way

$$\varnothing = \otimes + \succ\!\!\sim\!\!\prec \qquad (3)$$

The main part of the $N\Delta$ interaction is due to the pion exchange. Although information on local part of the $N\Delta$ interaction is rather scarce, one can deduce [16,19] that the corresponding LM parameters are essentially smaller then those for $NN$ interaction. Apart from that, at small transferred energies $\omega \ll m_\pi$ the $\Delta$-nucleon hole contribution is a smooth function of $\omega$ and $k$ in contrast to the nucleon–nucleon hole $(NN^{-1})$ contribution, which is not For this reason, and also for simplicity, we shell neglect the first graph on the right-hand side of (3).

A straightforward resummation of (1) in neutral channel yields [24,16]

$$\Gamma^R_{\alpha\beta} = \blacksquare = C_0\left(\mathcal{F}^R_{\alpha\beta} + \mathcal{Z}^R_{\alpha\beta}\boldsymbol{\sigma}_1\cdot\boldsymbol{\sigma}_2\right) + f^2_{\pi N}\mathcal{T}^R_{\alpha\beta}(\boldsymbol{\sigma}_1\cdot\boldsymbol{k})(\boldsymbol{\sigma}_2\cdot\boldsymbol{k}), \qquad (4)$$

with external indices $\alpha,\beta$ on the vertex.

$$\mathcal{F}_{\alpha\beta}^R = f_{\alpha\beta}\gamma(f_{\alpha\beta}), \quad \mathcal{Z}_{nn}^R = g_{nn}\gamma(g_{nn}), \quad \mathcal{Z}_{np}^R = g_{np}\gamma(g_{nn}), \quad \alpha, \beta = (n, p),$$
$$\mathcal{T}_{nn}^R = \gamma^2(g_{nn})D_{\pi^0}^R, \quad \mathcal{T}_{np}^R = -\gamma_{pp}\gamma(g_{nn})D_{\pi^0}^R, \quad \mathcal{T}_{pp}^R = \gamma_{pp}^2 D_{\pi^0}^R,$$
$$\gamma^{-1}(x) = 1 - 2xC_0 A_{nn}^R, \quad \gamma_{pp} = (1 - 4gC_0 A_{nn}^R)\gamma(g_{nn}), \tag{5}$$

$f_{nn} = f_{pp} = f + f'$, $f_{np} = f - f'$, $g_{nn} = g_{pp} = g + g'$, and $g_{np} = g - g'$, dimensional normalization factor is usually taken to be $C_0 = \pi^2/[m_N p_F(\varrho_0)] \simeq 300$ MeV$\cdot$ fm$^3 \simeq 0.77 m_\pi^{-2}$, $D_{\pi^0}^R$ is the full retarded Green function of $\pi^0$, $A_{\alpha\beta}$ is the corresponding $NN^{-1}$ loop (without spin degeneracy factor 2)

$$A_{\alpha\beta} = \underset{\alpha^{-1}}{\overset{\beta}{\bigcirc}}, \quad A_{nn}(\omega \simeq q) \simeq m_n^{*2}(4\pi^2)^{-1}\left(\ln\frac{1+v_{Fn}}{1-v_{Fn}} - 2v_{Fn}\right), \tag{6}$$

$A_{nn} \simeq -m_n^* p_{Fn}(2\pi^2)^{-1}$, for $\omega \ll qv_{Fn}, q \ll 2p_{Fn}$, and we for simplicity neglect proton hole contributions because of the small concentration of protons. Resummation of (1) in the charged channel yields

$$\widetilde{\Gamma}_{np}^R = \underset{p \quad p}{\overset{n \quad n}{\bowtie}} = C_0\left(\widetilde{\mathcal{F}}_{np}^R + \widetilde{\mathcal{Z}}_{np}^R \boldsymbol{\sigma}_1 \cdot \boldsymbol{\sigma}_2\right) + f_{\pi N}^2 \widetilde{\mathcal{T}}_{np}^R (\boldsymbol{\sigma}_1 \cdot \boldsymbol{k})(\boldsymbol{\sigma}_2 \cdot \boldsymbol{k}), \tag{7}$$

$$\widetilde{\mathcal{F}}_{np}^R = 2f'\widetilde{\gamma}(f'), \quad \widetilde{\mathcal{Z}}_{np}^R = 2g'\widetilde{\gamma}(g'), \quad \widetilde{\mathcal{T}}_{np}^R = \widetilde{\gamma}^2(g')D_{\pi^-}^R,$$
$$\widetilde{\gamma}^{-1}(x) = 1 - 4xC_0 A_{np}^R. \tag{8}$$

The LM parameters are rather unknown for isospin asymmetric nuclear matter and for $\varrho > \varrho_0$. Although some evaluations of these quantities have been done, much work is still needed in order to arrive at convincing results. Therefore for estimates we will use the values extracted from experiments on atomic nuclei. Using the argumentation of a relative locality of these quantities we will suppose the LM parameters to be independent on the density for $\varrho > \varrho_0$. One then can expect that the most uncertain will be the value of the scalar constant $f$ due to the essential role of the medium-heavy $\sigma$ meson in this channel. But this parameter does not enter the tensor force channel which is the most important for our purposes. Unfortunately, there are also essential uncertainties in the numerical values of some of the LM parameters even for atomic nuclei. These uncertainties are, mainly, due to attempts to get the best fit to experimental data in each concrete case by slightly modifying parameterization used for the residual part of the $NN$ interaction. E.g., calculations of [56] gave $f \simeq 0.25$, $f' \simeq 1$, $g \simeq 0.5$, $g' \simeq 1$ whereas [59–61], by including QP renormalization of the pre-factors, derived the values $f \simeq 0$, $f' \simeq 0.5 \div 0.6$, $g \simeq 0.05 \pm 0.1$, $g' \simeq 1.1 \pm 0.1$.

Typical energies and momenta entering $NN$ interaction of our interest are $\omega \simeq 0$ and $k \simeq p_{Fn}$. Then a rough estimate yields $\gamma(g_{nn}, \omega \simeq 0, k \simeq p_{Fn}, \varrho = \varrho_0) \simeq 0.35 \div 0.45$. For $\omega = k \simeq T$ which is typical for the weak processes with participation of $\nu\bar{\nu}$ one has $\gamma^{-1}(g_{nn}, \omega \simeq k \simeq T, \varrho = \varrho_0) \simeq 0.8 \div 0.9$.

## 2.2 Virtual Pion Mode

A resummation of diagrams yields the following Dyson equation for pions

$$\sim\sim = \sim\sim + \sim\!\bigcirc\!\sim + \sim\!\bigcirc\!\!\bullet\sim + \sim\boxed{\Pi^R_{res}}\sim \quad (9)$$

The $\pi N\Delta$ full-dot-vertex includes a background correction due to the presence of higher resonances, $\Pi^R_{res}$ is the residual retarded pion self-energy that includes the contribution of all the diagrams which are not presented explicitly in (9), such as $S$ wave $\pi NN$ and $\pi\pi$ scatterings (included by double-wavy line in (2)). The full vertex takes into account $NN$ correlations

$$\blacktriangleright\!\sim = \blacktriangleright\!\sim + \bigotimes\!\blacktriangleright\!\sim . \quad (10)$$

Due to this fact the nucleon particle-hole part of $\Pi_{\pi^0}$ is $\propto \gamma(g_{nn})$ and the nucleon particle-hole part of $\Pi_{\pi^\pm}$ is $\propto \gamma(g')$. The value of the $NN$ interaction in the pion channel is determined by the full pion propagator at small $\omega$ and $k \simeq p_{Fn}$, i.e. by the quantity $\widetilde{\omega}^2(k) = -(D^R_\pi)^{-1}(\omega = 0, k, \mu_\pi)$. Typical momenta of our interest are $k \simeq p_{Fn}$. Indeed the momenta entering the $NN$ interaction in MU and MMU processes are $k = p_{Fn}$, the momenta governing the MNB are $k = k_0$ [23] where the value $k = k_0 \simeq (0.9 \div 1)p_{Fn}$ corresponds to the minimum of $\widetilde{\omega}^2(k)$. The quantity $\widetilde{\omega} \equiv \widetilde{\omega}(k_0)$ has the meaning of the *effective pion gap*. It is different for $\pi^0$ and for $\pi^\pm$ since neutral and charged channels are characterized by different diagrams permitted by charge conservation, thus also depending on the value of the pion chemical potential, $\mu_{\pi^+} \neq \mu_{\pi^-} \neq 0$, $\mu_{\pi^0} = 0$. For $T \ll \varepsilon_{Fn}, \varepsilon_{Fp}$, one has $\mu_{\pi^-} = \mu_e = \varepsilon_{Fn} - \varepsilon_{Fp}$, as follows from equilibrium conditions for the reactions $n \to p\pi^-$ and $n \to pe\bar{\nu}$.

A change of the sign of $\widetilde{\omega}^2$ signals a phase transition to a pion condensate.

The typical density dependence of $\widetilde{\omega}^2$ is shown in Fig. 3. At $\varrho < (0.5 \div 0.7)\varrho_0$, one has $\widetilde{\omega}^2 = m_\pi^2 - \mu_\pi^2$. For such densities the value $\widetilde{\omega}^2(p_{Fn})$ essentially deviates from $m_\pi^2 - \mu_\pi^2$ tending to $m_\pi^2 + p_{Fn}^2 - \mu_\pi^2$ in small density limit.

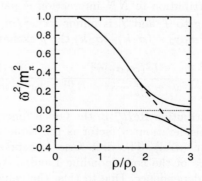

**Fig. 3.** Effective pion gap (for $\mu_{\pi^0} = 0$) versus baryon density, see [16].

At the critical point of the pion condensation ($\varrho = \varrho_{c\pi}$) the value $\widetilde{\omega}^2$ changes its sign when the $\pi\pi$ fluctuations are artificially switched off (dashed line in Fig. 3).

In reality $\pi\pi$ fluctuations are significant in the vicinity of the critical point and there occurs a first-order phase transition to a inhomogeneous pion-condensate state [62–64]. Therefore there are two branches (solid curves in Fig. 3) with positive and respectively negative values for $\widetilde{\omega}^2$. The calculations of [64] demonstrated that at $\varrho > \varrho_{c\pi}$ the free energy of the state with $\widetilde{\omega}^2 > 0$, where the pion mean field is zero, becomes larger than that of the corresponding state with $\widetilde{\omega}^2 < 0$ and a finite mean field. Therefore at $\varrho > \varrho_{c\pi}$ the state with $\widetilde{\omega}^2 > 0$ is metastable and the state with $\widetilde{\omega}^2 < 0$ and the pion mean field $\varphi_\pi \neq 0$ becomes the ground state.

The quantity $\widetilde{\omega}^2$ demonstrates how much the virtual (particle-hole) mode with pion quantum numbers is softened at a given density. The ratio $\alpha = D_\pi[\text{med.}]/D_\pi[\text{vac.}] \simeq 6$ for $\varrho = \varrho_0$, $\omega = 0$, $k = p_{FN}$ and for isospin symmetric nuclear matter. However this essential so called *"pion softening"* [57] does not significantly enhance the $NN$ scattering cross section because of a simultaneous essential suppression of the $\pi NN$ vertex by $NN$ correlations. Indeed, the ratio of the $NN$ cross sections calculated with FOPE and MOPE is

$$R = \frac{\sigma[\text{MOPE}]}{\sigma[\text{FOPE}]} \simeq \frac{\gamma^4(g',\omega \simeq 0, k \simeq p_{FN})(m_\pi^2 + p_{FN}^2)^2}{\widetilde{\omega}^4(p_{FN})}, \quad (11)$$

and for $\varrho = \varrho_0$ we have $R \lesssim 1$, whereas for $\varrho = 2\varrho_0$ we already get $R \sim 10$.

As follows from numerical estimates of different $\gamma$ factors entering (4) and (7), the main contribution to $NN$ interaction for $\varrho > \varrho_0$ is given by MOPE

$$\mathbf{I} \simeq \mathbf{\mathsf{X}} \quad (12)$$

whether this channel ($\mathcal{T} \propto (\boldsymbol{\sigma}_1 \cdot \boldsymbol{k})(\boldsymbol{\sigma}_2 \cdot \boldsymbol{k})$) of the reaction is not forbidden or suppressed by specific reasons as symmetry, small momentum transfer, etc.

The $\varrho$ meson contribution to $NN$ interaction is partially included in $g_{\alpha\beta}$, other part contributing to $\mathcal{T}$ and $\widetilde{\mathcal{T}}$ is minor ($\propto \widetilde{\omega}^2/m_\varrho^2$). Indeed, using that $(\boldsymbol{\sigma}_1 \times \boldsymbol{k})(\boldsymbol{\sigma}_2 \times \boldsymbol{k}) = k^2\boldsymbol{\sigma}_1\boldsymbol{\sigma}_2 - (\boldsymbol{\sigma}_1\boldsymbol{k}) \cdot (\boldsymbol{\sigma}_2\boldsymbol{k})$ the $\rho$ exchange can be cast as

$$\delta\Gamma^R_{N_1 N_2} = \left(\frac{f_\rho}{m_\rho}\right)^2 \left\{ \frac{k^2(\boldsymbol{\sigma}_1\boldsymbol{\sigma}_2)\gamma^2}{[\omega^2 - m_\rho^2 - k^2 - \Pi^R_\rho]} - \frac{(\boldsymbol{\sigma}_1\boldsymbol{k})(\boldsymbol{\sigma}_2\boldsymbol{k})\gamma^2}{[\omega^2 - m_\rho^2 - k^2 - \Pi^R_\rho]} \right\}. \quad (13)$$

We may omit the contribution $\text{Re}\Pi^R_\rho$ in the Green function since $\text{Re}\Pi^R_\rho \ll m_\rho^2$ for $\rho \sim m_\pi^3$ under consideration; $\gamma$ factor is the same as for the corresponding pion (charged or neutral). The first term is supposed to be included in phenomenological value of the corresponding Landau–Migdal parameter leading to its momentum dependence. Due to this, the value of $g'$ gets a 30% decrease (rather then an increase discussed in some works) with the momentum at $k = p_{FN}$ compared to the corresponding value at $k = 0$, see [65]. Thus one can think that $g'(0) > g'(p_{FN})$. Second term can be dropped since it is small ($\lesssim 1/30$ of MOPE contribution at $\rho = \rho_0$).

Another concern [66,67] was expressed in connection with the quasielastic polarization transfer experiment at LAMPF, EMC experiment and the Drell–Yan experiment at Fermi Lab, all of which did not observe, as predicted in several works, a pronounced pion excess in nuclei. First optimistic estimates demonstrated a pion excess at (15-30)% level which was ruled out by different models that analyzed the above mentioned experimental results including an artificially enhanced pion contribution and suppressing other important contributions. However it may well be that only in a very narrow vicinity of $\varrho_{c\pi}$ (far beyond $\varrho_0$) pion fluctuations grow at $T = 0$ and only at $T \neq 0$ pion excess must essentially grow in substantially wider density interval [63,64].

The physical reason of a contribution of virtual pions at $\omega < kv_F$ is clear. It is the well known Landau damping associated with possibility of the virtual pion decay to a nucleon particle-hole. As known from $\pi N$ scattering experiments, pions interact with nucleons. Thus there must be a contribution of virtual pions to the sum-rule for pion spectral function. There is no other way out. Therefore, to a question "Where virtual pions are?" we would still suggest an old fashioned answer: "They are hidden inside the matter". The only question which remains is: "What is the actual value of the pion excess?" To find a proper answer one needs to elaborate, besides the pionic contribution, all relevant specific contributions related to the phenomenon analyzed in the given concrete experiment. E.g., quasielastic polarization transfer data are reasonably fitted with a slightly increased value of the Landau–Migdal parameter $g'$ and with inclusion of $S$ wave repulsion at small energies [16,68,69], EMC experiment is reproduced with an inclusion of the soft pion mode and other relevant effects [70].

Thus instead of FOPE+$\varrho$ exchange, as a model of $NN$ interaction, which was used by [6] in their calculation of the emissivities of the two-nucleon reactions, one should use the full $NN$ interaction given by (4) and (7) or, in a simplified version, one should use its approximation by the MOPE part only.

## 2.3 Renormalization of the weak interaction

The full weak coupling vertex that takes into account the $NN$ correlations is determined by (10) where now the wavy line should be replaced by a lepton pair. Thus for the vertex of our interest, $N_1 \to N_2 l\bar{\nu}$, we obtain [24,16]

$$V_\beta = \frac{G}{\sqrt{2}} \left[ \widetilde{\gamma}(f') l_0 - g_A \widetilde{\gamma}(g') \boldsymbol{l}\boldsymbol{\sigma} \right], \tag{14}$$

for the $\beta$ decay and

$$V_{nn} = -\frac{G}{2\sqrt{2}} \left[ \gamma(f_{nn}) l_0 - g_A \gamma(g_{nn}) \boldsymbol{l}\boldsymbol{\sigma} \right], \quad V_{pp}^N = \frac{G}{2\sqrt{2}} \left[ \kappa_{pp} l_0 - g_A \gamma_{pp} \boldsymbol{l}\boldsymbol{\sigma} \right], \tag{15}$$

$$\kappa_{pp} = c_V - 2f_{np}\gamma(f_{nn})C_0 A_{nn}, \quad \gamma_{pp} = (1 - 4gC_0 A_{nn})\gamma(g_{nn}), \tag{16}$$

for processes on the neutral currents $N_1 N_2 \to N_1 N_2 \nu \bar{\nu}$, $V_{pp} = V_{pp}^N + V_{pp}^\gamma$, $G \simeq 1.17 \cdot 10^{-5}$ Gee$^{-2}$ is the Fermi weak coupling constant, $c_V = 1 - 4\sin^2\theta_W$, $\sin^2\theta_W \simeq 0.23$, $g_A \simeq 1.26$ is the axial-vector coupling constant, and $l_\mu = \bar{u}(q_1)\gamma_\mu(1 - \gamma_5)u(q_2)$ is the lepton current. The pion contribution $\sim q^2$ is small for typical $|q| \simeq T$ or $p_{Fe}$, and for simplicity is omitted. The in medium the value of $g_A$ (i.e. $g_A^*$) slightly decreases with the density which can be easily incorporated, cf. the Brown–Rho scaling idea [71]. Please notice that with a decrease of $g_A^*$, in particular, the value $\varrho_{c\pi}$ increases remaining however finite ($\simeq 2\varrho_0$ according to [72] where this decrease of $g_A^*$ up to 1 was discussed) due to attractive contribution of the $\Delta$ isobar in (9), whereas one would expect $\varrho_{c\pi} \to \infty$ for $g_A^* \to 1$ ignoring $\Delta$ contribution.

The $\gamma$ factors renormalize the corresponding vacuum vertices. These factors are essentially different for different processes involved. The matrix elements of the neutrino/antineutrino scattering processes $N\nu \to N\nu$ and of MNB behave differently in dependence on the energy-momentum transfer and whether $N = n$ or $N = p$ in the weak coupling vertex. Vertices

 (17)

are modified by the correlation factors (5) and (8). For $N = n$ these are $\gamma(g_{nn}, \omega, q)$ and $\gamma(f_{nn}, \omega, q)$ which lead to an enhancement of the cross sections for $\omega > qv_{Fn}$ and to a suppression for $\omega < qv_{Fn}$. The renormalization of the proton vertex (vector part of $V_{pp}^N + V_{pp}^\gamma$) is governed by the processes [24,73]

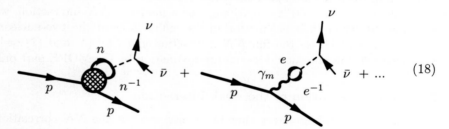 (18)

which are forbidden in the vacuum. For the systems with $^1S_0$ proton–proton pairing, $\propto g_A^2$ contribution to the squared matrix element (see (15)) is compensated by the corresponding contribution of the diagram with anomalous Green functions of protons. The vector current term is $\propto c_V^2$ in vacuum whereas it is $\propto \kappa_{pp}^2$ in medium (according to the first graph (18)). Thereby the corresponding vertices with participation of proton are enhanced in medium compared to their small vacuum value ($\propto c_V^2 \simeq 0.006$) which leads to an enhancement of the cross sections, up to $\sim 10 \div 10^2$ times for $1.5 \div 3\varrho_0$ depending on parameter choice. It does not contradict to the statement of [24] that correlations are rather suppressed in the weak interaction vertices. at $\varrho \leq \varrho_0$. The enhancement with the density comes from estimate (6) that directly follows from equation (2.30) of [24]. Also the enhancement factor (up to $\sim 10^2$) comes from the virtual in-medium photon ($\gamma_m$) whose propagator contains $1/m_\gamma^2 \sim 1/e^2$, where $m_\gamma$ is the effective spectrum gap, that compensates small $e^2$ factor from electromagnetic vertices, see [73]. It is included by means of the replacement $V_{pp}^N \to V_{pp}$. Other processes permitted in intermediate states like processes with $pp^{-1}$ and with the pion are suppressed by a small proton density and by $q^2 \sim T^2$ pre-factors, respectively. First diagram (18) was considered in [24], where the pPBF process was suggested, and then in [16,26], and it was shown that nPBF and pPBF processes may give contributions of the same order of magnitude. Finally, with electron and nucleon correlations included, we recover this statement and numerical estimate [24,26] (in spite of the mentioned misprints in the result [24]). Several subsequent papers [34,74,35] rediscovered pPBF process, however ignored the nucleon and electron correlation effects. (Although the authors were personally informed, they continued to insist [75] on their incorrect treatment, giving priority to their result.) Contribution of second diagram for pPFB process was recently incorporated in [76].

The paper [27] gives another example demonstrating that, although the vacuum branching ratio of the kaon decays is $\Gamma(K^- \to e^- + \nu_e)/\Gamma(K^- \to \mu^- + \nu_\mu) \approx 2.5 \times 10^{-5}$, in medium (due to $\Lambda p^{-1}$ decays of virtual $K^-$) it becomes of the order of unity. Thus we again see that, depending on which reaction channel is considered, in-medium effects may either strongly enhance the reaction rates or substantially suppress them. The ignorance of these effects may lead to misleading results; one may loose order of magnitude factors, while struggling for a numerical factors $\sim 1$.

## 2.4 Inconsistencies of FOPE model

Since FOPE model became the basis of the *"standard scenario"* of cooling simulations we would like first to demonstrate the principal inconsistencies of the model for the description of interactions in dense ($\varrho \gtrsim \varrho_0$) baryon medium. The only diagram in FOPE model which contributes to the MU and NB is

$$\tag{19}$$

Dots symbolize FOPE. This is the first non-zero Born approximation diagram, i.e. second order perturbative contribution in $f_{\pi N}$. In order to be theoretically consistent one should use perturbation theory to the very same second order in $f_{\pi N}$ for all the quantities. E.g., pion spectrum is then determined by pion polarization operator expanded up to the very same order in $f_{\pi N}$

$$\omega^2 \simeq m_\pi^2 + k^2 + \Pi_0^R(\omega, k, \varrho), \quad \Pi_0^R(\omega, k, \varrho) = \tag{20}$$

The value of $\Pi^0(\omega, k, \varrho)$ is easily calculated and the result contains no uncertain parameters. For $\omega \to 0$, $k \simeq p_F$ of interest to us and for isospin symmetric matter

$$\Pi_0^R \simeq -\alpha_0 - \mathrm{i}\beta_0\omega, \quad \alpha_0 \simeq \frac{2m_N p_F k^2 f_{\pi N}^2}{\pi^2} > 0, \quad \beta_0 \simeq \frac{m_N^2 k f_{\pi N}^2}{\pi} > 0. \qquad (21)$$

Substituting this expression in (20) we obtain a solution with $\mathrm{i}\omega < 0$ already for $\varrho > 0.3\varrho_0$, that would mean appearance of the pion condensation. Indeed, the mean field begins to increase with the time as $\varphi \sim \exp(-\mathrm{i}\omega t) \sim \exp(\alpha t/\beta)$ until repulsive $\pi\pi$ interaction will not stop its growth. But it is experimentally proven that there is no pion condensation in atomic nuclei, i.e. even at $\varrho = \varrho_0$. The puzzle is resolved as follows. The FOPE model does not work for such densities. One should replace the FOPE by the full $NN$ interaction given by (4), (7). The essential part of this interaction is due to MOPE with vertices corrected by $NN$ correlations. Also the $NN^{-1}$ part of the pion polarization operator is corrected by $NN$ correlations. Thus

$$\text{\raisebox{-0.3em}{\includegraphics[height=1.5em]{diagram}}} \simeq \Pi_0^R(\omega, k, \varrho)\gamma(g', \omega, k, \varrho) \qquad (22)$$

is suppressed by the factor $\gamma(g', \omega = 0, k \simeq p_F, \varrho \simeq \varrho_0) \simeq 0.35 \div 0.45$. The final solution of the dispersion relation (20), now with full $\Pi$ instead of $\Pi^0$, yields $\mathrm{i}\omega > 0$ for $\varrho = \varrho_0$ whereas the solution with $\mathrm{i}\omega < 0$, which shows the beginning of pion condensation, appears only for $\varrho > \varrho_{c\pi} > \varrho_0$.

## 3 Neutrino cooling of neutron stars

### 3.1 Emissivity of MMU process

Since DU process is forbidden up to sufficiently high density $\varrho_{cU}$, the main contribution for $\varrho < \varrho_{cU}$ and $T_{opac} > T > T_c$ comes from MMU processes schematically presented by the two diagrams of Fig. 1. MNB reactions give smaller contribution [23]. For densities $\varrho \ll \varrho_0$ the main part of the $NN$ interaction amplitude is given by the residual $NN$ interaction. In this case the $NN$ interaction amplitude can be better treated within the $T$ matrix approach which sums up the ladder diagrams in the particle-particle channel rather than by LM parameters. Calculations of MNB processes with the vacuum $T$ matrix [77] found essentially smaller emissivity than that given by the FOPE. Also the in-medium modifications of the $T$ matrix additionally suppress the rates of both MMU and MNB processes, see [78]. Thus even at small densities the FOPE model may give only a rough estimate of the emissivity of two nucleon processes. At $\varrho \gtrsim (0.5 \div 0.7)\varrho_0$ the reactions in particle-hole channel and more specifically those with participation of the soft pion mode begin to dominate.

The evaluations of [23,42,43,16] showed that the dominating contribution to MMU rate comes from the second diagram of Fig. 1, namely from contributions to it given by the first two diagrams of the series

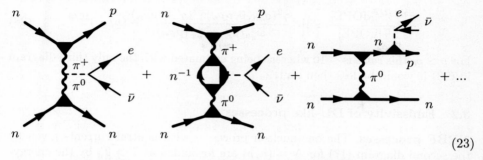

(23)

whereas the third diagram, which naturally generalizes the corresponding MU(FOPE) contribution, gives only a small correction for $\varrho \gtrsim \varrho_0$. The emissivity from the two first diagrams in a simplified notation [16,26] reads

$$\varepsilon^{MMU}[\text{MOPE}] \simeq 2.4 \cdot 10^{24}\, T_9^8 \left(\frac{\varrho}{\varrho_0}\right)^{10/3} \frac{(m_n^*)^3 m_p^*}{m_N^4} \left[\frac{m_\pi}{\alpha \widetilde{\omega}_{\pi^0}(p_{Fn})}\right]^4$$

$$\times \left[\frac{m_\pi}{\alpha \widetilde{\omega}_{\pi^\pm}(p_{Fn})}\right]^4 \Gamma^8\, F_1 \zeta(\Delta_n)\, \zeta(\Delta_p)\, \frac{\text{erg}}{\text{cm}^3\,\text{sec}}\,, \quad (24)$$

where $T_9 = T/10^9$ K is the temperature, $m_n^*$ and $m_p^*$ are the nonrelativistic effective neutron and proton masses, and the correlation factor $\Gamma^8$ is roughly

$$\Gamma^8 \simeq \gamma_\beta^2(\omega \simeq p_{Fe}, q \simeq p_{Fe}) \gamma^2(g_{nn}, \omega \simeq 0, k = p_{Fn}) \widetilde{\gamma}^4(g', 0, p_{Fn}),$$

$$\gamma_\beta^2(\omega, q) = \frac{\widetilde{\gamma}^2(f', \omega, q) + 3 g_A^2 \widetilde{\gamma}^2(g', \omega, q)}{1 + 3 g_A^2}\,, \quad (25)$$

while the second term in the prefactor,

$$F_1 \simeq 1 + \frac{3}{4 \widetilde{\gamma}^2(g', 0, p_{Fn}) \gamma_\beta^2(\omega \simeq p_{Fe}, q \simeq p_{Fe})} \left(\frac{\varrho}{\varrho_0}\right)^{2/3}, \quad (26)$$

is the contribution of the pion decay from intermediate states (first diagram (23)). The quantity $\Gamma$ effectively accounts for an appropriate product of the $NN$ correlation factors in different $\pi N_1 N_2$ vertices. For charged pions the value $\mu_\pi \neq 0$ is incorporated in the expression for the effective pion gap, for neutral pions $\mu_\pi = 0$. The value $\alpha \sim 1$ depends on condensate structure, $\alpha = 1$ for $\varrho < \varrho_{c\pi}$, and $\alpha = \sqrt{2}$ taking account of the new excitations on the ground of the charged $\pi$ condensate vacuum for $\varrho > \varrho_{c\pi}$. The factor

$$\zeta(\Delta_N) \simeq \begin{cases} \exp(-\Delta_N/T) & T \leq T_{cN}, \\ 1 & T > T_{cN}, \end{cases} \quad N = (n, p) \quad (27)$$

estimates the suppression caused by the $nn$ and $pp$ pairings. Deviation of these factors from simple exponents can be incorporated as in [35].

The ratio of the emissivities of MMU(MOPE) and MU(FOPE) is roughly

$$\frac{\varepsilon^{MMU}[\text{MOPE}]}{\varepsilon^{MU}[\text{FOPE}]} \simeq 10^3 \frac{\gamma^2(g_{nn},0,p_{Fn})\widetilde{\gamma}^2(g',0,p_{Fn})}{\widetilde{\omega}_{\pi^0}^4(p_{Fn})\widetilde{\omega}_{\pi^\pm}^4(p_{Fn})} (\varrho/\varrho_0)^{10/3}. \quad (28)$$

For $\varrho \simeq \varrho_0$ this ratio is $\sim 10$ whereas being estimated with the only third diagram in (23) it would be less then unit.

## 3.2 Emissivity of DU-like processes

**NPBF processes.** The one-nucleon processes with neutral currents given by the second diagram (17) for $N = (n,p)$ are forbidden at $T > T_c$ by the energy-momentum conservations but they can occur at $T < T_c$. Then physically the processes relate to NPBF, see Fig. 2. However they need special techniques to be calculated [32,24]. These processes $n \to n\nu\bar{\nu}$ and $p \to p\nu\bar{\nu}$ play very important role in the cooling of superfluid NS, see [24,25,16,26,36,35]. The emissivity for 3 types of neutrinos is given by [24,25]

$$\varepsilon[\text{nPBF}] = \frac{3 \cdot 4G^2 \left(\xi_1\gamma^2(f_{nn}) + \xi_2 g_A^{*2}\gamma^2(g_{nn})\right) p_{Fn} m_n^* \Delta_n^7}{15\pi^5} I\left(\frac{\Delta_n}{T}\right)$$

$$\simeq \zeta \cdot 10^{28} \left(\frac{\varrho}{\varrho_0}\right)^{1/3} \frac{m_n^*}{m_N} \left(\frac{\Delta_n}{\text{MeV}}\right)^7 I\left(\frac{\Delta_n}{T}\right) \frac{\text{erg}}{\text{cm}^3 \cdot \text{sec}}, \quad T < T_{cn}, \quad (29)$$

where $\xi_1 = 1$, $\xi_2 = 0$ for $S$-pairing and $\xi_1 = 2/3$, $\xi_2 = 4/3$ for $P$-pairing; compare with the result [35] where $\xi_1 = 1$, $\xi_2 = 2$ were obtained. I removed some misprints existed in [32,24,25,16]. For neutrinos $\omega(q) = q$, and therefore, the correlations are not so essential as it would be for $\omega \ll q$. Taking $\gamma^2 \simeq 1.3$ in the range of $S$-pairing we get $\zeta \simeq 5$ whereas for the $P$-pairing with $g_A^* \simeq 1.1$ we obtain $\zeta \simeq 9$, in agreement with numerical evaluations of [24] ($\zeta \simeq 7$) which were used later in the cooling simulations of [26]. Here $I(x) = \int_0^\infty \text{ch}^5 y dy/(\exp(x\text{ch}y)+1)^2$, with the asymptotics $I(x \gg 1) \simeq \exp(-2\Delta/T)\sqrt{\pi T/4\Delta}$, which provides also the appropriate asymptotic temperature behavior of (29). The emissivity of the process $p \to p\nu\bar{\nu}$ is given by [24]

$$\varepsilon(\text{pPBF}) = \frac{12G^2(\xi_1 \kappa_{pp}^2 + \xi_2 g_A^{*2}\gamma_{pp}^2 + \xi_3) p_{Fp} m_p^* \Delta_p^7}{15\pi^5} I\left(\frac{\Delta_p}{T}\right), \quad T < T_{cp}, \quad (30)$$

and $\xi_2 = 0$ for protons paired in $S$-state in NS matter, $\xi_3 \lesssim 1$ is due to the second diagram (18) and has a complicated structure [73,76]. For the process (30) the part of $NN$ and $ee$ correlations is especially important. One has $\kappa_{pp}^2 \sim 0.05 \div 1$ for $1 \div 3\varrho_0$ and $\xi_3 \sim 1$, and $\kappa_{pp}^2 + \xi_3 \sim 1$ instead of a small $c_V^2 \simeq 0.006$ value in absence of correlations; see the discussion in the subsection 2.3. Therefore, in agreement with [24,26,73,76], the emissivity of the process $p \to p\nu\bar{\nu}$ can be comparable with that for $n \to n\nu\bar{\nu}$ depending on the relative magnitudes of $\Delta_p$ and $\Delta_n$.

NPBF processes are very efficient for $T < T_c$ and are competing with MMU processes. The former win for not too massive stars. The analysis of the above processes supports also our general conclusion on the crucial role of in-medium effects in the cooling scenario.

**Pion (kaon) condensate processes.** The $P$ wave pion condensate can be of three types: $\pi_s^+$, $\pi^\pm$, and $\pi^0$ with different values of the critical densities $\varrho_{c\pi} = (\varrho_{c\pi^\pm}, \varrho_{c\pi_s^+}, \varrho_{c\pi^0})$, see [57]. Thus above the threshold density for the pion condensation of the given type, the neutrino emissivity of the MMU process (24) is to be supplemented by the corresponding PU processes

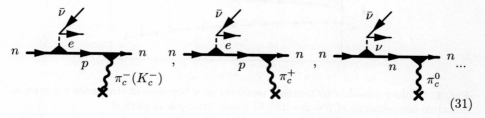

(31)

The emissivity of the charged pion condensate processes with inclusion of the $NN$ correlation effect (in a simplified treatment) renders, see [22,16],

$$\varepsilon[\text{PU}] \simeq 7 \cdot 10^{26} \frac{p_{Fn}}{m_\pi} \frac{m_n^* m_p^*}{m_N^2} \gamma_\beta^2(p_{Fe}, p_{Fe}) \, \widetilde{\gamma}^2(g', 0, p_{Fn}) T_9^6 \sin^2\theta \frac{\text{erg}}{\text{cm}^3 \text{sec}}. \quad (32)$$

Here $\varrho > \varrho_{c\pi}$ and $\sin\theta \simeq \sqrt{2|\widetilde{\omega}^2|/m_\pi^2}$ for $\theta \ll 1$. Of the same order of magnitude are the emissivities of other possible $\pi$ condensate reactions, e.g. for $n\pi_c^0 \to np e \bar{\nu}$ process at $\theta \ll 1$ the numerical factor is about two times larger. Since $\pi^\pm$ condensation probably reduces the energy gaps of the superfluid states by an order of magnitude, see [79], we may assume that superfluidity vanishes above $\varrho_{c\pi}$. Finally we note that although the PU processes have genuinely one-nucleon phase-space volumes, their contribution to the emissivity is suppressed relative to the DU by an additional $\widetilde{\gamma}^2(g', 0, p_{Fn})$ suppression factor due to existence of the extra $(\pi NN)$ vertex in the former case.

Fig. 4 compares the mass dependence of the neutrino cooling rates $L_\nu/C_V$, $L_\nu$ is the neutrino luminosity, $C_V$ is the heat capacity, for MMU(MOPE) and MU(FOPE) for non-superfluid matter. For the solid curves, the neutrino emissivity in pion-condensed matter is taken into account according to (32) and (24) with the parameter $\alpha = \sqrt{2}$. As a conservative estimate we took $\varrho_{c\pi} \simeq 3\varrho_0$. The dashed curves correspond to the model where no pion condensation is allowed. As one sees, the medium polarization effects when included in MMU may result in three order of magnitude increase of the cooling rate for the most massive stars. Even for stars of a rather low mass the cooling rate of MMU is still several times larger than for MU because even in this case the most efficient rate is given by the reactions shown in the right diagram in Fig. 1 (first two diagrams of (23)). The cooling rates for the NS of $M = 1.8 M_\odot$ with and without pion condensate differ only moderately (by a factor of 5 in this model). If we had used a smaller value of $\varrho_{c\pi}$ the ratio of PU emissivity to that of MMU would decrease and could even become $\lesssim 1$ in the vicinity of $\varrho_{c\pi}$. In contrary, the reaction rates for the FOPE model are rather independent of the star's mass for the stars with masses

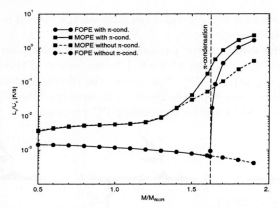

**Fig. 4.** Cooling rate due to neutrino emission as a function of star mass for a representative temperature of $T = 3 \times 10^8$ K. Superfluidity is neglected.

below the critical value $1.63 M_\odot$, at which transition into the pion condensed phase occurs after which they jump a value typical for PU. It is to be stressed that contrary to FOPE model, the MOPE model [23] consistently takes into account the pion softening effects for $\varrho < \varrho_{c\pi}$ and both the pion condensation and pion softening effects in the condensate for $\varrho > \varrho_{c\pi}$. For $\varrho > \varrho_{cK}$ the kaon condensate processes come into play. The most popular idea is the $S$ wave $K^-$ condensation (e.g. see [13]) which is allowed at $\mu_e > m^*_{K^-}$ due to possibility of the reaction $e \to K^- \nu$. Analogous condition for the pions, $\mu_e > m^*_{\pi^-}$, is not fulfilled owing to a strong $S$ wave $\pi NN$ repulsion [57,16] (again an in-medium effect!) otherwise $S$ wave $\pi^-$ condensation would occur at smaller densities than $K^-$ condensation. The neutrino emissivity of the $K^-$ condensate processes is given by equation analogous to (32) with a different $NN$ correlation factor and an additional suppression factor due to a small contribution of the Cabibbo angle. However qualitatively the scenario that permits kaon condensate processes is analogous to that with the pion condensate processes discussed above.

**Other resonance processes.** There are many other reaction channels allowed in the medium. For example any Fermi liquid permits propagation of zero sound excitations of different symmetry, which are related either to the pion or the quanta of a more local interaction represented here by $f_{\alpha,\beta}$ and $g_{\alpha,\beta}$. These excitations being present at $T \neq 0$ may also participate in the neutrino reactions. The most essential contribution comes from the neutral current processes [23] given by first two diagrams of the series

$$\tag{33}$$

Here the dotted line is a zero sound quantum of appropriate symmetry. These are the resonance processes (the second, of DU-type) analogous to those processes going on the condensates with the only difference that the rates of reactions with zero sounds are proportional to the thermal occupations of the corresponding spectrum branches whereas the rates of the condensate processes are proportional to the modulus squared of condensate mean field. The contribution of the resonance reactions is as a rule rather small due to small phase space volume ($q \sim T$) associated with the zero sounds. Please also bear in mind the analogy of the processes (33) with the corresponding phonon processes in the crust.

**DU processes.** The proper DU processes in matter, as $n \to pe^-\bar{\nu}_e$ and $pe^- \to n\nu_e$,

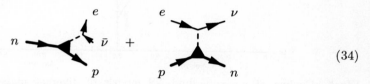 (34)

should also be treated with the full vertices. They are forbidden up to the density $\varrho_{cU}$ when triangle inequality $p_{Fn} < p_{Fp} + p_{Fe}$ begins to fulfill. For traditional EoS like that given by the variational theory [18] DU processes are permitted for $\varrho > 5\varrho_0$. The emissivity of the DU processes is

$$\varepsilon^{DU} \simeq 1.2 \cdot 10^{27} \frac{m_n^* m_p^*}{m_N^2} \left(\frac{\mu_l}{100\text{MeV}}\right) \gamma_\beta^2 \min\left[\zeta(\Delta_n), \zeta(\Delta_p)\right] T_9^6 \frac{\text{erg}}{\text{cm}^3 \text{ sec}}, \quad (35)$$

where $\mu_l = \mu_e = \mu_\mu$ is the chemical potential of the leptons in MeV. In addition to the usually exploited result [15], (35) contains $\gamma_\beta^2$ pre-factor (14) due to $NN$ correlations in the $\beta$ decay vertices, see [24,26].

It was realized in [80] that the softening of the pion mode in dense neutron matter could also give rise to a rearrangement of single-fermion degrees of freedom due to violation of Pomeranchuk stability condition for $\varrho = \varrho_{cF} < \varrho_{c\pi}$. It may result in a appearance of an extra Fermi sea for $\varrho_{cF} < \varrho < \varrho_{c\pi}$ and for small momenta $p < 0.2 p_{Fn}$, that in its turn opens a DU channel of neutrino cooling of NS from the corresponding layer. Due to the new feature of a temperature-dependent neutron effective mass, $m_n^* \propto 1/T$, we may anticipate an extra essential enhancement of the DU process, corresponding to a reduction in the power of the temperature dependence from $T^6$ to $T^5$. At early hot stage of NS thermal evolution this layer becomes opaque for the neutrinos thus slowing the neutrino transport from the massive NS core to the exterior.

Based on the Brown–Rho scaling idea [71], we argued in [81] that for the charged $\varrho$ meson condensation at a density relevant to NS ($\varrho_{c\varrho} \sim 3\varrho_0$ if $m_\varrho^*$ would drop to $m_\varrho/2$ at this density). If happened, it would open DU reaction already for $\varrho < \varrho_{c\varrho}$ and close it for $\varrho > \varrho_{c\varrho}$ due to an essential modification of the nuclear asymmetry energy.

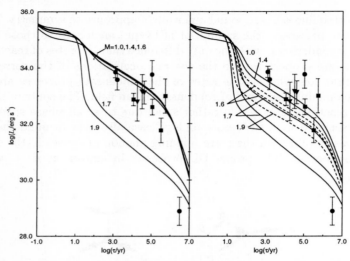

**Fig. 5.** Cooling of non-superfluid NS models of different masses constructed for the HV EOS [26]. Two graphs refer to cooling via MU(FOPE) +PU (left) and MMU(MOPE)+PU (right). In both cases, pion condensation is taken into account for the solid curves at $\varrho > \varrho_{c\pi}$, i.e. for $M > 1.6 M_\odot$ ($\varrho_{c\pi}$ is chosen to be $3\varrho_0$). The dashed curves in the right graph refer to a somewhat larger value of $\tilde{\omega}^2(p_{Fn})$, without pion condensation. The observed luminosities are labeled by dots. Possibilities of Fermi sea rearrangement and $\varrho^\pm$ condensation are ignored.

### 3.3 Comparison with soft $X$ ray data

The heat transport within the crust of NS establishes homogeneous density profile indexCrust at times $\lesssim (1 \div 10)$yr. After that time the subsequent cooling is determined by the simple relation $C_V \dot{T} = -L$, where $C_V = \sum_i C_{V,i}$ and $L = \sum_i L_i$ are the sums of the partial contributions to the heat capacity (specific heat integrated over the volume) and the luminosity (emissivity integrated over the volume).

The nucleon pairing gaps are rather purely known. Therefore one may vary them. The *"standard"* and *"nonstandard"* scenarios of the cooling of NS of several selected masses for suppressed gaps are demonstrated in the left panel of Fig. 5, [26]. Depending on the star mass, the resulting photon luminosities are basically either too high or too low compared to those given by observations. The situation changes, if the MMU process (24) is included. Now, the cooling rates vary smoothly with the star mass (see right panel of Fig. 5) such that the gap between *standard* and *non-standard* cooling scenarios is washed out. More quantitatively, by means of varying the NS mass between $(1 \div 1.6)$ $M_\odot$, one achieves an agreement with a large number of observed data points. This is true for a wide range of choices of the $\tilde{\omega}^2$ parameterization, independently whether pion (kaon) condensation can occur or not. Two parameterizations presented in Fig. 5 with pion condensation for $\varrho > 3\varrho_0$ and without differ only in the range which is covered by the cooling curves. The only point which does not fall in

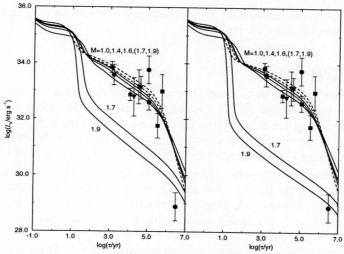

**Fig. 6.** Cooling of NS with different masses constructed for the HV EOS [26]. The cooling processes are MU-VS86, PU (only solid curves), NPBF, PPBF, and DU (only in the right graph). The dashed curves refer to the $M = 1.7$ and $1.9 M_\odot$ models without pion condensate. Possibilities of Fermi sea rearrangement and $\varrho^\pm$ condensation are ignored.

the range covered by the cooling curves correspond to the hottest known pulsar PSR 1951+32. The other three points which, as shown in the right panel of the Fig. 5, are also not fitted by the curves can be easily fitted by slight changes of the model parameters. The high luminosity of PSR 1951+32 may be due to internal heating processes, cf. [10].

We turn now to cooling simulations where the MU, NPBF, DU and PU take place simultaneously. Parameters of the pairing gaps are from Fig. 6 of [26]. Fig. 6 shows the cooling tracks of stars of different masses, computed for the HV EoS. Very efficient at $T < T_c$ become to be NPBF processes which compete with MMU processes. The former prevail for not too massive stars in agreement with estimation [24]. The DU process is taken into account in the right graph, whereas it is neglected in the left graph. The solid curves refer again to the $\tilde{\omega}^2$ parameterization with phase transition to pion condensate, the dashed curves to the one without phase transition (see Fig. 4). For masses in the range between 1.0 and 1.6 $M_\odot$, the cooling curves pass through most of the data points. We again recognize a photon luminosity drop by more than two orders of magnitude for the 1.7 $M_\odot$ mass star with pion condensate, due to suppression of the pairing gaps in this case. This drop is even larger if the DU is taken into account (right graph). This allows to account for the photon luminosity of PSR 1929+10.

Thus, comparison with the observed luminosities shows that one gets quite good agreement between theory and observations if one includes into consideration all available in-medium effects assuming that the masses of the pulsars are different. We point out that the description of these effects is constructed in essentially the same manner for all the hadronic systems as NS, atomic nuclei and heavy ion collisions, cf. [16,58].

## 3.4 Neutrino opacity

The importance of the in-medium effects for the description of neutrino transport at the initial stages of NS cooling was discussed in [23,42,43,16], where correlation effects, pion softening and pion condensation (at $\varrho > \varrho_{c\pi}$) were taken into account. The neutrino/antineutrino mean free paths can be evaluated from the corresponding kinetic equations via their widths $\Gamma_{\nu(\bar{\nu})} = -2 \mathrm{Im} \Pi^R_{\nu(\bar{\nu})}$, where $\Pi^R$ is the retarded self-energy, or within the QPA for the nucleons they can be also estimated via the squared matrix elements of the corresponding reactions. Hence the processes which most efficiently contribute to the emissivity are at early times (for $T \gtrsim 1$ MeV) also provide the dominant contribution to the opacity.

In the above *"nuclear medium cooling scenario"* at $T > T_c$ the most essential contribution was from MMU. Taking into account the $NN$ correlations in the strong coupling vertices of two-nucleon processes like MMU and MNB suppresses the rates, whereas the softening of the pion propagator essentially enhances them. For rather massive NS MOPE wins the competition. The mean free path of neutrino/antineutrino in MMU processes is determined from the same diagrams (23) as the emissivity. Its calculation (see (24)) with the two first diagrams yields

$$\frac{\lambda_\nu^{MMU}}{R} \simeq \frac{1.5 \cdot 10^5}{F_1(2\Gamma)^8 T_9^4} \left(\frac{\varrho_0}{\varrho}\right)^{10/3} \frac{m_N^4}{(m_n^*)^3 m_p^*} \left[\frac{\alpha \tilde{\omega}_{\pi^0}(p_{Fn})}{m_\pi}\right]^4 \left[\frac{\alpha \tilde{\omega}_{\pi^\pm}(p_{Fn})}{m_\pi}\right]^4 . \quad (36)$$

Using the relation $\lambda_\nu \simeq R$ one can evaluate $T_{opac}$. Including only the first diagram we get a simple estimate

$$T_9^{opac} \simeq 11 \frac{\varrho_0}{\varrho} \frac{\tilde{\omega}^2(p_{Fn}^2)}{[4\gamma(g_{nn}, 0, p_{Fn})\tilde{\gamma}(g', 0, p_{Fn})]^{1/2}} \frac{m_N}{m_N^*} . \quad (37)$$

For averaged value of the density $\varrho \simeq \varrho_0$ corresponding to a medium-heavy NS ($< 1.4 M_\odot$) with $\tilde{\omega}^2(p_{Fn}) \simeq 0.8 m_\pi^2$, $\tilde{\gamma} \simeq \gamma \simeq (0.3 \div 0.4)$ we get $T_{opac} \simeq (1 \div 1.5)$ MeV that is smaller then the value $T_{opac} \simeq 2$ MeV estimated with FOPE [6]. For $\varrho \simeq 2\varrho_0$ which corresponds to a more massive NS we estimate as $T_{opac} \simeq (0.3 \div 0.5)$ MeV. Thus pion softening results in a substantial decrease of neutrino/antineutrino mean free paths and the value of $T_{opac}$.

The diffusion equation determines the characteristic time scale for the heat transport of neutrinos from the hot zone to the star surface $t_0 \sim R^2 C_V \sigma^{-1} T^{-3} / \lambda_\nu$ ($\sigma$ is the Stefan–Boltzmann constant), for which we find the estimate $t_0 \sim 10$ min. at $T \simeq 10$ MeV and $\varrho \simeq \varrho_0$ for the values of parameters $\tilde{\omega}^2 \simeq 0.8 m_\pi$, $\Gamma \simeq 0.4$, $m_N^*/m_N \simeq 0.9$; $t_0$ becomes as large as several hours for $\varrho \simeq (2 \div 3)\varrho_0$. These estimates demonstrate that more massive NS cool down more slowly at $T > T_{opac}$ and faster at subsequent times than the less massive stars.

Due to the in-medium effects neutrino scattering cross section on the neutrons shown by the first diagram (17) requires the $NN$ correlation factor

$$\gamma_{n\nu}^2(\omega, q) = \frac{\gamma^2(f_{nn}, \omega, q) + 3g_A^2 \gamma^2(g_{nn}, \omega, q)}{1 + 3g_A^2}, \qquad (38)$$

as follows from Urca (15). This results in a suppression of the cross sections for $\omega < qv_{Fn}$ and in an enhancement for $\omega > qv_{Fn}$. Neutrino scattering cross sections on the protons are modified by

$$\gamma_{p\nu}^2(\omega, q) = \frac{\kappa^2(f_{np}, f_{nn}, \omega, q) + 3g_A^2 \gamma_{pp}^2(g_{nn}, \omega, q)}{1 + 3g_A^2}, \qquad (39)$$

that results in the same order of magnitude correction as given by (38).

Also there is a suppression of the $\nu N$ scattering and MNB reaction rates for soft neutrinos ($\omega \lesssim (3 \div 6)T$) due to multiple $NN$ collisions

$$\qquad (40)$$

(LPM effect). One may estimate these effects simply multiplying squared matrix elements of the $\nu N$ scattering and the MNB processes by the corresponding suppression pre-factors [28]. Qualitatively one may use a general pre-factor $C_0(\omega) = \omega^2/[\omega^2 + \Gamma_N^2]$, where $\Gamma_N$ is the nucleon width and the $\omega$ is the energy of $\nu$ or $\nu\bar{\nu}$ pair. In some works, see [82,83], correction factor, like $C_0$, was suggested at an ansatz level. Actually one does not need any ansatz type reductions. OTF, see [28], allows to calculate the rates using an exact sum rule. The modification of the charged current processes due to LPM effect is unimportant since the corresponding value of $\omega$ is $\simeq p_{Fe} \gg \Gamma_N$.

The main physical result we discussed is that in the medium the reaction rates are essentially modified. A suppression arises due to $NN$ correlations (for $\omega \ll kv_{FN}$) and infra-red pre-factors (coherence effects), and an enhancement due to the pion softening (and $NN$ correlations for $\omega \sim q$) and due to opening up of new efficient reaction channels. The pion softening demonstrates that already for densities $\varrho < \varrho_0$ the nucleon system begins to feel that it may have $\pi$ condensate phase transition for $\varrho > \varrho_{c\pi}$, although this $\varrho_{c\pi}$ value might be essentially larger than $\varrho_0$ or even not achieved.

## 4 The rate of radiation from dense medium. OTF

Perturbative diagrams are obviously irrelevant for calculation of in-medium processes and one should deal with dressed Green functions. The QPA for fermions is applicable if the fermion width is much less than all the typical energy scales essential in the problem ($\Gamma_F \ll \omega_{ch}$). In calculation of the emissivities of $\nu\bar{\nu}$ reactions the minimal scale is $\omega_{ch} \simeq 6T$, averaged $\nu\bar{\nu}$ energy for MNB reactions. For MMU $\omega_{ch} \simeq p_{Fe}$. For radiation of soft quanta of fixed energy $\omega < T$, $\omega_{ch} \simeq \omega$.

Within the QPA for fermions, the reaction rate with participation of the fermion and the boson is given by [22,23]

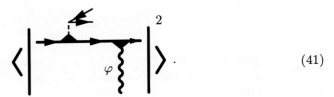 (41)

For equilibrium ($T \neq 0$) system there is the exact relation

$$(<\widehat{\varphi}_2^\dagger \widehat{\varphi}_1>)(p) = iD^{-+} + |\varphi_c|^2 = \frac{A_B}{\exp(\frac{\omega}{T})-1} + |\varphi_c|^2, \quad A_B = -2\mathrm{Im}D^R, \quad (42)$$

where $(<\widehat{\varphi}_2^\dagger \widehat{\varphi}_1>)(p)$ means the Fourier component of the corresponding non-equilibrium Green function and $\varphi_c$ is the mean field. Thus the rate of the reaction is related to the boson spectral function $A_B$ and the width ($\Gamma_B$) being determined by the corresponding Dyson equation, see Urca (9). $A_B$ is the delta-function at the spectrum branches related to resonance processes, like zero sound. The poles associated with the upper branches do not contribute at small temperatures due to a tiny thermal population of those branches. There is also a contribution to $\mathrm{Im}D^R$ proportional to $\mathrm{Im}\Pi^R$ given by the particle-hole diagram. Within the QPA taking Im part means the cut of the diagram. Thus we show [23] that this contribution is the same as that could be calculated with the help of the squared matrix element of the two-nucleon process

 (43)

This is precisely what one could expect using optical theorem. Thus unlimited series of all possible diagrams with in-medium Green functions (see (17), (31), (33), (34)) together with two-fermion diagrams (as given by (23)) and multiple-fermion diagrams (like (40)) would lead us to a double counting. The reason is that permitting the boson width effects (and beyond the QPA for fermions also permitting finite fermion widths) the difference between one-fermion, two-fermion and multiple-fermion processes in medium is smeared out. All the states are allowed and participate in production and absorption processes. Staying with the QP picture for fermions, the easiest way to avoid mentioned double counting is to calculate the reaction rates according to (41), i.e. with the help of the diagrams of the DU-like type, which already include all the contributions of the two-nucleon origin. Multiple $NN$ collision processes should be added separately. On the other hand, it is rather inconvenient to explicitly treat all one-nucleon processes dealing with different specific quanta instead of using of the full $NN$

interaction amplitude whenever it is possible. Besides, as we have mentioned, consideration of open fermion legs is only possible within the QPA for fermions since Feynman technique is not applicable if Green functions of ingoing and outgoing fermions have widths (that in another language means possibility of additional processes). Thus in [24,28] the idea was put forward to integrate over all in-medium states allowing all possible processes instead of specifying different special reaction channels.

In [24,28,27] it was shown that OTF in terms of full non-equilibrium Green functions is an efficient tool to calculate the reaction rates including finite particle widths and other in-medium effects. Applying this approach, e.g., to the antineutrino–lepton (electron, $\mu^-$ meson, or neutrino) production [24] we can express the transition probability in a direct reaction in terms of the evolution operator $S$,

$$\frac{d\mathcal{W}_{X\to\bar{\nu}l}^{\text{tot}}}{dt} = \frac{(1-n_l)dq_l^3 dq_{\bar{\nu}}^3}{(2\pi)^6 \, 4\,\omega_l\,\omega_{\bar{\nu}}} \sum_{\{X\}} \overline{<0|\,S^\dagger\,|\bar{\nu}l+X><\bar{\nu}l+X|\,S\,|0>}, \qquad (44)$$

where we presented explicitly the phase-space volume of $\bar{\nu}l$ states; lepton occupations of given spin, $n_l$, are put zero for $\nu$ and $\bar{\nu}$ which are supposed to be radiated directly from the system (for $T < T_{opac}$). The bar denotes statistical averaging. The summation goes over complete set of all possible intermediate states $\{X\}$ constrained by the energy-momentum conservation. It was also supposed that electrons/muons can be treated in the QPA, i.e. with zero widths. Then there is no need (although it is possible) to consider them in intermediate reaction states. Making use of the smallness of the weak coupling, we expand the evolution operator as $S \approx 1 - \mathrm{i}\int_{-\infty}^{+\infty} T\,\{V_W(x)\,S_{\text{nucl}}(x)\}dx_0$, where $V_W$ is the Hamiltonian of the weak interaction in the interaction representation, $S_{\text{nucl}}$ is the part of the $S$ matrix corresponding to the nuclear interaction, and $T\{...\}$ is the chronological ordering operator. After substitution into (44) and averaging over the arbitrary non-equilibrium state of a nuclear system, there appear chronologically ordered $(G^{--})$, anti-chronologically ordered $(G^{++})$ and disordered $(G^{+-}$ and $G^{-+})$ exact Green functions.

In graphical form the general expression for the probability of the lepton (electron, muon, neutrino) and anti-neutrino production is as follows

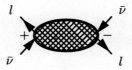

representing the sum of all closed diagrams $(-\mathrm{i}\Pi^{-+})$ containing at least one $(-+)$ exact Green function. The latter quantity is especially important. Various contributions from $\{X\}$ can be classified according to the number $N$ of $G^{-+}$ lines in the diagram

$$\frac{dW_{\bar{\nu}l}^{\text{tot}}}{dt} = \frac{(1-n_l)d^3q_{\bar{\nu}}dq_l^3}{(2\pi)^6 \, 4\omega_{\bar{\nu}}\omega_l} \left( \begin{array}{c} l \\ +\bigcirc N=1 \bigcirc - \\ \bar{\nu} \end{array} \begin{array}{c} \bar{\nu} \\ \\ l \end{array} + \begin{array}{c} l \\ +\bigcirc N=2 \bigcirc - \\ \bar{\nu} \end{array} \begin{array}{c} \bar{\nu} \\ \\ l \end{array} \cdots \right). \quad (45)$$

This procedure suggested in [24] is actually very helpful especially if the QPA holds for the fermions. Then contributions of specific processes contained in a closed diagram can be made visible by cutting the diagrams over the $(+-)$, $(-+)$ lines. In the framework of the QPA for the fermions $G^{-+} = 2\pi i n_F \delta(\varepsilon + \mu - \varepsilon_p^0 - \text{Re}\Sigma^R(\varepsilon + \mu, \mathbf{p}))$ ($n_F$ are fermionic occupations, for equilibrium $n_F = 1/[\exp((\varepsilon-\mu_F)/T)+1]$), and the cut eliminating the energy integral thus requires clear physical meaning. This way one establishes the correspondence between closed diagrams and usual Feynman amplitudes although in general case of finite fermion width the cut has only a symbolic meaning. Next advantage is that in the QPA any extra $G^{-+}$, since it is proportional to $n_F$, brings a small $(T/\varepsilon_F)^2$ factor to the emissivity of the process. Dealing with small temperatures one can restrict oneself to the diagrams of the lowest order in $(G^{-+}G^{+-})$, not forbidden by energy-momentum conservations, putting $T = 0$ in all $G^{++}$ and $G^{--}$ Green functions. Each diagram in (45) represents a whole class of perturbative diagrams of any order in the interaction strength and in the number of loops.

Proceeding further we may explicitly decompose the first term in (45) as

$$+\bigcirc N=1 \bigcirc - \;=\; +\bigcirc\!\bigcirc\bigcirc - \;+\; +\bigcirc\!\bigcirc\bigcirc - \quad . \quad (46)$$

The full vertex in the diagram (46) of given sign is irreducible with respect to the $(+-)$ and $(-+)$ nucleon–nucleon hole lines. This means it contains only the lines of given sign, all $(--)$ or $(++)$. Second diagram with anomalous Green functions exists only for systems with pairing. In the framework of the QPA in fact these diagrams determine the proper DU and also NPBF processes calculated in [24,25] within OTF. In the QP picture the contribution to DU process vanishes for $\varrho < \varrho_{cU}$. Then the second term (45) comes into play which within the same QP picture contains two-nucleon processes with one $(G^{-+}G^{+-})$ loop in intermediate states, etc.

The full set of diagrams for $\Pi^{-+}$ can be further explicitly decomposed as series [28] (from now on using brief notations)

$$-\!\!\bigcirc{-i\Pi}\bigcirc\!\!- \;=\; -\!\!\bigcirc\!\bigcirc\!\!- \;+\; -\!\!\bigcirc\!\!\bigcirc\!\!- \;+\; -\!\!\bigcirc\!\!\bigcirc\!\!\bigcirc\!\!- \;+\cdots$$

$$+\; -\!\!\bigcirc\!\!\bigcirc\!\!- \;+\; -\!\!\bigcirc\!\!\times\!\!\bigcirc\!\!- \;+\; -\!\!\bigcirc\!\!\bigcirc\!\!\bigcirc\!\!- \;+\;\cdots \quad (47)$$

Full dot is the weak coupling vertex including all the diagrams of one sign, $NN$ interaction block is the full block also of one sign diagrams. The lines are full Green functions with the widths. The most essential term is the one-loop diagram (see (46)), which is positive definite, and including the fermion width corresponds to the first term of the classical Langevin result, for details see [28]. Other diagrams represent interference terms due to rescatterings. In some simplified representations (e.g., as we used within Fermi liquid theory) the 4-point functions (blocks of $NN$ interaction of given sign diagrams) behave like intermediate bosons (e.g. zero-sounds and dressed pions). In general it is not necessary to consider different quanta dealing instead with the full $NN$ interaction (all diagrams of given sign). For particle propagation in an external field, e.g. infinitely heavy scattering centers (proper LPM effect), only the one-loop diagram remains, since one deals then with a genuine one-body problem. In the quasiclassical limit for fermions (with small occupations $n_F$) all the diagrams given by first line of series (47) with arbitrary number of "$-+$" lines are summed up leading to the diffusion result, for details see [28]. For small momenta $q$ this leads to a suppression factor of the form $C = \omega^2/(\omega^2 + \Gamma_x^2)$, $\Gamma_x$ incorporates rescattering processes. In general case the total radiation rate is obtained by summation of all diagrams in (47). The value $-i\Pi^{-+}$ determines the gain term in the generalized kinetic equation for $G^{-+}$, see [84,85], that allows to use this method in non-equilibrium problems, like for description of neutrino transport in semi-transparent region of the neutrino-sphere of supernovas, as we may expect.

In the QP limit diagrams 1, 2, 4 and 5 of (47) correspond to the MMU and MNB processes related to a single in-medium scattering of two fermionic QP and can be symbolically expressed as Feynman amplitude (48a)

(48)

The one-loop diagram in (47) is particular, since its QP approximation in many cases vanishes as we have mentioned. However the full one-loop includes QP graphs of the type (48b), which survive to the same order in $\Gamma_N/\omega_{ch}$ as the other diagrams (therefore in [24] where QP picture was used this diagram was considered as allowed diagram). In QP series such a term is included in second diagram of (45) although beyond the QPA it is included as the proper self-energy insertion to the one-loop result, i.e. in first term (45) [28]. In fact it is positive definite and corresponds to the absolute square of the amplitude (48a). The other diagrams 2, 4 and 5 of (47) describe the interference of amplitude (48a) either with those amplitudes where the weak coupling quantum ($l\bar{\nu}$ pair) couples to another leg or with one of the exchange diagrams. For neutral interactions diagram (47:2) is more important than diagram 4 while this behavior reverses for charge exchange interactions (the latter is important, e.g., for gluon radiation from quarks in QCD transport due to color exchange interactions). Diagrams like 3 describe the interference terms due to further rescatterings of the source fermion with others as shown by (48c). Diagram (47:6) describes the production from intermediate states and relates to the Feynman graph (48d). For photons in the soft limit ($\omega \ll \varepsilon_F$) this diagram (48d) gives a smaller contribution to the photon production rate than the diagram (48a), where the normal bremsstrahlung contribution diverges like $1/\omega$ compared to the $1/\varepsilon_F$-value typical for the coupling to intermediate fermion lines. For $\nu\bar{\nu}$ bremsstrahlung (48d) gives zero due to symmetry. However in some cases the process (48d) might be very important even in the soft limit. This is indeed the case for the MMU process considered above. Some of the diagrams which are not presented explicitly in (47) give more than two pieces, if being cut, so they never reduce to the Feynman amplitudes. However in the QPA they give zero contribution [28].

With $\Gamma_F \sim \pi^2 T^2/\varepsilon_F$ for Fermi liquids, the criterion $\Gamma_F \ll \omega_{ch} \sim T$ is satisfied for all thermal excitations $\Delta\varepsilon \sim T \ll \varepsilon_F/\pi^2$. However with the application to soft radiation this concept is no longer justified. Indeed series of QP diagrams is not convergent in the soft limit and there is no hope to ever recover a reliable result by a finite number of QP diagrams for the production of soft quanta. With *full Green functions*, however, one obtains a description that uniformly covers both the soft ($\omega \ll \Gamma_F$) and the hard ($\omega \gg \Gamma_F$) regimes. In the vicinity of $\varrho_{c\pi}$ the quantity $\Gamma_F$ being roughly estimated in [62,64] as $\Gamma_F \propto \pi^2 \Gamma^2 T m_\pi/\widetilde{\omega}$, and coherence effects come into play.

In order to correct QP evaluations of different diagrams by the fermion width effects for soft radiating quanta one can simply multiply the QP results by different pre-factors [28]. E.g., comparing the one-loop result at non-zero $\Gamma_F$ with the first non-zero diagram in the QPA ($\Gamma_F = 0$ in the fermion Green functions) we get

$$\cdots \bigcirc \cdots = C_0(\omega) \left\{ \cdots \bigcirc \cdots \right\}_{\text{QPA}} , \qquad (49)$$

at small momentum $q$. For the next order diagrams we have

$$\cdots \bigcirc \cdots = C_1(\omega) \left\{ \cdots \bigcirc \cdots \right\}_{\text{QPA}} , \quad C_1(\omega) = \omega^2 \frac{\omega^2 - \Gamma_F^2}{(\omega^2 + \Gamma_F^2)^2} ,$$

$$\cdots \bigcirc\!\bigcirc \cdots = C_0(\omega) \left\{ \cdots \bigcirc\!\bigcirc \cdots \right\}_{\text{QPA}} , \quad C_0(\omega) = \frac{\omega^2}{\omega^2 + \Gamma_F^2} . \qquad (50)$$

where factors $C_0$, $C_1$, ... cure the defect of the QPA for soft $\omega$. The factor $C_0$ complies with the replacement $\omega \to \omega + i\Gamma_F$. A similar factor is observed in the diffusion result, where however the macroscopic relaxation rate $\Gamma_x$ enters, due to the resummation of all rescattering processes.

Finally, we demonstrated how to calculate the rates of different reactions in dense equilibrium and non-equilibrium matter and compared the results derived in closed diagram technique with those obtained in the standard technique of computing of the squared matrix elements.

## 5 Cooling of Hybrid Neutron Stars

Let us in addition review how the color superconducting quark matter, if exists in interiors of massive NS, may affect the neutrino cooling of HNS, [53]. A detailed discussion of the neutrino emissivity of quark matter without taking into account of the possibility of the color superconductivity has been given first in ref. [86]. In this work the quark direct Urca (QDU) reactions $d \to ue\bar{\nu}$ and $ue \to d\nu$ have been suggested as the most efficient processes. In the color superconducting matter the corresponding expression for the emissivity modifies as

$$\epsilon_\nu^{\text{QDU}} \simeq 9.4 \times 10^{26} \alpha_s (\varrho/\varrho_0) Y_e^{1/3} \zeta_{\text{QDU}} \, T_9^6 \text{ erg cm}^{-3} \text{ s}^{-1}, \tag{51}$$

where due to the pairing the emissivity of QDU processes is suppressed by a factor, very roughly given by $\zeta_{\text{QDU}} \sim \exp(-\Delta_q/T)$. At $\varrho/\varrho_0 \simeq 2$ the strong coupling constant is $\alpha_s \approx 1$ decreasing logarithmically at still higher densities, $Y_e = \varrho_e/\varrho$ is the electron fraction. If for somewhat larger density the electron fraction was too small ($Y_e < Y_{ec} \simeq 10^{-8}$), then all the QDU processes would be completely switched off [87] and the neutrino emission would be governed by two-quark reactions like the quark modified Urca (QMU) and the quark bremsstrahlung (QB) processes $dq \to uqe\bar{\nu}$ and $q_1 q_2 \to q_1 q_2 \nu \bar{\nu}$, respectively. The emissivities of QMU and QB processes are smaller than that for QDU being suppressed by factor $\zeta_{\text{QMU}} \sim \exp(-2\Delta_q/T)$ for $T < T_{\text{crit},q} \simeq 0.4 \, \Delta_q$. For $T > T_{\text{crit},q}$ all the $\zeta$ factors are equal to unity. The modification of $T_{\text{crit},q}(\Delta_q)$ relative to the standard BCS formula is due to the formation of correlations as, e.g., instanton- anti-instanton molecules. The contribution of the reaction $ee \to ee\nu\bar{\nu}$ to the emissivity is very small [88], $\epsilon_\nu^{ee} \sim 10^{12} Y_e^{1/3} (\varrho/\varrho_0)^{1/3} T_9^8$ erg cm$^{-3}$ s$^{-1}$, but it can become important when quark processes are blocked out for large values of $\Delta_q/T$ in superconducting quark matter.

For the quark specific heat [53] used expression of [86] being however suppressed by the corresponding $\zeta$ factor due to color superfluidity. Therefore gluon-photon and electron contributions play important role.

The heat conductivity of the matter is the sum of partial contributions $\kappa = \sum_i \kappa_i$, $\kappa_i^{-1} = \sum_j \kappa_{ij}^{-1}$, where $i, j$ denote the components (particle species). For quark matter $\kappa$ is the sum of the partial conductivities of the electron, quark and gluon components $\kappa = \kappa_e + \kappa_q + \kappa_g$, where $\kappa_e \simeq \kappa_{ee}$ is determined by electron-electron scattering processes since in superconducting quark matter the partial contribution $1/\kappa_{eq}$ (as well as $1/\kappa_{gq}$) is additionally suppressed by a $\zeta_{\text{QDU}}$ factor, as for the scattering on impurities in metallic superconductors. Due to very small resulting value of $\kappa$ the typical time necessary for the heat to reach the star surface is large, delaying the cooling of HNS.

EoS used in ref. [53] included a model of the hadronic matter, region of mixed phase and of pure quark matter. A hard EoS for the hadronic matter was used, finite size effects were disregarded in description of the mixed phase, and the bag constant $B$ was taken to be rather small that led to the presence of a wide region of mixed and quark phases already for the HNS of the mass $M = 1.4\ M_\odot$. On the other hand, absence of a dense hadronic region within this EoS allowed to diminish uncertainties in description of in-medium effects in hadronic matter suppressing them according to that used in the "standard scenario".

With the above inputs ref. [53] solved the evolution equation for the temperature profile. In order to demonstrate the influence of the size of the diquark and nucleon pairing gaps on the evolution of the temperature profile solutions were performed with different values of the quark and nucleon gaps. Comparison of the cooling evolution ($\lg T_s$ vs. $\lg t$) of HNS of the mass $M = 1.4\ M_\odot$ is given in Fig. 7. The curves for $\Delta_q \gtrsim 1$ MeV are very close to each other demonstrating typical large gap behavior. The behavior of the cooling curve for $t \leq 50 \div 100$ yr is in straight correspondence with the heat transport processes. The subsequent time evolution is governed by the processes in the hadronic shell and by a delayed transport within the quark core with a dramatically suppressed neutrino emissivity from the color superconducting region. In order to demonstrate this feature a calculation was performed with the nucleon gaps ($\Delta_i(n)$, $i = n, p$) being artificially suppressed by a factor 0.2. Then up to $\lg(t[\mathrm{yr}]) \lesssim 4$ the behavior of the cooling curve is analogous to the one would be obtained for pure hadronic matter. The curves labelled "MMU" show the cooling of hadronic matter with inclusion of appropriate medium modifications in the $NN$ interaction. These effects have an influence on the cooling evolution only for $\lg(t[\mathrm{yr}]) \lesssim 2$ since the specific model EoS used does not allow for high nucleon densities in the hadronic phase at given example of HNS of $M = 1.4\ M_\odot$. The effect would be much more pronounced for larger star masses, a softer EoS for hadronic matter and smaller values of the gaps in the hadronic phase. Besides, incorporation of finite size effects within description of the mixed phase reducing its region should enlarge the size of the pure hadronic phase.

The unique asymptotic behavior at $\lg(t[\mathrm{yr}]) \geq 5$ for all the curves corresponding to finite values of the quark and nucleon gaps is due to a competition between normal electron contribution to the specific heat and the photon emissivity from the surface since small exponentials switch off all the processes related to paired particles. This tail is very sensitive to the interpolation law $T_s = f(T_m)$ used to simplify the consideration of the crust. The curves coincide at large times due to the uniquely chosen relation $T_s \propto T_m^{2/3}$.

The curves for $\Delta_q = 0.1$ MeV demonstrate an intermediate cooling behavior between those for $\Delta_q = 50$ MeV and $\Delta_q = 0$. Heat transport becomes not efficient after first $5 \div 10$ yr. The subsequent $10^4$ yr evolution is governed by QDU processes and quark specific heat being only moderately suppressed by the gaps and by the rates of NPBF processes in the hadronic matter (up to $\lg(t[\mathrm{yr}]) \leq 2.5$). At $\lg(t[\mathrm{yr}]) \geq 4$ begins the photon cooling era.

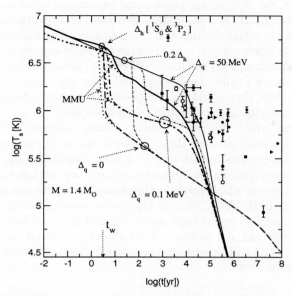

**Fig. 7.** Evolution of the surface temperature $T_s$ of HNS with $M = 1.4 M_\odot$ for $T_s = (10 T_m)^{2/3}$, where $T$ is in K, see [8]. Data are from [26] (full symbols) and from [35] (empty symbols), $t_w$ is the typical time which is necessary for the cooling wave to pass through the crust.

The curves for normal quark matter ($\Delta_q = 0$) are governed by the heat transport at times $t \lesssim 5$ yr and then by QDU processes and the quark specific heat. The NPBF processes are important up to $\lg(t[\text{yr}]) \leq 2$, the photon era is delayed up to $\lg(t[\text{yr}]) \geq 7$. For times smaller than $t_w$ (see Fig. 7) the heat transport is delayed within the crust area [15]. Since for simplicity this delay was disregarded in the heat transport treatment, for such small times the curves should be interpreted as the $T_m(t)$ dependence scaled to guide the eye by the same law $\propto T_m^{2/3}$, as $T_s$.

For the CFL phase with large quark gap, which expected to exhibit the most prominent manifestations of color superconductivity in HNS and QCNS, [53] thus demonstrated an essential delay of the cooling during the first $50 \div 300$ yr (the latter for QCNS) due to a dramatic suppression of the heat conductivity in the quark matter region. This delay makes the cooling of HNS and QCNS not as rapid as one could expect when ignoring the heat transport. In HNS compared to QCNS (large gaps) there is an additional delay of the subsequent cooling evolution which comes from the processes in pure hadronic matter.

In spite of that we find still too fast cooling for those objects compared to ordinary NS. Therefore, with the CFL phase of large quark gap it seems rather difficult to explain the majority of the presently known data both in the cases of the HNS and QCNS, whereas in the case of pure hadronic stars the available data are much better fitted even within the same simplified model for the hadronic matter. For 2SC (3SC) phases one may expect analogous behavior to that demonstrated by $\Delta_q = 0$ since QDU processes on unpaired quarks are then allowed, resulting in a fast cooling. It is however not excluded that new observations may lead to lower surface temperatures for some supernova remnants and will be better consistent with the model which also needs further improvements. On the other hand, if future observations will show very large temperatures for young compact stars they could be interpreted as a manifestation of large gap color superconductivity in the interiors of these objects.

**Concluding**, the *"nuclear medium cooling scenario"* allows easily to achieve agreement with existing data. However there remains essential uncertainty in quantitative predictions due to a poor knowledge, especially, of the residual interaction treated above in an economical way within a phenomenological Fermi liquid model which needs further essential improvements. As for color superconductivity in HNS, QCNS, and QS, characterizing by large diquark pairing gaps, we did not find an appropriate fit of existing $X$ ray data. Situation will be changed if more cold, old objects will be observed. Also, if young hot objects ( on scale $\sim 10^2$ yr.) will be observed it could be interpreted as a signature of CFL phase in these objects.

**Acknowledgement**

Author appreciates the hospitality and support of GSI Darmstadt and ECT* Trento. He thanks E. Kolomeitsev for discussions.

# References

1. S. Shapiro, S.A. Teukolsky: *Black Holes, White Dwarfs and Neutron Stars: The Physics of Compact Objects*. ( Wiley, New York 1983), chapter 11
2. D. Pines, R. Tamagaki, S. Tsuruta (eds.): *Neutron Stars*. (Addison-Weseley, New York 1992)
3. Ed. by M. A. Alpar, Ü. Kiziloglu, J. van Paradijs (eds.): *The Lives of the Neutron Stars* (NATO ASI Ser. C, 450; Dordrecht: Kluwer 1995)
4. J.N. Bahcall, R.A. Wolf: Phys. Rev. B **140**, 1445 (1965)
5. S. Tsuruta, A.G.W. Cameron: Canad. J. Phys. **43**, 2056 (1965)
6. B. Friman, O.V. Maxwell: Ap. J. **232**, 541 (1979)
7. O.V. Maxwell: Ap. J. **231**, 201 (1979)
8. S. Tsuruta: Phys. Rep. **56**, 237 (1979)
9. K. Nomoto, S. Tsuruta: Ap. J. Lett. **250**, 19 (1981)
10. Ch. Schaab, F.Weber, M.K. Weigel, N.K. Glendenning: Nucl. Phys. A **605**, 531 (1996)
11. G. Flowers, P.G. Sutherland, J.R. Bond: Phys. Rev. D **12**, 315 (1975)
12. O. Maxwell, G.E. Brown et al.: Ap. J. **216**, 77 (1977)
13. G.E. Brown, K. Kubodera, D. Page, P. Pizzochero: Phys. Rev. D **37**, 2042 (1988)
14. T. Tatsumi: Prog. Theor. Phys. **88**, 22 (1988)
15. J.M. Lattimer, C.J. Pethick, M. Prakash, P. Haensel: Phys. Rev. Lett. **66**, 2701 (1991)

16. A.B. Migdal, E.E. Saperstein, M.A. Troitsky, D.N. Voskresensky: Phys. Rep. **192**, 179 (1990)
17. T. Takatsuka, R. Tamagaki: Prog. Theor. Phys. **97**, 1 (1997)
18. A. Akmal, V.R. Pandharipande, D.G. Ravenhall: Phys. Rev. C **58**, 1804 (1998)
19. T. Suzuki, H. Sakai, T. Tatsumi: nucl-th/9901097
20. G.E. Brown, C.H. Lee, M. Rho, V. Thorsson: Nucl. Phys. A **567**, 937 (1994)
21. E.E. Kolomeitsev, D.N. Voskresensky, B. Kämpfer: Nucl. Phys. A **588**, 889 (1995)
22. D.N. Voskresensky, A.V. Senatorov: JETP Lett. **40**, 1212 (1984)
23. D.N. Voskresensky, A.V. Senatorov: JETP **63**, 885 (1986)
24. D.N. Voskresensky, A.V. Senatorov: Sov. J. Nucl. Phys. **45**, 411 (1987)
25. A.V. Senatorov, D.N. Voskresensky: Phys. Lett. B **184**, 119 (1987)
26. Ch. Schaab, D. Voskresensky, et al.: Astron. Astrophys. **321**, 591 (1997)
27. E.E. Kolomeitsev, D.N. Voskresensky: Phys. Rev. C **60**, 034610 (1999)
28. J. Knoll, D. N. Voskresensky: Phys. Lett. B **351**, 43 (1995); Ann. Phys. (New York) **249**, 532 (1996)
29. N.K. Glendenning: Phys. Rev. D **46**, 1274 (1992)
30. T. Tatsumi: Prog. Theor. Phys. **69**, 1137 (1983)
31. H. Umeda, K. Nomoto, et al: Ap.J. **431**, 309 (1994)
32. E. Flowers, M. Ruderman, P. Sutherland: Ap. J. **205**, 541 (1976)
33. K.P. Levenfish, D.G. Yakovlev: Astron. Reports, **38**, 247 (1994)
34. D.G. Yakovlev, A.D. Kaminker, K.P. Levenfish: Astron. Astrophys. **343**, 650 (1999)
35. D.G. Yakovlev, K.P. Levenfish, Yu.A. Shibanov: Phys. Usp. **42**, 737 (1999)
36. D. Page: *Many Faces of Neutron Stars*, ed. by R. Buccheri, J. van Paradijs, M.A. Alpar (Dordrecht, Kluwer, 1998) p. 538
37. R.F. Sawyer, A. Soni: Ap. J. **230**, 859 (1979)
38. R.F. Sawyer: Ap. J. **237**, 187 (1980)
39. R.F. Sawyer, A. Soni: Ap. J. **216**, 73 (1977)
40. S. Reddy, M. Prakash: Ap.J. **423**, 689 (1997)
41. J.A. Pons, S. Reddy, et al.: Ap.J. **513**, 780 (1999)
42. D.N. Voskresensky, A.V. Senatorov, B. Kämpfer, H. Haubold: Astrophys. Space Sci. **138**, 421 (1987)
43. H. Haubold, B. Käempfer, A.V. Senatorov, D.N. Voskresensky: Astron. Astrophys. **191**, L22 (1988)
44. A. Sedrakian, A. Dieperink: Phys. Lett. B **463**, 145 (1999); Phys.Rev. **D62**, 083002 (2000)
45. S.Yamada: Nucl. Phys. **A662**, 219 (2000)
46. A.B. Migdal: ZhETF. **61**, 2209 (1971) (in Engl.: JETP. **34**, 1184 (1972))
47. D.N. Voskresensky, G.A. Sorokin, A.I. Chernoutsan: JETP. Lett **26**, 465 (1977)
48. N.K. Glendenning: *Compact Stars* ( New York: Springer- Verlag 1997)
49. E. Witten: Phys. Rev. D **30**, 272 (1984)
50. M. Alford, K. Rajagopal, F. Wilczek: Phys. Lett. B **422**, 247 (1998)
51. R. Rapp, T. Schäfer, E.V. Shuryak, M. Velkovsky: Phys. Rev. Lett. **81**, 53 (1998)
52. D. Blaschke, T. Klähn, D.N. Voskresensky: Ap.J. **533**, 406 (2000)
53. D. Blaschke, H. Grigorian, D.N. Voskresensky: astro-ph/0009120, Astron. Astrophys., in press (2001)
54. D. Page, M. Prakash, J.M. Lattimer, A. Steiner: Phys. Rev. Lett. **85**, 219 (2048)
55. L.D. Landau: Sov. JETP. **3**, 920 (1956)
56. A.B. Migdal: *Theory of Finite Fermi Systems and Properties of Atomic Nuclei* (Willey and Sons, New York 1967; second. ed. (in Rus.), Nauka, Moscow, 1983)
57. A.B. Migdal: Rev. Mod. Phys. **50**, 107 (1978)

58. D. N. Voskresensky: Nucl. Phys. A **555**, 293 (1993)
59. E.E. Saperstein, S.V. Tolokonnikov: JETP Lett. **68**, 553 (1998)
60. S.A. Fayans, D. Zawischa: Phys. Lett. B **363**, 12 (1995)
61. I.N. Borzov, S.V. Tolokonnikov, S.A. Fayans: Sov. J. Nucl. Phys. **40**, 732 (1984)
62. A.M. Dyugaev: Pisma v ZhETF. **22**, 181 (1975)
63. D.N. Voskresensky, I.N. Mishustin: JETP Lett. **34**, 303 (1981); Sov. J. Nucl. Phys. **35**, 667 (1982)
64. A.M. Dyugaev: ZhETF. **83**, 1005 (1982); Sov. J. Nucl. Phys. **38**, 680 (1983)
65. A.P. Platonov, E.E, Saperstein, S.V. Tolokonnikov, S.A. Fayans: Phys. At. Nucl., **58**, 556 (1995)
66. G.F. Bertsch, L. Frankfurt and M. Strikman: Science. **259**, 773 (1993)
67. G.E. Brown, M. Buballa, Zi Bang Li and J. Wambach: Nucl. Phys. A **593**, 295 (1995)
68. M. Ericson: Nucl. Phys. A **577** (1994) 147c
69. J. Delorm, M. Ericson: Phys. Rev. C **49** (1994) 1763
70. E. Marco. E. Oset and P. Fernandez de Cordoba: Nucl. Phys. A **611**, 484 (1996)
71. G.E. Brown, M. Rho: Phys. Rev. Lett. **66**, 2720 (1991); Phys. Rep. **269**, 333 (1996)
72. G. Baym, D. Campbell, R. Dashen, J. Manassah: Phys. Lett. B **58**, 304 (1975)
73. D.N. Voskresensky, E.E. Kolomeitsev, B. Kämpfer: JETP. **87**, 211 (1998)
74. A.D. Kaminker, P. Hansel, D.G. Yakovlev: Astron. Astroph. **345**, L14 (1999)
75. D.G. Yakovlev, A.D. Kaminker, O.Y. Gnedin, P. Hansel: astro-ph/0012122
76. L.B. Leinson: Phys. Lett. B **473**, 318 (2000)
77. C. Hanhart, D. R. Phillips, and S. Reddy: nucl-th/0003445
78. D. Blaschke, G. Röpke, et all.: MNRAS. **273**, 596 (1995)
79. T. Takatsuka, R. Tamagaki: Progr. Theor. Phys. **64**, 2270 (1980)
80. D.N. Voskresensky, V. A. Khodel, M. V. Zverev, J.W. Clark: astro-ph/0003172; Ap. J. **533**, L127 (2000)
81. D.N. Voskresensky: Phys. Lett. B **392**, 262 (1997)
82. G. Raffelt, D. Seckel: Phys. Rev. D **52**, 1780 (1995)
83. S. Hannestad, G. Raffelt: astro-ph/971132
84. Yu.B. Ivanov, J. Knoll, D.N. Voskresensky: Nucl. Phys. A **672**, 314 (2000)
85. Yu.B. Ivanov, J. Knoll, H.van Hees, D.N. Voskresensky: nucl-th/0005075
86. N. Iwamoto: Ann. Phys. **141**, 1 (1982)
87. R.C. Duncan, S.L. Shapiro, I. Wasserman: ApJ **267**, 358 (1983)
88. A.D. Kaminker, P. Haensel: Acta Phys. Pol. B **30**, 1125 (1999)

# Books for Further Study

### Observation

R. D. Blandford, A. Hewish, A. G. Lyne, L. Mestel (Eds.)
*Pulsars as Physics Laboratories*
(Oxford University Press; 197 pages, 1992)
ISBN 0-198-53983-5

Francis Graham-Smith, Andrew G. Lyne
*Pulsar Astronomy*
(Cambridge University Press; 275 pages, 1998)
ISBN 0-521-59413-8

Richard N. Manchester, Joseph H. Taylor
*Pulsars*
(W. H. Freeman & Co.; 217 pages, 1977)
ISBN 0-716-70358-0

### Theory

M. Baldo (Ed.)
*Nuclear Methods and the Nuclear Equation of State*
(World Scientific; 512 pages, 1999)
ISBN 981-02-2165-7

V.S. Beskin, A.V. Gurevich, Ya.N. Istomin
*Physics of the Pulsar Magnetosphere*
(Cambridge University Press; 432 pages, 1993)
ISBN 0-521-41746-5

Norman K. Glendenning
*Compact Stars: Nuclear Physics, Particle Physics, and General Relativity*
(Springer Verlag; 1st edition, 390 pages, 1996; 2nd edition, 536 pages, 2000)
ISBN 0-387-98977-3

Vladimir M. Lipunov
*Astrophysics of Neutron Stars*
(Springer Verlag, 322 pages, 1987)
ISBN 3-540-53568-3

Peter Meszaros
*High-Energy Radiation from Magnetized Neutron Stars*
(University of Chicago Press; 531 pages, 1992)
ISBN 0-226-52094-3


F. Curtis Michel
*Theory of Neutron Star Magnetospheres*
(University of Chicago Press; 517 pages, 1991)
ISBN 0-226-52330-6


G. S. Saakian
*Equilibrium Configurations of Degenerate Gaseous Masses*
(John Wiley & sons; 311 pages, 1973)
ASIN 0-470-74805-2


Stuart L. Shapiro and Saul A. Teukolsky
*Black Holes, White Dwarfs and Neutron Stars: The Physics of Compact Objects*
(John Wiley & Sons; 672 pages, 1st edition, 1983)
ISBN 0-471-87316-0


Jean Louis Tassoul
*Stellar Rotation*
(Cambridge University Press; 272 pages, 2000)
ISBN 0-521-77218-4


Fridolin Weber
*Pulsars as Astrophysical Laboratories for Nuclear and Particle Physics*
(Institute of Physics Publishing, 682 pages, 1999)
ISBN 0-750-30332-8


Ia. B. Zeldovich and I. D. Novikov
*Stars and Relativity*
(Dover Publications; 522 pages, reissue edition, 1997)
ISBN 0-486-69424-0

# Index

Accretion 152
- disk 315
- donor mass 323
- epoch 317
- onto neutron star 306, 313
- pathway from canonical to ms pulsar 293, 321
- rate 155, 297, 316
- x-rays from 313

Babu-Brown model 42, 44
Baryon octet 219, 312
BCS phase 212, 244
BCS theory 30, 50, 99, 206
Bethe-Brueckner-Goldstone theory 1, 2, 10, 473
Bethe-Faddeev equations 9, 11
Bethe-Salpeter equation 219
Binaries
- donor mass 318
- x-ray 159, 305, 306, 316, 318, 325
Black hole 75, 366, 397
Boltzmann equation 343, 368
- equilibrium diffusion approximation 370
- neutrino transport 345
Bose condensate 74, 79, 367
Braking index 70, 292, 309
- as function of stellar spin 309, 313
- during phase transition 292, 312, 313
Bremsstrahlung 468, 482
Brueckner-Hartree-Fock approximation 2, 37, 41, 47, 138
- three-nucleon correlations 8
- relativistic 138

Canonical pulsar 306
Chandrasekhar limit 54, 58, 311
Charge neutrality 327, 374

Chiral density wave *see* LOFF
Chiral symmetry 175, 235
- breaking 175
- effective interactions 178
- lagrangian 183
- restoration 177
Coherence length 99
Collective excitations 42, 380
- RPA 381, 394
Color superconductivity 203, 221, 235, 243, 474
- crystalline(LOFF) 243
Color-flavor locking 206, 236
Color-spin locking 207
Condensate
- chiral 235
- diquark 236
- kaon 175
- strange diquark 240
Conservation
- of baryon number 325, 374, 379
- of charge 325, 379
Conservation laws
- for phases in equilibrium 328
- in hydrodynamics 103, 104
Conserved charge 327
Cooling scenario 402
- non-standard (rapid) 398, 468, 488
- standard (slow) 367, 398, 468, 488
Cooper pairs 30, 45, 57, 76, 79, 99, 100, 221, 235
Coulomb lattice 328
Crab pulsar 305
- power output 307
- properties 308
Critical temperature 30
Crust 56, 67, 73, 127
- accreted 152, 171
- crystalline lattice 128

- deformation   165
- density inversion   171
- elastic strain   62, 164
- elastic stress   128
- electrical conductivity   127, 170
- impurities   169
- non-equilibrium processes   128
- plate drift   69
- shear modulus   166, 168
- thermal conductivity   127, 170

Crustquakes   62, 68
Crystalline lattice   328, 470

Deconfined hadronic matter   *see* Quark matter
Deleptonization   24, 364, 394
Diquark
- condensate   221
- correlations   218

Dispersion by interstellar electrons   306
Dynamical screening   205, 238
Dyson equation   31
Dyson-Schwinger equation   219

Effective interactions   15, 31, 45, 135, 138, 148, 470, 474
Electron chemical potential
- effect of hyperonization   197

Entrainment effect   59, 88, 98
Equation of state   178, 191, 193, 200, 255, 319, 329, 440
- crust   128, 160
- first order phase transition   329
- models   405, 412
- of dense matter   18, 25, 161, 364, 404, 405
- of neutrino trapped matter   371
- polytropes   406, 410
- softening   306, 311, 375, 401, 406

Euler equation   80, 84, 105, 125

Fermi-liquid theory   469, 474
Fluctuations
- of density   44
- of spin   44

Gamma-ray bursts   275
Gap
- equation   45
- function   30, 31, 42, 399

Gibbs
- energy   132, 153
- phase equilibrium   192, 327
- rules   374, 400

Ginzburg-Landau theory   82, 100, 101
Global color model   219
Gluon propagator model   224
Gluons   176
Goldstone boson   176, 209, 224, 236
Gorkov equations   31
Gravitational collapse   138, 365
Gravitational waves   73, 98, 125, 128, 357
Green's functions   31
Guitar nebula pulsar   424

Hamiltonian   2
- dynamics   76

Hartree-Fock approximation   134, 135
Heavy-ion collisions   203, 238, 253, 285
Hyperonization   195, 311
Hyperons   21, 24, 372, 401, 403
- effect on kaon condensate   195
- in neutron star   24, 191, 195, 329
- interactions   21, 25

Instantons   212
Interacting Fermi systems   31
- superfluid   31

Interstellar medium   440
- dispersion of radio signal   306, 320

Kaon
- in neutron star   190
- in nuclear matter   178, 179, 186
-- with hyperon   198
- in nuclear matter   178
- optical potential   178

Kaon condensate   24, 178, 209, 365, 373, 397, 468, 473, 485
- geometrical structure   195

Kaon-nucleon interaction   178
Kaonic atom   178, 188
Kick
- neutron star   424
- velocity   424

Lagrangian density   75, 76, 80, 83, 85, 90, 372, 384
Landau parameters   44
Lattice QCD   175

Lindhard function  40
Liquid drop model  129, 143, 146

Maclaurin spheroid  115
Magnetic field  *see* Neutron Star
Magneto-hydrodynamics  98
Meissner effect  242
Millisecond pulsar  239, 306, 440
MIT bag model  226, 258, 330, 376, 408, 473
Mutual friction  65, 102

Nambu-Gorkov propagator  206, 222
Nambu-Jona-Lasinio model  212, 219
Neutrino  364
– differential cross section  381, 385, 389
– diffusion  352, 366
– emission  152, 239, 367, 468
– heating  336
– luminosity  371, 392
– mean free path  366, 380, 382, 386, 389, 394, 474
– opacity  364, 380
– transport  369
– trapping  365
Neutron star  1, 20, 54, 55, 305
– accretion  152, 306, 313, 314, 316, 321, 323, 324, 440, 441
– binding energy  23, 264, 414
– census in the Milky Way  460
– conserved charges  325, 327
– conversion to strange star  275
– crust  127, 152
– – liquid phase  152
– crystalline lattice  328
– ejector phase  441
– evolution  440
– evolutionary path  313, 321, 323, 324, 451
– glitches  60, 97, 128, 413
– – in quark matter  246
– hybrid  289, 312, 329
– in low-mass x-ray binary  306, 321, 440, 451
– initial spin  434
– kaons  190
– Kepler frequency  27
– kick  424
– – and supernova assymetry  424
– – velocity  424

– large space velocities  424
– magnetic field  316, 440
– – evolution  242, 321, 444
– mass-radius relationship  26, 263, 269, 405, 408, 416
– maximum mass  28, 198, 260, 311, 365, 366
– moment of inertia  59, 290, 306, 308, 310, 311, 316, 412
– oscillations  98, 128
– population synthesis  440, 456
– propeller phase  441
– quark matter core  311, 312, 317, 318
– rotating  288
– – moment of inertia  267
– – perturbation expansion  288
– – population clustering  298, 305, 318
– – rapidly  265
– – slowly  288
– spin  440, 441
– – distribution of accretors  317
– – evolution  287, 291, 317
– – properties  305
– structure  25, 26, 259, 289, 321, 322
– thermal evolution  30, 127, 398, 401, 467, 468
– thermal radiation  97, 398, 408
– two-component model  60
– x-ray accretor  298, 306, 313–316
– – spin clustering of the population  298, 305, 313, 320
– x-ray bursters  269
NJL model  400
– Lagrangian  376
Nuclear matter  1, 138, 312, 372, 374
– compressibility  129, 372, 392, 406, 412
– equation of state  1, 8, 13, 23
– in-medium effects  380, 394, 469
– mixed phase  374, 375, 378
– neutronization  132
– polarization tensor  381
– saturation  13
– – properties  330
– symmetry energy  24, 372, 373, 392, 414
Nuclei  130
– Coulomb energy  141, 144, 147
– experimental masses  131, 132
– Hartree-Fock calculations  135, 136
– non-spherical  138, 145, 148

- shell effects   133, 136
- surface energy   146, 153
- surface tension   140, 144, 147
Nucleon resonances   475
Nucleon-nucleon interaction   2, 5, 129, 135, 138, 358, 468, 475
- meson exchange   468, 479, 482
- partial waves   12, 19, 34
- phase shifts   36
- potentials   2, 8, 14, 30, 143
- three-body forces   13, 16, 24, 25, 135, 143

Ohmic decay   448
Oppenheimer - Volkoff bound   55
Optical potential
- for kaons   374, 397
Oscillations   99
- non-radial   99, 120, 128

Pairing theory   30
- for nuclei   30, 132
Partition function   372
Pauli-Gürsey symmetry   224
Phase diagram
- critical line   286
- for QCD   235
- for rotating stars   286
Phase transition   145, 150, 170, 175, 178, 290, 379
- and moment of inertia   291, 308
- Coulomb lattice   328
- deconfinement   200, 306, 311, 313, 318, 329
- - signal   287, 294, 305
- degrees of freedom   326
- first order   148, 149, 162, 325, 384
- geometrical structure in mixed phase   194, 199, 328
- global conservation law   327
- in neutron star   305, 306
- kaon condensate   192
- phase equilibrium in multi-component substance   193, 327
- second order   133
Pion condensate   209, 365, 469, 473, 485
Polarization effects   30, 40, 44
Proto-neutron star   241, 364
- collapse   396, 397
- convection   346, 352

- convective overturn   346
- evolution   368, 392
Pulsar   54, 97
- canonical   305
- Crab   61, 62, 68, 71
- evolution from canonical to ms pulsar   321
- glitches   64, 67
- graveyard   324
- millisecond   159, 305, 306
- recycled   313
- timing   55
- Vela   61, 62, 71, 368

Quantum chromodynamics   14, 175
- vacuum   177
- partition function   176
- phase diagram   235
Quark   175
- confinement   226
- deconfinement   287, 311
- light flavors   330
- strange   198
Quark matter   200, 305, 306, 309–312, 327, 330, 365, 376, 397, 473
- mixed phase   310, 380
- polarization tensor   387
- strange   248, 255
- superconductivity   57, 235, 387, 404

R-mode instability   241
Rayleigh-Taylor instability   171
Reactions   158, 380, 383
- capture   154, 155, 171, 417
- picnonuclear   152
Relativistic mean-field theory   139, 149, 330, 372, 373, 400
Reynolds number   112
Roche lobe   323

Scattering matrix   38, 493
- G-matrix   2, 3, 7, 10, 41, 43
- T-matrix   11, 41, 482
Schwarzschild radius   54
Self-energy   30, 31, 45
- Brueckner-Hartree-Fock   2, 5, 8, 37, 47
- Hartree-Fock   5, 33
Skyrme force   *see* Effective interactions
Spacetime metric   74
Strange

– matter
–– equation of state 255
– quark
–– stars 259
Strange matter 366, 407, 473
Strange star 54, 58
Strangelets 279
Superconductivity 72, 399
– proton matter 57, 59
Superfluid 65
– interacting Fermi systems 31
– Magnus force 65, 91
– moment of inertia 64
– order parameter 100
– phase 81
– pinning 64, 67
– relaxation times 63, 64, 67
– rotation 63
– vortices 73, 100, 102
Superfluid Helium 73, 79, 80, 82, 86, 103
Superfluid hydrodynamics 72, 99
Superfluidity 54, 56, 58, 72, 73, 97, 98, 367, 399
– hyperons 404
– Landau model 82, 84, 86
– neutron matter 30, 54, 57, 59
– relativistic treatment 72, 74, 86
Supernova
– anisotropies 351
– asymmetry 424
– clumping 351
– core bounce 365
– core collapse 128, 138, 343, 425
– delayed mechanism 343, 426
– detection limits 395
– explosion 138, 335, 364, 365, 473
–– Rayleigh-Taylor instability 345
–– revival 336
– instability
–– Quasi-Ledoux criterion 355
–– Solberg-Høiland criterion 357
– neutrino 241
–– driven 338, 426
–– signal 359
– neutron star kick 358, 424
– nucleosynthesis 349
– prompt shock 346
– radioactive elements 349

– remnants 467
– shock 340, 365
–– revival 342
– SN 1987A 351, 395
Surface tension 194, 329

Tensor
– momentum-energy 78, 90
Tensor virial method 98
Thermal evolution 404
Thermodynamical potential 205, 226
Thomas-Fermi approximation 135, 137, 138, 151
Tolman–Oppenheimer–Volkoff equations 259

Urca process 240, 367, 398, 402, 484
– modified 367, 399, 402, 468, 470, 482
– on hyperons 403, 468
– on quarks 403

Variational principle 80
Vertex corrections 10, 30, 469, 471, 474, 479, 482
Virial equations 105, 109
Virial theorem 98, 142, 147
Vortex 65, 73, 86, 100, 102
– friction force 65, 102
– lattice 66
– pinning 67, 68, 247

White dwarf 54, 56
Wigner-Seitz approximation 131, 141, 147, 153, 384

X-ray
– accretor 306
– binaries (LMXBs) 239, 270, 293, 300, 305
– binary 159
– bursts 170, 270, 320
– neutron star 306
– photosphere 127
– quasi periodic oscillations 273, 293, 300, 320
– satellite 467
– thermal emission 367
X-rays 97

# Lecture Notes in Physics

For information about Vols. 1–542
please contact your bookseller or Springer-Verlag

Vol. 543: H. Gausterer, H. Grosse, L. Pittner (Eds.), Geometry and Quantum Physics. Proceedings, 1999. VIII, 408 pages. 2000.

Vol. 544: T. Brandes (Ed.), Low-Dimensional Systems. Interactions and Transport Properties. Proceedings, 1999. VIII, 219 pages. 2000

Vol. 545: J. Klamut, B. W. Veal, B. M. Dabrowski, P. W. Klamut, M. Kazimierski (Eds.), New Developments in High-Temperature Superconductivity. Proceedings, 1998. VIII, 275 pages. 2000.

Vol. 546: G. Grindhammer, B. A. Kniehl, G. Kramer (Eds.), New Trends in HERA Physics 1999. Proceedings, 1999. XIV, 460 pages. 2000.

Vol. 547: D. Reguera, G. Platero, L. L. Bonilla, J. M. Rubí(Eds.), Statistical and Dynamical Aspects of Mesoscopic Systems. Proceedings, 1999. XII, 357 pages. 2000.

Vol. 548: D. Lemke, M. Stickel, K. Wilke (Eds.), ISO Surveys of a Dusty Universe. Proceedings, 1999. XIV, 432 pages. 2000.

Vol. 549: C. Egbers, G. Pfister (Eds.), Physics of Rotating Fluids. Selected Topics, 1999. XVIII, 437 pages. 2000.

Vol. 550: M. Planat (Ed.), Noise, Oscillators and Algebraic Randomness. Proceedings, 1999. VIII, 417 pages. 2000.

Vol. 551: B. Brogliato (Ed.), Impacts in Mechanical Systems. Analysis and Modelling. Lectures, 1999. IX, 273 pages. 2000.

Vol. 552: Z. Chen, R. E. Ewing, Z.-C. Shi (Eds.), Numerical Treatment of Multiphase Flows in Porous Media. Proceedings, 1999. XXI, 445 pages. 2000.

Vol. 553: J.-P. Rozelot, L. Klein, J.-C. Vial Eds.), Transport of Energy Conversion in the Heliosphere. Proceedings, 1998. IX, 214 pages. 2000.

Vol. 554: K. R. Mecke, D. Stoyan (Eds.), Statistical Physics and Spatial Statistics. The Art of Analyzing and Modeling Spatial Structures and Pattern Formation. Proceedings, 1999. XII, 415 pages. 2000.

Vol. 555: A. Maurel, P. Petitjeans (Eds.), Vortex Structure and Dynamics. Proceedings, 1999. XII, 319 pages. 2000.

Vol. 556: D. Page, J. G. Hirsch (Eds.), From the Sun to the Great Attractor. X, 330 pages. 2000.

Vol. 557: J. A. Freund, T. Pöschel (Eds.), Stochastic Processes in Physics, Chemistry, and Biology. X, 330 pages. 2000.

Vol. 558: P. Breitenlohner, D. Maison (Eds.), Quantum Field Theory. Proceedings, 1998. VIII, 323 pages. 2000

Vol. 559: H.-P. Breuer, F. Petruccione (Eds.), Relativistic Quantum Measurement and Decoherence. Proceedings, 1999. X, 140 pages. 2000.

Vol. 560: S. Abe, Y. Okamoto (Eds.), Nonextensive Statistical Mechanics and Its Applications. IX, 272 pages. 2001.

Vol. 561: H. J. Carmichael, R. J. Glauber, M. O. Scully (Eds.), Directions in Quantum Optics. XVII, 369 pages. 2001.

Vol. 562: C. Lämmerzahl, C. W. F. Everitt, F. W. Hehl (Eds.), Gyros, Clocks, Interferometers...: Testing Relativistic Gravity in Space. XVII,507 pages. 2001.

Vol. 563: F. C. Lázaro, M. J. Arévalo (Eds.), Binary Stars. Selected Topics on Observations and Physical Processes. 1999.IX, 327 pages. 2001.

Vol. 564: T. Pöschel, S. Luding (Eds.), Granular Gases. VIII, 457 pages. 2001.

Vol. 565: E. Beaurepaire, F. Scheurer, G. Krill, J.-P. Kappler (Eds.), Magnetism and Synchrotron Radiation. XIV, 388 pages. 2001.

Vol. 566: J. L. Lumley (Ed.), Fluid Mechanics and the Environment: Dynamical Approaches. VIII, 412 pages. 2001.

Vol. 567: D. Reguera, L. L. Bonilla, J. M. Rubí (Eds.), Coherent Structures in Complex Systems. IX, 465 pages. 2001.

Vol. 568: P. A. Vermeer, S. Diebels, W. Ehlers, H. J. Herrmann, S. Luding, E. Ramm (Eds.), Continuous and Discontinuous Modelling of Cohesive-Frictional Materials. XIV, 307 pages. 2001.

Vol. 569: M. Ziese, M. J. Thornton (Eds.), Spin Electronics. XVII, 493 pages. 2001.

Vol. 570: S. G. Karshenboim, F. S. Pavone, F. Bassani, M. Inguscio, T. W. Hänsch (Eds.), The Hydrogen Atom: Precision Physics of Simple Atomic Systems. XXIII, 293 pages. 2001.

Vol. 571: C. F. Barenghi, R. J. Donnelly, W. F. Vinen (Eds.), Quantized Vortex Dynamics and Superfluid Turbulence. XXII, 455 pages. 2001.

Vol. 572: H. Latal, W. Schweiger (Eds.), Methods of Quantization. XI, 224 pages. 2001.

Vol. 573: H. M. J. Boffin, D. Steeghs, J. Cuypers (Eds.), Astrotomography. XX, 434 pages. 2001.

Vol. 574: J. Bricmont, D. Dürr, M. C. Galavotti, G. Ghirardi, F. Petruccione, N. Zanghi (Eds.), Chance in Physics. XI, 288 pages. 2001.

Vol. 575: M. Orszag, J. C. Retamal (Eds.), Modern Challenges in Quantum Optics. XXIII, 405 pages. 2001.

Vol. 576: M. Lemoine, G. Sigl (Eds.), Physics and Astrophysics of Ultra High Energy Cosmic Rays. X, 318 pages. 2001.

Vol. 577: N. Thomas, I. P. Williams (Eds.), Solar and Extra-Solar Planetary Systems. X, 266 pages. 2001.

Vol. 578: D. Blaschke, N. K. Glendenning, A. Sedrakian (Eds.), Physics of Neutron Star Interiors. XI, 509 pages. 2001.

Vol. 579: R. Haug, H. Schoeller (Eds.), Interacting Electrons in Nanostructures. X, 227 pages. 2001.

Vol. 580: K. Baberschke, M. Donath, W. Nolting (Eds.), Band-Ferromagnetism: Ground-State and Finite-Temperature Phenomena.XI, 346 pages. 2001.

Vol.581: J. M. Arias, M. Lozano (Eds.), An Advanced Course in Modern Nuclear Physics. XI, 362 pages. 2001.

# Monographs

For information about Vols. 1–27 please contact your bookseller or Springer-Verlag

Vol. m 28: O. Piguet, S. P. Sorella, Algebraic Renormalization. IX, 134 pages. 1995.

Vol. m 29: C. Bendjaballah, Introduction to Photon Communication. VII, 193 pages. 1995.

Vol. m 30: A. J. Greer, W. J. Kossler, Low Magnetic Fields in Anisotropic Superconductors. VII, 161 pages. 1995.

Vol. m 31 (Corr. Second Printing): P. Busch, M. Grabowski, P.J. Lahti, Operational Quantum Physics. XII, 230 pages. 1997.

Vol. m 32: L. de Broglie, Diverses questions de mécanique et de thermodynamique classiques et relativistes. XII, 198 pages. 1995.

Vol. m 33: R. Alkofer, H. Reinhardt, Chiral Quark Dynamics. VIII, 115 pages. 1995.

Vol. m 34: R. Jost, Das Märchen vom Elfenbeinernen Turm. VIII, 286 pages. 1995.

Vol. m 35: E. Elizalde, Ten Physical Applications of Spectral Zeta Functions. XIV, 224 pages. 1995.

Vol. m 36: G. Dunne, Self-Dual Chern-Simons Theories. X, 217 pages. 1995.

Vol. m 37: S. Childress, A.D. Gilbert, Stretch, Twist, Fold: The Fast Dynamo. XI, 406 pages. 1995.

Vol. m 38: J. González, M. A. Martín-Delgado, G. Sierra, A. H. Vozmediano, Quantum Electron Liquids and High-Tc Superconductivity. X, 299 pages. 1995.

Vol. m 39: L. Pittner, Algebraic Foundations of Non-Com-mutative Differential Geometry and Quantum Groups. XII, 469 pages. 1996.

Vol. m 40: H.-J. Borchers, Translation Group and Particle Representations in Quantum Field Theory. VII, 131 pages. 1996.

Vol. m 41: B. K. Chakrabarti, A. Dutta, P. Sen, Quantum Ising Phases and Transitions in Transverse Ising Models. X, 204 pages. 1996.

Vol. m 42: P. Bouwknegt, J. McCarthy, K. Pilch, The W3 Algebra. Modules, Semi-infinite Cohomology and BV Algebras. XI, 204 pages. 1996.

Vol. m 43: M. Schottenloher, A Mathematical Introduction to Conformal Field Theory. VIII, 142 pages. 1997.

Vol. m 44: A. Bach, Indistinguishable Classical Particles. VIII, 157 pages. 1997.

Vol. m 45: M. Ferrari, V. T. Granik, A. Imam, J. C. Nadeau (Eds.), Advances in Doublet Mechanics. XVI, 214 pages. 1997.

Vol. m 46: M. Camenzind, Les noyaux actifs de galaxies. XVIII, 218 pages. 1997.

Vol. m 47: L. M. Zubov, Nonlinear Theory of Dislocations and Disclinations in Elastic Body. VI, 205 pages. 1997.

Vol. m 48: P. Kopietz, Bosonization of Interacting Fermions in Arbitrary Dimensions. XII, 259 pages. 1997.

Vol. m 49: M. Zak, J. B. Zbilut, R. E. Meyers, From Instability to Intelligence. Complexity and Predictability in Nonlinear Dynamics. XIV, 552 pages. 1997.

Vol. m 50: J. Ambjørn, M. Carfora, A. Marzuoli, The Geometry of Dynamical Triangulations. VI, 197 pages. 1997.

Vol. m 51: G. Landi, An Introduction to Noncommutative Spaces and Their Geometries. XI, 200 pages. 1997.

Vol. m 52: M. Hénon, Generating Families in the Restricted Three-Body Problem. XI, 278 pages. 1997.

Vol. m 53: M. Gad-el-Hak, A. Pollard, J.-P. Bonnet (Eds.), Flow Control. Fundamentals and Practices. XII, 527 pages. 1998.

Vol. m 54: Y. Suzuki, K. Varga, Stochastic Variational Approach to Quantum-Mechanical Few-Body Problems. XIV, 324 pages. 1998.

Vol. m 55: F. Busse, S. C. Müller, Evolution of Spontaneous Structures in Dissipative Continuous Systems. X, 559 pages. 1998.

Vol. m 56: R. Haussmann, Self-consistent Quantum Field Theory and Bosonization for Strongly Correlated Electron Systems. VIII, 173 pages. 1999.

Vol. m 57: G. Cicogna, G. Gaeta, Symmetry and Perturbation Theory in Nonlinear Dynamics. XI, 208 pages. 1999.

Vol. m 58: J. Daillant, A. Gibaud (Eds.), X-Ray and Neutron Reflectivity: Principles and Applications. XVIII, 331 pages. 1999.

Vol. m 59: M. Kriele, Spacetime. Foundations of General Relativity and Differential Geometry. XV, 432 pages. 1999.

Vol. m 60: J. T. Londergan, J. P. Carini, D. P. Murdock, Binding and Scattering in Two-Dimensional Systems. Applications to Quantum Wires, Waveguides and Photonic Crystals. X, 222 pages. 1999.

Vol. m 61: V. Perlick, Ray Optics, Fermat's Principle, and Applications to General Relativity. X, 220 pages. 2000.

Vol. m 62: J. Berger, J. Rubinstein, Connectivity and Superconductivity. XI, 246 pages. 2000.

Vol. m 63: R. J. Szabo, Ray Optics, Equivariant Cohomology and Localization of Path Integrals. XII, 315 pages. 2000.

Vol. m 64: I. G. Avramidi, Heat Kernel and Quantum Gravity. X, 143 pages. 2000.

Vol. m 65: M. Hénon, Generating Families in the Restricted Three-Body Problem. Quantitative Study of Bifurcations. XII, 301 pages. 2001.

Vol. m 66: F. Calogero, Classical Many-Body Problems Amenable to Exact Treatments. XIX, 749 pages. 2001.

Vol. m 67: A. S. Holevo, Statistical Structure of Quantum Theory. IX, 159 pages. 2001.

Vol. m 68: N. Polonsky, Supersymmetry: Structure and Phenomena. Extensions of the Standard Model. XV, 169 pages. 2001.

Vol. m 69: W. Staude, Laser-Strophometry. High-Resolution Techniques for Velocity Gradient Measurements in Fluid Flows. XV, 178 pages. 2001.